Sustainable Finance

Series Editors

Karen Wendt, CEO. Eccos Impact GmbH
President of SwissFinTechLadies, President Sustainable-Finance
Cham, Switzerland

Margarethe Rammerstorfer, Professor for Energy Finance and Investments
Institute for Finance, Banking and Insurance WU Vienna
Vienna, Austria

Sustainable Finance is a concise and authoritative reference series linking research and practice. It provides reliable concepts and research findings in the ever growing field of sustainable investing and finance, SDG economics and Leadership with the declared commitment to present the theories, methods, tools and investment approaches that can fulfil the United Nations Sustainable Development Goals and the Paris Agreement COP 21/22 alongside with de-risking assets and creating triple purpose solutions that ensure the parity of profit, people and planet through choice architecture passion and performance. The series addresses market failure, systemic risk and reinvents portfolio theory, portfolio engineering as well as behavioural finance, financial mediation, product innovation, shared values, community building, business strategy and innovation, exponential tech and creation of social capital. Sustainable Finance and SDG Economics series helps to understand keynotes on international guidelines, guiding accounting and accountability principles, prototyping new developments in triple bottom line investing, cost benefit analysis, integrated financial first plus impact first concepts and impact measurement. Going beyond adjacent fields (like accounting, marketing, strategy, risk management) it integrates the concept of psychology, innovation, exponential tech, choice architecture, alternative economics, blue economy shared values, professions of the future, leadership, human and community development, team culture, impact, quantitative and qualitative measurement, Harvard Negotiation, mediation and complementary currency design using exponential tech and ledger technology. Books in the series contain latest findings from research, concepts for implementation, as well as best practices and case studies for the finance industry. This book series is indexed in SCOPUS

Chi-Wei Su • Oana-Ramona Lobonț
Ana-Cristina Nicolescu • Nicoleta Mihaela Doran
Editors

Green Finance and the Challenges of Climate Change

Policies, Practices, and Global Perspectives

Springer

Editors
Chi-Wei Su
Capital Market Research Institute, School of Economics, Department of Finance
Qingdao University
Qingdao, China

Oana-Ramona Lobonţ
Faculty of Economics and Business Administration, Department of Finance, Business Information Systems and Modelling
West University of Timişoara
Timişoara, Timiş County, Romania

Ana-Cristina Nicolescu
Faculty of Economics and Business Administration, Department of Finance, Business Information Systems and Modelling
West University of Timişoara
Timişoara, Timiş County, Romania

Nicoleta Mihaela Doran
Faculty of Economics and Business Administration
University of Craiova
Craiova, Romania

ISSN 2522-8285 ISSN 2522-8293 (electronic)
Sustainable Finance
ISBN 978-3-032-07292-4 ISBN 978-3-032-07293-1 (eBook)
https://doi.org/10.1007/978-3-032-07293-1

This Springer imprint is published by the registered company Springer Nature Switzerland AG
The registered company address is: Gewerbestrasse 11, 6330 Cham, Switzerland

Acknowledgments

This work was supported by a grant from the Romanian Ministry of Research, Innovation and Digitalization, the project with the title "*Economics and Policy Options for Climate Change Risk and Global Environmental Governance*" (CF 193/28.11.2022, Funding Contract no. 760078/23.05.2023), within Romania's National Recovery and Resilience Plan (PNRR)—Pillar III, Component C9, Investment I8 (PNRR/2022/C9/MCID/I8)—Development of a program to attract highly specialized human resources from abroad in research, development, and innovation activities.

Disclaimer

The editors listed in this publication have contributed equally to the preparation, development, and coordination of the content. Their roles and responsibilities were shared equitably throughout the editorial process.

Contents

Contributors

Firu Mihai Alexandru Faculty of Economics and Business Administration, Doctoral School in Economic Sciences "Eugeniu Carada", University of Craiova, Craiova, Romania

Bianca-Raluca Băditoiu Faculty of Economics and Business Administration, Department of Management and Entrepreneurship, West University of Timisoara, Timisoara, Romania

Diana Balan Doctoral School of Economics and Business Administration, Faculty of Economics and Business Administration, West University of Timişoara, Timişoara, Timiş County, Romania

Ioana Birlan Doctoral School of Economic Cybernetics and Statistics, The Bucharest University of Economic Studies, Bucharest, Romania

Gherţescu Claudia Doctoral School in Economic Sciences "Eugeniu Carada", Faculty of Economics and Business Administration, University of Craiova, Craiova, Romania

Robert-Ştefan Constantin Faculty of Cybernetics, Statistics and Economic Informatics, The Bucharest University of Economic Studies, Bucharest, Romania

Andreea-Florentina Crăciun Doctoral School of Economics and Business Administration, West University of Timişoara, Timişoara, Romania

Mirela Cristea Faculty of Economics and Business Administration, Department of Finance, Banking and Economic Analysis, University of Craiova, Craiova, Romania

Cristina Criste Faculty of Economics and Business Administration, Doctoral School of Economics and Business Administration, West University of Timisoara, Timişoara, Romania

Doran Marius Dalian Doctoral School in Economic Sciences "Eugeniu Carada", Faculty of Economics and Business Administration, University of Craiova, Craiova, Romania

Mihiţ Lavinia Daniela Doctoral School in Economic Sciences "Eugeniu Carada", Faculty of Economics and Business Administration, University of Craiova, Craiova, Romania

Adriana AnaMaria Davidescu Department of Statistics and Econometrics, The Bucharest University of Economic Studies, Bucharest, Romania

Department of Education, Training and Labour Market, National Scientific Research Institute for Labour and Social Protection, Bucharest, Romania

Rongrong Dong School of Economics, Qingdao University, Qingdao, China

Nicoleta Mihaela Doran Faculty of Economics and Business Administration, University of Craiova, Craiova, Romania

Anastasia Doraş Lisnic Doctoral School of Economics and Business Administration, Faculty of Economics and Business Administration, West University of Timişoara, Timişoara, Timiş County, Romania

Berceanu Dorel Faculty of Economics and Business Administration, University of Craiova, Craiova, Romania

Costea Florin Doctoral School of Economics and Business Administration, West University of Timişoara, Timişoara, Romania

Manta Liviu Florin Faculty of Automation, Computers and Electronics, University of Craiova, Craiova, Romania

Badareu Gabriela Faculty of Economics and Business Administration, University of Craiova, Craiova, Romania

Pirtea Marilen Gabriel Faculty of Economics and Business Administration, Finance, Business Information Systems and Modelling Department, West University of Timisoara, Timişoara, Romania

Hadasa Ligia Pricopiuc Gag Faculty of Economics and Business Administration, Doctoral School of Economics and Business Administration, West University of Timisoara, Timişoara, Romania

Manta Alina Georgiana Faculty of Economics and Business Administration, University of Craiova, Craiova, Romania

Omer Serkan Gulal Nevsehir Haci Bektas Veli University, Nevsehir, Türkiye

Alina Ionaşcu Doctoral School of Economics and Business Administration, Faculty of Economics and Business Administration, East-European Center for Research in Economics and Business, West University of Timisoara, Timisoara, Romania

Rahil Irfan Department of Business Administration, Air University Multan Campus, Multan, Pakistan

Oana-Ramona Lobonţ Faculty of Economics and Business Administration, Department of Finance, Business Information Systems and Modelling, West University of Timişoara, Timişoara, Timiş County, Romania

Poenaru Maria Magdalena Faculty of Horticulture, University of Craiova, Craiova, Romania

Bădîrcea Roxana Maria Faculty of Economics and Business Administration, University of Craiova, Craiova, Romania

Andru Mazilescu Faculty of Economics and Business Administration, West University of Timisoara, Timişoara, Romania

Olaru Mihail Doctoral School in Economic Sciences "Eugeniu Carada", Faculty of Economics and Business Administration, University of Craiova, Craiova, Romania

Nicoleta-Claudia Moldovan Faculty of Economics and Business Administration, Department of Finance, Business Information Systems and Modelling, West University of Timişoara, Timişoara, Timiş County, Romania

Ariana-Denisa Moţ Faculty of Economics and Business Administration, Finance, Organisational Governance and Tax Consulting Master Program, West University of Timişoara, Timişoara, Romania

Ana-Cristina Nicolescu Faculty of Economics and Business Administration, Department of Finance, Business Information Systems and Modelling, West University of Timişoara, Timişoara, Timiş County, Romania

Graţiela Georgiana Noja Department of Marketing, International Business and Economics, Faculty of Economics and Business Administration, West University of Timişoara, Timişoara, Romania

Lobonţ Oana-Ramona Finance, Business Information Systems and Modelling Department, Faculty of Economics and Business Administration, West University of Timişoara, Timişoara, Romania

Salih Ozdemir Department of International Trade and Logistics, Kayseri University, Kayseri, Türkiye

Meng Qin School of Marxism, Qingdao University, Qingdao, China

Pîrvu Ramona Faculty of Economics and Business Administration, University of Craiova, Craiova, Romania

Adnan Safi School of Economics, Qingdao University, Qingdao, Shandong, China

Chi-Wei Su Capital Market Research Institute, School of Economics, Department of Finance, Qingdao University, Qingdao, China

Ran Tao Doctoral School of Economics and Business Administration, West University of Timisoara, Timişoara, Romania

Alexandra-Mădălina Ţăran Faculty of Economics and Business Administration, Department of Finance, Information Systems and Business Modelling, West University of Timisoara, Timisoara, Romania

Ebru Topcu Alanya Alaaddin Keykubat University, Alanya, Antalya, Türkiye

Mert Topcu Department of Economics, Alanya Alaaddin Keykubat University, Alanya, Antalya, Türkiye

Muhammad Umar Department of Finance and Accounting, Adnan Kassar School of Business, Lebanese American University, Beirut, Lebanon

Ana-Elena Varadi Doctoral School of Economics and Business Administration, West University of Timişoara, Timişoara, Romania

Sorana Vătavu Faculty of Economics and Business Administration, Department of Finance, Business Information Systems and Modelling, West University of Timisoara, Timişoara, Timiş County, Romania

Kaihua Wang Qingdao University, Qingdao, Shandong Province, China

Wenqian You Qingdao University, Qingdao, Shandong Province, China

Manal Yunis Department of Information Technology and Operations Management, Adnan Kassar School of Business, Lebanese American University, Beirut, Lebanon

Li Zhengzheng School of Economics, Qingdao University, Qingdao, China

The Green Finance Ecosystem: Conceptual Foundations and Theoretical Frameworks

Oana-Ramona Lobonţ, Andreea-Florentina Crăciun, Nicoleta-Claudia Moldovan, and Ariana-Denisa Moţ

1 Introduction

Contemporary challenges caused by environmental climate change have transformed sustainable development into a mandatory target, and green finance (GF) has become a point of reference in today's economies. While much has been written on the practical tools of green finance, such as green bonds, ESG investing, or sustainable banking, a gap exists in the conceptual and structural understanding of the broader green finance ecosystem (GFE). This paper addresses that gap by analysing the theoretical foundations, historical evolution, and key institutional frameworks that shape this emerging field.

Our study is grounded in a multi-layered qualitative and bibliometric methodology. The investigation begins with a content analysis focused on tracing an axis of historical development for the concept of green finance, identifying key moments and intellectual shifts that contributed to its current understanding. This is followed by a science mapping and performance analysis using VosViewer software, which enabled us to analyse 2317 publications from 2007 to 2025 relevant to the topics of

O.-R. Lobonţ · N.-C. Moldovan
Faculty of Economics and Business Administration, Department of Finance, Business Information Systems and Modelling, West University of Timişoara, Timişoara, Timiş County, Romania
e-mail: oana.lobont@e-uvt.ro; nicoleta.moldovan@e-uvt.ro

A.-F. Crăciun (✉)
Doctoral School of Economics and Business Administration, West University of Timişoara, Timişoara, Romania
e-mail: andreea.craciun@e-uvt.ro

A.-D. Moţ
Faculty of Economics and Business Administration, Finance, Organisational Governance and Tax Consulting Master Program, West University of Timişoara, Timişoara, Romania
e-mail: ariana.mot02@e-uvt.ro

C.-W. Su et al. (eds.), *Green Finance and the Challenges of Climate Change*, Sustainable Finance, https://doi.org/10.1007/978-3-032-07293-1_1

"green finance," "financial ecosystem," and "sustainable development" to identify dominant perspectives, research clusters, and knowledge gaps. In addition, a diversity area analysis was conducted to map the thematic breadth of the literature and assess how different academic disciplines contribute to the construction of the green finance ecosystem. These insights were then contextualised through a documentary and public policy review, highlighting institutional strategies, international standards, and governance mechanisms that support or hinder the development of GFE.

The need to address this topic results from its novelty. At the time of writing this article, only three documents are available in the Web of Science database that present "green finance ecosystem" as a concept. The essence of this paper is to contribute to building a theoretical basis for the idea by integrating multiple quantitative and qualitative analyses of documents related to this subject.

This paper brings several original contributions to the literature. First, it provides a structured conceptual framework of the green finance ecosystem, synthesising diverse academic sources, theoretical currents, and policy documents. Second, the study outlines a clear historical trajectory of the GF concept, identifying its transformation from a normative ideal to a structured policy and research domain. Third, by combining bibliometric mapping with documentary and policy analysis, we offer a unique meta-perspective that bridges fragmented discourses and sets the stage for a more integrated research agenda. Our findings provide valuable insights for scholars, policymakers, and institutions seeking to understand and position themselves within the evolving architecture of green finance. They also lay the groundwork for future studies that may build on the defined theoretical base.

The structure of this paper reflects the layered and interdisciplinary nature of our investigation into the green finance ecosystem. Following the introduction, the literature review in the paper's second section aims to build the historical framework for developing the concept of green finance by capturing the most relevant information through content analysis. Section 3 outlines the data and methodology, detailing the sources of information used in the study. This includes a comprehensive bibliographic dataset extracted from the Web of Science Core Collection and a curated selection of policy documents and strategic frameworks relevant to green finance at the international and national levels. Section 4 presents the main findings, structured into three complementary subsections: the bibliometric analysis (including science mapping and performance analysis), the research area diversity analysis, and the documentary and public policy review, each offering distinct yet interconnected insights into the development and current landscape of green finance. Finally, Section 5 provides the conclusions, summarising the paper's key contributions and outlining implications for future research and policy development in sustainable finance.

2 Literature Review

To better understand how the concept of "Green Finance" has evolved and how it has come to play an essential role in the transition to a sustainable economy, it is necessary to mark its main historical stages of development. For this, we considered graphically representing the key periods and moments in history that shaped the analysed concept, the time axis being represented in Fig. 1.

The first signs of the importance of integrating environmental issues into the economic sphere appeared in the context of the ecological and energy crises of the 1970s, such as the oil crisis of 1973, which was the first energy crisis faced by the Western world after World War II (Issawi, 1978). Environmental movements began to gain momentum in society, and Western governments introduced the first environmental regulations. In this context, the idea that financial activities could support environmental protection was still in its infancy. The Brundtland Report, published in 1987, was also a key moment, introducing the concept of "sustainable development" into the international debate as a balance between economic growth, social inclusion, and environmental protection (Hajian & Kashani, 2021). Even though there were no formalised green financial products at the time, the theoretical foundations for subsequent developments were laid.

The 1990s saw the first stage of institutionalisation of environmental concerns in the financial sector. The UNEP ("United Nations Environment Programme") Finance Initiative was among the first global initiatives to recognise the role of the banking sector in supporting sustainable development (Desai, 2020). However, the turning point came with the entry into force of the Kyoto Protocol (1997) (Miyamoto & Takeuchi, 2019), which went into effect in 2005, paving the way for market-based instruments such as emissions trading, which directly linked financial markets to the fight against climate change.

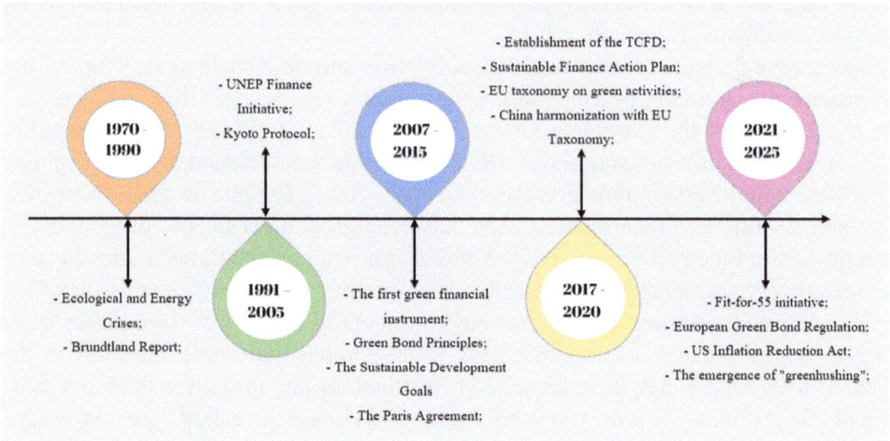

Fig. 1 Axis of Historical Development of Green Finance. (Source: Own elaboration in Microsoft Word)

The emergence of green instruments on the financial market marked the period 2007–2015. The European Investment Bank issued the first Climate Awareness Bond in 2007, followed in 2008 by the World Bank, which issued the first traditional green bond (Banga, 2018). This step triggered a shift that channelled capital towards projects targeting green, renewable energy and energy efficiency. The emergence of the Green Bond Principles in 2014, under the auspices of the ICMA (De Assis et al., 2022), was essential for standardising the market. At the same time, the Sustainable Development Goals (SDGs) adopted in 2015 provided a broad global framework for aligning investments with sustainability (Halisçelik & Soytas, 2019). Another key milestone in the evolution of green finance was the signing of the Paris Agreement (2015), which catalysed the mobilisation of capital to combat climate change (Erickson & Brase, 2019). This agreement recognised the importance of the financial sector's involvement in limiting global warming to below 2°C and triggered a series of national and transnational initiatives. Institutions such as the European Central Bank, the Bank of England and the People's Bank of China have started to integrate climate risks into monetary and financial supervision policies.

The post-2015 period has been characterised by an acceleration of regulation and the more systematic integration of sustainability into financial governance. Initiatives such as the Task Force on Climate-related Financial Disclosures (TCFD) (Chua et al., 2022) have guided companies and investors towards transparency on climate risks. The European Union has become a leader in regulation, publishing the Action Plan on Sustainable Finance in 2018 (Busch et al., 2021) and launching the EU Taxonomy in 2020 (Kooths, 2022), which clearly defines which economic activities are "green." Since 2018, green finance has expanded rapidly outside European markets. China has become a global leader in green bond issuance, promoting its standards and gradually aligning with international ones (Larsen, 2022). Even though the European Bank launched the first green bond in 2007, by 2020 the green bonds market in Europe still did not occupy a significant part of the listed instruments (Frydrych, 2021).

Recent years have brought both consolidation and new challenges. The EU has continued its regulatory process, launching initiatives such as "Fit-for-55" (Schlacke et al., 2022) and the European Green Bond Regulation (EuGB) (Maragopoulos, 2023). The Inflation Reduction Act (IRA) 2022 in the US included the largest public investment in the green transition (Bistline et al., 2023). Despite progress in regulating and developing green finance, there has been growing criticism, particularly in Anglo-Saxon financial markets, of the lack of consistency and transparency in communicating environmental responsibility. The practice known as "greenwashing" — misleadingly presenting activities as environmentally friendly — has attracted the attention of regulators such as the SEC in the United States and the FCA in the United Kingdom, which have launched investigations and initiatives to clarify ESG standards. At the same time, a reverse reaction has emerged, called "greenhushing," whereby some companies choose to promote no longer their sustainable goals or actions for fear of controversy or reputational consequences (Dias et al., 2025). This dynamic reflects the increasingly visible tensions between the ethical imperatives of the green transition and the economic and political pressures surrounding this emerging field.

The scientific literature brings together concepts such as “green economy”, “sustainability”, “energy efficiency”, “combating climate change” or “green bonds” in many documents (Cai & Zhang, 2009; Sohail et al., 2025; Dadabada, 2025; Ozili, 2022), but there is a gap in defining the green financial ecosystem. This term is mentioned in three articles, but the focus is on other main themes. Mukherjee et al. (2024) investigate how the financing of green innovation start-ups interacts with investors, the state and the green finance ecosystem. Even though the term was used in the abstract and, subsequently, in one of the hypotheses addressed, the green finance ecosystem is approached as the green entrepreneurial finance ecosystem. In this sense, a definition of the concept aims to examine the interactions between government institutions, academic institutions, business incubators, and private entities that support the development and implementation of clean technologies, including through innovation, government funding, and public policies. Huang (2025) approaches financial technology (fintech) as a pillar of sustainable development, conducting the research through a bibliographic review analysis. The study’s main results mark fintech as an engine for sustainable development by supporting green finance through technological innovation, digital services and public policies targeting investments in renewable energy and reducing carbon emissions. In the work of the authors Zhang and Umair (2023), the most relevant information related to our research topic can be found, namely that the main elements of the green finance ecosystem are green bonds, renewable energy stocks and carbon markets, the political decision-makers and the state being a secondary part of this system.

Over more than five decades, green finance has moved from theory to practice, developing from a niche area of environmental activism into an essential segment of the global financial system. Currently, the major challenges relate to harmonising standards, avoiding regulatory duplication, integrating systemic climate risks, and, last but not least, attracting private capital to projects with real impact. The success of the green transition will depend on the financial system’s ability to remain flexible and transparent, without losing sight of the objectives of equity and global inclusion. Moreover, to pursue sustainable economic development and achieve relevant growth targets, it is necessary to define the green finance ecosystem and set this concept as a benchmark for good practices in economic activity, government decisions related to the environment and economy, and society’s response to climate change.

3 Data and Methodology

In a global context characterised by ecological imbalances, economic uncertainty, and rapid social change, green finance is becoming an essential element in rethinking sustainable development strategies. Climate change, new international regulations, and pressure to reduce carbon emissions have significantly transformed the global financial system (Low, 2023). This transformation highlights the need for a theoretical and practical framework integrating sustainability principles into economic decisions.

This paper aims to highlight the structure, dynamics and central points, both conceptual and institutional, of the green finance field, through a comprehensive analysis using the "Web of Science Core Collection" database. To retrieve the data necessary for the study, we used key terms such as "financial ecosystem" and "sustainable development" or "green finance", resulting in a number of 2317 results at a global level, published between 2007 and 2025 that address our research topic.

Considering the complexity and vastness of the specialised literature regarding the topic addressed, three complex methodologies were used to observe the significant contributions in this field. The first methodology used was bibliometric analysis techniques, which begins with Performance Analysis, which emphasises the main features of the data sample and science mapping, which includes five perspectives, such as studying keyword co-emergence networks, co-authoring, bibliographic coupling, co-citation per sources and international co-operation between countries (Bernatović et al., 2021), all done using VOSviewer software. The main aim is twofold: on the one hand, identifying the central themes and theories underlying this field, and on the other hand, observing the geographical distribution of expertise and research networks that contribute to the expansion and strengthening of green funding.

The database was uploaded to VOSviewer to allow for a structured and detailed analysis of the data, according to a step-by-step methodology:

(a) Web of Science—Clarivate (Analytic database):

- The subject of the research: "Financial Ecosystem", and "Sustainable Development" or "Green Finance";
- Period: 2007–2025;

(b) VOSviewer Instruments:

- Recorded content for VOSviewer: Full text and cited references.

(c) The analysis was based on:

- keywords;
- citations/countries;
- citations/authors.
- bibliographic coupling/countries
- co-citation/sources

Further, we focused the analysis on assessing the thematic diversity of the identified green finance articles, using statistical indicators established in the literature (Palaghianu, 2016). Three indicators were used in the diversity analysis to capture this aspect: the Shannon Index, the Berger-Parker Index and the Simpson Index. Each indicator provides a different perspective on how themes are distributed within the analysed sample. To analyse the thematic diversity, we used data extracted from the Web of Science database to identify the main research directions in the field of green finance. Initially, 2317 articles published between 2007 and 2025 were identified. However, following the processing process—converting the ".txt" generated by the platform into a compatible format (.xlsx) for bibliometric analysis—we found that only 2051 articles contained complete keyword information. Thus, the diversity analysis was based on this final subset of articles, as only for these was it possible to classify the thematic classification needed to calculate the diversity indicators.

The Shannon index was the first indicator, a tool often used in ecology and bibliometric analysis to measure information diversity. It considers both the total number of subject categories and the proportions in which they occur. In other words, this index reflects how many themes there are and how well they are represented. The Formula (1) is presented below:

$$H' = \Sigma\left(p_i \times \ln\left(p_i\right)\right) \tag{1}$$

where p_i is the proportion of articles in a given theme out of all articles. This index helps detect both richness and balance in the thematic structure. A higher Shannon value indicates a broader and more evenly distributed thematic landscape, whereas a lower value suggests thematic concentration or dominance by a few categories. The Shannon index proves particularly valuable for assessing interdisciplinary research areas, as it highlights the extent of diversity or concentration within the scholarly output. It also enables comparative evaluations across time periods or geographic regions, offering a perspective on how thematic focus has broadened or narrowed over time.

The second indicator applied was the Berger-Parker index, which measures the dominance of a single category concerning the whole set analysed. In other words, this index focuses on identifying the theme with the highest dominance and expresses this dominance by a simple ratio (2):

$$d = \frac{N_{\max}}{N} \tag{2}$$

where N_{max} is the number of articles in the most frequent theme and N is the total number of articles. A high Berger-Parker index indicates a strong literature concentration around a single prevailing theme, whereas a lower score reflects a more evenly distributed thematic structure.

This metric is particularly useful for detecting thematic dominance or highlighting the disproportionate presence of a specific topic. It offers a clear and accessible measure of thematic centralisation, helping to assess whether a research area is still developing, with multiple topics coexisting, or already consolidated around a core subject.

The last indicator applied was the Simpson index, which assesses the likelihood that two randomly chosen articles deal with the same theme. It helps understand the degree of thematic focus and is sensitive to dominant themes. The Formula (3) is:

$$D = \Sigma\left(p_i^2\right) \tag{3}$$

where p_i is the proportion of articles in each theme. The Simpson index captures the prevalence of dominant themes within a dataset by placing greater emphasis on the most frequently occurring topics. A higher value of D indicates a stronger thematic concentration, where only a limited number of themes are prominently featured. This metric is well-suited for evaluating how thematically focused a body of literature is. It is especially effective in revealing situations where a small number of topics dominate the discourse, offering perspective on the degree of specialisation or maturity in a research area. In this way, it is a useful complement to the Shannon index, prioritising concentration over diversity.

After calculating the three indicators, we added a frequency analysis of the words within the abstracts for the documents considered in the study to delimit a framework of terms directly connected to the green finance ecosystem theme. In this regard, we worked with the MAXQDA analysis software for frequency analysis, eliminating connecting words and limiting the calculation to nouns only.

The last methodology applied in this research was Policy Review, focusing on a sample of six relevant policies or initiatives at the national, European, and global levels. Each document was analysed based on common criteria: Document/Policy Name, Level of Intervention (National, EU or Global), Actors Involved, Key Objectives, Instruments Applied, Type of Intervention, and Obstacles Identified. The data was extracted from official sources such as EU legislative texts, government websites, institutional reports, etc., and the matrix for this type of analysis was completed on the basis of these sources. A policy review systematically examines and interprets public policy documents, strategies, and regulations relevant to a specific topic or sector. Its purpose is to understand the objectives, instruments, actors, and implementation mechanisms involved, while also identifying gaps, overlaps, or areas for improvement in the existing policy landscape.

4 Results

4.1 Bibliometric Analysis

The preliminary stage of bibliometric analysis begins with performance analysis, which includes publication and citation numbers. This type of analysis, carried out using the Web of Science database, highlights the general characteristics of the selected sample (Cucari et al., 2022), such as the number of publications over various periods, the number of citations, and the types of documents, as shown in Table 1:

Over the past few years, it has been observed that the COVID-19 pandemic has accelerated the transition to sustainable and resilient economic models (Morrissey & Heidkamp, 2022) by exposing the "fragility" of the global financial system, but

Table 1 Performance analysis data sample

Unit of analysis	Results
The most productive year	2024: 668 Documents
The least productive year	2007 and 2013: Average of ≈4 documents p.y.
Type of documents	98.50%—Articles, Early Access and Book Chapters
Number of authors	4889
Number of countries	99
Number of organisations	2053
Number of keywords	5383
Research Areas	58
The most productive Research Area	Environmental Sciences Ecology—1034

Source: Data processed by the author

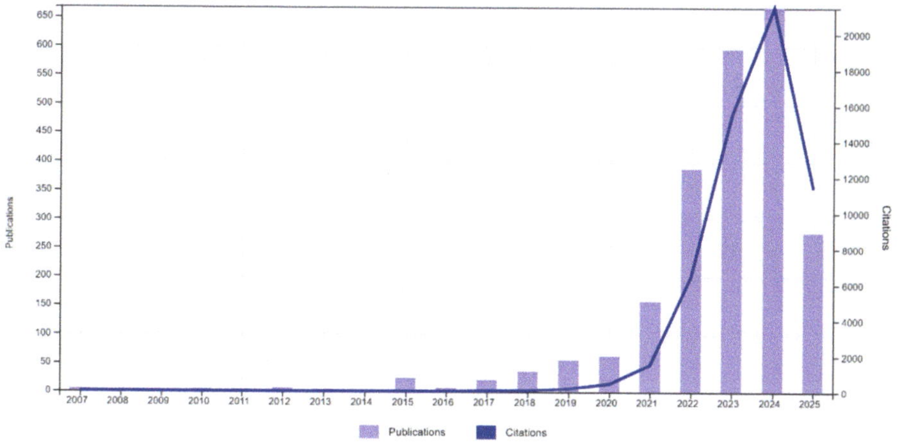

Fig. 2 Publications and Citations Numbers distributed by years. (Source: author's processing in Web of Science)

also by highlighting the need for economic and ecological resilience. Thus, more and more researchers are focusing their analysis on the interactions between financial and sustainable ecosystems, in search of possible solutions that are more resilient to future shocks. This increased interest in the post-pandemic period in concepts such as "green finance," "sustainable development," and "financial ecosystem" can also be seen in Fig. 2.

It is possible to see how interest in this research topic grew between 2021 and 2024, when it peaked with 668 publications, compared to the first year of analysis, 2007, when only four articles were published. On the other hand, after 2020, many governments and international institutions launched economic recovery plans, including investments in green infrastructure, renewable energy, and sustainable development (Ruddock & Ruddock, 2022). These policies fueled academic interest in the impact of green financial instruments on the global financial system.

As regards document citations, these can also be seen in Fig. 2. The total number of citations referring to the sample of documents analysed is 57,304, which is the total number of citations for all items in the results set, while the average number of citations per article is 24.73. After analysing the dataset, we noticed that 436 articles have no citations, which is justified by the fact that they were published recently (2024–2025). Thus, this lack of citations can be explained by the "natural" delay in citations, as scientific papers need time to become visible and relevant in the scientific community, especially those research papers that address emerging or innovative fields. Thus, these publications may be among the first to study certain recent cases, which could be a foundation for future research.

Next, we will detail the second part of the bibliometric analysis—science mapping, which includes five types of analysis. The first type of map is the keyword analysis (Fig. 3), which includes 174 terms that meet the minimum threshold of 15 co-occurrences, organised into several distinct clusters, differentiated by colour. The size of the nodes reflects the frequency of the terms, and their proximity indicates the strength of the semantic links between concepts.

At the centre of the network analysed is the concept of "green finance", the thematic core around which all the other concepts revolve. Four major clusters, each representing a distinct research sub-domain, emerge around it. The first, highlighted in red, includes terms such as 'digital finance', 'green technology innovation', 'performance' and 'management'. This cluster suggests a research orientation towards economic efficiency, technological innovation and the application of sustainable policies in corporate governance. At the same time, it can be observed how green

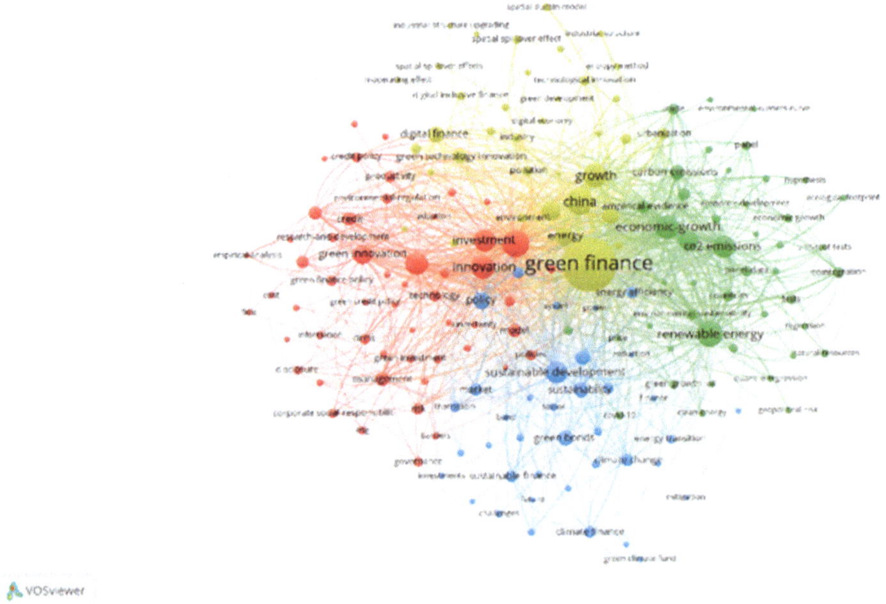

Fig. 3 Keyword Network. (Source: author's processing in VOSviewer)

finance is seen as a factor contributing to the competitiveness of companies and their adaptation to new environmental requirements, in a context in which ESG principles and corporate social responsibility are becoming increasingly relevant.

The second cluster, marked by the colour green, revolves around concepts such as “renewable energy”, “CO_2 emissions”, “economic growth” and “natural resources”. This set of terms reflects the close link between green finance and the global goals of reducing emissions and promoting sustainable development. The theme is closely linked to energy and climate transition policies, particularly those implemented by the European Union, China or other emerging economies. From a theoretical point of view, this cluster builds on the foundations of green modernisation theory and green growth theory, where financial mechanisms are designed to counteract the negative effects of traditional economic development by providing more balanced and sustainable alternatives.

The yellow cluster revolves around terms such as “technological innovation”, “growth”, “China” and “carbon emissions”, emphasising the importance of Asia, and China in particular, in the process of integrating green finance into macroeconomic strategies. The high frequency of the notions of innovation, urbanisation and digitalisation suggests a dual model, where economic development and sustainability are no longer treated as separate dimensions, but as complementary elements, connected through technological progress. This approach reflects the direction of public policies promoted by some Asian governments, such as using green credits or initiatives to support the green transition, marking a shift from reactive interventions to proactive and integrated strategies.

As for the blue cluster, it emphasises terms such as ‘sustainable development’, ‘green bonds’, ‘climate finance’ and ‘ESG’ (environmental, social and corporate governance), outlining an institutional and policy-focused analysis framework. This group highlights the concern for building a sustainable financial infrastructure, where tools such as the EU green taxonomy, green banks and climate funds contribute to enhancing transparency and efficiency in the governance of the green transition. From a theoretical perspective, the influence of concepts such as stakeholder theory and the theory of market failures, which provide justifications for public intervention to sustain collective goods and correct economic imbalances, is notable here.

An interesting aspect is the frequent presence of the terms ‘policy’, ‘impact’, ‘transition’ and ‘mitigation’ at the intersection of all clusters, which gives them the role of linking elements between the economic, technological, environmental and institutional dimensions of green finance (Khemiri, 2024). This points to a clear trend towards an interdisciplinary and cross-sectoral approach, where future research will increasingly seek to integrate these components coherently in policy formulation and practical implementation of sustainable development.

The second type of analysis is the network of international collaborations in green finance research (Fig. 4), which was created based on a minimum threshold of 10 documents per country and at least 30 citations. Out of 99 countries analysed, only 46 met these criteria, suggesting a geographical concentration of expertise and scientific production in the field.

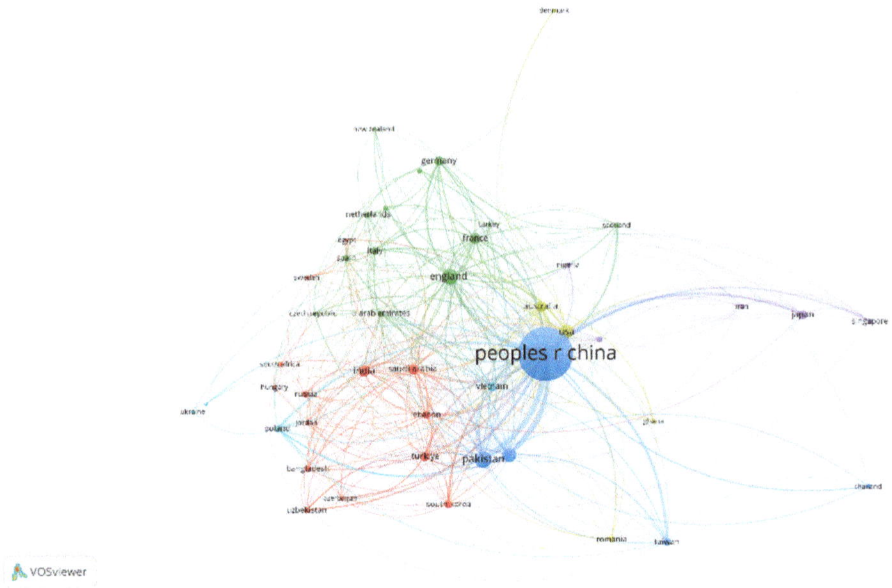

Fig. 4 Scientific co-authorship network, based on the countries where the documents were published. (Source: author's processing in VOSviewer)

At the centre of the map of the scientific co-authorship network is China (People's Republic of China), which stands out with the most significant number of publications and academic links globally. This dominant position reflects China's strategic role in the global energy transition and its substantial investment in research, technological innovation and sustainable financing. Its central role suggests an extensive network of partnerships with countries in regions as diverse as South and South-East Asia, Europe, the Middle East, Africa and North America, which contributes to a remarkable scientific influence in the field of green finance.

Another important point on the map is Pakistan, which, although less present in the established international academic literature, is gaining more and more relevance in regional research on green finance. This increased visibility can be explained by collaborations with China on major initiatives such as the Belt and Road (Pich, 2021), where the green infrastructure component has become increasingly important. Also, cooperation with neighbouring South Asian countries—such as India, Bangladesh or Uzbekistan—has prompted the emergence of an Asian regional focus on financial sustainability studies.

"The European cluster" is well-defined and is dominated by countries such as Germany, France, the Netherlands, Italy and the United Kingdom. They form a dense network of collaboration with each other and other regions of the world. These countries stand out as important centres of academic and institutional research in areas such as climate policy and ESG (Environmental, Social and Governance), reflecting the high standards promoted by the European Union on green taxonomy and investment transparency. Therefore, it is unsurprising that these European countries are closely connected with other major global players such as China, the United States, and Australia, indicating a dynamic global network of scientific and policy exchanges.

On the other hand, countries such as Iran, Japan, Singapore, and South Korea appear in a separate cluster (marked in purple), in a more peripheral position but representative of technologically advanced Asia (Sun, 2025). Even if their number of publications and citations is lower compared to the main players, these countries contribute significantly to research in areas such as digital innovation, green bonds, and carbon neutrality goals.

The United States occupies an intermediate position in the network, with multiple transcontinental connections with Asia, Europe, and Oceania. Still, it does not have the dominant role that it does in other research areas. This positioning may reflect a more decentralised approach to environmental funding in the American context, in contrast to the more coordinated strategies adopted by the European Union or China. On the African side, countries such as South Africa, Nigeria and Ghana form an expanding cluster, indicating a growing interest in sustainability and green finance in emerging economies. Although collaborative networks are still limited, recent trends suggest a progressive integration into global dynamics, notably through partnerships with Europe and China.

The third network is the collaborative dimension, which can be seen in Fig. 5. It is explored through an analysis of the scientific co-authorship network based on a minimum of 5 documents/authors, 15 co-appearances, and at least 50 citations. This network shows geographical fragmentation but also increasingly pronounced thematic cohesion.

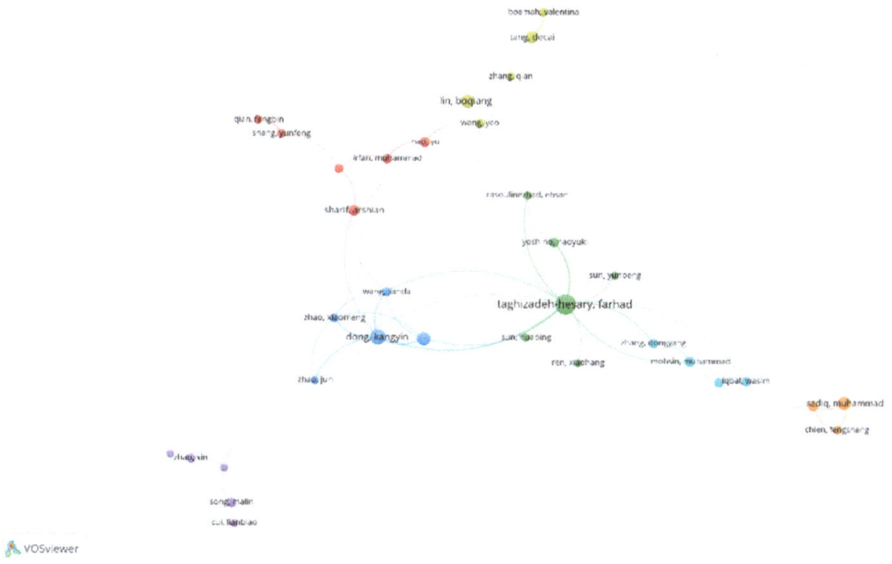

Fig. 5 Network of co-authors, based on the number of documents per author. (Source: author's processing in VOSviewer)

The seven clusters identified outline several regional research centres, particularly in East Asia, South Asia and the Middle East, where analytical paradigms around ESG (environmental, social and corporate governance) investments, energy efficiency, green policies and banking sector regulation are increasingly developing. Collaborations between authors suggest a functional polarisation within the network: some clusters focus on market dynamics, such as green bonds or carbon pricing, while others focus more on the institutional and regulatory dimensions of the transition to a sustainable economy.

The bibliometric analysis highlights the complexity and dynamism of the field of green finance, built around a dense network of concepts, institutions and international collaborations. The evolution of key terms, their organisation into thematic clusters and the structure of co-authorship networks reflect not only the variety of sub-domains covered—from technological innovation and climate policy to financial mechanisms and ESG (Environmental Social Governance) regulation, but also an increasing trend towards closer links between disciplines (Spulbar & Birau, 2019). A mature scientific field with a global footprint is emerging, with China, Europe, and Southeast Asia holding key positions in volume of contributions and strategic influence on research directions. There is also a gradual increase in academic interest in emerging regions such as Africa and South Asia, suggesting an accelerating globalisation of research related to the transition to a sustainable economy.

The presented Fig. 6 illustrates a bibliographic coupling network across countries (minimum 10 publications/country, minimum 30 citations, resulting a threshold of 46 registrations), which highlights the geographical distribution of research, emphasising relevant contributions from a national perspective, thus revealing the institutional and conceptual networks that shape the field of study. The network's

Fig. 6 Bibliographic Coupling per Country. (Source: author's processing in VOSviewer)

core is dominated by China ('Peoples R China'), which stands out as the central pole of influence in terms of the volume of publications and its connections with other countries, reflected in its sizeable node and the density of links. There is also a strong connectivity between China and countries such as Pakistan, India, Malaysia and Saudi Arabia, suggesting active collaborations and common research directions in green finance.

Countries from Western Europe (such as England, Germany and France), as well as the USA and Australia, are also found in the network, but with relatively more fragmented activity, with the colours indicating earlier years of publication (2022), in contrast to more recent contributions from emerging countries (2023–2024). Thus, the colour palette allows the temporal evolution of each country's involvement to be tracked, providing a synchronic and diachronic picture of research dynamics in the field. This map, therefore, provides a comprehensive picture of the global distribution of expertise and emerging trends, facilitating the identification of key actors and axes of collaboration in sustainable development and green finance.

A final scientific map is reflected in Fig. 7, representing a co-citation analysis applied on the analysed sample, using a minimum threshold of 200 citations for each included source. This scientific mapping technique allows the identification of frequently co-cited publications, suggesting thematic proximity and common conceptual relevance. The visual map generated with VOSviewer highlights several well-defined thematic clusters, representing structural sub-domains of research: the green cluster dominated by the Journal of Cleaner Production (j clean prod) indicates a research focus on sustainability, while the blue cluster, centred on publications such as Energy Policy and Renewable & Sustainable Energy Reviews, reflects

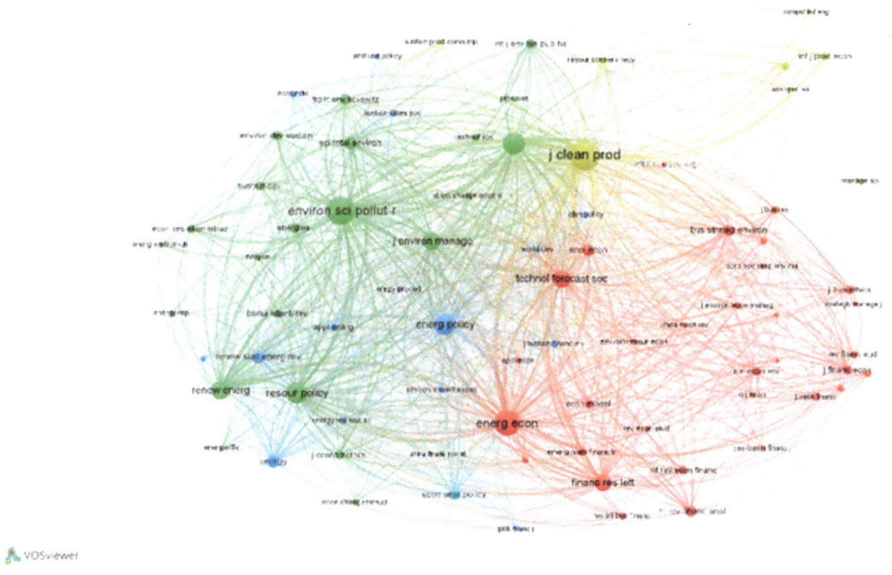

Fig. 7 Co-citation per Sources. (Source: author's processing in VOSviewer)

interest in energy policy and renewable resources. The red cluster, represented by journals such as Energy Economics, Finance Research Letters or Technological Forecasting and Social Change, emphasises the convergence between economics, finance and technological innovation. The high connectivity between these clusters suggests an epistemic interdependence between green finance's economic, technological and environmental components. Co-citation analysis, therefore, allows the identification of seminal publications in the field and a clearer understanding of how the main research directions have been shaped and consolidated, as well as the positioning of central, peripheral and linking actors in the scientific network.

These findings confirm that green finance is no longer a peripheral field, but a central element in shaping the economic and environmental policies of the future. Given these developments, future research directions should focus on strengthening international cooperation, regulatory alignment and the development of integrated methodologies to assess the practical impact of green finance instruments. In addition, it would be helpful to extend the research to topics such as financial inclusion, climate risk management and digitisation of sustainable financial processes. In the longer term, such a bibliometric approach provides a solid basis for identifying gaps in the current literature and for targeting research towards areas with the greatest potential to support the development of a green, equitable and resilient economy.

4.2 Research Area Diversity

A quantitative analysis was conducted on a sample of 2051 academic articles with complete keyword information to understand better the intellectual structure and thematic focus of green finance research. Table 2 represents the main themes addressed in the 2051 articles. To be able to realise this representation, we have divided the scientific articles according to the main keywords; as can be observed, the dominant theme is "Green Finance", accounting for almost 40% of the total number of articles. This majority presence reflects the constant interest of

Table 2 Research topics and frequency of occurrence in the analysed corpus

Category	No. of items
"Green Bond"	53
"Green Finance"	804
"ESG / CSR"	16
"Climate Change & Risk"	55
"Carbon & Emission"	78
"Renewable Energy"	105
"Digital/Fintech"	159
"Sustainable Development"	70
"Green Innovation/Technology"	55
"Supply Chain/Banking"	56

Source: Data processed by authors

researchers in understanding and developing sustainable financial mechanisms that can support climate policies and green innovation in an efficient and scalable way.

This thematic predominance may be linked to the increasing institutional focus on green financing tools, such as EU Taxonomy regulations or the proliferation of green bonds, which have captured scholarly attention due to their potential to realign capital flows towards sustainable objectives. Moreover, the maturity of "Green Finance" as a research field may also explain its centrality, with well-established methodologies, data sources, and policy linkages attracting sustained academic engagement.

In the general interpretation, a higher Shannon index indicates a high diversity, suggesting that the articles are relatively evenly distributed among the different topics. In our case, the value obtained (Table 3) was approximately 1.38, indicating a low level of diversity. This low Shannon score suggests the existence of a dominant theme occupied by the term "green finance" with over 39% of the total, and the other 61% of the total is occupied by the different themes, some of which are insignificant in proportion. The analysis results place the themes "digital/fintech" and "renewable energy" in the following two places, but with 7.7% and 5.1% of the total. Referring to this result, we can state that the concept of "green finance ecosystem" cannot yet be defined through research fields, having to analyse the words used in the document abstracts to create a field of terms that are directly associated with this concept, and implicitly to outline a definition. This low score also suggests researchers branching into adjacent topics beyond financial mechanisms, such as renewable energy systems, digital platforms, and carbon markets. While the field

Table 3 Shannon index, Berger-Parker index, and Simpson index

		Shannon Index			Simpson Index	
Category	no. of items	(p_i)	$\ln(p_i)$	$p_i \times \ln(p_i)$	(p_i)	$(p_i)^2$
Green bond	53	0.02584	−3.65579	−0.09447	0.02584	0.00066
Green finance	804	0.39200	−0.93648	−0.36711	0.39200	0.15366
ESG/CSR	16	0.00780	−4.85349	−0.03786	0.00780	0.00006
Climate change & risk	55	0.02681	−3.61875	−0.09704	0.02682	0.00071
Carbon & emission	78	0.03803	−3.26937	−0.12434	0.03803	0.00144
Renewable energy	105	0.05119	−2.97212	−0.15216	0.05119	0.00262
Digital/fintech	159	0.07752	−2.55718	−0.19824	0.07752	0.00601
Sustainable development	70	0.0341	−3.37759	−0.11528	0.03413	0.00116
Green innovation/technology	55	0.02681	−3.61875	−0.09704	0.02682	0.00071
Supply chain/banking	56	0.02730	−3.60073	−0.09831	0.02730	0.00074
Others	600	0.29254	−1.22915	−0.35958	0.02584	0.00066
Total	2051					
Shannon Index	1.38184					
Berger-Parker Index[a]	0.392					
Simpson Index					0.1678	

Source: Data processed by authors

[a]Calculation method: number of articles on the topic of green finance divided by the total number of articles

gravitates strongly around finance, the integration of environmental and technological dimensions illustrates a cross-sectoral engagement vital for the real-world application of sustainable strategies and for developing a green finance ecosystem.

In our case, the dominant theme was "Green Finance", with 804 articles out of a total of 2051, leading to a Berger-Parker index of 0.392 (Table 3). Interpretation of this result shows that almost 40% of the analysed research focuses on a single theme, which indicates a rather strong dominance and relatively low diversity according to this indicator. Thus, although there is variety, one central topic monopolises the researchers' attention. The Berger-Parker index underscores the asymmetrical nature of research focus in this field. Such dominance may signal a mature and deeply explored area or a bottleneck in innovation, depending on how the research evolves over time. This could prompt an inquiry into whether funding mechanisms and academic incentives are disproportionately favoring financial over technical or social themes within sustainability.

Regarding the Simpson Index, the result obtained in our analysis was 0.1678 (Table 3), a relatively low value. Since in the interpretation of this index a lower value suggests a higher diversity, we can say that despite the obvious dominance of one theme, the rest of the topics are sufficiently diverse and well represented to maintain a reasonable balance in the distribution of the survey. The Simpson Index provides a useful corrective to the Berger-Parker score by emphasizing the non-negligible role played by smaller categories. In this case, themes such as "Digital/ Fintech" and "Carbon & Emission", while not dominant, still command enough attention to contribute meaningfully to thematic richness. This implies that the field is not entirely monopolized, and there is space for growth and diversification.

By comparing these three indices, we get a more complete picture of how green finance research is structured. While the Berger-Parker index highlights the intense concentration around the topic of "Green Finance", the Shannon and Simpson indices manage to highlight the significant contribution of other topics such as "Digital/ Fintech", "Renewable Energy" or "Carbon & Emissions", which, although having smaller weights, support an essential degree of diversity. Taken together, these indicators show that the research ecosystem surrounding green finance is both centralised and fragmented. Centralised in the sense that one concept dominates the discourse, but fragmented enough to indicate active exploration in emerging areas. The interplay between concentration and diversity may indicate a maturing field, where foundational knowledge in green finance sets the stage for more niche or interdisciplinary research to flourish.

As a conclusion for this analysis, the thematic diversity of the green finance literature is balanced but with a clearly defined central theme. This structure reflects the steadily growing interest in sustainable finance topics, but it also shows an openness to exploring other dimensions of the green transition, such as technology, renewable energy, or sustainable supply chains. Looking ahead, such diversity can stimulate interdisciplinary integration and the development of a more complex research framework more adapted to contemporary challenges.

4.3 Word Frequency

Through word frequency analysis, we aimed to identify the main terms directly associated with this research topic, the central goal being to define the term "green finance ecosystem". The choice of the abstract as part of the word frequency analysis is motivated precisely by its definition, namely, it is that part of the paper that contains the summary of the key points of the paper (Bahadoran et al., 2020). Moreover, the large number of articles analysed led us to consider only one part of the document for the efficiency of the analysis. The first result generated by the MAXQDA software is shown in Fig. 8 and represents the 50 most frequent noun words with a minimum frequency of 50 occurrences in the data sample. Compared to the keyword occurrence analysis, we can identify individual words that build the body of scientific literature and describe the main results of the studies.

The centrality of a word is associated with its size, indicating high frequency. Thus, "finance" becomes the most frequent word in the abstract documents, followed by "development," "energy," "innovation," and "financing." To outline a definition of the green finance ecosystem, we selected the first 20 words found in Table 4 to include as large a part of the body of specialised literature as possible.

The frequency value depends on the number of occurrences in a document and the number of occurrences in each document. In terms of the number of occurrences in total documents, "finance" still ranks first, followed by "development" and "impact", all three being key words in the definition construction. Referring to the words in the table, we can say that "green finance ecosystem" is a complex model of economic activity, in which financing, development and innovation are oriented towards efficiency, reduced environmental impact and sustainable growth,

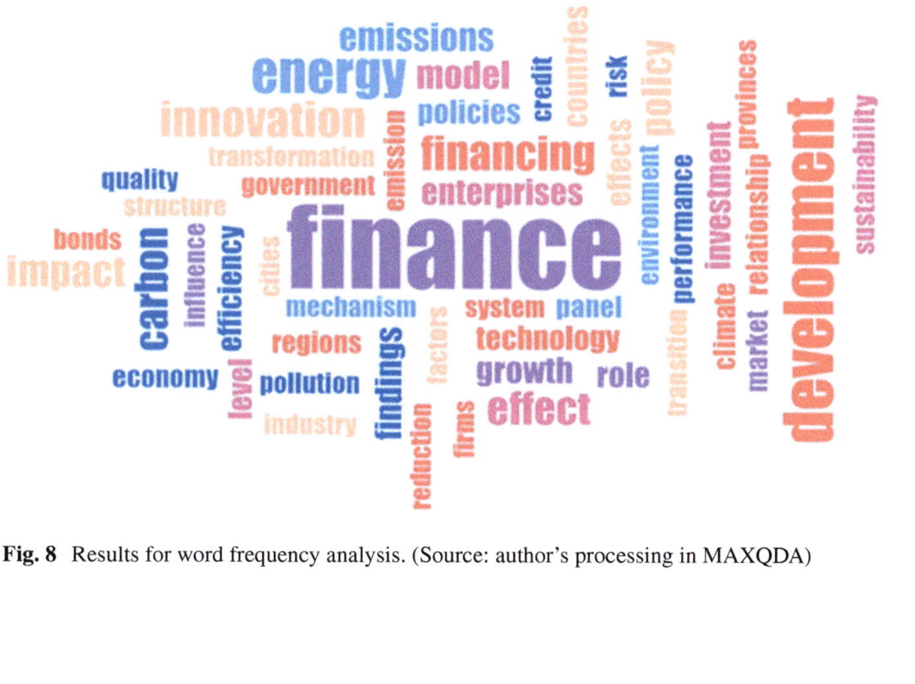

Fig. 8 Results for word frequency analysis. (Source: author's processing in MAXQDA)

Table 4 Results for word frequency analysis

Word	Word length	Frequency	%	Rank	Documents	Documents %
Finance	7	6861	5.56	1	1718	74.08
Development	11	3547	2.87	2	1331	57.40
Energy	6	2660	2.16	3	706	30.44
Innovation	10	2095	1.70	4	739	31.87
Financing	9	1967	1.59	5	735	31.69
Carbon	6	1752	1.42	6	569	24.54
Effect	6	1740	1.41	7	868	37.43
Impact	6	1613	1.31	8	1011	43.60
Policy	6	1489	1.21	9	759	32.73
Model	5	1207	0.98	10	797	34.37
Emissions	9	1181	0.96	11	455	19.62
Findings	8	1114	0.90	12	893	38.51
Enterprises	11	1043	0.85	13	344	14.83
Growth	6	1034	0.84	14	548	23.63
Investment	10	959	0.78	15	492	21.22
Role	4	868	0.70	16	663	28.59
Technology	10	825	0.67	17	390	16.82
Efficiency	10	820	0.66	18	323	13.93
Countries	9	800	0.65	19	431	18.59
Effects	7	784	0.64	20	524	22.60

Source: author's processing in MAXQDA

integrating carbon emission reduction policies with the development of efficient technologies and investments in green energy.

4.4 *Policy Review*

The Policy Review methodology aims to enable a uniform and valid comparison between policies with similar objectives but different structures and efficiencies. The first policy analysed in Table 5 is the European Green Deal (The European Green Deal, 2021), the European Union's framework strategy for the transition to a green and sustainable economy, with the primary objective of achieving climate neutrality by 2050. The leading actors are the European Commission, supported by the European Parliament, the Council of the EU, and specialised agencies such as the European Environment Agency or the European Investment Bank. The instruments used include European funds such as InvestEU and the Just Transition Fund, regulations (such as the "Fit for 55" package), green taxation schemes, and cross-cutting energy, agriculture, and transport policies. The type of intervention is strategic and integrated, targeting all sectors of the economy. Among the main limitations are institutional complexity—decisions are slow and difficult to coordinate between

Table 5 Policy review matrix

No.	Document/ policy	Level (National/ EU/Global)	Main actors	Key objectives	Instruments used
1.	"European Green Deal" (2019)	EU	European Commission	Climate neutrality by 2050	European funds, Regulations, Green Taxes
2.	EU Taxonomy Regulation (2020/852)	EU	European Parliament	Uniform classification of sustainable economic activities	Technical eligibility criteria, reporting obligations
3.	Sovereign green bonds (RO, 2024)	National (Romania)	Ministry of Finance, National Bank of Romania	Financing green projects	EUR bond issue with green prospectus
4.	Energy efficiency programs—KfW (Efficiency House Standards) (2009)	National (Germany)	Ministry of Economy and Climate (BMWK), KfW Bankengruppe	Improving the energy performance of buildings, reducing CO_2 emissions	Subsidised loans and grants for home modernisation according to "Efficiency House" levels (55, 70, 85, etc.)
5.	GFANZ (Glasgow Financial Alliance for Net Zero) (2021)	Global	Banks, investment funds, and insurers	Aligning financial assets with the objectives of the Paris Agreement	Voluntary commitments, decarbonization plans, reports
6.	NGFS (Network for Greening the Financial System) (2017)	Global	Central banks, regulatory authorities	Integrating climate risks into the financial system	Guides, reports, climate scenarios, recommendations

Source: Data processed by the author

states—and opposition from certain governments concerned about the economic impact. The main observation is that this policy provides the general framework for most green policies in the EU, but implementation depends on the will and capacity of member states.

The second policy selected is the EU Taxonomy Regulation (EU Taxonomy for Sustainable Activities, n.d.), an essential reference document for green financing in the European Union, as it provides a unified system for classifying sustainable economic activities. Adopted by the European Parliament, in collaboration with the European Commission and the Council of the EU, the regulation aims to increase transparency in the financial sector and prevent greenwashing. The key instruments are technical eligibility criteria, reporting obligations for large companies and financial actors, and interconnection with other regulations (e.g. SFDR, CSRD). It is a regulatory intervention with a significant impact on investment decisions. However, its application is hampered by the lack of quality ESG Environmental, Social and Governance) data, difficulties interpreting the criteria, and the slow process of updating the annexes. It is worth noting that, although it is a powerful tool, its effectiveness depends on how quickly and comprehensively the criteria are updated and the quality of national monitoring.

At the national level, we have adopted two documents/policies, namely Sovereign green bonds (Romania) and Energy efficiency programs—KfW (Efficiency House Standards) (Germany). The issuance of green bonds by the Romanian state in 2024 is a national initiative to finance projects with a positive environmental impact, coordinated by the Ministry of Finance and supported by the National Bank of Romania. The main objective is to attract international capital to develop green infrastructure. The type of intervention is financial, and its success depends on investor confidence. Obstacles include the immaturity of Romania's green financial market, the lack of an ESG investment culture, and dependence on external markets. Although this is a significant step, Romania needs a more robust institutional ecosystem to transform this initiative into a sustainable green financing mechanism. As for the programmes implemented in Germany by the Federal Ministry for Economic Affairs and Climate Action (BMWK) and the state-owned bank KfW (Kolb, 2017), they aimed to increase the energy efficiency of buildings and reduce CO_2 emissions. The main instruments used were low-interest loans and grants for building renovation, conditional on meeting specific energy standards (e.g., "Efficiency House 55"). The intervention is financial and informative, involving mandatory technical advice for beneficiaries. Although the program was successful (it led to renovating over 240,000 homes and a reduction of 5.9 Mt CO_2 annually), its limitations included unequal access between regions, administrative complexity, and high initial costs for beneficiaries.

Globally, we have selected as a document/policy GFANZ (Glasgow Financial Alliance for Net Zero) and NGFS (Network for Greening the Financial System). GFANZ is a global initiative that brings together banks, investment funds, insurance companies, and other financial institutions to align financial portfolios with the Paris Agreement. (Glasgow Financial Alliance for Net Zero, 2025). The leading players are international financial entities such as HSBC, BlackRock, and Allianz, which cooperate with the United Nations and other global entities. The instruments are voluntary commitments, decarbonization plans, and the publication of annual progress reports. Although it has substantial symbolic value and creates pressure on financial markets, the lack of sanctions and homogeneity among members limit its real impact. Despite the impressive capital mobilisation, without mandatory regulations or independent verification mechanisms, these commitments risk remaining at the declarative level. The NGFS is a global network of central banks and financial regulators that promotes the integration of climate risks into financial system supervision (NGFS Short-term Climate Scenarios for Central Banks and Supervisors | Network for Greening the Financial System, n.d.). The tools include guidelines, climate scenarios, best practice reports, and financial supervision and regulation recommendations. However, the standards are not mandatory, and implementation varies significantly between regions. The NGFS plays a key role in raising awareness of climate risks in the banking sector, but its effectiveness depends on the transposition of recommendations into national legislation.

In conclusion, this analysis shows that the best structured and coherent documents/policies are the Taxon's own EU Green Deal, the most structured and cohesive of all the EU's Green Papers/policies. This is as clear, concrete, coherent, and accompanied by performance indicators. The GFANZ initiative is based on voluntary participation, which reduces and limits its scope. The concrete rules on the eligibility of funding are clearly visible at the European and national levels, and the European support schemes are being developed and implemented in an obvious and transparent way, with clear eligibility criteria and precise mechanisms for funding. As far as the priorities of the European Union's institutions are concerned, Germany is proving to be an active leader, with the involvement of the Investment Bank and the KfW, its institution in Romania, and in the field of decentralization, with a very consistent, but still modest, approach: the Ministry of Finance and the NBR are making progress, but without a consolidated institutional network. Among the four main shortcomings are the absence of national regulators (in the case of Romania), the low level of enforcement of the GFA Z, and the lack of sanctions or a unitary regulatory framework (NGFS).

The ecosystem of green financing works piecemeal: at the European level, there is a solid, stable, and widely varying degree of ambition between Member States, and at the global level, a lack of local regulatory frameworks leads to inefficiencies. Amongst a few others, some outstanding regulations (e.g., GE tax, CSRD) are in a constant development process and provide a model of good practice in clean energy.

5 Conclusions

Green finance has emerged as a central element in the reconstruction of economic paradigms in a global context characterised by environmental crises, social pressures and the urgent need for a sustainable transition. Our enquiry has brought the complexity and potential of this burgeoning field to the fore. The present study conceptualised the green finance ecosystem, combining a rigorous theoretical analysis with an extensive bibliometric approach, a Research Area Diversity analysis and a review of relevant public policies.

By quantitatively analysing more than 2300 scientific papers and using the VOSviewer software, we were able to map scientific networks, identify major thematic clusters and assess the diversity of the field. A clear picture of a maturing field has emerged, where concepts such as ESG, green bonds, digitisation and green growth theory converge in an increasingly complex financial architecture. Green Finance is thus not just a theoretical construct but an emerging reality in public policy and global markets.

The results also revealed significant geographical and institutional fragmentation. Europe remains the leader in regulation, while Asia, particularly China, stands out in terms of research volumes and practical applicability. Therefore, harmonising the international regulatory framework and strengthening cooperation between financial, political and academic actors is a key challenge. The research also found a contrast between developed regions with clear regulatory frameworks (such as the EU or China) and emerging regions where the lack of a consolidated ecosystem limits the practical applicability of green instruments. Analysing public policies—from the European Green Deal to sovereign green bonds—reveals multiple but often fragmented efforts towards global sustainable financing. The lack of a unified assessment and reporting framework remains a major obstacle to the effectiveness of these initiatives.

The analysis of the frequency of words in the abstracts of the documents allowed us to outline a definition of the concept of "green finance ecosystem", thus contributing to the development of the specialised literature. In addition, the low scores for the indices within the research area diversity analysis indicate an almost unidirectional orientation of the research, which means that researchers need to integrate elements of interdisciplinarity in their works to encompass everything that the green finance ecosystem means. As we stated in the definition, GFE is not limited to green finance; it also considers green consumption, sustainable development, environmental protection mechanisms, efficient policies, and digital transformation.

At the same time, current reality shows us that progress is not linear: phenomena such as greenwashing or greenhushing can distort perceptions and jeopardise the credibility of the whole system. This is why future directions for research and action must be anchored in a logic of transparency, institutional coherence and international collaboration. Green Finance cannot operate in isolation; it must be integrated into a broad mechanism that connects financial actors, public policies and scientific research.

Linking bibliometric findings with policy analysis shows that the success of Green Finance depends on the coherent integration of financial instruments with social and environmental objectives. Thus, a systemic approach that harnesses the potential of both technological innovation and responsible governance is essential. At the same time, phenomena such as 'greenwashing' or 'greenhushing', which risk undermining public confidence and the effectiveness of interventions, must be avoided.

Looking to the future, the green finance ecosystem has the potential to redefine global economic architecture. Still, its success will depend on our collective ability to implement it coherently, equitably, and in a way that adapts to local realities. A clear direction towards standardisation, digitisation, financial education and openness to innovation is needed so that the green transition is not only possible, but sustainable over time.

Acknowledgments This work was supported by a grant from the Romanian Ministry of Research, Innovation and Digitalization, the project with the title "*Economics and Policy Options for Climate Change Risk and Global Environmental Governance*" (CF 193/28.11.2022, Funding Contract no. 760078/23.05.2023), within Romania's National Recovery and Resilience Plan (PNRR)—Pillar III, Component C9, Investment I8 (PNRR/2022/C9/MCID/I8)—Development of a program to attract highly specialised human resources from abroad in research, development, and innovation activities.

Competing Interests The authors have no conflicts of interest to declare relevant to this chapter's content.

References

Bahadoran, Z., Mirmiran, P., Kashfi, K., & Ghasemi, A. (2020). The principles of biomedical scientific writing: Abstract and keywords. *International Journal of Endocrinology and Metabolism, 18*(1). https://doi.org/10.5812/ijem.100159

Banga, J. (2018). The green bond market: a potential source of climate finance for developing countries. *Journal of Sustainable Finance & Investment, 9*(1), 17–32. https://doi.org/10.1080/20430795.2018.1498617

Bernatović, I., Gomezel, A. S., & Černe, M. (2021). Mapping the knowledge-hiding field and its future prospects: a bibliometric co-citation, co-word, and coupling analysis. *Knowledge Management Research & Practice, 20*(3), 394–409. https://doi.org/10.1080/14778238.2021.1945963

Bistline, J. E. T., Brown, M., Domeshek, M., Marcy, C., Roy, N., Blanford, G., Burtraw, D., Farbes, J., Fawcett, A., Hamilton, A., Jenkins, J., Jones, R., King, B., Kolus, H., Larsen, J., Levin, A., Mahajan, M., Mayfield, E., McFarland, J., … Zhao, A. (2023). Power sector impacts of the Inflation Reduction Act of 2022. *Environmental Research Letters, 19*(1), 014013. https://doi.org/10.1088/1748-9326/ad0d3b

Busch, D., Ferrarini, G., & Van Den Hurk, A. (2021). The European Commission's Sustainable Finance Action Plan and other international initiatives. In *EBI studies in banking and capital markets law* (pp. 19–59). https://doi.org/10.1007/978-3-030-71834-3_2

Cai, D. C., & Zhang, Q. A. (2009). General Introduction about the development of sustainable development theory in China. In *Dynamics of urban agglomeration in China: Preferences of energy-saving and environment-friendly society* (pp. 11–15).

Chua, W. F., James, R., King, A., Lee, E., & Soderstrom, N. (2022). Task force on climate-related financial disclosures (TCFD) implementation: An overview and insights from the Australian accounting standards board dialogue series. *Australian Accounting Review, 32*(3), 396–405. https://doi.org/10.1111/auar.12388

Cucari, N., Tutore, I., Montera, R., & Profita, S. (2022). A bibliometric performance analysis of publication productivity in the corporate social responsibility field: Outcomes of SciVal analytics. *Corporate Social Responsibility and Environmental Management, 30*(1), 1–16. https://doi.org/10.1002/csr.2346

Dadabada, P. K. (2025). Analyzing the impact of ESG integration and FinTech innovations on green finance: A comparative case studies approach. *Journal of the Knowledge Economy, 16*, 7959–7978. https://doi.org/10.1007/s13132-024-02197-0

De Assis, T. F., Ricci, L. M., Monteiro, T. G. M., De Abreu, V. H. S., De Almeida D'Agosto, M., & Santos, A. S. (2022). Sustainable transport indicators and mitigation actions applied to the Green Bond principles. In *Environmental footprints and eco-design of products and processes* (pp. 139–169). https://doi.org/10.1007/978-981-19-7226-3_6

Desai, B. H. (2020). 14. United Nations Environment Programme (UNEP). *Yearbook of International Environmental Law, 31*(1), 319–325. https://doi.org/10.1093/yiel/yvab060

Dias, L. P. C., Bagatini, F. Z., & Perin, M. G. (2025). Green "under-communication:" a systematic literature review on greenhushing. *ReMark – Revista Brasileira De Marketing, 24*(1), 387–414. https://doi.org/10.5585/2025.26718

Erickson, L. E., & Brase, G. (2019). *Paris Agreement on climate change* (pp. 11–22). CRC Press eBooks. https://doi.org/10.1201/9781351116589-2

EU Taxonomy for Sustainable Activities. (n.d.). *Finance*. https://finance.ec.europa.eu/sustainable-finance/tools-and-standards/eu-taxonomy-sustainable-activities_en. Date of access June 2025.

Frydrych, S. (2021). Green bonds as an instrument for financing in Europe. *Ekonomia I Prawo, 20*(2), 239–255. https://doi.org/10.12775/eip.2021.014

Glasgow Financial Alliance for Net Zero. (2025, April 22). *Glasgow Financial Alliance for Net Zero*. https://www.gfanzero.com/. Date of access June 2025.

Hajian, M., & Kashani, S. J. (2021). Evolution of the concept of sustainability. From Brundtland Report to sustainable development goals. In *Sustainable resource management* (pp. 1–24). Elsevier eBooks. https://doi.org/10.1016/b978-0-12-824342-8.00018-3

Halisçelik, E., & Soytas, M. A. (2019). Sustainable development from millennium 2015 to sustainable development goals 2030. *Sustainable Development, 27*(4), 545–572. https://doi.org/10.1002/sd.1921

Huang, Y. (2025). Fintech as a catalyst for sustainable development: a bibliographic review of drivers, technologies, econometrics and regional insights. *Journal of Accounting Literature*. https://doi.org/10.1108/JAL-02-2025-0094

Issawi, C. (1978). The 1973 oil crisis and after. *Journal of Post Keynesian Economics, 1*(2), 3–26. https://doi.org/10.1080/01603477.1978.11489099

Khemiri, M. A. (2024). Examining the relationship between climate risk, economic policy uncertainty, and credit risk: evidence from MENA banks toward sustainable economic development. In *Palgrave studies in impact finance* (pp. 579–593). Springer International Publishing. https://doi.org/10.1007/978-3-031-65756-6_21

Kolb, C. (2017). *KfW efficiency houses & swarming – time for energy transition 2.0. KfW stories.* https://www.kfw.de/stories/kfw/stories/environment/energy-efficiency/energieeffizienzhaus/. Date of access June 2025.

Kooths, S. (2022). EU taxonomy: Mission impossible. *The Economists Voice, 19*(2), 243–249. https://doi.org/10.1515/ev-2022-0028

Larsen, M. L. (2022). Driving global convergence in green financial policies: China as policy pioneer and the EU as standard setter. *Global Policy, 13*(3), 358–370. https://doi.org/10.1111/1758-5899.13105

Low, N. (2023). Climate transformation and social transformation. In *Social transformation for climate change* (pp. 3–9). Routledge. https://doi.org/10.4324/9781003382133-2

Maragopoulos, N. (2023). Toward a European Green Bond Standard: a European initiative to promote sustainable finance. In *EBI studies in banking and capital markets law* (pp. 21–50). https://doi.org/10.1007/978-3-031-38692-3_2

Miyamoto, M., & Takeuchi, K. (2019). Climate agreement and technology diffusion: Impact of the Kyoto Protocol on international patent applications for renewable energy technologies. *Energy Policy, 129*, 1331–1338. https://doi.org/10.1016/j.enpol.2019.02.053

Morrissey, J., & Heidkamp, P. (2022). Sustainability after COVID-19: pillars for a just transition. *Environmental Sustainability, 5*(2), 261–269. https://doi.org/10.1007/s42398-022-00231-y

Mukherjee, A., Owen, R., Scott, J. M., & Lyon, F. (2024). Financing green innovation startups: a systematic literature review on early-stage SME funding. *Venture Capital*. https://doi.org/10.1080/13691066.2024.2410730

NGFS Short-term Climate Scenarios for central banks and supervisors | Network for Greening the Financial System. (n.d.). Network for Greening the Financial System. https://www.ngfs.net/en/publications-and-statistics/publications/ngfs-short-term-climate-scenarios-central-banks-and-supervisors. Date of access June 2025.

Ozili, P. K. (2022). Green finance research around the world: a review of literature. *International Journal of Green Economics, 16*(1), 56. https://doi.org/10.1504/ijge.2022.125554

Palaghianu, C. (2016). *A tool for computing diversity and consideration on differences between diversity indices*. arXiv (Cornell University). https://doi.org/10.48550/arxiv.1602.04005

Pich, C. (2021). The belt and road initiatives and the master plan on ASEAN connectivity 2025: Connecting the dots, integrating the region. In *Series on Asian Regional Cooperation Studies* (pp. 17–29). World Scientific. https://doi.org/10.1142/9789811234316_0002

Ruddock, L., & Ruddock, S. (2022). Investment in infrastructure as a key to sustainable economic recovery: the role of the building industry. *International Journal of Strategic Property Management, 26*(6), 439–449. https://doi.org/10.3846/ijspm.2022.18430

Schlacke, S., Wentzien, H., Thierjung, E., & Köster, M. (2022). Implementing the EU climate law via the 'Fit for 55' package. *Oxford Open Energy, 1*. https://doi.org/10.1093/ooenergy/oiab002

Sohail, M. T., Ullah, S., Ozturk, I., & Sohail, S. (2025). Energy justice, digital infrastructure, and sustainable development: A global analysis. *Energy, 319*. https://doi.org/10.1016/j.energy.2025.134999

Spulbar, C., & Birau, R. (2019). Determinants of bank cost efficiency in transition economies: evidence for Latin America, Central and Eastern Europe, and South-East Asia. *IGI Global Scientific Publishing*. https://doi.org/10.4018/978-1-5225-9269-3.ch001

Sun, J. (2025). Sustainable FinTech for sustainable tourism, inclusive partnership, and smart city in low-carbon transition: alipay+ ecosystem in Singapore (Changi Pay, EZ-Link Wallet, ComfortDelgro, GrabPay, OCBC Digital), South Korea (Kakao Pay, NAVER Pay, Toss) and Japan (Paypay). In *Sustainable development goals series* (pp. 467–512). Springer Nature Singapore. https://doi.org/10.1007/978-981-96-3580-1_17

The European Green Deal. (2021, July 14). *European Commission*. https://commission.europa.eu/strategy-and-policy/priorities-2019-2024/european-green-deal_en. Date of access June 2025.

Zhang, Y. F., & Umair, M. (2023). Examining the interconnectedness of green finance: an analysis of dynamic spillover effects among green bonds, renewable energy, and carbon markets. *Environmental Science and Pollution Research, 30*, 77605–77621. https://doi.org/10.1007/s11356-023-27870-w

Green Finance and Social Equity: Bridging the Gap Between Sustainability and Inclusivity

Mert Topcu and Salih Ozdemir

1 Introduction

Sustainability has emerged as the defining imperative of the twenty-first century, encompassing environmental protection, economic resilience, and social equity as its three core pillars. As the adverse impacts of climate change, biodiversity loss, and environmental degradation intensify, achieving sustainability is no longer a matter of choice but a fundamental condition for long-term human and ecological survival (IPCC, 2023). Climate change, in particular, has become a pervasive risk factor, affecting developed and developing nations alike and threatening the stability of financial systems by exposing banks, insurance companies, corporations, households, and governments to mounting environmental threats (European Commission, 2024). Addressing these intertwined risks necessitates a profound transformation in financial decisions, with sustainability principles guiding investments and policymaking at every level.

Green finance is at the heart of this transformation, a powerful policy and market-based lever for steering capital toward projects with clear environmental and social benefits, such as clean energy, sustainable agriculture, low-carbon infrastructure, and climate adaptation. Green financial instruments, including green bonds, ESG funds, and sustainability-linked credit, have helped mainstream environmental considerations into financial decision-making, internalise ecological risks, and stimulate innovation in green technologies (Flammer, 2021; Bhutta et al., 2022; Fu et al., 2023). Sustainable finance, by explicitly incorporating environmental, social, and

M. Topcu (✉)
Department of Economics, Alanya Alaaddin Keykubat University, Alanya, Antalya, Türkiye
e-mail: mert.topcu@alanya.edu.tr

S. Ozdemir
Department of International Trade and Logistics, Kayseri University, Kayseri, Türkiye
e-mail: salihozdemir@kayseri.edu.tr

C.-W. Su et al. (eds.), *Green Finance and the Challenges of Climate Change*, Sustainable Finance, https://doi.org/10.1007/978-3-032-07293-1_2

governance (ESG) criteria, provides a framework to ensure that long-term investments promote environmental integrity and social well-being (European Commission, 2025). However, a central question remains: can green finance deliver sustainability in an environmentally effective and socially inclusive manner?

Inequality plays a crucial and often underexamined role in shaping the outcomes and effectiveness of sustainability efforts. While sustainability is typically understood in environmental terms, it fundamentally reflects a normative framework rooted in intergenerational justice, inclusive prosperity, and the equitable distribution of resources and risks (United Nations, 2015). Social inequality is not merely a parallel challenge to environmental degradation; it is a structural factor that shapes how environmental interventions unfold in practice (Berthe & Elie, 2015). Within the green finance framework, social considerations such as inequality, inclusiveness, labour relations, investment in human capital, community development, and human rights are just as essential as environmental and governance factors (European Commission, 2025). Policies designed to address climate and environmental pressures must, therefore, be responsive to the social structures within which they are embedded; otherwise, they risk entrenching existing inequalities or generating new forms of exclusion.

While the literature on green finance has expanded significantly, much of it has focused on the environmental dimension, with comparatively less attention devoted to its social dimension, particularly issues of inequality. The existing research increasingly emphasises the mediation role of sustainable finance in advancing environmental and economic outcomes. Yet, it consistently neglects the social equity dimension, overlooking how income inequality, access disparities, and institutional inclusivity shape who ultimately benefits from green finance interventions (Guo et al., 2022; Wang & Ma, 2022; Wang et al., 2023; Zhang & Jin, 2024; Meng et al., 2025). Emerging studies highlight several thematic clusters within the green finance domain, including green investments, green economic growth, environmental performance, green innovation, corporate social responsibility, climate risk management, and financial development (Desai & Patel, 2025; Hu & Gan, 2025). However, these works often assume the uniform effectiveness of green finance mechanisms, paying insufficient attention to the contextual and distributional factors determining who benefits from these initiatives and under what conditions. Specifically, the moderating role of income inequality within the green finance–sustainability relationship remains underexplored, particularly in comparative, cross-national research.

This chapter addresses this critical gap by investigating whether and how income inequality influences the effectiveness of green finance in achieving sustainability goals. The central research question we raised is whether income inequality moderates the relationship between green finance and sustainability, and if so, how does it affect the inclusivity of sustainability outcomes? We hypothesise that while green finance positively contributes to sustainability performance, its effectiveness in delivering inclusive sustainability outcomes is significantly diminished in contexts marked by high income inequality. In unequal societies, barriers to access, participation, and institutional responsiveness limit the extent to which green financial flows can generate system-wide, socially just environmental improvements. Conversely, more egalitarian contexts are better positioned to translate green investments into broad-based sustainability gains, as inclusivity is embedded in their financial and institutional frameworks. We argue that models overlooking the moderating role of income inequality risk overestimate the transformative potential of green finance, thereby producing policies that fall short of achieving both environmental targets and social inclusivity, ultimately undermining the goals of a just and equitable ecological transition.

To empirically investigate these relationships, we conduct a cross-country panel analysis covering 24 developed and developing countries from 2015 to 2022. We apply the Driscoll-Kraay technique, which is particularly well-suited for addressing heteroskedasticity in panel data, thus providing robust estimates of the conditional effectiveness of green finance across diverse socio-economic settings. Through this approach, we aim to contribute to the literature on green finance and environmental economics and broader debates on sustainable development, social justice, and the design of inclusive policy frameworks.

The remainder of this chapter is organised as follows: Section 2 reviews the relevant literature on green finance, inequality, and sustainability, identifying key debates and research gaps. Section 3 presents the data and modelling approach, detailing the variables, sources, and empirical strategy employed. Section 4 reports the main empirical findings, examining the interaction between green finance and income inequality. Finally, Section 5 provides concluding remarks, discussing the policy implications of the results, offering insights into how green finance mechanisms can be better designed to promote inclusivity alongside environmental effectiveness.

2 Literature Review

Green finance has increasingly been viewed as an essential instrument in facilitating global environmental goals, particularly in mitigating climate change, promoting low-carbon technologies, and investing in sustainable infrastructure. The literature generally agrees on the capacity of long-term green finance projects to stimulate environmental innovation and support green transition pathways for sustainable and inclusive development across various economies (Bhattacharyya, 2022; Tariq & Hassan, 2023). It includes tools such as green bonds, ESG funds, and sustainability-linked credit mechanisms to reorient capital flows toward environmentally beneficial outcomes (Fu et al., 2023). However, while green finance has demonstrated strong environmental performance, its effectiveness in advancing social equity remains ambiguous. Much of the existing research emphasises macroeconomic or environmental indicators but does not adequately examine the distributional consequences of green financial mechanisms. In addition to commonly cited instruments like ESG funds and green bonds, recent reviews have emphasised the role of systemic actors in shaping green finance ecosystems. For instance, Cui et al. (2020)

developed an evolutionary game model incorporating governments, financial institutions, enterprises, and consumers, finding that system integrity enhances both sustainable development and clean production. Similarly, Liu and Wu (2023) highlight that rising environmental awareness has increased demand for sustainability disclosure, though assurance and consistency challenges persist.

A growing strand of the literature has begun to explore this complexity through mediation and moderation frameworks that reveal the underlying mechanisms and contextual conditions influencing green finance effectiveness. Several studies emphasise mediation effects, where green finance either acts as a transmission channel or is shaped by broader economic, institutional, or organisational processes. Zhou et al. (2022) find that fintech innovation improves green growth through the mediating role of green finance. Similarly, Wang and Ma (2022) demonstrate that green finance contributes to CO_2 emissions reduction via industrial restructuring and economic growth, while Li et al. (2025) show that green finance policies reduce regional carbon emissions by promoting green technological innovation and optimizing industrial structure. In a complementary finding, Guo et al. (2022) highlight the mediating role of technological innovation in linking green finance policies to emission reduction, emphasising the importance of coordinated innovation alliances and regional policy support. In a corporate governance context, Wang et al. (2023) reveal that corporate social responsibility mediates the relationship between green governance, green finance, and sustainability performance. While these studies uncover how green finance mediates various sustainability mechanisms, others focus on the conditional nature of its effectiveness by examining moderating influences. In contrast, a smaller but growing body of work has adopted moderation-based models to investigate the conditional nature of green finance outcomes. Iqbal et al. (2024) show that employee green values and knowledge sharing significantly moderate the influence of bank policies on green financing sustainability. Zhang and Jin (2024) identify that industrial structure optimisation enhances the impact of green finance on improving energy consumption structures. These studies underscore that green finance is not universally effective, and its performance depends heavily on institutional, behavioural, and structural conditions.

Beyond these methodological contributions, a number of studies also raise conceptual concerns about green finance's inclusiveness. For example, Dong et al. (2024) highlight that while green finance has fostered inclusive development in some Chinese cities, the benefits are uneven and more pronounced in areas with lower baseline inclusivity. At the sectoral level, Zheng et al. (2021) document how green finance enhances social, economic, and environmental performance in the banking sector, supporting Sustainable Development Goals. Moreover, Chen et al. (2024) explored the relationship between digital inclusive finance and corporate green innovation in China, A-share listed companies spanning 2014–2019, through a multidimensional fixed-effect model. The findings reveal that digitally inclusive finance markedly enhances firms' capacity for green innovation. Another study, Van Niekerk (2024), views that green finance can only bridge employment, opportunity, and wealth gaps when deliberately designed with inclusion in mind. Yet, empirical findings on the social impacts of green finance remain mixed. Simionescu and

Gavurová (2023) highlight that income inequality influences pollution patterns in newer EU member states, underscoring the role of distributive dynamics in sustainability transitions. In a study on OECD countries, Han and Gao (2024) find that green finance improves environmental outcomes, but its influence on social inclusion is statistically insignificant. Bhutta et al. (2022) expand this discussion by underlining the critical role of green bonds in financing sustainable projects and drawing attention to regulatory and disclosure-related challenges. Reviews such as those by Bhutta et al. (2022) and Fu et al. (2023) emphasise regulatory gaps and disclosure limitations as persistent obstacles, especially regarding green bond issuance. Using a bibliometric approach, Khan et al. (2024) further call for developing standardised performance indices and integrating with fintech to improve environmental and financial outcomes. Meanwhile, Meng et al. (2025) highlight that China's Green Finance Reform and Innovation Policy (GFRI) reduced emissions through improved resource allocation and governance, particularly when coupled with inclusive financial mechanisms.

While these studies significantly advance the field by clarifying both direct and mediated pathways, they often overlook deeper structural conditions that may alter the effectiveness of green finance initiatives. Despite these efforts, income inequality remains almost absent from the literature as a moderating factor in the green finance–sustainability relationship. Existing studies tend to treat green finance as exogenously beneficial, without interrogating the structural socio-economic context that may amplify or constrain its impacts. As outlined in the introduction, inequality has the potential to substantially weaken or even reverse the environmental benefits of green finance, making it a critical variable that remains both empirically underexplored and conceptually undertheorized. The present study directly addresses this gap by testing the moderating role of income inequality in shaping the green finance–sustainability nexus, contributing to the development of a more equity-oriented and context-sensitive framework for assessing the true transformative potential of green finance.

3 Data and Model

This chapter utilises an unbalanced panel dataset covering both developed and developing economies over the period 2015–2022. The sample includes the following 24 countries: Australia, Austria, Brazil, Canada, China, Denmark, Finland, France, Germany, India, Italy, Japan, Republic of Korea, Luxemburg, Mexico, Netherlands, New Zealand, Norway, Spain, Sweden, Switzerland, United Arab Emirates, United Kingdom, and United States.

From an empirical perspective, this chapter aims to investigate the relationship between green finance and sustainability performance, while explicitly considering the role of social inequality as a potential moderator. As shown in function (1), we describe sustainability as a function of green finance, social inequality, and a set of control variables.

$$sustainability = f\left(green\ finance,\ social\ inequality,\ controls\right) \quad (1)$$

In this setup, sustainability is represented using the Sustainable Development Index (SDI), which measures each country's overall sustainability performance by combining environmental, social, and economic dimensions. On the right-hand side, green bond issuances are used as a key indicator of green finance, representing each country's annual volume of green bonds, capturing the scale of financial resources mobilised toward environmentally sustainable projects. The traditional Gini coefficient based on disposable income is used to capture social inequality. In addition, the model also incorporates two control variables: GDP per capita (in constant 2015 USD) as a proxy of quantitative development and voice and accountability as a proxy of governance quality.

Detailed descriptions of each variable, including their definitions, units of measurement, and data sources, are provided in Table 1. This setup is expected to provide a comprehensive framework to explore how green financial flows interact with social and institutional factors to shape sustainability performance across countries.

To empirically examine the relationship between green finance, inequality, and sustainability performance, we specify the following panel data model:

$$SDI_{it} = \beta_0 + \beta_1 \ln GDP_{it} + \beta_2 gq_{it} + \beta_4 greenfin_{it} + \beta_5 inq_{it} + \varepsilon_{it} \quad (2)$$

where SDI stands for sustainability, *lnGDP* stands for the natural logarithm of the per capita income, *gq* stands for the governance quality, *greenfin* stands for the green finance, and *inq* stands for the social inequality.

As discussed earlier, the benefits of green finance are not uniformly distributed but are shaped by a country's underlying socio-economic structure. In particular, income inequality may condition the extent to which green finance contributes to sustainability. In more unequal societies, the flow of green investments may disproportionately benefit wealthier groups or regions, thereby reducing their environmental reach and systemic effectiveness. Therefore, we extend Eq. (2) with the

Table 1 Variable description and data sources

Heading level	Explanation	Measure/unit	Source
SDI	Sustainable development index	Index (0–1)	Sustainable Development Index.org, Hickel (2020)
GDP	Gross Domestic Product per capita, constant 2015 prices	USD (constant 2015)	World Bank
gq	Voice and accountability	Estimate (−2.5 to +2.5)	World Development Indicators
grrenfin	Green bond issuances	Billion US Dollars	International Monetary Fund
inq	Gini index, disposable income	Index (0–100)	Standardised World Income Inequality Database (SWIID), Solt (2020)

inclusion of an interaction term to capture the conditional nature of green finance effectiveness.

$$SDI_{it} = \beta_0 + \beta_1 \ln GDP_{it} + \beta_2 gq_{it} + \beta_4 greenfin_{it} + \beta_5 inq_{it} + \beta_6 interaction_{it} + \varepsilon_{it} \quad (3)$$

With the inclusion of the interaction term,[1] estimation of the slope parameters defined in Eq. (3) allows for an explicit test of whether the impact of green finance on sustainability performance varies systematically with inequality levels.

Table 2 summarises the descriptive statistics of the variables. Notice that the highest standard deviation comes from the green finance variable, followed by the inequality variable, indicating that green finance levels vary widely across the sample countries and over time. This substantial variation reflects the heterogeneous capacity of countries, particularly across developed and developing economies, to mobilise financial resources for environmental purposes. The high dispersion validates the inclusion of a diverse sample, as it captures the unequal distribution of green finance and the differing institutional and socio-economic contexts under which it operates. Such diversity is essential to empirically assess whether income inequality moderates the effectiveness of green finance in achieving sustainability goals.

Table 3 shows the correlation matrix among the variables. Notice that all covariates, except for the inequality variable, provide a positive correlation with the sustainability performance, underscoring the importance of considering income

Table 2 Descriptive stats

Variable	Obs	Mean	Std. dev.	Min	Max
SDI	192	0.416	0.204	0.112	0.781
lngdp	192	10.432	0.869	7.367	11.612
gq	192	0.965	0.837	−1.660	1.780
greenfin	183	9.775	15.587	0.038	99.428
inq	184	32.494	5.524	25.500	47.400

Table 3 Correlation matrix

	SDI	lngdp	gq	greenfin	inq
SDI	1.000				
lngdp	0.671	1.000			
gq	0.3578	0.674	1.000		
greenfin	0.042	0.023	0.248	1.000	
inq	−0.526	−0.788	−0.709	−0.042	1.000

[1] We follow a demeaning procedure while constructing the interaction term. Note that demeaning the variables before constructing the interaction term ensures consistency with the fixed effects transformation and mitigates multicollinearity. It improves the interpretability of coefficients and allows the interaction term to accurately reflect the within-unit moderating relationship. Without demeaning, the model may conflate within and between variations or yield biased estimates due to incorrect interpretation of interaction effects under fixed effects estimation.

inequality not just as a control variable, but as a potential moderator that could alter the relationship between green finance and sustainability.

4 Methods and Findings

In panel data econometrics, analyses where the cross-sectional dimension exceeds the time dimension ($N > T$) often face various diagnostic issues, particularly heteroskedasticity. To address these problems, this study employs the robust standard errors estimator developed by Driscoll and Kraay (1998), which yields consistent results even when the error terms exhibit heteroskedasticity, serial correlation, and cross-sectional dependence. The Driscoll-Kraay standard errors can be implemented with pooled OLS, fixed effects, or random effects estimators. To determine the appropriate model, the F-test is used to compare the fixed effects model with the Pooled Ordinary Least Squares (POLS) model, while the Lagrange Multiplier (LM) test is used to compare the random effects model with the POLS model. The POLS model is preferred if the null hypothesis is not rejected in both tests. The Hausman specification test is a widely used approach to compare fixed and random effects models. When the null hypothesis of no correlation between the individual effects and the regressors is rejected, the fixed effects model is preferred (Das, 2019). Table 4 provides the results of the three tests applied to the panel OLS estimation[2] of Eq. (3). Empirical results indicate that the fixed effects model is the most appropriate among the estimators.

Having found the appropriate estimator, the slope coefficients in Eq. (3) can be estimated using the robust standard errors estimator by Driscoll and Kraay (1998). Estimation results reported in Table 5 indicate that i) a 1% increase per capita income increases sustainability by 0.080 unit, ii) a 1-unit increase in governance quality increases sustainability index by 0.651 unit, iii) a 1-unit increase in green finance increases sustainability index by 0.055 unit, iv) a 1-unit increase in inequality decreases sustainability by 0.015 unit and, v) a 1-unit increase in interaction term decreases sustainability index by 0.001 unit.

Table 4 Estimator determination

Tests	Probabilies
F-test	$p < 0.01$
Breusch-Pagan LM test	$p < 0.01$
Hausman test	$p < 0.01$

[2]Depending on the OLS estimation, we also calculate the VIF scores of the independent variables with a maximum of 3,15 and a mean of 2,39, indicating that there is no multicollinearity among the covariates.

Table 5 Regression results

Variables	Coefficients
lngdp	8.0027* (2.397)
gq	0.651* (0.234)
greenfin	0.055** (0.010)
inq	−0.015** (0.001)
interaction	−0.001** (0.0003)
constant	−33.610** (11.406)
R-squared	0.22
F-test	$p < 0.01$

The maximum number of lags to be considered in the autocorrelation structure is 2
** and * indicate significance at the 1% and 5% levels, respectively
The values in parentheses are robust standard errors
F is a test for joint model significance

5 Conclusion

This chapter examined whether income inequality moderates the relationship between green finance and sustainability. Empirical findings based on a cross-country panel covering 24 developed and developing economies from 2015 to 2022 confirm that while green finance improves sustainability performance generally, its impact becomes negative in the presence of high-income inequality. This validates our hypothesis and further indicates that inequality is not merely a limiting factor but a dominant structural constraint that can fully override the beneficial effects of green finance.

The findings of this chapter carry important implications for both policymakers and researchers. Our empirical results demonstrate that while green finance has a statistically significant and positive effect on sustainability performance, income inequality systematically moderates this relationship. Specifically, as inequality rises, the positive impact of green finance on sustainability outcomes not only diminishes but reverses direction, turning into a net negative effect. This implies that green financial instruments, in isolation, are not sufficient to generate system-wide environmental improvements, particularly in socio-economic contexts marked by exclusion and disparity. These findings underscore the urgent need to integrate social equity considerations into designing, implementing, and evaluating green finance mechanisms.

While the estimated effect of the combined influence of green finance and income inequality may appear relatively small in magnitude, its reversal in direction signals a meaningful structural distortion that should not be overlooked. In the context of sustainability policy, even modest adverse effects can accumulate over time and lead to disproportionately large environmental and social costs, especially in countries where persistent inequality is deeply rooted. This small average effect may mask more severe localised consequences, particularly for disadvantaged populations. Therefore, the finding should be interpreted cautiously: in unequal societies, green financial flows may inadvertently undermine sustainability goals rather than support them. Recognising and addressing this risk is essential for designing green finance frameworks that are both effective and equitable.

From a policy standpoint, the results suggest that green finance must be embedded in inclusive institutional and regulatory frameworks to reach its full transformative potential. Policymakers should avoid the assumption that financial tools alone will yield inclusive environmental outcomes. Instead, governments and development finance institutions should adopt a dual-focus approach that links green investment promotion with targeted redistributive and capacity-building policies. This could include:

- Broadening access to green finance instruments through inclusive financial services, such as community green bonds, subsidised credit lines for low-income households and small enterprises, and mobile-based green microfinance platforms.
- Strengthening institutional accountability and participation, ensuring marginalised communities have a voice in determining green investment priorities through participatory budgeting or local sustainability councils.
- Integrating social conditionality clauses into green bond issuance or green public procurement contracts, requiring that a share of project benefits (e.g., clean energy access, green jobs) directly serve disadvantaged groups.
- Coupling green finance with human capital investment, particularly in education and green skills training, to enhance the ability of underrepresented populations to engage in and benefit from the sustainability transition.

These policy recommendations highlight that green finance cannot substitute for broader structural reforms. In countries with high income inequality, the underlying distribution of power, resources, and opportunities must be addressed concurrently with the expansion of green finance. Without such structural changes, green financial mechanisms may reinforce patterns of elite capture, deepen spatial disparities, and generate "green enclaves" of development that fail to trigger systemic change.

From a methodological perspective, the findings of this chapter also offer key insights for the academic community. First, including interaction terms between financial and socio-economic variables, such as green finance and income inequality, enables a more context-sensitive understanding of sustainability outcomes. Future research should continue to explore nonlinear and conditional relationships within green transitions, moving beyond average treatment effects to unpack who benefits, under what conditions, and to what extent. Second, the use of robust panel estimation techniques such as the Driscoll-Kraay estimator proves valuable for addressing cross-sectional dependence and heteroskedasticity prevalent in cross-country datasets, particularly those involving policy variables.

Researchers should also consider expanding the conceptual framework of green finance to include multi-dimensional equity, incorporating income-based inequality and gender, spatial, and generational disparities. Moreover, future empirical work could benefit from disaggregating green finance into specific components, such as mitigation versus adaptation finance, public versus private flows, or domestic versus international sources, to identify where inclusivity bottlenecks are most acute.

In conclusion, this chapter reinforces that green finance can only catalyse inclusive sustainability if aligned with broader equity-enhancing strategies. As governments, investors, and international organisations seek to scale up green finance in response to the climate crisis, they must not overlook the social terrain upon which these policies unfold. Designing for impact requires recognising that sustainability without inclusivity is not truly sustainable, and that closing the equity gap is central to closing the sustainability gap. Ultimately, green finance should not be treated as a standalone solution but as a policy tool that requires complementary equity-enhancing reforms to be effective.

References

Berthe, A., & Elie, L. (2015). Mechanisms explaining the impact of economic inequality on environmental deterioration. *Ecological Economics, 116*, 191–200.

Bhattacharyya, R. (2022). Green finance for energy transition, climate action and sustainable development: Overview of concepts, applications, implementation and challenges. *Green Finance, 4*(1), 1–35.

Bhutta, U. S., Tariq, A., Farrukh, M., Raza, A., & Iqbal, M. K. (2022). Green bonds for sustainable development: Review of literature on development and impact of green bonds. *Technological Forecasting and Social Change, 175*, 121378.

Chen, C., Fan, M., & Fan, Y. (2024). The role of digital inclusive finance in green innovation. *PLoS One, 19*(12), e0315598.

Cui, H., Wang, R., & Wang, H. (2020). An evolutionary analysis of green finance sustainability based on multi-agent game. *Journal of Cleaner Production, 269*, 121799.

Das, P. (2019). *Econometrics in theory and practice: Analysis of cross section, time series and panel data with Stata 15.1*. Springer.

Desai, R., & Patel, S. (2025). Analysing the research development in green finance: Performance analysis and directions for future research. *FIIB Business Review*. https://doi.org/10.1177/23197145241306838

Dong, K., Zhao, C., Dong, X., & Taghizadeh-Hesary, F. (2024). Can green finance promote inclusive development? Empirical evidence from China. *Sustainability Science*, 1–15.

Driscoll, J. C., & Kraay, A. C. (1998). Consistent covariance matrix estimation with spatially dependent panel data. *Review of Economics and Statistics, 80*(4), 549–560.

European Commission. (2024, May 3). *Report on the monitoring of climate-related risk to financial stability*. Retrieved from https://finance.ec.europa.eu/publications/report-monitoring-climate-related-risk-financial-stability_en

European Commission. (2025, May 7). *Overview of sustainable finance*. Retrieved from https://finance.ec.europa.eu/sustainable-finance/overview-sustainable-finance_en

Flammer, C. (2021). Corporate green bonds. *Journal of Financial Economics, 142*(2), 499–516.

Fu, C., Lu, L., & Pirabi, M. (2023). Advancing green finance: A review of sustainable development. *Digital Economy and Sustainable Development, 1*, 20.

Guo, C. Q., Wang, X., Cao, D. D., & Hou, Y. G. (2022). The impact of green finance on carbon emission – Analysis based on mediation effect and spatial effect. *Frontiers in Environmental Science, 10*, 844988.

Han, J., & Gao, H. (2024). Green finance, social inclusion, and sustainable economic growth in OECD member countries. *Humanities and Social Sciences Communications, 11*, 140.
Hickel, J. (2020). The sustainable development index: Measuring the ecological efficiency of human development in the anthropocene. *Ecological Economics, 167*, 106331.
Hu, D., & Gan, C. (2025). Green finance development and its origin, motives, and barriers: An exploratory study. *Environment. Development and Sustainability*, 1–25.
IPCC. (2023, May 5). *AR6 synthesis report: Climate change 2023*. Retrieved from https://www.ipcc.ch/report/sixth-assessment-report-cycle/
Iqbal, Z., Akram, M., & Rao, Z. U. R. (2024). Unlocking green potential: A mediation-moderation analysis of bank policies-related practices and green financing sustainability in Pakistan. *International Journal of Ethics and Systems*. https://doi.org/10.1108/IJOES-02-2024-0057
Khan, H. H. A., Ahmad, N., Yusof, N. M., & Chowdhury, M. A. M. (2024). Green finance and environmental sustainability: A systematic review and future research avenues. *Environmental Science and Pollution Research, 31*(6), 9784–9794.
Li, H., Yu, Z., Chen, G., & Nie, Y. (2025). Research on the impact of green finance on regional carbon emission reduction and its role mechanisms. *Scientific Reports, 15*(1), 1–17.
Liu, C., & Wu, S. S. (2023). Green finance, sustainability disclosure and economic implications. *Fulbright Review of Economics and Policy, 3*(1), 1–24.
Meng, S., Duan, Y., & Zhou, W. (2025). How does green finance policy reduce the carbon emissions of polluting enterprises? A mediation effect analysis of inclusive finance. *Research in International Business and Finance, 102914*, 102914.
Simionescu, M., & Gavurová, B. (2023). Pollution, income inequality and green finance in the new EU member states. *Humanities and Social Sciences Communications, 10*(1), 1–11.
Solt, F. (2020). Measuring income inequality across countries and over time: The standardized world income inequality database. *Social Science Quarterly, 101*(3), 1183–1199.
Tariq, A., & Hassan, A. (2023). Role of green finance, environmental regulations, and economic development in the transition towards a sustainable environment. *Journal of Cleaner Production, 413*, 137425.
United Nations. (2015, May 4). *Transforming our world: The 2030 agenda for sustainable development*. Retrieved from https://docs.un.org/en/A/RES/70/1
Van Niekerk, A. J. (2024). Economic inclusion: Green finance and the SDGs. *Sustainability, 16*(3), 1128.
Wang, J., & Ma, Y. (2022). How does green finance affect CO2 emissions? Heterogeneous and mediation effects analysis. *Frontiers in Environmental Science, 10*, 931086.
Wang, L., Ur Rehman, A., Xu, Z., Amjad, F., & Ur Rehman, S. (2023). Green corporate governance, green finance, and sustainable performance nexus in Chinese SMES: A mediation moderation model. *Sustainability, 15*(13), 9914.
Zhang, H., & Jin, J. (2024). Assessing the effect of green finance, energy consumption structure and environmental sustainable development: A moderated mediation model. *Economic Change and Restructuring, 57*(2), 87.
Zheng, G. W., Siddik, A. B., Masukujjaman, M., & Fatema, N. (2021). Factors affecting the sustainability performance of financial institutions in Bangladesh: The role of green finance. *Sustainability, 13*(18), 10165.
Zhou, G., Zhu, J., & Luo, S. (2022). The impact of fintech innovation on green growth in China: Mediating effect of green finance. *Ecological Economics, 193*, 107308.

Artificial Intelligence and ESG Disclosure: Measuring Transparency in AI-Driven Firms

Ioana Birlan, Adriana AnaMaria Davidescu, and Robert-Ştefan Constantin

1 Introduction

As the economy transitions towards sustainability and digitalisation, ESG (Environmental, Social, and Governance) reporting transparency has become a major marker of corporate responsibility and non-financial risks. At the same time, with many businesses now incorporating artificial intelligence (AI) into their companies, new questions arise surrounding the legitimacy and accuracy of sustainability disclosures. While some warn of the implications of "digital greenwashing," others argue that the integration of artificial intelligence can help companies enhance reporting that will be more representative of actual performance. This research seeks to answer one main question: How does AI adoption influence ESG reporting transparency among large publicly traded companies? The study approaches this inquiry through a rigorous methodological framework grounded in causal inference

I. Birlan (✉)
Doctoral School of Economic Cybernetics and Statistics, The Bucharest University of Economic Studies, Bucharest, Romania
e-mail: ioana.birlan@csie.ase.ro

A. A. Davidescu
Department of Statistics and Econometrics, The Bucharest University of Economic Studies, Bucharest, Romania

Department of Education, Training and Labour Market, National Scientific Research Institute for Labour and Social Protection, Bucharest, Romania
e-mail: adriana.alexandru@csie.ase.ro

R.-Ş. Constantin
Faculty of Cybernetics, Statistics and Economic Informatics, The Bucharest University of Economic Studies, Bucharest, Romania
e-mail: constantinrobert21@stud.ase.ro

C.-W. Su et al. (eds.), *Green Finance and the Challenges of Climate Change*, Sustainable Finance, https://doi.org/10.1007/978-3-032-07293-1_3

and machine learning-based variable selection. To evaluate the degree to which ESG disclosure scores align with actual ESG performance, firms that adopt AI are compared with firms that do not, using a standardardized transparency measure. This research aims to estimate the causal effect of AI adoption on the relationship between ESG reporting and performance, positioning ESG disclosure practices as a proxy for greenwashing risk.

This research brings several innovations that strengthen both its stientific relevance and practical value. First, it introduces a hybrid approach for identifying AI-adopting firms, combining rule-based filters with zero-shot classification using transformer models. Moreover, this approach makes it possible to detect AI adoption even when companies do not explicitly mention it in public reports. Second, the study defines a composite ESG disclosure indicator as the standardized difference between a company's ESG Disclosure Score (Bloomberg) and its ESG Performance Score (Refinitiv). By centering this measure at the industry level (TRBC), the analysis allows for clearer testing of greenwashing behaviour. Third, it brings together causal inference and machine learning through the Post-Double LASSO procedure, which selects relevant covariates while avoiding bias from overfitting or weak predictors, ensuring more reliable causal analysis. Fourth, three comparative treatment effect estimation methods: Nearest Neighbor Matching, Entropy Balancing, and Covariate Balancing Propensity Score (CBPS) are applied to test the robustness of the findings. A Rosenbaum-type sensitivity analysis further examines how results hold under potential unobserved confounding. Finally, this study contributes to current debates concerning ESG transparency and the responsible use of AI, emphasizing the importance of intergrating digitalization indicators into future ESG scoring frameworks.

This research offers a threefold contribution to the existing literature. Empirically, it provides robust evidence that AI adoption is associated with a stronger alignment between ESG disclosure and actual ESG performance. Methodologically, it integrates methods from NLP and causal inference into an applied framework for evaluating corporate sustainability. Normatively, it highlights the need to revise ESG standards and criteria to account for the role of technological factors in shaping transparency and mitigating the risk of broad or inconsistent interpretations of greenwashing.

The paper is organised as follows. The introduction presents the topic's relevance and introduces the reader to the main objectives of the research, together with the innovation elements. Section 2. is dedicated to the literature review and presents the most relevant publications surrounding AI, ESG performance, and greenwashing risk. Section 3 is dedicated to data and methodology and includes identification of AI firms, indicator construction, and treatment effect estimation. Section 4 reports reports the empirical results and provides a detailed examination of the differences between AI and non-AI adopting firms. The last section has been dedicated to the discussion of the main empirical findings and it ends with conclusions and future research directions.

2 Literature Review

Environmental, Social, and Governance (ESG) criteria have long served as a key measure of corporate performance and risk. Yet, rapid technological advancements are introducing new challenges to this framework. Although ESG is intended to be a holistic measure that ties together a plethora of metrics, its interpretation often remains complex due to inconsistent metrics, fragmented regulations, and varying methodologies. These blurred lines in evaluating ESG performance can lead to greenwashing, a practice in which companies portray themselves as more environmentally or socially responsible than they truly are. As as tep toward more transparent ESG standards, Ekaristi et al. (2025) propose applying machine learning (ML) and deep learning methods to detect anomalies, assess ESG transparency, and create an avenue of real-time verification. However, they conclude that interdisciplinary, collaborative input is key to an improved recognized and standard ESG framework. Such standards would then identify risks related to not only potential greenwashing but also "under-communication," meaning that the real performance of the company exceeds reported performance.

Strong ESG performance is not simply about developing a more environmentally and socially friendly image of the company; it is also a new investment opportunity (Xie & Wu, 2025). The use of AI in this space has a double purpose; it goes beyond an assessment of the legitimacy of a firm's ESG to allow a firm to tie it to performance and perform more internal audits in a timely and efficient manner. Backing this up, Xie and Wu (2025) show that the use of AI generates considerable positive impacts, especially in companies that are digitally advanced, which indicates that the firm's level of digitalization is a critical factor in the green transition. They indicate that the greatest potential benefit of AI can be realised in the environmental dimension through the development of greener technologies, and in the social dimension through strengthened philanthropic responsibility. However, governance can be a challenge, as certain applications of AI may inadvertently weaken internal controls and overall performance, ultimately harming a firm's ESG standing.

Importantly, the impact of AI on ESG varies widely and is often be moderated by factors such as digitalization and market dynamics. In fact, the benefits of AI tend to be most pronounced among technology-intensive or highly competitive firms where innovation plays a central role (Xie & Wu, 2025). This reflects a broader understanding that AI excels at analyzing large volumes of information (Lu & Li, 2025; Zhou et al., 2025; Wang & Chen, 2025; Yu et al., 2025), which allows firms to extract better insights—such as identifying supply chain risks in real time or optimizing ESG investment portfolios through improved diversification. To further illustrate thes importance of context, Weng (2025) proposes that, for ESG purposes, AI is most useful in non-state-owned companies as they feel more pressure to respond quickly to market and regulatory changes influenced by new technologies for competitive advantage. Taken together, these findings suggest that the role of AI in ESG should be viewed through the interplay of technological innovation, market dynamics, and corporate structure.

The integration of AI into corporate strategies aimed at enhancing Environmental, Social, and Governance performance is far from universal. AI adoption is context-dependent and can be influenced by organizational characteristics such as industry and firm size, which suggest that tailored implementations generate the most value in organizations. This aligns with the resource-based view (RBV), which holds that specific AI capabilities are a strategic asset that provides better value when aligned with an organization's unique resources and context (Li et al., 2025; Yu et al., 2025).

The effective use of AI to achieve ESG-oriented objectives primarily depends on the firm's organizational maturity. This dependency can be better understood through concepts from organizational theory, particularly stakeholder theory and organizational learning theory (OLT). As Li et al. (2025) explain, stakeholder theory and OLT are grounded in the reciprocal relationship between a company and its stakeholders and the ability of the firm to learn, adapt and innovate (Li et al., 2025). For a technology that has the capacity to be as disruptive as AI, it is important to have a high level of trust from stakeholders to mitigate potential pushback and some level of learning orientation in place in order to navigate the complexities of implementation. This includes the firm's ability to mitigate operational and ethical risks, like algorithmic bias, data privacy infringement, and regulatory non-compliance. This ultimately creates the following hypothesis: To effectively translate AI capabilities into improved ESG outcomes, a firm must possess both strong stakeholder trust and internal capacity to manage AI-related risk.

The relationship between AI and enhanced ESG performance tends to be indirect, unfolding through a series of intermediary mechanisms. Li et al. (2025) identify green innovation as a key pathway, through which firms leverage AI to develop environmentally sustainable technologies Similarly, Yu et al. (2025) outlined that AI improves production and supply chain efficiencies, which in turn strengthens resource reallocation toward long-term ESG oriented initiatives.

The strength and longevity of this relationship are moderated by a range of contextual factors. For example, organizational resilience reinforces this relationship, as resilient firms are better positioned to leverage technological capabilities in pursuit of sustainability goals (Li et al., 2025). External contextual environmental factors also play a key role; there is greater realization of value from AI when used in more competitive industries, while high environmental uncertainty diminishes effect, as firms revert to short term stability (Yu et al., 2025). It is also difficult to summarize the context of ownership structure, as the literature is conflicting on the relationship overall. Some studies show there is strong and positive association between AI and ESG in state owned enterprises (SOEs) as they have greater resource access and are aligned with public policy goals (Li et al., 2025). Other studies have shown that the competitive pressures facing non-state owned enterprises (non-SOEs) allow them to reap the benefits of AI engagement, thus driving ESG opportunities by adapting to technology at a more rapid pace (Weng, 2025). Overall, complex interactions are at play, and thus it is essential to consider having a standardised reporting framework, such as the AI-ESG protocol from Sætra (2022), to allow firms to more systematically assess and transparently disclose these dynamics.

3 Methodology

The objective of this research is to investigate whether firms identified as adopting artificial intelligence differ in their ESG transparency than similar firms that do not adopt AI. This problem is tackled through a multi-stage causal inference framework that combines variable selection, covariate balancing, and matching estimators to estimate the effect.

The construction of the treatment indicator used to identify AI-adopting firms relies on a hybrid approach that leverages rule-based filtering, firm identity recognition, and transformer-based zero-shot classification. The approach has three steps. First, a list of well-known AI-core companies was compiled, and firms with publicly-acknowledged AI specialization or activities such as NVIDIA, Palantir and Snowflake were labeled acc. Second, key-word matching was applied to the business description field to detect specific AI-activities, such as machine learning, deep learning, neural network, or AI-powered, among others. The purpose of this step was to identify firms with strong textual signals of AI integration.

To reduce the risk of false negatives and better identify cases where AI activity is not explicitly stated such as when it is embedded within broader technology operations, a transformer-based large language model (facebook/bart-large-mnli) was used for zero-shot classification. This approach allows the model to assess whether a firm's business description supports the statement: "This company is an AI company," even without prior labeled examples. It treats the classification task as one of textual entailment, which is particularly useful when AI involvement is ambiguous, multi-layered, or still emerging which are common issues in real-world corporate disclosures (Moreno-Garcia et al., 2023; Zhang et al., 2025; Chae & Davidson, 2025). The model assigns a probability indicating how strongly a firm's description aligns with the AI or non-AI category. To ensure reliability, a threshold of 0.75 was used: only firms scoring above this cutoff were labeled as AI adopters. Those with lower confidence scores or unclear signals were conservatively classified as non-AI to reduce the chance of overestimating AI adoption.

Despite the multi-layered methodology there are limitations. For instance, Adobe and Electronic Arts (EA) were characterized as AI due to their generative media capabilities or predictive modeling tools being AI-like features, but not central to their business revenue. Likewise, a "No AI" classification does not necessarily mean that AI was completely absent; instead, it indicates that the firm's business description did not include dominant AI signalling. A manual review for borderline cases was undertaken to confirm model outputs and to ensure the tagging process was consistent with the conceptual intent of this study, which was isolating firms that had a significant and identity-defining capability concerning AI.

One of the constraints of this analysis is that it only examines firms within the S&P 500 index. This choice is a matter of practicality and a methodological issue. Companies in this benchmark index typically have more complete and reliable ESG disclosures with consistent financial reporting. Limiting the sample in this way helps reduce firm size variations, international exposure, and required disclosure levels, which may add noise to the analysis. Nonetheless, the S&P 500 still provides a diversified representation of sectors and technologies, allowing room to observe salient differences between firms. Moreover, using a well-defined universe helps to reduce biases and complications that are often attached to studying firms from different countries or listing segments.

The final dataset is comprised of firm-level data supplied by Bloomberg and Refinitiv, both commonly used financial data providers, consisting of 503 firms that were originally listed in the S&P 500 index. Bloomberg provides access to the ESG Disclosure Score, which reflects how much environmental, social, and governance-related information a firm reports in its sustainability communication. Disclosure Score reflects scope and frequency of reporting, not the underlying sustainability performance of the firm. The ESG Disclosure Score is expressed relative on a scale of 0–100.

The dimension on ESG performance was calculated by combining Refinitiv's ESG Combined Score. The ESG Combined Score summarises how well a firm managed ESG factors based on indicator-level data and event-level negative incidents categorised as ESG Controversies. To examine differences that may exist between a firm's self-presentation and its ESG performance, a transparency metric was operationalised in the same manner detailed by Yu et al. (2020).

The transparency metric was operationalised using the difference between the standard ESG Disclosure Score from Bloomberg and the standard ESG Performance Score from Refinitiv, which were normalised at the TRBC industry level. The transparency metric is the peer-relative measure and represents the outcome variable in this study as it reflects how aligned the firms' sustainability communications are, relative to actual ESG behaviours. A positive score reflects firms that disclose more than they perform, suggesting that these firms may overemphasise their sustainability communication, and vice versa for negative scores.

The standardized values for disclosure and performance were obtained via z-score normalization within each TRBC (Thomson Reuters Business Classification) industry group, ensuring that comparisons account for sector-specific reporting norms and ESG expectations. Let D_{jkt} and P_{jkt} denote the ESG disclosure and performance scores, respectively, for firm j in industry k at time t. Let $\underline{D_k}$ and σ_{D_k} be the mean and standard deviation of disclosure scores within industry k, and analogously $\underline{P_k}$, σ_{P_k} for performance scores. The standardized (peer-relative) values are given by:

$$\hat{D}_{jkt} = \frac{D_{jkt} - \overline{D}_k}{\sigma_{D_k}}, \hat{P}_j = \frac{P_{jkt} - \overline{P}_k}{\sigma_{P_k}} \tag{1}$$

The transparency score GW_{jkt} is then constructed as the difference between standardised disclosure and standardised performance.

$$GW_{jkt} = \hat{D}_{jkt} - \hat{P}_{jkt} \tag{2}$$

To ensure comparability, disclosure and performance scores were initially scaled to the [0,1] interval by dividing raw scores by 100 before normalisation. Firms missing either component were excluded from the analysis to avoid biased estimates from imputation or structural zeros.

In order to enhance the interpretability of the outcome variable, the transparency metric is interpreted as a proxy for the exposure of greenwashing risk, as conceptualised in the framework put forth by Yu et al. (2020). In their model, greenwashing is framed not as a dichotomous (yes/no) condition, but as a continuous dimension that captures the difference between how much a firm reports on ESG issues and how it performs. If a firm's standardised ESG Disclosure Score is greater than its standardised ESG Performance Score, then within its peer group, it will indicate that the firm engages in more ESG talk than in action. Such scenarios are interpreted as revealing greenwashing risk, especially if the gap is large and persistent.

Conversely, if a firm's standardised ESG Performance Score is greater than its standardised ESG Disclosure Score, it is interpreted as not enough ESG talk to match its performance, which might reflect a conservative disclosure policy, or a lack of strategic emphasis on leveraging its commitment to ESG in its public communications. If the transparency value is relatively close to zero, we can interpret it as ESG talk aligned with their performance, i.e. their reported operational reality matches its operational reality. This metric does not explicitly observe the intent of a firm to mislead through deception, only how much misalignment is present relative to peers in terms of the gap between disclosure and action.

In order to isolate the effect of AI adoption on the level of transparency in ESG disclosure, a set of covariates was included in the models to control for observed firm aspects that might confound the treatment assignments and outcome. These covariates represent the firm's distinct financial, operational, and structural aspects, and they all originate from the Refinitiv Database. The financial variables include total assets, turnover, and operating income; all expressed in USD, with logarithmic transformations where necessary to decrease skewness and heteroscedasticity. In addition to total assets, turnover, and operating income as performance measures, we include the number of full-time employees and return on assets (ROA) as interrelated measures of scale and efficiency; ROA is calculated as total assets divided by operating income.

The debt-to-equity ratio, as a percentage, measures capital structure. As financial control variables, the objective is to capture size and profitability, leverage and debt, essentially (the intensity of labour costs will also vary among firms).

Aside from the continuous covariates, categorical covariates represent the firm's industrial and regional context within the dataset. Industry affiliation was constructed using the TRBC industry categorization system, which was retained at lower levels but collapsed to higher-order aggregate sectors to avoid a multitude of very sparse categories. Country of headquarters was included to account for national level institutional differences which might affect national-level obligations to disclose, or which may afford or impose disclosure norms and regulatory pressure.

Both categorical covariates were transformed via one-hot encoding, which allowed the matching and balancing phase to include variables that account for heterogeneity in terms of sectoral and regional exposure, while adding extra controls. Both industry and country groups that had fewer than five firms were grouped into an "Other" category to limit the influence of rare categories. Missing values were imputed using the median to avoid systematic bias while keeping sample size intact.

Due to high dimensionality and the high number of potential confounders (many of which are correlated with one another, and are weakly associated with treatment and/or outcome), a penalised regression was used to identify a sparse set of covariates that was still informative. Specifically, the Post-Double LASSO method, which ensures valid post-selection inference in a causal framework (Belloni et al., 2014). The benefit of this method is based on the sparsity assumption which supposes that the proper set of confounding variables is small relative to the set of observed covariates.

The double-selection technique thus applies two regularized regressions. The first regression is a logistic LASSO model that models the treatment indicator (AI adoption). This first step determines which variables are predictive of treatment assignment via an L1-penalized likelihood estimator using cross-validation to select the optimal regularization parameter. The logistic LASSO solves the following optimization problem:

$$\hat{\beta}_{treat} = argmin_{\beta} \left\{ -l(\beta;, T;, X) + \lambda_1 \sum_{p}^{j=1} \left| \beta_j \right| \right\} \tag{3}$$

where $l(\beta; T, X)$ is the log-likelihood for the binary treatment variable T, X is the matrix of covariates, and λ_1 is a penalty parameter selected via cross-validation.

Second, a LASSO linear regression is used to predict the outcome variable, ESG transparency. The outcome model solves:

$$\hat{\gamma} = argmin_{\gamma} \left\{ Y - X_{\gamma 2}^{2} + \lambda_2 \gamma_1 \right\} \tag{4}$$

$$\hat{Y} = X\hat{\gamma} \tag{5}$$

where Y is the vector of transparency scores, X is the covariate matrix, and λ_2 is a tuning parameter selected via cross-validation. This is the canonical LASSO objective described in Tibshirani (1996) and reviewed in Hastie et al. (2009).

The set of selected covariates is defined as the union of those selected in either model:

$$\hat{S} = \left\{ j : \hat{\beta}_{treat,j} \neq 0 \right\} \cup \left\{ j : \hat{\gamma}_j \neq 0 \right\} \tag{6}$$

This double selection approach guarantees the inclusion of variables that are pertinent to predicting either treatment or outcome, mitigating omitted variable bias (Belloni et al., 2014; Chernozhukov et al., 2018). Lastly, to make a final selection, variance inflation factors (VIFs) are calculated, and covariates will be removed with the largest multicollinearity (VIF > 10) until the remaining covariates have acceptable tolerance for collinearity (Kutner et al., 2005).

This method produces a parsimonious set of confounders resulting in the number of confounders to balance the limit of covariate adjustment versus the risk of overfitting (Cilliers et al., 2024). Doing this adds rigor to the use of a machine learning-based variable selection into a traditional casual inference pipeline making it more plausible that the conditional independence assumption used in valid propensity score estimation, and eventual treatment effect identification is valid (Chernozhukov et al., 2018; Farrell et al., 2021).

Estimating Average Treatment Effect on the Treated (ATT) was accomplished with multiple propensity score-based estimation techniques with different strengths in balancing covariates and isolating causal effects, under differing assumptions. The estimators nearest neighbor matching (NNM), entropy balancing (EB), and covariate balancing propensity score (CBPS) followed by a formal Rosenbaum-type sensitivity analysis were use on the reduced set of covariates selected for use from the previous sampling using post double LASSO selection.

The MatchIt package in R (Ho et al., 2024) performed nearest neighbour matching without replacement, matching the closest untreated unit to the treated unit for each treated unit, based on the propensity score. To reduce poor matches, the best practice for matching utilized a caliper of 0.2 standard deviations of the logit of the propensity score (Austin, 2011). This method estimates the ATT by comparing the outcomes between matched pairs of treated and control firms. The propensity score for firm i is defined as:

$$\mathrm{e_i} = \Pr\left(Z_i = 1 | X_i\right) \tag{7}$$

where $Z_i \in \{0, 1\}$ indicates treatment assignment, and X_i is a vector of observed covariates.

The matching process was designed to minimize mean differences in covariates between treated and control groups. Balance was assessed using standardised mean differences (SMD), and thresholds of 0.1 were used to determine acceptable balance (Ho et al., 2007). Matching diagnostics were conducted pre- and post-matching to evaluate balance improvement in covariates:

$$SMD = \frac{\overline{\mathrm{X}}_\mathrm{T} - \overline{\mathrm{X}}_\mathrm{C}}{\sqrt{\frac{s_C^2 + s_C^2}{2}}} \tag{8}$$

where $\overline{\mathrm{X}}_\mathrm{T}$ and $\overline{\mathrm{X}}_\mathrm{C}$ are the sample means and s_T^2 and s_C^2 are the sample variances of the treated and control groups.

Entropy balancing (using the ebal package) constructs covariate distributions that are unit-specific weights for the control group, so that the reweighted covariate distribution matches the covariate moments (mean and variance) of the treated group. This technique does not estimate a propensity score explicitly and achieves exact balance on moments of the covariate distribution (Hainmueller, 2012). For a continuous variable x_j, constraints are imposed as first moment and second moment:

$$\sum_{n_t}^{i=1} w_i \left(x_{ij} - \mu_j\right) = 0, \tag{9}$$

$$\sum_{n_t}^{i=1} w_i \left(x_{ij} - \mu_j\right)^2 = 0 \tag{10}$$

where μ_j is the mean of the trained group. For a categorical covariate represented via GLM parameterization with level l proportion constraint:

$$\sum_{n_t}^{i=1} w_i x_{il} = m_r \tag{11}$$

The optimisation problem minimises the Kullback–Leibler divergence (relative entropy) between the new weights q_i:

$$d(w) = \sum_{n_t}^{i=1} w_i \log\left(\frac{w_i}{q_i}\right) \tag{12}$$

subject to balance constraints and normalizing conditions:

$$\sum w_i = 1, w_i \geq 0 \tag{13}$$

The optimal solution w^* can be derived from the dual form:

$$w_i^* = \frac{q_i \exp\left(-\sum_{r=1}^{R} \lambda_r^* c_{ri}(x)\right)}{\sum_{i=1}^{n_t} q_i \exp\left(-\sum_{r=1}^{R} \lambda_r^* c_{ri}(x)\right)} \tag{14}$$

where λ^* are the lagrange multipliers. This dual problem is solved using Newton–Raphson optimization (Hainmueller, 2012; SAS PSMATCH Documentation, 2023).

This method is valuable when overlap between treatment and control groups is moderate to poor, as it provides a way to avoid discarding observations, while utilizing optimization to reweight controls to look like the treated sample (Källberg & Waernbaum, 2023). It was used to estimate ATT, by applying outcome regression on the reweighted sample.

The Covariate Balancing Propensity Score (CBPS) method jointly optimizes covariate balance and the modeling of the treatment assignment (Imai & Ratkovic, 2014). We used the CBPS package in R (Fong et al., 2025), which estimates a propensity score in a way that balances covariates, while also maximizing predictive accuracy of the treatment assignment model. CBPS demonstrates strong performance in finite samples, including scenarios where the treatment assignment model may be misspecified. Formally, CBPS seeks $\hat{\beta}$ such that:

$$\sum_{i=1}^{N}\left[Z_i - \frac{\exp\left(X_i^T\beta\right)}{1+\exp\left(X_i^T\beta\right)}\right]X_i = 0 \tag{15}$$

This ensures that the residuals from the treatment assignment model are othogonal to the covariates, improving finite sample balance. The model simultaneously satisfies the score condition for the logistic likelihood:

$$\sum_{i=1}^{N}\left[Z_i - \pi\left(X_i;\ \beta\right)\right]X_i = 0 \tag{16}$$

and the balance condition:

$$\sum_{N}^{i=1} Z_i X_i = \sum_{N}^{i=1} \pi\left(X_i;\ \beta\right) X_i \tag{17}$$

In the ATT framework, weights are constructed as:

$$w_i = \begin{cases} 1, & if\ Z_i = 0 \\ \dfrac{\pi\left(X_{i;}\ \hat{\beta}\right)}{1-\pi\left(X_{i;}\ \hat{\beta}\right)}, & if\ Z_i = 0 \end{cases} \tag{18}$$

Estimation is performed via the Generalized Method of Moments (GMM). In this analysis, CBPS weights were utilized to derive estimates of ATT through weighted regression of the outcome on the treatment indicator. SMD plots and balance tables were used to check covariate balance. The recommendations for making balance assessments and achieving optimal covariate alignment are based on foundational works (Rosenbaum & Rubin, 1983; Rubin, 2001).

To evaluate the sensitivity of the ATT estimates to potential unobserved confounding factors, a Rosenbaum bounds sensitivity analysis was performed using the rbounds package. The Rosenbaum bounds procedure estimates the extent of an unmeasured covariate's effect on the results, viewed in terms of the odds characteristics of the unmeasured variable. The analysis indicates a threshold point for the sensitivity parameter Gamma (Γ) value at which the treatment effect would no longer be considered significant (Rosenbaum, 2002; Heng & Small, 2021):

$$= \frac{\pi_{ij}(1-\pi_{ij'})}{\pi_{ij'}(1-\pi_{ij})} \tag{19}$$

where π_{ij} and $\pi_{ij'}$ are the probabilities of treatment assignment for two units within the same matched set. This bound measures the maximum possible odds ratio between treatment assignments due to unmeasured confounding. Where a value of 1 means there is no hidden bias, larger values indicate varying levels of potential (unrecorded) confounding. The threshold value where the treatment effect was no longer statistically significant was recorded.

The importance of running this sensitivity analysis for the eventual interpretation of causal claims in observational studies is important for understanding unmeasured confounding, specifically when AI use could be a proxy variable for disclosure strategies not accounted for in the observed covariates (Langworthy et al., 2023).

4 Empirical Results

The empirical investigation begins with a descriptive assessment of the outcome variable: the peer-relative greenwashing score (GW), constructed to capture firm-level misalignment between ESG communication and ESG performance. As established in the methodological framework, the GW score is derived by standardizing both Bloomberg's ESG Disclosure Score and Refinitiv's ESG Performance Score within each TRBC industry group and then computing their difference. This transformation ensures comparability across sectors with heterogeneous disclosure norms and ESG expectations. The resulting variable reflects a firm's relative position within its industry in terms of the gap between reported and realized sustainability performance, where higher values suggest greater emphasis on ESG-related claims than on verifiable outcomes.

Figure 1 displays the distribution of GW scores across the full sample of S&P 500 firms. The distribution is approximately symmetric and centered near zero, with most firms falling within one standard deviation of the mean. This pattern indicates that, for the majority of firms, ESG disclosure levels are moderately aligned with observed ESG performance. A smaller number of firms extend into the upper tail, where disclosure exceeds peer-normalized performance, while the left tail contains firms that appear to understate ESG outcomes. These patterns reflect underlying differences in reporting behavior and provide initial validation that the GW score captures meaningful variation in transparency practices across the sample.

An examination of the relative distributions between the AI-adopter group and the non-AI firms suggests a slight divergence in their profiles of transparency. There is a peak for both of the distributions around zero. This position indicates that both groups of companies are generally professional in how they report on ESG disclosure levels, roughly aligned to the ESG performance they have documented. However, the AI-adopter comparison group appears to have more mass on the left side of their distribution, indicating that AI-adopter companies have a higher, relative, concentration of negative GW values, i.e., lower ESG transparency, than the

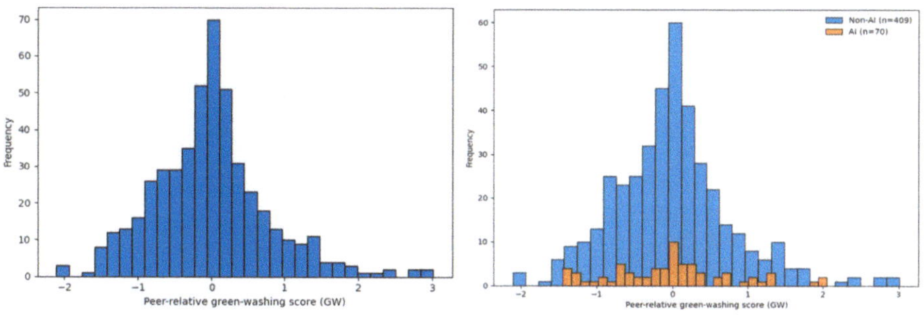

Fig. 1 Distribution of peer-relative green-washing scores for the full sample (left) and by AI adoption status (right)

non-AI group. This indicates, on average, that AI firms are reporting lower levels of ESG disclosure than expected given their ESG performance levels.

This distinction establishes the argument for the analysis to substantiate the central research hypothesis of the paper: firms classified as adopting artificial intelligence have systematically different transparency profiles compared with comparable non-adopting firms. Given that the GW score is created as a continuous, industry-relative, and standardised reporting-based construct, what happens to the GW adoption behaviour in the sample frame is important for the eventual interpretation of the later causal treatment effects. The aggregate shape and width of the distributions imply that the measure reflects distinct differences among firms in ESG disclosure behaviour, and that extreme outlier scores do not mostly drive the GW adoption scores, nor do sector predominately cluster distortions. Thus the GW score adds credibility as a credible basis for conducting a causal analysis.

To continue measuring how ESG disclosure and performance diverge at the firm level, Fig. 2 is a two-way scatter representation of ESG performance against NGO transparency, classified according to AI-adopter versus non-AI firms. This representation of data builds on the prior distribution analysis for actually representing the positional relationship of the underlying ESG disclosure and its associated performance elements to the GW score. The represented 45-degree line could be considered the demarcation point at which a firm's ESG transparency equals its realized ESG performance. Firms above their non-AI firms on the transparency relative to performance indicate that the firms may be overstating ESG performance illustrated in sustainability reporting. Conversely, firms with lower GW ratings than the transparency line indicate companies demonstrating greater ESG performance in real terms as compared to what they publicly referenced in their disclosure.

The quadrant visualization in Fig. 3 provides a more structured view of how firms distribute across combinations of ESG disclosure and performance. Each firm is positioned according to its raw ESG Disclosure Score (horizontal axis) and ESG Performance Score (vertical axis), with the sample split at the median to create four interpretive zones: high disclosure/high performance (HD/HP), high disclosure/low performance (HD/LP), low disclosure/high performance (LD/HP), and low disclosure/low performance (LD/LP).

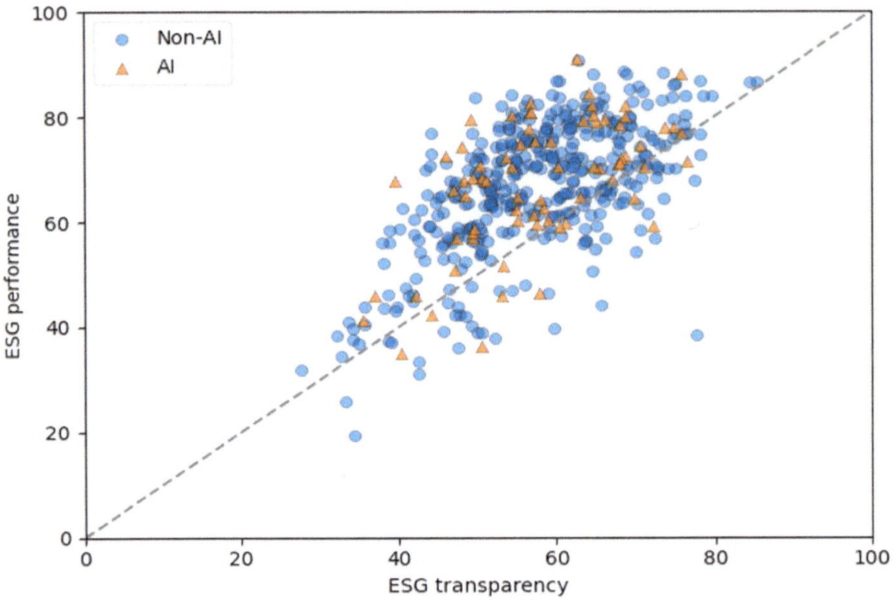

Fig. 2 ESG Performance and ESG transparency by AI adoption status

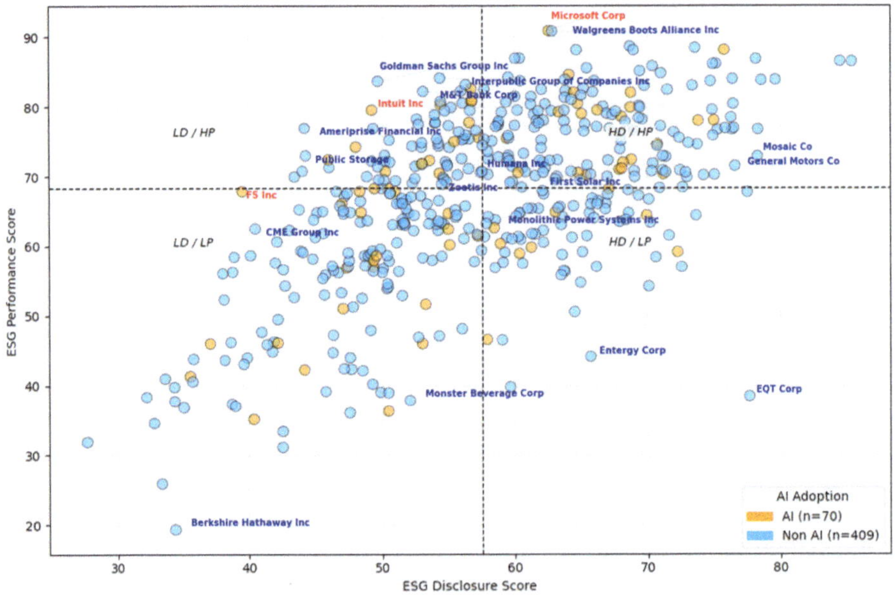

Fig. 3 ESG Disclosure vs. Performance: Quadrant Positioning of AI and Non-AI Firms

Most firms fall within the HD/HP and LD/LP quadrants, but a few consistently perform well in both areas. Microsoft and Walgreens Boots Alliance are examples of firms with consistent strong ESG scores on both dimensions. For comparison, firms such as EQT Corp, and Entergy Corp source relatively high ESG reporting versus their actual performance. Company examples such as Berkshire Hathaway and Monster Beverage represent firms with relatively weak performance on both dimensions. In order to understand how AI and non-AI firms differ in these ESG dimensions, the numbers of firms were analysed by quadrant. Most of the 70 AI firms we reviewed were in the LD/LP quadrant with 24, followed by 23 in HD/HP, 14 in LD/HP, and 9 in HD/LP. For comparison, the non-AI firms appeared mostly in the HD/HP quadrant, with 143 and in the LD/LP quadrant, with 141. When adjusting for size there were roughly one-third of each category represented within LD/LP; there were some differences, however, AI firms populated the LD/HP versus non-AI firms at a higher share of 20% vs. 14.7% respectively, while AI firms appeared in the HD/LP as a smaller share of 12.9–15.9% respectively suggesting that some AI firms are less likely to report their ESG efforts even when they have strong performance.

This suggests potential differences in disclosure behaviour when comparing AI applicants against non-AI applicants. AI adopters appear somewhat more likely to maintain a lower disclosure profile even when they have solid ESG outcomes, whereas non-AI firms appeared more often in the HD/LP quadrant, where disclosure migrated above performance.

To provide more context of the relative differences in disclosure behaviour between AI-associated firms and non-AI-associated firms, Fig. 4 estimates average greenwashing scores in industry categories, separating AI-adopting firms from their non-AI counterparts. Viewing the behaviour at an industry level emphasizes the extent that disclosure-performance misalignment varies not just by technology adoption but more distinctly by industry. Although most industries demonstrated relatively modest differences, the differences in average behaviour in several sectors are more substantial.

Several trends emerge. AI-adopting companies have significantly lower greenwashing (GW) scores than non-AI-adopting firms. For example, the trends are especially noticeable in the Financial Technology (Fintech), Pharmaceuticals, and Advanced Medical Equipment industries, where the average GW scores for AI adopting firms are noticeably strongly negative. A negative GW score indicates that firms in these industries have better ESG performance than publicly disclosed. This implies that they have a more conservative or understated disclosure strategy when their ESG performance is above average when considering their respective industry. These firms may simply not wish to signal their sustainability efforts unless they can substantiate it through some operational action. Non-AI adopting firms in these same industries tend toward near-zero or positive GW scores in the average, indicating they disclose in alignment with their ESG performance or more than what can be justified by their actual ESG behavior. For example, in the fintech industry, non-AI-adopted firms have the highest positive average GW score in the sample,

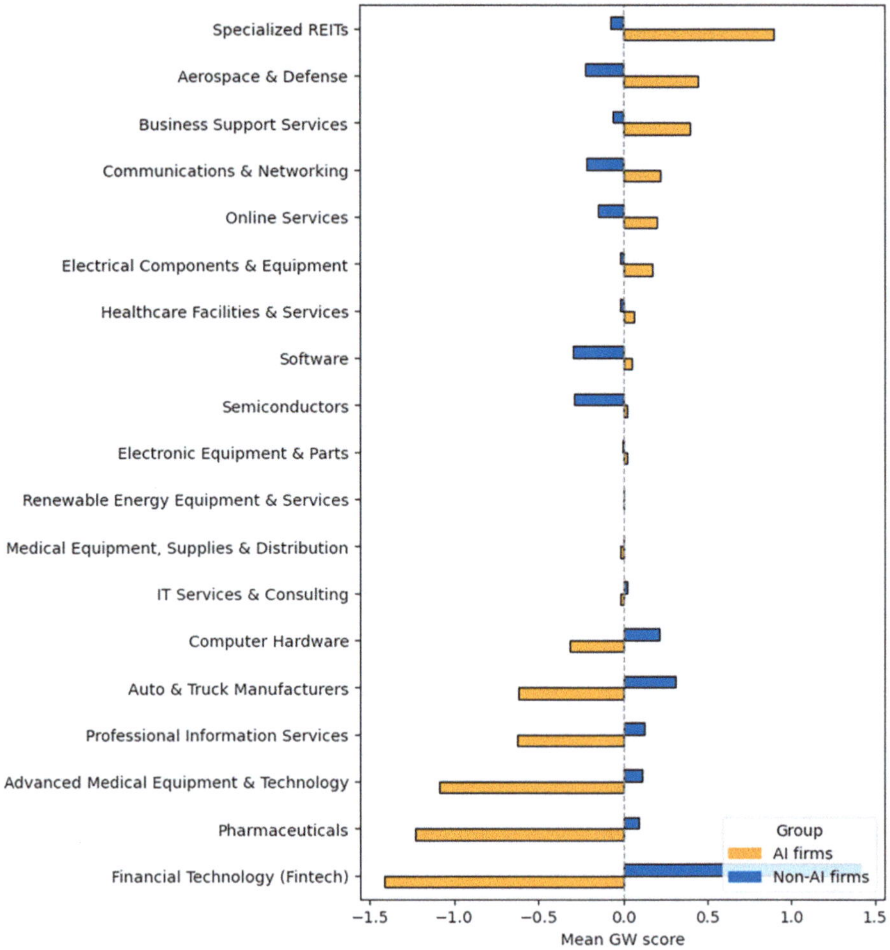

Fig. 4 Mean Peer-Relative Greenwashing Scores by Industry and AI-Adoption Status

suggesting that the extent of their sustainability communication is overstated relative to their peers' ESG outcomes when peer-adjusted.

Figure 5 presents the relationship between emissions scores and greenwashing behaviour across both AI and non-AI firms. The emissions score comes from Refinitiv and identifies a firm's commitment and success in reducing environmental emissions through production and operations. Higher emissions scores indicate stronger environmental performance. On the vertical axis, the GW score depicts the alignment of reported ESG claims—performance at the industry level.

This data does not present a strong linear pattern, some trends are worth examining. First, firms with low emissions scores tend to cluster around zero or slightly negative GW values, indicating modest or conservative disclosure practices when firms perform poorly on re-eventing emissions. There was a more widely dispersed

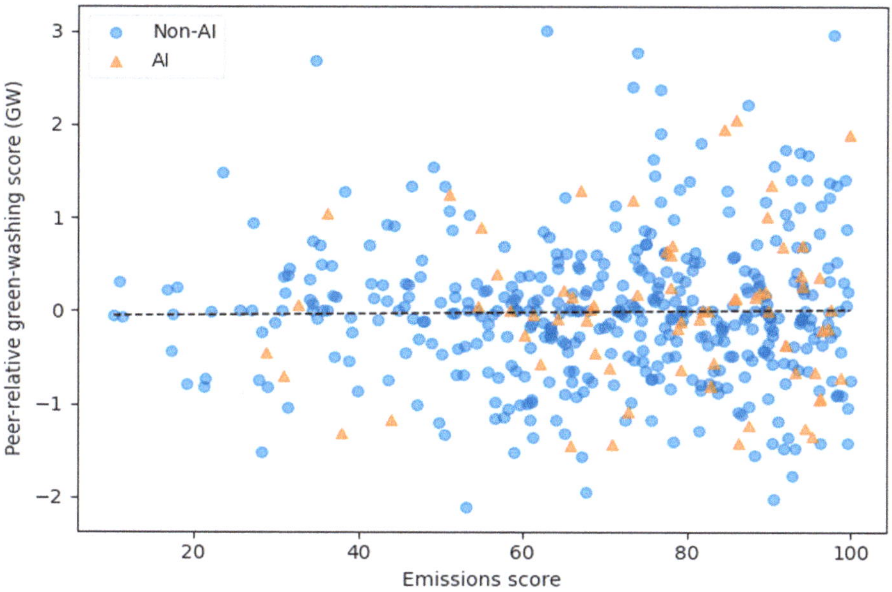

Fig. 5 Relationship Between Emissions Scores and Greenwashing Scores

range of GW scores toward the upper range of emissions scores. Some firms with strong emissions scores also had strong GW scores, which suggests that strong environmental performance is paired with disproportionately positive disclosure 0 and creates a situation where firms may be over-communicating relative to their peer group.

Building on the earlier findings that AI-adopting firms tend to communicate differently on ESG matters, we now turn to examine how ESG disclosure aligns with actual ESG performance across the environmental, social, and governance pillars (Fig. 6). Specifically, we assess whether this alignment is moderated by AI adoption status, using interaction models that account for firm size. The results are summarised in Tables 1, 2, and 3.

In the environmental pillar (Table 1), we find higher levels of disclosure strongly associated with better performance ($\beta = 0.679$, $p < 0.001$), with larger firms also performing better. The interaction between AI and disclosure is not statistically significant ($p = 0.246$), indicating that adoption of AI does not change how well environmental reporting relates to environmental performance. In short, it seems that environmental reporting is sufficiently similar across firms regardless of the adoption of AI.

We see a different picture for the social pillar (Table 2). AI firms perform better overall ($\beta = 0.159$, $p = 0.005$), but the negative interaction term ($\beta = -0.211$, $p = 0.045$) shows that higher levels of disclosure do not lead to better performance for social. This could point to AI firms being less transparent or having more challenges in accurately reporting on social matters.

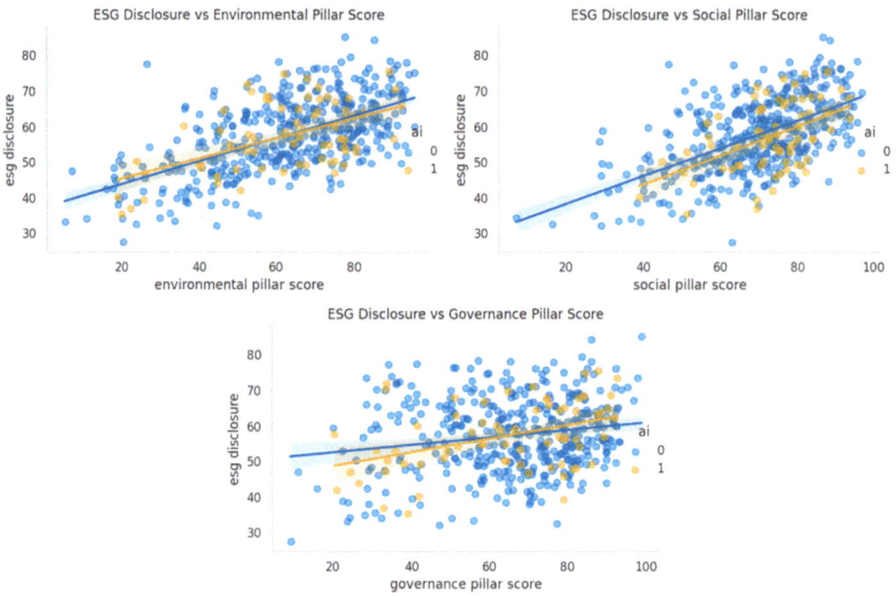

Fig. 6 Alignment Between ESG Disclosure and Environmental, Social, and Governance Pillars: AI vs Non-AI Firms

Table 1 Environmental pillar

Variable	Coefficient	Std. Error	t-Statistic	p-Value
Intercept	−0.4038	0.147	−2.753	0.006
AI	0.0823	0.071	1.159	0.247
Disclosure	0.6789	0.047	14.599	0.000
AI × Disclosure	−0.1520	0.131	−1.162	0.246
Firm Size	0.0283	0.006	4.646	0.000

Table 2 Social pillar

Variable	Coefficient	Std. Error	t-Statistic	p-Value
Intercept	0.0675	0.118	0.574	0.067
AI	0.1591	0.057	2.792	0.005
Disclosure	0.5311	0.037	14.231	0.000
AI × Disclosure	−0.2108	0.105	−2.007	0.045
Firm Size	0.0143	0.005	2.918	0.004

Table 3 Governance pillar

Variable	Coefficient	Std. Error	t-Statistic	p-Value
Intercept	0.2506	0.176	1.427	0.154
AI	−0.1848	0.085	−2.174	0.030
Disclosure	0.1970	0.056	3.538	0.000
AI × Disclosure	0.3277	0.157	2.092	0.037
Firm Size	0.0116	0.007	1.588	0.113

For governance (Table 3), AI firms start with a lower score ($\beta = -0.185$, $p = 0.030$), but governance performance tends to improve at higher levels of disclosure ($\beta = 0.328$, $p = 0.037$). This implies that when AI firms disclose more on governance, it tends to indicate real improvement, possibly because they are under greater scrutiny and trying to build trust through transparency of governance. AI and environmental reporting look fairly robust across all firms. However, social reporting looks weaker for AI firms and shows social impact. Where governance reporting might be more meaningful in the same AI firms, it is increasingly important to look closely at the relationship between disclosure and actual performance as AI becomes more prevalent.

To further elucidate the performance-oriented outcomes, we examine the distribution of ESG controversy scores for both AI and non-AI firms in Fig. 7. The environmental, social, and governance scores are standardised across the sample and analysed to convey deviations from the firm's average controversy record. Both segments (AI, Non-AI) present a bimodal distribution. Furthermore, we see that the Non-AI group have more concentration around higher z-scores, suggesting there are fewer controversial incidents within this group. AI firms display a relatively flat distribution and slightly more density on the lower tail of the distribution, indicating a higher average potential for an ESG incident. These implications indicate that AI adopters may face more scrutiny or have more difficulty avoiding controversies altogether. This also provides additional depth in understanding the perception of their ESG behaviour in the market.

The volcano plot in Fig. 8 shows the magnitude and significance of differences between AI and non-AI firms across various variables. Most variables cluster around zero effect size, suggesting limited differences overall. However, several features have high standardised differences and statistical significance. Notably, AI firms report higher scores for emissions, social practices, and specific financial indicators like turnover and operating income. These variables appear in the top-right quadrant of the plot, indicating that they are not only meaningfully different but also statistically reliable. This pattern suggests that AI-adopting firms may outperform non-AI counterparts in selected ESG and financial dimensions.

A post-double LASSO procedure was applied to identify appropriate covariates for the propensity score estimation. The first step involved fitting an L1-penalised logistic regression to model the probability of AI adoption. Results indicate that AI adoption is primarily concentrated in deep-tech sectors. The largest positive coefficients were associated with sectoral dummies for Semiconductors, Semiconductor Equipment, Software, IT Services, and Communications Hardware (Fig. 9).

On the contrary, Food Processing and Electric Utilities negatively adjusted to AI, suggesting that traditional industries had minimal representation. Financial variables had an effect as well; EBITDA was positive for AI tagging ($\beta \approx +0.64$), while log(Assets) and total assets were negative ($\beta \approx -0.53$ and -0.37); this adjusted for profitability and indicated that larger firm size was not an indicator of AI once profitability was considered. Location only contributed a little explanatory power, and the only location dummy to be retained was the Ireland dummy, potentially as it is

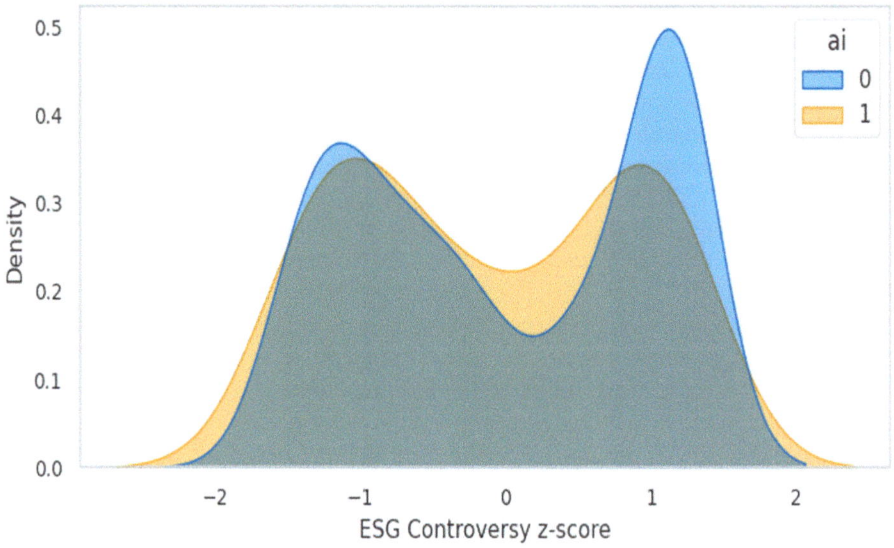

Fig. 7 Distribution of standardised ESG Controversy Scores between AI and non-AI firms

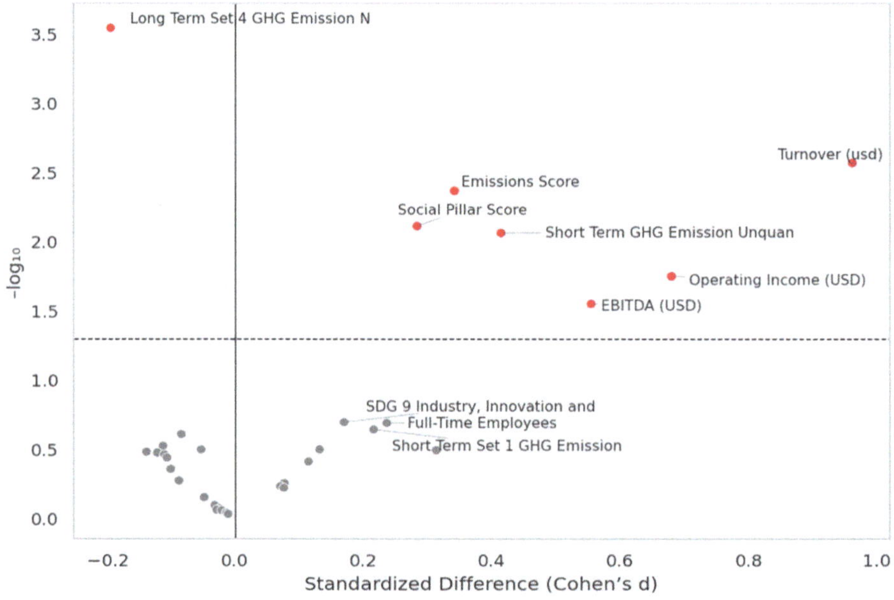

Fig. 8 Volcano plot of AI vs Non-AI firms by ESG and Financial Indicators

home to some tax-efficient AI hubs. ESG-related predictors included emissions score (positive effect) and ESG controversies score (negative effect). At the same time, several sector dummies and firm-level operational indicators such as turnover and full-time employees also contributed.

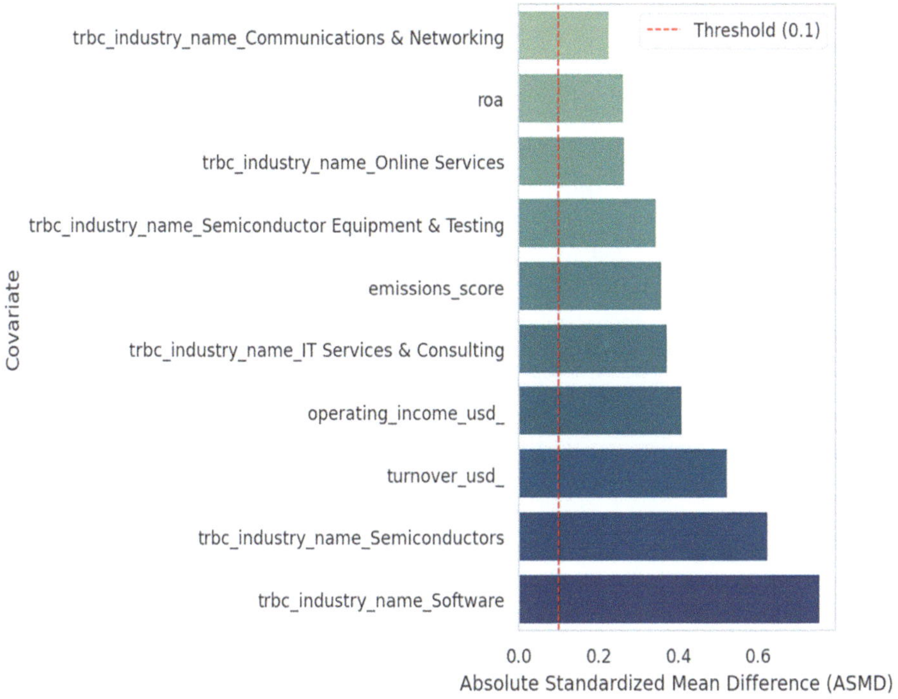

Fig. 9 Pre-Treatment Imbalances between AI and Non-AI firms

In the second stage, LASSO was applied to the outcome model, where the only selected predictor of the greenwashing gap (GW2) was the environmental innovation score, with a small negative coefficient ($\beta \approx -0.06$). This result suggests that observable firm characteristics do not strongly explain greenwashing variation and remain largely in the residual component. This means that emphasis is placed on accurately modelling the treatment assignment, as the outcome appears less structurally determined.

Regularisation parameters selected through cross-validation confirmed a moderate penalty in the treatment model (logit $C \approx 1.62$) and a mild penalty in the outcome model (LASSO $\alpha \approx 0.05$). A variance inflation factor (VIF) check was applied to avoid multicollinearity, and covariates with VIF > 10 were excluded.

The final set of covariates used for propensity score estimation includes firm-level financial indicators: log of total assets, operating income, turnover, return on assets, total reported assets, and number of full-time employees; ESG-related measures: emissions score, environmental controversies, environmental innovation score, and ESG controversies score; a geographic control for firms based in Ireland; sectors represented, dummy variables for Aerospace & Defense, Business Support Services, Communications & Networking, Computer Hardware, Electrical Components & Equipment, Electronic Equipment & Parts, Financial Technology (Fintech), Healthcare Facilities & Services, IT Services & Consulting, Medical

Equipment & Supplies, Online Services, Professional Information Services, Renewable Energy Equipment & Services, Semiconductor Equipment & Testing, Semiconductors, and Software.

The set of covariates identified through post-double LASSO and filtered for multicollinearity formed the basis for propensity score estimation. These covariates capture firm-level financial structure, ESG-related exposure, geographic positioning, and detailed sectoral affiliation. This stage aimed to construct an appropriate counterfactual by ensuring that, conditional on observed characteristics, the probability of AI adoption was balanced across treated and control groups.

Propensity scores were estimated via a logistic regression model wherein the treatment indicator was regressed on the included covariates. The propensity scores represent the estimated probability that a firm will adopt AI, conditional on its covariate profile. Means were compared pre-matching using standardized mean differences (SMDs) for each covariate between the AI and non-AI groups, finding significant differences, particularly regarding sector affiliation and financial indicators, including operating income and turnover, with several SMDs being greater than 0.2, implying that if both groups were compared directly, the direct comparisons would be obscured by structural differences.

To address this imbalance, nearest neighbour matching (NNM) was conducted using the logit of the estimated propensity score, applying a calliper of 0.2 standard deviations to improve match quality and reduce poor pairings. This procedure successfully matched 57 of the 70 treated firms to non-AI firms with comparable covariate profiles. Post-matching diagnostics showed substantial reductions in standardised mean differences across all covariates, with nearly all falling below 0.1. This improvement in balance indicates that the matching process yielded a more valid counterfactual group.

Entropy balancing was also used to create weightings for the covariate-adjusted control group such that the covariate distribution's first and second moments matched the treated group. Unlike matching, reweighting retains the original sample size, while matching uses reweighted comparisons and removes control units. An exact balance was obtained on means and variances of covariates, which confirms that the entropy balancing methodology was able to produce an equivalent matched counterfactual in the control group.

The covariate balancing propensity score (CBPS) approach offered a third approach that estimated propensity scores and balanced covariates. The CBPS method estimates propensity scores in direct conjunction with covariate balancing. CBPS performs well under a model misspecification assumption, providing an unbiased estimate of propensity scores when all observations can be used. Balance statistics from the CBPS approach showed improvements similar in quality to NNM, confirming the reliability of the reweighted comparisons.

Estimates of the average treatment effect on the treated (ATT) obtained from these three methods are presented in Table 4. The results reveal consistent and statistically meaningful differences in ESG transparency scores across all estimators. Nearest neighbour matching based on logit scores produced an ATT of 0.1652, with a nearly identical result using the treatment-on-the-treated estimator. The probit-based version yielded a slightly lower ATT of 0.1286. These estimates suggest that AI-adopting firms have transparency scores between 0.13 and 0.17 standard deviations higher than their matched non-AI counterparts, conditional on observable characteristics. This pattern is visually confirmed in Figs. 10 and 11, illustrating how the distribution of propensity scores and their logit transformations shift from imbalance in the full sample to substantial overlap after matching, supporting the validity of the resulting ATT estimates.

The CBPS analysis yielded an average treatment effect on the treated (ATT) of 0.1345, further supporting the pattern seen in matching techniques. By comparison, the estimate from entropy balancing was more conservative, yielding an ATT

Table 4 Estimated average treatment effects on the treated (ATT) using propensity score methods

Estimator	Estimated ATT	Matched treated units
NNM (Logit, caliper)	0.1652	57/70
NNM (Logit, TT estimate)	0.1658	57/70
NNM (Probit)	0.1286	57/70
Covariate Balancing Propensity Score (CBPS)	0.1345	All
Entropy Balancing (EB)	0.0807	All

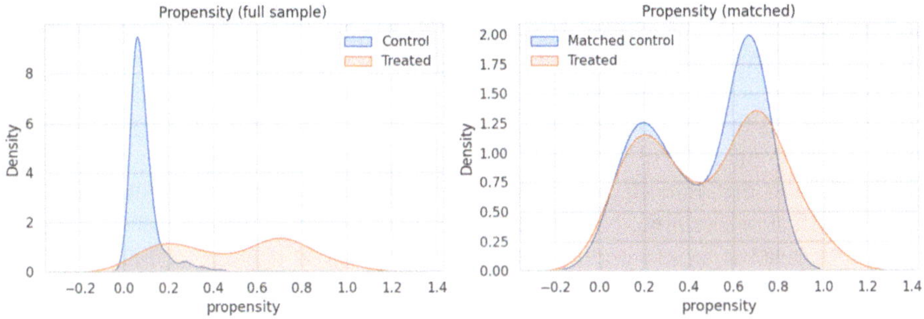

Fig. 10 Propensity Score Distributions Before and After Nearest Neighbour Matching

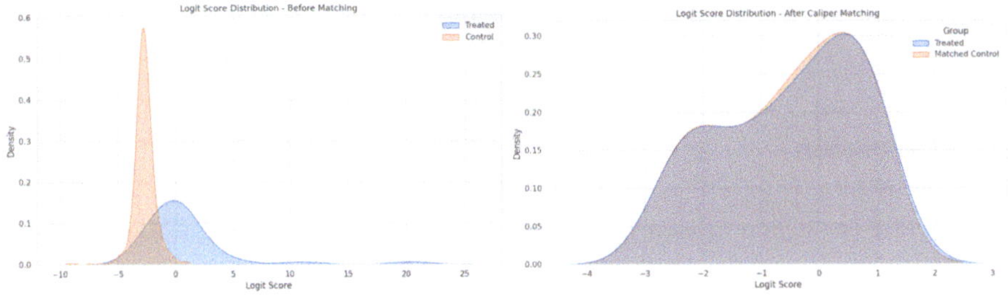

Fig. 11 Logit Score Distributions Before and After Matching

estimate of 0.0807, but the relationship and significance remained positive. We believe these results imply that firms adopting AI have a stronger relationship between their ESG disclosures and their ESG performance, unencumbered by firm-level heterogeneity. The patterns of results were held across estimation strategies, and the size of the effects was small to moderate based on the strength of the prior adjustment methodology. A Rosenbaum sensitivity analysis was conducted to assess the robustness of these results. Results showed that the treated effects would remain statistically significant only where the unobserved biases were large. This indicates that the results not only hold across different methods but also for reasonable deviations from the unconfoundedness assumption.

These results further suggest that raw differences in transparency can overshadow behavioural realities. AI adopters appeared less transparent prior to adjustment. However, once differences in size, sector, profitability, and ESG characteristics are controlled for, proportionally to non-adopters, AI-adopting firms present better alignment between what they are reporting on and what they are delivering. These findings stress the value of causal inference techniques to differentiate signals from structural confounding as ESG takes shape in the academy and, hence, in practice.

The differences in ATT method estimates can be attributed to how each method achieves balance on observed characteristics between the treated and control groups. Nearest neighbour matching (NNM) connects each AI firm with the most similar non-AI firm based on propensity scores, and will exclude any firms that are not closely matched. This will provide the greatest balance between the groups, but excludes firms from the sample size, resulting in a smaller sample size and, therefore, larger effect sizes. In comparison, entropy balancing and CBPS use the whole sample size and reweight control firms when comparisons are made, instead of dropping them from the sample. Therefore, both methods provide approximate or exact balance on observed characteristics but produce model-based conservative ATT estimates that smooth over more variation in the data. In statistical terms, this is a tradeoff. For example, NNM may offer clearer treatment contrasts (if matches are available), but would undoubtedly be compromised if many or poor matches were located. There is also a strong argument for weighting with the analysis of an experimental evaluation, as weighting methods provide coherence and accuracy through maintaining a larger portion of data—the downside is that it could also diminish the estimated treatment effect. Lastly, the agreement of direction across all methods suggests confidence, while the difference in size demonstrates, along with other factors, the interpretative nature of methods.

5 Discussion

The empirical results provide new evidence on the relationship between artificial intelligence adoption and ESG transparency, offering a refined perspective on whether AI engagement correlates with greater or lesser alignment between disclosure and performance. Previous studies have raised concerns about technological opacity and greenwashing risks among digitally advanced firms (e.g. Floridi & Cowls, 2022; Chen, 2024), suggesting that AI adopters may obscure environmental and social shortcomings behind complex innovations. However, the present findings indicate the opposite. Once structural differences in size, sector, financial structure, and ESG maturity are accounted for using causal inference methods, AI-adopting firms display higher disclosure-performance alignment than comparable non-AI firms.

This aligns with emerging research suggesting that AI can improve ESG data quality and reporting consistency (Siqueira et al., 2024), especially when deployed for sustainability analytics and environmental monitoring. It also supports the argument that AI adoption may signal institutional capacity and a higher baseline of organisational sophistication, characteristics often linked to improved ESG governance. The results, therefore, nuance the popular narrative around AI-driven greenwashing and suggest that the observed scepticism toward AI adopters may partly be driven by confounding structural features rather than behavioural differences.

At a methodological level, this study contributes to recent work on causal ESG evaluation by combining a hybrid classification of AI exposure with rigorous matching and weighting estimators. The robustness of the findings across nearest neighbour matching, entropy balancing, and CBPS estimators suggests that the effect is not method-dependent but reflects a systematic pattern in the data. Visual diagnostics confirmed that covariate balance improved significantly across methods, and sensitivity analysis further indicated that results are unlikely to be driven by moderate levels of unobserved confounding.

Several limitations should be acknowledged. First, the study relies on firm-level ESG data that may reflect measurement limitations, particularly in the social and governance pillars, where transparency is often weaker. Second, the classification of AI adopters, though methodologically comprehensive, relies on publicly disclosed text and may miss subtler or internal deployments of AI. Third, unobserved confounders remain a threat despite rigorous selection and adjustment; factors such as corporate culture, internal compliance structures, or board-level commitment to sustainability may influence AI adoption and transparency but remain unmeasured. Finally, the outcome variable focuses on disclosure-performance alignment, a proxy for transparency but not a direct measure of stakeholder impact or environmental outcomes.

Future research could extend this framework in several directions. One avenue is to explore heterogeneity in treatment effects across industries or regions, particularly given regulatory variations in ESG disclosure. Another is to incorporate time dynamics by examining whether AI adoption increases transparency over time, possibly using panel matching or event study designs. Integrating firm-level data on AI implementation intensity or type, such as predictive analytics versus automation, could offer more granular insights. Lastly, expanding the analysis to include outcomes related to real ESG performance, such as emissions reduction or social impact metrics, would provide a complete picture of how AI relates to sustainability beyond disclosure.

Taken together, the findings challenge assumptions that AI engagement increases greenwashing and instead suggest that, conditional on firm characteristics, AI adopters may be more committed to credible ESG communication. These insights are particularly relevant for regulators, investors, and ESG rating agencies seeking to differentiate between genuine transparency and strategic disclosure in an increasingly digitised corporate landscape.

6 Conclusion

This study offers a robust quantitative assessment of how artificial intelligence adoption influences ESG transparency, applying a rigorous causal inference framework that integrates hybrid firm classification, post-double LASSO variable selection, and multiple matching and weighting estimators. By combining statistical precision with interpretability, the analysis isolates the effect of AI adoption on the alignment between ESG disclosures and actual performance, helping to distinguish genuine transparency from structural reporting patterns. Across all methods employed, AI-adopting firms demonstrated a consistently higher degree of disclosure-performance alignment, which counters common narratives that associate digital sophistication with greater greenwashing risk.

The findings make an important contribution to the empirical ESG literature and the emerging field of responsible AI governance. They suggest that when embedded within firms' operational models, AI capabilities may facilitate stronger reporting discipline and accountability. This has clear implications for ESG rating agencies, institutional investors, and regulators who are increasingly called upon to assess the credibility of sustainability claims in a digitised corporate environment. This evidence supports the view that AI adoption, under certain structural conditions, can enhance and not obscure the reliability of ESG communication.

At a policy level, this calls for more nuanced frameworks that integrate technological indicators such as AI integration levels into ESG scoring methodologies and disclosure regulations. It also reinforces the importance of transparency protocols considering AI deployment's content and context. Future research should build on this framework by incorporating longitudinal data, sector-specific dynamics, and more granular indicators of AI intensity. As the regulatory landscape evolves, particularly with forthcoming legislation on AI ethics and corporate sustainability reporting, empirical tools like those developed in this study will be essential for evidence-based policymaking.

Acknowledgments This research was conducted within the Business Economics Data Science Lab and supported by the project ASE-RISE—'Promoting Excellence in Research through Interdisciplinarity, Digitalisation, and Open Science Integration to Enhance International Visibility' (CNFIS-FDI-2025-F-0457).

References

Austin, P. C. (2011). Optimal caliper widths for propensity-score matching when estimating differences in means and differences in proportions in observational studies. *Pharmaceutical Statistics, 10*(2), 150–161. https://doi.org/10.1002/pst.433

Belloni, A., Chernozhukov, V., & Hansen, C. (2014). High-dimensional methods and inference on structural and treatment effects. *Journal of Economic Perspectives, 28*(2), 29–50. https://doi.org/10.1257/jep.28.2.29

Chae, Y., & Davidson, T. (2025). Large language models for text classification: From zero-shot learning to instruction-tuning. *Sociological Methods & Research.* https://doi.org/10.1177/00491241251325243

Chen, S. (2024). The influence of artificial intelligence and digital technology on ESG reporting quality. *International Journal of Global Economics and Management, 3*(1), 301–310. https://doi.org/10.62051/IJGEM.v3n1.36

Chernozhukov, V., Chetverikov, D., Demirer, M., Duflo, E., Hansen, C., Newey, W., & Robins, J. (2018). Double/debiased machine learning for treatment and structural parameters. *The Econometrics Journal, 21*(1), C1–C68. https://doi.org/10.1111/ectj.12097

Cilliers, J., Elashmawy, N., & McKenzie, D. (2024). *Using post-double selection Lasso in field experiments*. Policy Research Working Paper 10931, World Bank. http://hdl.handle.net/10986/42206

Ekaristi, C. Y. D., Utomo, D. C., & Rohman, A. (2025). Big data and AI in ESG performance measurement: A bibliometric analysis. *Edelweiss Applied Science and Technology, 9*(5), 2732–2749. https://doi.org/10.55214/25768484.v9i5.7587

Farrell, M. H., Liang, T., & Misra, S. (2021). Deep neural networks for estimation and inference. *Econometrica, 89*(1), 181–213. https://doi.org/10.3982/ECTA16901

Floridi, L., & Cowls, J. (2022). A unified framework of five principles for AI in society. In S. Carta (Ed.), *Machine learning and the city: Applications in architecture and urban design* (pp. 535–545). Hoboken, NJ: John Wiley & Sons Ltd. https://doi.org/10.1002/9781119815075.ch45

Fong, C., Ratkovic, M., & Imai, K. (2025). *CBPS: Covariate balancing propensity score (Version 0.23) [R package].* https://CRAN.R-project.org/package=CBPS

Hainmueller, J. (2012). Entropy balancing for causal effects: A multivariate reweighting method to produce balanced samples in observational studies. *Political Analysis, 20*(1), 25–46. https://doi.org/10.1093/pan/mpr025

Hastie, T., Tibshirani, R., & Friedman, J. (2009). *The elements of statistical learning: Data mining, inference, and prediction* (2nd ed.). Springer.

Heng, S., & Small, D. S. (2021). *Sharpening the Rosenbaum sensitivity bounds to address concerns about interactions between observed and unobserved covariates*. arXiv Preprint. https://arxiv.org/abs/2004.00766

Ho, D. E., Imai, K., King, G., & Stuart, E. A. (2007). Matching as nonparametric preprocessing for reducing model dependence in parametric causal inference. *Political Analysis, 15*(3), 199–236. https://doi.org/10.1093/pan/mpl013

Ho, D. E., Imai, K., King, G., & Stuart, E. A. (2024). *MatchIt: Nonparametric preprocessing for parametric causal inference (Version 4.5.4) [R package].* https://cran.r-project.org/package=MatchIt

Imai, K., & Ratkovic, M. (2014). Covariate balancing propensity score. *Journal of the Royal Statistical Society: Series B (Statistical Methodology), 76*(1), 243–263. https://doi.org/10.1111/rssb.12027

Källberg, D., & Waernbaum, I. (2023). Large sample properties of entropy balancing estimators of average causal effects. *Econometrics and Statistics.* https://doi.org/10.1016/j.ecosta.2023.11.004

Kutner, M. H., Nachtsheim, C. J., Neter, J., & Li, W. (2005). *Applied linear statistical models* (5th ed.). McGraw-Hill/Irwin.

Langworthy, B., Wu, Y., & Wang, M. (2023). An overview of propensity score matching methods for clustered data. *Statistical Methods in Medical Research, 32*(4), 641–655. https://doi.org/10.1177/09622802221133556

Li, J., Wu, T., Hu, B., Pan, D., & Zhou, Y. (2025). Artificial intelligence and corporate ESG performance. *International Review of Financial Analysis, 102*, 104036. https://doi.org/10.1016/j.irfa.2025.104036

Lu, G., & Li, B. (2025). Artificial intelligence and green collaborative innovation: An empirical investigation based on a high-dimensional fixed effects model. *Sustainability, 17*, 4141. https://doi.org/10.3390/su17094141

Moreno-Garcia, C. F., Jayne, C., Elyan, E., & Aceves-Martins, M. (2023). A novel application of machine learning and zero-shot classification methods for automated abstract screening in systematic reviews. *Decision Analytics Journal, 6*, 100162. https://doi.org/10.1016/j.dajour.2023.100162

Rosenbaum, P. R. (2002). *Observational studies* (2nd ed.). Springer.

Rosenbaum, P. R., & Rubin, D. B. (1983). The central role of the propensity score in observational studies for causal effects. *Biometrika, 70*(1), 41–55. https://doi.org/10.1093/biomet/70.1.41

Rubin, D. B. (2001). Using propensity scores to help design observational studies: Application to the tobacco litigation. *Health Services and Outcomes Research Methodology, 2*(3–4), 169–188. https://doi.org/10.1023/A:1020363010465

Sætra, H. S. (2022). The AI ESG protocol: Evaluating and disclosing the environment, social, and governance implications of artificial intelligence capabilities, assets, and activities. *Sustainable Development, 31*(2), 1027–1037. https://doi.org/10.1002/sd.2438

SAS Institute Inc. (2023). *Entropy balancing method in PSMATCH procedure*. SAS Documentation. https://documentation.sas.com/doc/ja/sasstudiocdc/v_036/pgmsascdc/statug/statug_psmatch_details29.htm

Siqueira, M. B., dos Santos, V. M., Diniz, E. & Paula, A. P. (2024). Artificial intelligence for sustainability: A systematic literature review in information systems. *Revista de Gestão Social e Ambiental, 18*(3), 1–20. https://doi.org/10.24857/rgsa.v18n3-178

Tibshirani, R. (1996). Regression shrinkage and selection via the lasso. *Journal of the Royal Statistical Society: Series B (Methodological), 58*(1), 267–288. http://www.jstor.org/stable/2346178

Wang, L., & Chen, Y. (2025). Artificial intelligence and corporate investment efficiency: Evidence from China. *Emerging Markets Review, 68*, 101314. https://doi.org/10.1016/j.ememar.2025.101314

Weng, M. (2025). Green innovation through artificial intelligence technology: Enhancing environmental, social, and governance performance. *Finance Research Letters, 75*, 106921. https://doi.org/10.1016/j.frl.2025.106921

Xie, H., & Wu, F. (2025). Artificial intelligence technology and corporate ESG performance: Empirical evidence from Chinese-listed firms. *Sustainability, 17*, 420. https://doi.org/10.3390/su17020420

Yu, E. P. Y., Luu, B. V., & Chen, C. H. (2020). Greenwashing in environmental, social and governance disclosures. *Research in International Business and Finance, 52*, 101192. https://doi.org/10.1016/j.ribaf.2020.101192

Yu, X., Fan, L., & Yu, Y. (2025). Artificial intelligence and corporate ESG performance: A mechanism analysis based on corporate efficiency and external environment. *Sustainability, 17*(9), 3819. https://doi.org/10.3390/su17093819

Zhang, K., Zhang, Q., Wang, C. -C., & Jang, J. -S. R. (2025). Comprehensive study on zero-shot text classification using category mapping. *IEEE Access, 13*, 23526–23541. https://doi.org/10.1109/ACCESS.2025.3538103

Zhou, X., Li, G., Wang, Q., Li, Y., & Zhou, D. (2025). Artificial intelligence, corporate information governance and ESG performance: Quasi-experimental evidence from China. *International Review of Financial Analysis, 102*, 104087. https://doi.org/10.1016/j.irfa.2025.104087

Low-Carbon Transition: The Economic Implications of Climate Change

Andru Mazilescu, Sorana Vătavu, and Ana-Cristina Nicolescu

1 Introduction

Climate change has become one of the most pressing global challenges, with wide-ranging implications not only for environmental systems but also for economic structures and policy design. The scientific consensus emphasises that the accumulation of greenhouse gases—particularly carbon dioxide (CO_2)—resulting from human activities is driving rapid changes in global temperatures, precipitation patterns, and ecological stability (Stocker et al., 2013). In response, the concept of transitioning to a low-carbon economy has emerged as a strategic imperative for governments, industries, and international bodies alike.

A low-carbon economy is centred on reducing greenhouse gas emissions through cleaner energy systems, greener technologies, and more efficient resource use. However, this transition is not simply a technical challenge—it is also an economic transformation. As pointed out by Acemoglu et al. (2012), transitioning to a low-carbon growth path necessitates reorienting technological change and investment strategies toward green innovation, which may entail short-term economic trade-offs, particularly in sectors heavily dependent on fossil fuels. These trade-offs include transitional job losses, industrial restructuring, and the reallocation of capital, which can affect GDP growth and sectoral competitiveness (Popp, 2019).

A. Mazilescu (✉)
Faculty of Economics and Business Administration, West University of Timisoara, Timişoara, Romania
e-mail: andru.mazilescu99@e-uvt.ro

S. Vătavu · A.-C. Nicolescu
Faculty of Economics and Business Administration, Department of Finance, Business Information Systems and Modelling, West University of Timisoara, Timişoara, Timiş County, Romania
e-mail: sorana.vatavu@e-uvt.ro; cristina.nicolescu@e-uvt.ro

C.-W. Su et al. (eds.), *Green Finance and the Challenges of Climate Change*, Sustainable Finance, https://doi.org/10.1007/978-3-032-07293-1_4

Theoretically, the low-carbon transition challenges the traditional model of growth that is closely linked to carbon-intensive production. While environmental economics has long emphasised the externalities of pollution (Nordhaus, 1991), more recent studies have explored the potential for "decoupling" economic performance from emissions. For instance, Stern and Stiglitz (2017) argue that well-designed carbon pricing, coupled with investment in clean technology, can foster both environmental sustainability and economic efficiency.

Empirical evidence on the economic consequences of climate mitigation is mixed. Some studies suggest that the adoption of renewable energy, while environmentally beneficial, can impose short-term costs on GDP due to high upfront investment and lower energy return on investment (Apergis & Payne, 2014; Burke & Stephens, 2018). Others emphasise the long-term productivity and employment gains from green sectors, especially when complemented by innovation and supportive policy frameworks (Popp et al., 2020). The heterogeneity of these outcomes across countries highlights the importance of context-specific analysis that considers structural economic differences, industrial composition, and institutional capacity.

In the context of the European Union—a region actively pursuing climate neutrality—the transition is particularly salient. EU member states differ significantly in their reliance on fossil fuels, energy imports, and industrial competitiveness, resulting in uneven economic effects of climate policy. Research by Tagliapietra et al. (2019) highlights that decarbonization policies can have distributional impacts, both within and between countries, depending on how industries and labour markets adjust to the change.

This study contributes to the ongoing debate by empirically analysing the relationship between economic growth, industrial structure, CO_2 emissions, and renewable energy adoption in EU countries from 2001 to 2023. By employing panel data methods, the research controls for major economic shocks—such as the 2007–2009 financial crisis and the 2020–2021 COVID-19 pandemic—while assessing how these variables interact over time. Its contribution to the scientific literature highlights how renewable energy adoption may temporarily dampen economic output due to transitional inefficiencies and reallocation costs. The model also tests the hypothesis that the transition to a low-carbon economy involves measurable trade-offs in terms of GDP, but may also offer resilience against future economic and environmental risks. Accordingly, rather than assuming full transformation, the study reveals how sustainability pressures a restructure in industrial systems while the traditional sectors continue to fuel GDP growth.

In sum, the low-carbon transition represents a fundamental shift in how economies function. It challenges established production models, calls for new investment patterns, and redefines the metrics of economic success. By grounding this inquiry in both theory and data, the present study aims to inform the broader academic and policy conversation on how to reconcile economic development with environmental sustainability. Overall, the study offers clear economic evidence for the need to invest in technological innovation to mitigate energy market instability and facilitate smoother transition pathways.

The remainder of this chapter is organised as follows. The next section provides a comprehensive review of the existing literature, outlining the current state of research on the economic dimensions of the low-carbon transition, particularly studies that assess the impact of renewable energy adoption and carbon emissions on economic growth. This is followed by the data and methodology section, where we present the dataset used for our empirical analysis and justify the selection of variables and indicators. The empirical strategy employs a panel data tested with multiple regression methods to account for unobserved heterogeneity across countries and assess the robustness of our results. The following section presents and discusses the empirical findings, interpreting them in the context of broader theoretical and policy debates. Finally, the chapter concludes with a summary of key insights, implications for policymaking, and suggestions for future research avenues in the field of climate economics.

2 Literature Review

An important part of the scientific research examines whether shifting to renewable energy fosters or hinders economic growth. Although the current focus is on how sustainability imperatives try to reshape the traditional sectors, the persistence of emission-intensive industries remains, despite policy shifts. Further on, we will present various studies focused on the relationship between carbon emissions and economic growth and the growth benefits of renewables, with a focus on their empirical approach.

We start by presenting research evidence that renewable energy adoption can serve as both a catalyst for economic expansion and a strategic solution to environmental degradation. Many panel data studies find a growth impact of renewables. For example, Apergis and Payne (2010) analysed 20 OECD countries, including many EU member states, through panel cointegration techniques and overviewing the 1985–2005 period. They identified a long-run equilibrium where renewable energy consumption has a positive and significant coefficient in the GDP equation, alongside capital and labour. Moreover, their causality tests indicated bidirectional causality between renewable energy use and GDP in both the short and long run, supporting a feedback hypothesis that renewable energy adoption and economic growth reinforce each other. Similarly, focusing on major renewable energy-consuming countries worldwide, Saidi and Omri (2020) employed a panel fully modified OLS (FMOLS) and a Vector Error Correction Model. They report that renewable energy is effective in boosting economic growth while also reducing CO_2 emissions. These findings align with those of other global panels (e.g., Omri et al., 2015; Kahia et al., 2021), which generally conclude that increasing the share of renewables does not harm growth and can even stimulate GDP through the creation of new industries and improved energy security.

However, evidence is not uniform because although renewable energy has long-term benefits, studies suggest that its growth-enhancing effects can vary significantly depending on national context, energy mix, and policy design. Some studies focusing on Europe suggest the growth benefits of renewables may be limited or context dependent. Menegaki (2011), using a panel of 27 EU countries (1997–2007) with a random-effects model, found no significant causal relationship between renewable energy consumption and GDP, an outcome consistent with the "neutrality hypothesis". Menegaki's model included controls such as total energy use, emissions, and employment; yet the coefficient for renewable energy remained insignificant, implying that during that period, Europe's renewables expansion did not significantly drive GDP. The author partly attributes this to uneven and still insufficient renewable energy deployment across EU states. This contrasts with the positive results of Apergis and Payne, which could be due to methodological differences (static panel vs. cointegration approach) or sample differences. Such divergence points to a debate in the literature: whether renewables inherently spur growth or whether the effect materialises only under certain conditions (e.g., sufficient scale, supportive policies, or in combination with other growth drivers).

Several studies have extended this analysis using different methods. Tugcu et al. (2012), for instance, compared the impacts of renewable and non-renewable energy on GDP in G7 countries using both panel and time-series techniques. They and others often find mixed causal patterns—some countries exhibit energy-led growth while others show growth-led energy demand—reinforcing that the renewable-growth nexus can vary by country context. Overall, the EU panel evidence suggests a small but positive long-term impact of renewable energy adoption on GDP. However, short-run effects may be negligible or strongly improve growth, but might require complementary factors (such as innovation, investment, etc.).

The relationship between carbon emissions and economic growth has been extensively studied, often through the lens of the Environmental Kuznets Curve (EKC) hypothesis. In EU economies, numerous panel studies have identified a nonlinear relationship where emissions initially rise with GDP and then decline once higher income levels are achieved. For example, Balsalobre-Lorente et al. (2018) examined five large EU countries (EU-5) from 1985 to 2016. Using fixed-effects panel models, they confirmed an inverted-U followed by a second upswing ("N-shaped") relationship between per capita GDP and CO_2 emissions. This suggests that beyond a certain income level, emissions may rise again unless further structural changes or policies are implemented to mitigate this trend. They also found that increased renewable electricity usage and energy innovation helped reduce emissions, suggesting that environmental policies can decouple economic growth from carbon emissions. Consistent with this, a recent panel analysis of 27 EU countries by Onofrei et al. (2022) found a cointegrating relationship between GDP and emissions, with a 1% increase in GDP resulting in approximately a 0.07% rise in CO_2 in the long run. Their results suggest that economic growth in the EU continues to modestly increase emissions, underscoring the challenge of achieving absolute decoupling from emissions. Nevertheless, higher-income EU nations also show increasing demand for environmental quality, leading to stronger climate policies. This dynamic aligns with the EKC concept—growth enables investment in cleaner technology and regulation, which eventually curtails emissions.

Globally, findings on the growth–emissions nexus go beyond Europe. Dong et al. (2018) overviewed 50 highly globalised countries (with a system-GMM estimator), confirming that GDP per capita has an inverted-U relationship with CO_2, consistent with the EKC, and that greater renewable energy usage significantly lowers carbon intensity. On the other hand, studies focusing on developing countries often find that the turning point of the EKC occurs at higher income levels or reveal a predominantly positive GDP–CO_2 link within their sample, indicating that emissions continue to rise with output (e.g., Sharif et al., 2019, on emerging Asia). These discrepancies fuel debate on whether growth inevitably helps environmental outcomes or if deliberate policy intervention is required. Notably, panel studies with fixed effects that control for country heterogeneity generally support the existence of a long-run growth-emissions relationship (positive or inverted U). In contrast, some time-series studies have failed to find cointegration in individual cases. For instance, earlier work by Soytas and Sari (2009) for the U.S. and Turkey showed that while energy use drives CO_2 emissions, CO_2 emissions did not feed back into GDP, suggesting that moderating emissions might not harm growth in the short run. In summary, the literature suggests that unchecked economic growth tends to increase CO_2 emissions in the absence of green policies; however, the highest-income economies (many of which are in the EU) are beginning to experience a weakening of this link as renewable adoption and efficiency improvements take effect.

Although economies with robust industrial sectors often experience a higher impact on GDP, they face greater challenges in decarbonisation due to higher energy and emissions intensity. Accordingly, another factor influencing GDP is the composition of a country's economy—the relative size of agriculture, industry, and services, as well as the degree of high-value activities. Classic development theories posit that structural transformation, such as shifting labour from low-productivity agriculture to higher-productivity industry, is a key driver of growth. In the EU context, attention has shifted to deindustrialisation and whether a decline in manufacturing share affects growth potential. Peneder (2003) conducted a cross-country panel analysis and found that differences in industrial structure significantly correlate with different growth rates, lending support to the idea that specific sectors (manufacturing and knowledge-intensive industries) act as engines of growth. In Europe's advanced economies, however, the picture is nuanced. Havlik (2015) examined structural change in new EU member states from Central and Eastern Europe during 1995–2011 using a shift-share analysis (a decomposition technique). He found that countries which experienced greater reallocation towards manufacturing and trade enjoyed faster aggregate GDP growth, especially in Northern and Central Europe. By contrast, some Southern European economies showed less structural change and correspondingly lower growth, despite being high-income—a pattern Havlik identifies as a North–South divide in the EU. This suggests that continued industrial upgrading (e.g., expanding modern manufacturing and high-value services) contributed to better growth performance in the 2000s. In contrast, regions that stagnated in their economic structure saw sluggish growth.

Major macroeconomic crises—such as financial meltdowns or recessions—have immediate adverse effects on GDP, but studies debate the duration and magnitude of their impact. Panel econometric studies enable researchers to isolate crisis effects by controlling for country- and time-fixed effects, as well as other covariates. The consensus is that crises significantly dent short-term growth and can have lasting output costs. Reinhart and Rogoff's historical analysis (2009) documented that severe financial crises often lead to deep and protracted recessions. More recent panel evidence refines this view. Madeira (2024) analyses 40 years of crises across 102 countries with industry-level panel data. Using a differences-in-differences approach within a fixed-effects framework, he finds that banking crises reduce overall manufacturing output growth by about 2.7 percentage points on average. In contrast, currency crises and sovereign debt crises cut manufacturing growth by roughly 6% and 1%, respectively. These are sizeable impacts, highlighting that crises typically cause a significant drop in GDP relative to trend. Crucially, Madeira proved that the effect is even larger for industries dependent on external finance (credit), and that the 2008–2009 global financial crisis had powerful negative effects on output in both advanced and emerging economies. Such findings reinforce that a systemic crisis can push economies below their previous growth trajectory.

In the EU, the 2008 sovereign debt and banking crisis led to a double recession in several countries. Studies focusing on Europe (often using panel data for EU members) find heterogeneous effects: nations with more robust financial systems or policy responses recovered faster, whereas others suffered persistent GDP losses. Havlik (2015) also assessed the impact of the 2008–2011 crisis in his study of EU structural change, finding that despite different pre-crisis growth patterns, both new and old member states experienced significant slowdowns and required years to recover lost output. This aligns with IMF analyses, which noted that many EU economies did not return to their pre-crisis GDP trend even several years after the crisis. On the other hand, some research using panel regressions (e.g., Romer & Romer, 2017, who re-examine advanced economy crises) suggests that when controlling for crisis severity, the long-run impact may not be as permanently scarring as previously thought, with growth resuming albeit from a lower base. The debate here centers on whether crises cause permanent level shifts in GDP (with output never catching up to the old path) or just temporary deviations.

The prevailing view from multi-country panel studies is that financial crises have statistically significant negative effects on GDP growth and investment in the medium term, and policy measures, such as post-2008 macroprudential policies, can moderate but not eliminate these losses. These are important reasons to consider the effects of crisis in our empirical analysis. On one side, the emissions could be temporarily reduced due to lower economic activity, and, on the other side, green investments could face a delay due to fiscal constraints or policy shifts.

By bringing together all these diverse ideas from scientific literature we observe several consistent patterns. First, renewable energy adoption is generally found to support economic growth or at least to not inhibit it, especially over the long run, which is encouraging for green transition policies. Many panel studies confirm the positive GDP elasticity of renewable energy. However, few such as Menegaki (2011) found neutral effects in the EU, highlighting that results can depend on the sample and method. Second, CO_2 emissions and GDP are strongly intertwined: growing economies tend to emit more CO_2 in the absence of interventions; however, high-income countries (many in the EU) exhibit signs of an emissions slowdown or reversal at advanced stages of economic growth. The literature debates the exact shape of this relationship (inverted-U vs. N-shaped). There is a consensus that environmental constraints eventually require a shift in energy and industrial practices to sustain growth. Third, the structure of the economy influences growth performance: panel evidence from Europe and beyond indicates that economies oriented toward higher-productivity sectors achieve faster GDP growth, whereas those that over-rely on low-productivity or diminishing sectors may stagnate. Finally, macroeconomic crises are acute shocks to GDP, with panel estimates quantifying sizable negative impacts and suggesting that output can remain below its pre-crisis trend for many years. This highlights the importance of resilience and effective policy in mitigating crises.

Notably, the methodologies used also contribute to points of divergence. Panel OLS with fixed effects is valued for controlling unobserved heterogeneity, but it assumes homogeneous coefficients across all observations. Some studies find differences when using random effects (Menegaki, 2011) or more sophisticated estimators (e.g. system-GMM for endogeneity, as in some globalisation-CO_2 studies). Time-series analyses at the country level sometimes yield contrasting causality directions (for example, one country might exhibit GDP to renewables causality but no renewables to GDP effect), while panel studies seek an overarching trend. Cointegration and error-correction models often detect long-run equilibrium relationships that static panels miss, explaining why Apergis and Payne found a positive link where Menegaki did not. Structural models (like CGE simulations or structural VARs) complement this literature by exploring scenario-based outcomes (such as how a shift to 100% renewables or a hypothetical crisis would affect GDP), generally corroborating the empirical findings that gradual transitions have minor growth penalties, and big shocks have significant short-term costs.

Despite consensus on the necessity of gas emission reduction, the form of the emissions-growth curve remains unsettled. While many studies confirm the growth elasticity of renewables, conflicting results underline that national context, policy design, and methodological approaches can significantly alter the economic impact of renewable adoption. Therefore, there is a need for country-level or sector-specific analyses that account for institutional, structural, and technological heterogeneity across EU member states. Accordingly, our empirical focus will be on several high-income economies, to observe contrasts based on industrial outlines.

3 Data and Methodology

This study examines the economic implications of the low-carbon transition by analysing the relationship between economic growth and key environmental and structural indicators. The empirical analysis focuses on five European Union member states—Belgium, Denmark, France, Germany, and Italy—chosen for their diverse energy mixes, industrial structures, and policy approaches toward sustainability. All these are high-income economies but with contrasting industrial profiles: manufacturing-heavy (Germany, Italy), finance and services (Belgium, France), and clean-technology innovation (Denmark). Although pursuing distinct national strategies, these countries are sharing overarching climate targets within the European Green Deal. While acknowledging some countries' diversity, we consider these five countries for their common goals in terms of the low-carbon transition landscape, which strengthens the analysis of structural indicators like employment, investment, and carbon productivity. The period under analysis spans over more than two decades (2001–2023), subject to data availability for all indicators collected. These are presented in Table 1.

Due to differences in the scales of the indicators, we considered including them in an econometric analysis by creating a new variable consisting of GDP growth, computed based on the level in the previous year, and a new CO_2 emissions variable, set up as the natural logarithm form.

To ensure the validity of the econometric analysis, the variables were tested for stationarity. Unit root tests indicated the presence of non-stationarity for the Share of energy from renewable sources, CO_2 emissions, and Gross value added and income by main industry variables. Therefore, we employed the first-order difference and reached stationarity with these indicators. The results of the unit-root tests are presented in Table 2.

The descriptive statistics are presented in Table 3 for the indicators initially collected from open-access databases (listed in Table 1) and for the indicators prepared for the regression models.

From the descriptive statistics, we observe higher volatility, as indicated by a larger standard deviation, for GDP growth, the Share of energy from renewable sources (first difference), and Gross value added and income by main industry (first difference). However, differences across countries could be a cause, which is why we will also employ the fixed effects model to observe whether these differences may significantly affect regression results.

To estimate the relationship between economic growth and the selected explanatory variables, we employ a Panel Ordinary Least Squares (OLS) approach, followed by Fixed Effects (FE) and Random Effects (RE) models, to account for unobserved heterogeneity across countries that might bias or confound the estimates. While the Fixed Effects models control for time-invariant country-specific characteristics, the Random Effects models assume that these effects are uncorrelated with the explanatory variables. The FE model is selected to eliminate bias due to time-invariant country-specific factors, such as geography or institutional background, which may correlate with the explanatory variables. By controlling these effects, FE provides more robust within-country estimates. To explore variance across countries while retaining more degrees of freedom, RE assumes country-specific effects which are random and uncorrelated with the regressors. This is useful when these assumptions are held and when we want to generalise results beyond our sample.

Table 1 Variable description

Variable	Description	Unit of measure	Source of data	Type of variable
GDP	Measures the total monetary value of all goods and services produced within a country over a specific period. It is a key indicator of a country's economic performance and growth.	euros	World Bank	Dependent variable
Industry as % of GDP	Represents the contribution of the industrial sector (manufacturing, mining, construction, etc.) to a country's total Gross Domestic Product (GDP). It indicates the level of industrialisation and economic structure.	%	Eurostat	Independent variable
Share of energy from renewable sources	Measures the proportion of gross final energy consumption that comes from renewable energy sources such as wind, solar, hydro, biomass, and geothermal. It is a key indicator for monitoring progress toward EU renewable energy targets.	%	Eurostat	Independent variable
CO_2 emissions	Represents the total amount of carbon dioxide released into the atmosphere from fossil fuel combustion and industrial processes. It is a key environmental indicator linked to climate change.	tons	Our World in Data	Independent variable
Gross value added and income by main industry	Measures the contribution of each industrial sector to the economy by calculating the difference between output and intermediate consumption. It provides an economic assessment of industry performance, productivity, and profitability, being measured out of the GDP.	%	Eurostat	Control variable
Dummy	A binary variable indicating economic distress periods. It takes the value 1 for years impacted by major global crises—namely the 2007–2009 financial crisis and the 2020–2021 COVID-19 pandemic—and 0 for all other years.	0/1		Control variable

To determine the appropriate specification for our sample, we perform a Hausman test, which statistically evaluates whether the unique errors are correlated with the regressors, guiding the choice between fixed or random effects. The general regression model is presented below:

$$GDPgrowth_{it} = a_i + \beta_1 * Ind_{it} + \beta_2 * RenewEn_{it} + \beta_3 * CO_{2_{it}} + \varepsilon_{it} \tag{1}$$

To observe any changes in the results obtained, we will further test the model with two control variables: one related to the gross value added and income by main

Table 2 Unit root tests

Variable	Levin-Lin-Chu test Adjusted t* (p-value)	Im–Pesaran–Shin test Z-t-tilde-bar (p-value)
GDP growth	−4.6776 (0.0000)	−4.6580 (0.0000)
Industry as % of GDP	−2.1576 (0.0155)	−1.5663 (0.0586)
Share of energy from renewable sources	−0.3096 (0.3784)	3.3533 (0.9996)
Share of energy from renewable sources (first order difference)	−4.6586 (0.0000)	−5.9893 (0.0000)
CO_2 emissions (ln values)	2.4160 (0.9922)	4.3538 (1.0000)
CO_2 emissions (first order difference in ln values)	−6.5456 (0.0000)	−6.2914 (0.0000)
Gross value added and income by main industry	2.3101 (0.9896)	1.6166 (0.9470)
Gross value added and income by main industry (first order difference)	0.9974 (0.8407)	−4.9871 (0.0000)

Table 3 Descriptive statistics

Variable	Obs	Mean	Std. dev.	Min	Max
GDP growth	115	4.0844	8.219	−14.75	23.49
Industry as % of GDP	115	21.505	2.894	16.087	26.843
Share of energy from renewable sources	115	14.575	8.623	1.6	44.396
Share of energy from renewable sources (first difference)	115	6.635	8.043	−14.66	31.981
CO_2 emissions (ln values)	115	19.204	1.0677	17.121	20.634
CO_2 emissions (first difference in ln values)	115	−0.106	0.27	−0.644	0.809
Gross value added and income by main industry	115	88.897	1.364	84.9	91.3
Gross value added and income by main industry (first difference)	115	0.0575	0.3846	−0.932	1.332
Dummy	115	0.174	0.3807	0	1

industry, and a dummy specific to the COVID-19 pandemic. Accordingly, the regression model becomes:

$$GDPgrowth_{it} = a_i + \beta_1 * Ind_{it} + \beta_2 * RenewEn_{it} + \beta_3 * CO_{2_{it}} + \beta_4 * GVA_{it} + \beta_5 * dummy_t + \varepsilon_{it} \tag{2}$$

where: i represents the countries considered in our analysis (i = 1..5), t is the year observed (t = 2001..2023), α is the intercept, the βs are the regression coefficients associated with the explanatory variables (independent and control variables), and ε is the error term.

Given the structure of our panel, including five countries (N = 5) over a 23-year period (2001–2023) we take careful steps to ensure robust inference in the presence of potential econometric issues, applying robust standard errors to account for heteroskedasticity across countries and therefore, use panel-corrected standard errors. To capture potential endogeneity and dynamic effects, we extend our analysis using the Generalized Method of Moments (GMM) estimator, specifically the Arellano-Bond dynamic panel GMM. This methodological framework allows us to combine static and dynamic perspectives, control for country-specific fixed effects while correcting for common panel issues and improves estimator consistency—yielding more reliable insights into the drivers of economic growth.

4 Results and Discussions

This section presents the empirical findings derived from the regression models employed, including ordinary least squares (OLS), fixed effects (FE), random effects (RE) models, and the Hausman test. These results provide insights into the relationship between economic growth and the explanatory variables: industrial structure, the share of energy from renewable sources, the level of carbon dioxide emissions, the gross value added and income by primary industry, and the influence of the COVID-19 crisis. Results are presented in a comparative manner in Table 4.

Results indicate a statistically significant influence from the industry as a percentage of GDP (representing the contribution of the industrial sector) and CO_2

Table 4 Regression results

	OLS	FE	RE	OLS (control var.)	FE (control var.)	RE (control var.)
Ind	0.448*	2.385***	0.448*	0.504*	2.817***	0.504*
	(0.263)	(0.559)	(0.263)	(0.265)	(0.566)	(0.265)
RenewEn	−0.0177	−0.0400	−0.0177	0.00584	0.00346	0.00584
	(0.0947)	(0.0935)	(0.0947)	(0.0958)	(0.0924)	(0.0958)
CO_2	5.943**	4.203	5.943**	4.962*	2.655	4.962*
	(2.818)	(2.747)	(2.818)	(2.915)	(2.776)	(2.915)
GVA				−3.657*	−5.66***	−3.657*
				(2.220)	(2.112)	(2.220)
dummy				2.122	4.294**	2.122
				(2.191)	(2.092)	(2.191)
Constant	−4.806	−46.4***	−4.806	−6.429	−56.6***	−6.429
	(5.711)	(12.07)	(5.711)	(5.798)	(12.32)	(5.798)
Observations	115	115	115	115	115	115
R-squared (within)	0.1051	0.186	0.1051	0.1367	0.245	0.1367
F / Wald test	8.34**	8.15***	8.34**	11.23**	6.81***	11.23**
Hausman test chi sq. (p-value)		14.92 (0.0019)			21.17 (0.0008)	

Standard errors in parentheses
*** $p < 0.01$, ** $p < 0.05$, * $p < 0.1$

emissions on GDP growth, both in a direct manner. Moreover, it seems that the gross value added and income by main industry has an indirect influence on GDP growth. The share of energy from renewable sources was not statistically significant. It changed its coefficient signs, leading us to believe that it should not be considered a decisive factor influencing economic growth. Lastly, the dummy for the COVID-19 pandemic has a direct influence on GDP growth, as indicated by the fixed effects model results.

Based on the regression results, we will further discuss the implications of the suggested relationships. The CO_2 emissions are positively associated with GDP, with coefficients ranging from 2.65 to 5.94, depending on the model, reinforcing the persistent linkage between output and emissions observed in the Environmental Kuznets Curve (EKC) literature. This highlights the challenge of separating growth from environmental degradation (Onofrei et al., 2022).

The industrial sector, consistently demonstrating a positive and significant association with GDP growth, showed regression coefficients ranging from 0.5 to 2.8, suggesting that an increase in industrial activity will positively influence economic growth. This aligns with the literature that emphasises the role of manufacturing and high-value industries in driving productivity and output (Peneder, 2003; Havlik, 2015).

The negative influence of Gross Value Added (GVA) and income by main industry (GVA) on GDP growth may be explained by deindustrialisation, which reduces the high-productivity sector's output, or fiscal policy shifts related to significant changes in tax or subsidies, which can distort the link between GVA and GDP. Similar indirect influence was found by Karami et al. (2019) during periods of structural transition, when investments, often tied to capital-intensive sectors, were negatively associated with economic growth. Considering the sample and the countries analysed, we mention some specificities: in Denmark hi-tech industries and other niche sectors can disproportionately affect GDP; in France and Italy there is a low productivity growth in services (there is a heavy reliance on the services sector, compared to a weak industrial performance) and over-reliance on public sector employment; Germany was vulnerable in trade shocks related to energy prices or supply chain disruptions such as those from the automotive sector or manufacturing, which represents an important share of GDP.

Finally, with the COVID-19 crisis period modelled, results suggest that economic recoveries and policy responses are likely to play a role in mitigating the long-term negative effects on the environment. These findings contribute to a broader understanding of the economic implications of the low-carbon transition and provide a nuanced basis for evaluating the impacts of climate policy across the diverse European economies.

Across all model specifications the statistically significant results seem robust, supporting several key hypotheses (e.g. the Environmental Kuznets Curve) or policies about the low-carbon transition and its economic implications: the carbon pricing, a policy tool that assigns a cost to CO_2 emissions, aiming to internalise the environmental cost of greenhouse gas emissions; the Green Investment lag, which refers to the delay in investment flows into green technologies and infrastructure needed to meet the climate change goals.

The Hausman test yields significant p-values, confirming that the Fixed Effects model is more appropriate for our sample. Thus, differences across the countries analysed are important and influential in defining the relationship between the variables employed in the analysis.

Further on, we test cross-sectional dependence in panels and based on these results we employ a new regression model, corrected (CORR) for heteroskedasticity and auto-correlation and adequate for a sample with fewer cross-sections ($i = 5$) than years ($t = 23$) overviewed, followed by a dynamic model (GMM) to test the relevance of GDP growth from previous years. Based on the Pesaran test of cross-sectional independence, the fixed effects model with control variables returns a value of 12.66 with a p-value of 0.000, which makes us reject the null hypothesis of no cross-sectional dependence.

Considering the complex effects of pollution and industry, we expect better confirmation of regression results through the dynamic model. Results are presented in Table 5.

The latest results confirm the positive influence of industry and CO_2 emissions. However, the latter's statistical significance is limited, as well as the indirect influence of gross value added and income by main industry on GDP growth. Within the dynamic model, employing the Arellano-Bond estimation, the dummy variable has

Table 5 Regression results for corrected and dynamic models

	CORR	GMM
L1.GDP growth		0.0138
		(0.095)
Ind	0.504*	3.466***
	(0.258)	(0.645)
RenewEn	0.00584	−0.0514
	(0.0932)	(0.0988)
CO_2	4.962*	3.233
	(2.838)	(2.826)
GVA	−3.657*	−4.768**
	(2.161)	(2.157)
Dummy	2.122	3.874*
	(2.133)	(2.111)
Constant	−6.429	
	(5.645)	
Observations	115	115
Wald test	11.85**	42.69***
Sargan test: chi sq. (p-value)		95.82 (0.3999)

Standard errors in parentheses
*** $p < 0.01$, ** $p < 0.05$, * $p < 0.1$

a positive and statistically significant coefficient, reflecting a direct effect on GDP growth. The level of GDP from the previous years does not appear statistically significant in the dynamic model tested.

To summarise the empirical evidence, the results support the idea that the industrial sector remains a critical engine of growth, while the adoption of renewable energy introduces short-term economic frictions. The strong positive relationship between GDP and CO_2 emissions reaffirms the urgency of structural decoupling strategies, which refer to long-term changes in the economic system that enable countries to grow their GDP while reducing environmental pressures. These findings contribute to a broader understanding of the economic implications of the low-carbon transition and provide a nuanced basis for evaluating the impacts of climate policy across diverse European economies.

Structural decoupling strategies are essential to improve efficiency and transform the underlying economic architecture. Carbon pricing, emissions trading schemes, and green tax reforms must be designed to maintain investment incentives and avoid regressive effects. As governments should support industrial modernisation, especially in sectors with high value-added potential, public policies must incentivise cleaner production methods related to carbon capture or circular economy models. By these means, industrial policy could balance competitiveness with sustainability, avoiding carbon leakage while supporting innovation. The transition towards renewables might disrupt certain labour markets related to traditional energy, automotive or heavy industry, and therefore governments could ensure combining retraining programs, regional development funds, temporary subsidies for affected sectors, somehow covering for energy price volatility and distributional impacts.

5 Conclusions

This research examines the economic aspects of the low-carbon transition in some of the most developed economies within the European Union, utilising empirical panel data from five countries spanning the period 2001–2023. By employing different regression models, we assess the relationship between industrial structure, renewable energy integration, carbon emissions and GDP. The findings reveal complex relationships, particularly considering current involvement in climate-related economic transformation.

The analysis confirms the continued importance of the industrial sector as a driver of economic growth in the EU, by confirming a strong and positive link between industrial output and GDP, reinforcing the role of manufacturing and high-productivity sectors in sustaining economy. Conversely, the negative influence suggested from renewable energy consumption on GDP growth emphasises a transition towards cleaner energy sources, which imposes short-term economic costs and transitional frictions linked to investment and labor market shifts. The persistent positive correlation between CO_2 emissions and GDP growth underscores the challenge of distinguishing between economic activities and environmental degradation. Findings underscore a tension between environmental goals and economic stability, especially during transition periods, and the need for targeted green investment strategies. Accordingly, these relationships also reflect the necessity for adjustments in infrastructure shifts and capital reallocation for companies. The sample analysed tackles economic structures through gross value added and income by main industry, because energy policies and climate exposure vary across EU countries. Moreover, renewable energy share may suffer from inconsistent reporting standards, which is why the sample is limited to few countries with full data available. While fixed effects models control time-invariant country-specific factors, they cannot consider dynamic shocks or policy changes over time. Therefore, the dynamic model is more appropriate to capturing time-varying effects and using lagged variables as instruments.

The significant impact of the COVID-19 crisis on GDP growth highlights the need for resilience strategies that mitigate economic downturns and ensure that climate transition policies remain on track during periods of financial stress.

These findings highlight a key policy challenge in reconciling environmental goals with economic stability, particularly during periods of transition. While the long-term benefits of a low-carbon economy—including energy security, innovation, and climate risk reduction—are widely acknowledged, the empirical evidence suggests that short-run GDP effects may not be uniformly positive. The fact that the analysis is limited to five EU countries may restrict the generalisation across the EU, especially in less developed or more energy-dependent economies. However, a successful transition requires targeted interventions related to green investment incentives and innovation-friendly regulations.

This study contributes to the broader climate economics literature by providing a country-specific and empirically grounded assessment of the trade-offs inherent in the decarbonisation process. Future research should consider sectoral analyses and dynamic modelling frameworks to better understand the full spectrum of transition costs and benefits. For example, policy variables related to carbon pricing, subsidies or innovation grants could be included in the analysis to assess the effectiveness in mitigating transition costs. A sustainable future is not only an environmental imperative but also a way to reach economic balance. Therefore, recognising and addressing transitional challenges is essential to ensure that climate policies achieve their goals without compromising economic performance or social cohesion. For it, indicators of inequality or employment shifts might be added in future research to assess the social sustainability of climate policies.

Acknowledgments This work was supported by a grant from the Romanian Ministry of Research, Innovation and Digitalization, the Project with the title "Economics and Policy Options for Climate Change Risk and Global Environmental Governance" (CF 193/28.11.2022, Funding Contract no. 760078/23.05.2023), within Romania's National Recovery and Resilience Plan (PNRR)—Pillar III, Component C9, Investment I8 (PNRR/2022/C9/MCID/I8)—Development of a program to attract highly specialised human resources from abroad in research, development, and innovation activities.

We are grateful to Nikola Markovic (Master Student and Project team member) for his contributions, which greatly aided the development of this work.

Competing Interests The authors have no conflicts of interest to declare relevant to this chapter's content.

References

Acemoglu, D., Aghion, P., Bursztyn, L., & Hemous, D. (2012). The environment and directed technical change. *American Economic Review, 102*(1), 131–166. https://www.nber.org/papers/w15451

Apergis, N., & Payne, J. E. (2010). Renewable energy consumption and economic growth: Evidence from a panel of OECD countries. *Energy Policy, 38*(1), 656–660. https://ideas.repec.org/a/eee/enepol/v38y2010i1p656-660.html

Apergis, N., & Payne, J. E. (2014). Renewable energy, output, CO_2 emissions, and fossil fuel prices in Central America: Evidence from a panel of countries. *Energy Economics, 42*, 226–232. https://doi.org/10.1016/j.eneco.2014.01.003

Balsalobre-Lorente, D., Shahbaz, M., Roubaud, D., & Farhani, S. (2018). How economic growth, renewable electricity and natural resources contribute to CO_2 emissions? *Energy Policy, 113*, 356–367. https://doi.org/10.1016/j.enpol.2017.10.050

Burke, M. J., & Stephens, J. C. (2018). Political power and renewable energy futures: A critical review. *Energy Research & Social Science, 35*, 78–93. https://doi.org/10.1016/j.erss.2017.10.018

Dong, K., Sun, R., Jiang, H., & Zeng, X. (2018). CO_2 emissions, economic growth, and the environmental Kuznets curve in China: What roles can nuclear energy and renewable energy play? *Journal of Cleaner Production, 196*, 51–63. https://doi.org/10.1016/j.jclepro.2018.05.271

Havlik, P. (2015). Patterns of structural change in the new EU member states. *DANUBE: Law, Economics and Social Issues Review, 6*(3), 133–157. https://doi.org/10.1515/danb-2015-0009

Kahia, M., Omri, A., & Jarraya, B. (2021). Green energy, economic growth and environmental quality nexus in Saudi Arabia. *Sustainability, 13*(3), 1264. https://doi.org/10.3390/su13031264

Karami, M., Elahinia, N., & Karami, S. (2019). The effect of manufacturing value added on economic growth: Empirical evidence from Europe. *Journal of Business, Economics and Finance, 8*(2), 133–146. https://doi.org/10.17261/Pressacademia.2019.1044

Madeira, C. (2024). The impact of financial crises on industrial growth: Lessons from the last 40 years. *BIS Working Paper No. 1214*. Bank for International Settlements. https://doi.org/10.1515/danb-2015-0009

Menegaki, A. N. (2011). Growth and renewable energy in Europe: A random effect model with evidence for neutrality hypothesis. *Energy Economics, 33*(2), 257–263. https://doi.org/10.1016/j.eneco.2010.10.004

Nordhaus, W. D. (1991). To slow or not to slow: The economics of the greenhouse effect. *Economic Journal, 101*(407), 920–937. https://doi.org/10.2307/2233864

Omri, A., Daly, S., Rault, C., & Chaibi, A. (2015). Financial development, environmental quality, trade and economic growth: What causes what in MENA countries. *Energy Economics, 48*, 242–252. https://doi.org/10.1016/j.eneco.2015.01.008

Onofrei, M., Vatamanu, A. F., & Cigu, E. (2022). The relationship between economic growth and CO_2 emissions in EU countries: A cointegration analysis. *Frontiers in Environmental Science, 10*, 934885. https://doi.org/10.3389/fenvs.2022.934885

Peneder, M. (2003). Industrial structure and aggregate growth. *Structural Change and Economic Dynamics, 14*(4), 427–448. https://doi.org/10.1016/S0954-349X(02)00052-8

Popp, D. (2019). Environmental policy and innovation: A decade of research. *International Review of Environmental and Resource Economics, 13*(3–4), 265–337. https://doi.org/10.1561/101.00000111

Popp, D., Vona, F., Marin, G., & Chen, Z. (2020). *The employment impact of green technologies*. NBER Working Paper No. 27321. https://www.nber.org/papers/w27321

Reinhart, C. M., & Rogoff, K. S. (2009). The aftermath of financial crises. *American Economic Review, 99*(2), 466–472. https://doi.org/10.1257/aer.99.2.466

Romer, C. D., & David, H. R. (2017). New evidence on the aftermath of financial crises in advanced countries. *American Economic Review, 107*(10), 3072–3118. https://doi.org/10.1257/aer.20150320

Saidi, K., & Omri, A. (2020). The impact of renewable energy on carbon emissions and economic growth in 15 major renewable energy-consuming countries. *Environmental Research, 186*, 109567. https://doi.org/10.1016/j.envres.2020.109567

Sharif, A., Raza, S. A., Ozturk, I., & Afshan, S. (2019). The dynamic relationship of renewable and nonrenewable energy consumption with carbon emission: A global study with the application of heterogeneous panel estimations. *Renewable Energy, 133*, 685–691. https://doi.org/10.1016/j.renene.2018.10.052

Soytas, U., & Sari, R. (2009). Energy consumption, economic growth, and carbon emissions: Challenges faced by an EU candidate member. *Ecological Economics, 68*, 1667–1675. https://doi.org/10.1016/j.ecolecon.2007.06.014

Stern, N., & Stiglitz, J. (2017). *Report of the high-level commission on carbon prices*. World Bank Group. https://www.researchgate.net/publication/318284315_Report_of_the_High-Level_Commission_on_Carbon_Prices.

Stocker, T. F., Qin, D., Plattner, G.-K., et al. (2013). *Climate change 2013: The physical science basis*. Cambridge University Press. https://doi.org/10.1017/CBO9781107415324

Tagliapietra, S., Zachmann, G., Edenhofer, O., Glachant, J.-M., Linares, P., & Loeschel, A. (2019). The European Union energy transition: Key priorities for the next five years. *Energy Policy, 132*, 950–954. https://doi.org/10.1016/j.enpol.2019.06.060

Tugcu, C. T., Ozturk, I., & Aslan, A. (2012). Renewable and non-renewable energy consumption and economic growth: Evidence from G7 countries. *Energy Policy, 47*, 194–205. https://doi.org/10.1016/j.eneco.2012.08.021

Integrating Climate Risk into Financial Stability: A Systemic Perspective

Ana-Elena Varadi, Oana-Ramona Lobonţ, Graţiela Georgiana Noja, and Sorana Vătavu

1 Introduction

Climate change has emerged as a critical systemic risk to global financial stability. The increasing frequency and severity of physical climate events, alongside the profound economic adjustments necessitated by the transition to a low-carbon economy, pose multifaceted challenges to financial institutions and markets worldwide. Accordingly, our research investigates the evolving integration of climate-related financial risks—both physical and transitional—into financial stability evaluations, whilst clarifying the fundamental mechanisms through which these risks are conveyed into the financial system, affecting credit, market, liquidity, and operational risk categories.

Climate-related impacts are becoming consistent, influencing various segments of the financial system concurrently, and are marked by non-linearities and tipping points that may result in sudden and significant repercussions. A consensus exists among central banks and regulators that these risks significantly impact their financial stability objectives; the possibility of irreversible environmental impacts

A.-E. Varadi (✉)
Doctoral School of Economics and Business Administration, West University of Timişoara, Timişoara, Romania
e-mail: ana.varadi@e-uvt.ro

O.-R. Lobonţ · S. Vătavu
Faculty of Economics and Business Administration, Department of Finance, Business Information Systems and Modelling, West University of Timişoara, Timişoara, Timiş County, Romania
e-mail: oana.lobont@e-uvt.ro; sorana.vatavu@e-uvt.ro

G. G. Noja
Department of Marketing, International Business and Economics, Faculty of Economics and Business Administration, West University of Timişoara, Timişoara, Romania
e-mail: gratiela.noja@e-uvt.ro

C.-W. Su et al. (eds.), *Green Finance and the Challenges of Climate Change*, Sustainable Finance, https://doi.org/10.1007/978-3-032-07293-1_5

heightens the urgency, and the increasing awareness is that present financial asset valuations may be insufficiently accounting for inherent climate-related financial risks. This mispricing engenders a latent vulnerability: an abrupt market re-evaluation of climate risks may instigate price adjustments, resulting in cascading adverse effects on financial institutions and markets. The complex interconnections within the financial system indicate that climate-related shocks are not standalone occurrences, given that they can be intensified.

For this reason, our paper offers a comprehensive, multidimensional perspective on climate-induced risks, positioning them as systemic threats to global financial stability rather than isolated environmental concerns. Its main contributions lie in integrating institutional analysis with econometric evidence and highlighting differentiated vulnerabilities across advanced and fossil-fuel-dependent economies, while proposing empirically grounded macroprudential recommendations. The main objective is to analyse physical and transition risks, outlining their transmission mechanisms and systemic consequences for the financial sector, thus underscoring the formulation of pragmatic strategies for regulators, policymakers, and financial institutions to bolster resilience against climate-induced financial instability. The analysis emphasises the need to reform conventional risk mitigation procedures for addressing the interconnected and non-linear characteristics of climate risks, providing insights into how global financial systems can enhance their preparedness for and mitigation of these emerging threats. This paper contributes to the sustainable finance literature through a multi-dimensional and evidence-based perspective, bridging gaps between theory, regulation, and empirical risk assessment. It employs a dual methodological approach integrating qualitative and quantitative methods through the employment of regression and mediation analysis, facilitating the identification of climate-financial transmission mechanisms. Crucially, this research reconceptualises climate change not merely as an environmental concern but as a systemic threat to global financial and economic stability, calling for a fundamental reassessment of traditional financial risk and regulatory paradigms.

A primary goal is to provide proper insights customised for regulators, policymakers, and financial institutions. This entails synthesising existing knowledge and identifying and addressing empirical gaps. Special emphasis is placed on the intricacies of climate-related shocks' transmission within financial systems and the notable discrepancies in institutional practices across various jurisdictions. The analysis aims to compare and contrast methodologies in developed economies with those in emerging market and fossil-fuel-dependent economies, which frequently encounter unique vulnerabilities and capacities by examining key international frameworks, including those proposed by the Network for Greening the Financial System (NGFS), the Task Force on Climate-related Financial Disclosures (TCFD) and its development into the IFRS S2 standard, and the Sustainable Finance Disclosure Regulation (SFDR).

This paper is structured as follows: Section 2 comprehensively analyses the current literature, identifying the fundamental theoretical frameworks and empirical studies that support this research. The review rigorously assesses the existing knowledge base, pinpointing deficiencies and controversies that this study aims to tackle. Section 3 delineates the data and methodology utilised in this analysis, specifying the research design, data collection techniques, and analytical strategies to guarantee robust and reliable outcomes. Section 4 presents and rigorously analyses the empirical results. This section analyses the findings in relation to the current literature and the study's research objectives, addressing their implications, significance, and any unforeseen outcomes. The paper concludes by synthesising the principal findings, emphasising their contributions to the field, and addressing the study's limitations and potential for future research.

2 Literature Review

2.1 Defining Physical and Transition Risks from a Systemic Perspective

Climate change poses significant threats to financial systems through two primary channels: physical risks and transition risks. Physical risks encompass the economic and financial losses arising from acute events— cyclones, floods, and wildfires— and chronic climatic changes, including sea level rise, prolonged droughts, and shifts in long-term precipitation patterns. These risks can disrupt asset values, impair business operations, and destabilise financial markets. Such risks cause damage to property, land, and infrastructure, disrupt business activities and pose risks to health and productivity. From the perspective of financial institutions, physical risks result in increased credit risk due to loan defaults from impacted households and businesses, decreased collateral value, particularly for real estate in vulnerable regions, and increased claims on insurance. For example, banks holding significant mortgage portfolios in flood or drought-prone areas face substantial risks to income from rents and depreciation of property values (Network for Greening the Financial System, 2019). Furthermore, transition risks arise from adjusting to a low-carbon economy. These risks are driven by changes in climate policy (e.g., implementation of carbon pricing mechanisms, emissions standards, or phase-out of fossil fuel subsidies), technological innovation (e.g., rapid development and deployment of renewable energy technologies displacing carbon-intensive alternatives), and shifts in market sentiment and societal preferences (e.g., growing consumer demand for sustainable products, investor pressure for decarbonization). Transition risks can lead to significant revaluation of assets, particularly the "stranding" of assets in carbon-intensive sectors that become economically unviable or face premature write-downs. Understanding these transmission channels is fundamental to assessing their potential influence on financial stability.

The materiality of these risks—determining which risks are important to which sorts of investors and how and when they may affect investments—is of primary concern. In accordance with the estimations of the Network for Greening the Financial System (2024a), both physical and transition risks are projected to increase in the coming decades: transition risks are expected to escalate with intensified global efforts to mitigate carbon emissions, while physical risks will become more pronounced as global temperatures continue to rise. A critical consideration is the interplay between these two risk types: delayed or insufficient action on mitigation will inevitably lead to more severe and widespread physical impacts in the future. Such projections suggest that after 2050, physical risks are expected to dominate if climate action falls short, although transition risks will persist if mitigation goals are not met. Therefore, managing climate financial risk requires an integrated strategy that simultaneously addresses physical and transition dimensions, recognising their evolving materiality over different time horizons.

The Financial Stability Board (2025) acknowledges climate risk as a possible source of systemic risk to the global financial system due to the physical and transition concerns associated with climate change. This systemic nature arises from climate change's broad and correlated impacts, affecting numerous economic agents and sectors, often in a non-linear and unpredictable manner. Financial stability concerns are scrutinised when asset prices undergo rapid adjustments in response to unexpected materialisations of sudden shifts in market perception regarding these risks (Aglietta & Espagne, 2016). Such climate-related shocks do not occur independently as they interact and amplify pre-existing financial vulnerabilities, such as high levels of corporate or sovereign debt, asset mispricing, or liquidity mismatches within the financial system. The inherent interconnectedness of the modern financial system implies that climate-related losses originating in one sector or region can propagate rapidly across institutions and markets. The potential for a "climate Minsky moment"—a sudden, sharp collapse in asset prices triggered by a collective shift in investor sentiment about unpriced climate risks—represents a critical systemic threat (Carney, 2015). Understanding these mechanisms and amplification channels is crucial for developing effective financial stability monitoring and macroprudential responses.

2.2 *Pivotal International Frameworks and Integration Measures*

The global financial landscape is increasingly grappling with the systemic implications of climate change. Several international frameworks and integration measures have emerged to guide the financial sector in understanding, managing, and disclosing climate-related financial risks. These initiatives are crucial in shaping institutional responses and regulatory expectations. Table 1 below provides a detailed overview of the key frameworks and measures, highlighting their objectives, the actors involved, and the inherent challenges in their implementation.

Nonetheless, despite the adopted frameworks for mitigating and controlling climate change, integrating climate-related risks into financial stability assessments involves diverse actors, each with distinct roles, responsibilities, and levels of

Table 1 International climate-related financial risk frameworks

Framework	Establishment	Key Objectives/ Contributions	Primary Actors Involved	Challenges/ Limitations
Network for Greening the Financial System (NGFS)	Established in 2017	Facilitates the exchange of best practices and advances climate risk management within the financial sector. Develops and regularly updates common reference climate scenarios, which are widely utilised for stress testing and risk assessment by both public authorities and private financial institutions. Provides high-level recommendations for integrating climate considerations into financial stability monitoring, micro-prudential supervision, and central banks' portfolio management.	Central banks, financial supervisors	Difficulty in downscaling global or regional scenario outputs to the granular level required for specific institutions or assets. Acknowledged limitations in fully capturing climate tipping points, the impacts of acute physical risks, nature-related risks beyond climate change, or complex socio-economic feedback loops (e.g., climate-induced migration). Constrained precision for exact institutional risk quantification, necessitating careful interpretation of results.
Task Force on Climate-related Financial Disclosures (TCFD)	Established in 2015 Mandate completed in October 2023	Developed voluntary recommendations to enhance climate-related financial disclosures' consistency, comparability, and quality. Structured around four thematic pillars: Governance, Strategy, Risk Management, and Metrics and Targets. Served globally as the foundational framework for numerous mandatory and voluntary disclosure regimes.	Financial Stability Board (FSB), companies, financial institutions, investors	The voluntary nature presented limitations in universal adoption. Its mandate transitioned to the IFRS Foundation in 2024, shifting towards more prescriptive global standards.

(continued)

Table 1 (continued)

Framework	Establishment	Key Objectives/ Contributions	Primary Actors Involved	Challenges/ Limitations
IFRS Foundation (IFRS S1 & S2 Standards)	Assumed responsibility from TCFD in 2024	Oversees global disclosure practices and promotes alignment with its IFRS S1 (general requirements for disclosure of sustainability-related financial information) and IFRS S2 (climate-related disclosures) standards. Aims to establish a harmonised global baseline for climate disclosure, embedding climate considerations more deeply within standard financial reporting.	IFRS Foundation, companies, financial institutions, investors, regulators	Critical dependence on consistent adoption and enforcement across diverse jurisdictions to prevent a fragmented landscape. Requires developing and implementing robust assurance mechanisms to enhance reliability and credibility. Significant capacity-building efforts are essential for both preparers and users, particularly in emerging market and developing economies (EMDEs) with limited data and technical expertise. Challenges in quantifying the financial impacts of climate risks from scenario analysis and assessing risks across complex global value chains.
Sustainable Finance Disclosure Regulation (SFDR) (EU)	Establishment 2021	Enhances transparency in the EU financial sector regarding sustainability-related risks and impacts. Mandates disclosures by financial market participants (FMPs) and financial advisers on integrating sustainability risks into investment decision-making processes. Requires assessing and reporting adverse sustainability impacts arising from investment activities.	Financial market participants (FMPs), financial advisers within the EU	Potential for a complex and fragmented reporting landscape for multinational corporations and financial institutions operating across multiple jurisdictions due to variations in national IFRS S2 adoptions or overlapping regional requirements.

Source: Authors' compilation

preparedness (Wu et al., 2023). For instance, central banks and financial supervisors are at the forefront of efforts to embed climate risk considerations into the financial system. Their responses are evolving along several dimensions. In micro-supervision, authorities increasingly set expectations for financial institutions regarding governance, risk management, strategy, and disclosure of climate-related risks. This includes encouraging or mandating climate stress tests and scenario analyses to assess firm-level resilience. The ECB (2021) has articulated comprehensive expectations for banks within the European Union to integrate climate and environmental risks as key drivers within traditional prudential risk categories, including credit, market, operational, and liquidity risks. Concurrently, member jurisdictions of the Network for Greening the Financial System are advancing climate-related risk assessment frameworks and developing analytical tools to identify, evaluate, and mitigate such risks. These efforts also support the formulation of policy recommendations to strengthen the role of central banks in addressing the challenges posed by climate change.

Beyond institutional-level measures, there is a growing focus on regulatory frameworks for addressing the systemic nature of climate risks. These could include sectoral capital requirements, concentration limits for exposures to high-risk sectors, adjustments to risk weights, or countercyclical capital buffers if climate risks are deemed to contribute to systemic cyclical risk. The FSB (2022) is developing a forward-looking analytical toolkit to help authorities monitor and assess climate-related vulnerabilities from a systemic perspective. However, the design and calibration of such tools are complex. For this reason, a gradual, targeted, and scalable approach is often advocated, allowing policies to adapt as understanding and data improve.

Financial institutions are heterogeneous in their response to climate risks. Banks are particularly exposed through their lending and investment activities. Physical risks can impair mortgage portfolios and commercial loans, while transition risks affect exposures to carbon-intensive sectors. Insurers face dual threats: on the liability side from increased claims due to more frequent and severe extreme weather events, and on the asset side from the impact of climate risks on their investment portfolios. The potential for insurance to become unaffordable or unavailable in high-risk regions poses a broader societal and financial stability challenge. Asset managers and institutional investors, such as pension funds and mutual funds, are exposed to physical and transition risks through their vast holdings of equities and bonds. However, many institutions lack preparedness. Common challenges include data availability and quality, methodological complexity in quantifying climate risks and conducting scenario analysis, the difficulty of integrating long-term climate considerations into shorter-term business and capital planning cycles, and a lack of professional local expertise.

For businesses, adapting to climate change means embedding climate considerations into their core operations, including how they plan long-term investments, manage their supply chains, and develop new products and services (Lobonț et al., 2025). This involves concrete actions such as strengthening resilience to physical climate impacts, improving energy efficiency, transitioning to renewable energy, innovating with green products, and creating credible plans to achieve net-zero emissions.

However, companies encounter significant hurdles in this strategic shift. These commonly include the substantial costs associated with transitioning to sustainable practices, the complexity of accurately quantifying the financial implications of climate risks, the challenge of managing global supply chains vulnerable to diverse climate hazards, and difficulties in securing affordable "green" financing. Increasingly, climate risk management is no longer a secondary compliance or CSR concern but a fundamental strategic imperative. Failing to adjust business models to the physical realities of a changing climate and the economic transition towards decarbonization poses severe financial risks and existential threats to long-term viability. Conversely, those organisations that effectively adapt will uncover opportunities for innovation and gain a competitive edge.

2.3 Global Divergences and Commonalities

While the recognition of climate-related financial risks is becoming extensive, the approaches to managing these risks, the pace of regulatory development, and institutional preparedness exhibit significant global divergences, particularly between developed and emerging economies.

Advanced economies have generally led the development and implementation of regulatory frameworks addressing climate-related financial risks. The European Union, for example, has introduced a comprehensive sustainable finance agenda, including the Sustainable Finance Disclosure Regulation (SFDR) and the Corporate Sustainability Reporting Directive (CSRD), which mandates detailed sustainability reporting. Additionally, many developed countries were among the first to adopt the Task Force on Climate-related Financial Disclosures recommendations (2021) and are now progressing toward alignment with the IFRS S2 standards. Supervisory authorities in these jurisdictions also tend to be more advanced in conducting climate stress testing and in embedding climate-related risks into their supervisory and regulatory practices.

Although they are the most vulnerable to the physical impacts of climate change due to geographical factors, reliance on climate-sensitive sectors like agriculture, and lower adaptive capacity, the developed and emerging economies face different climate-induced challenges. The climate finance needs of developed and emerging economies for mitigation and adaptation are immense, yet actual financial flows fall far short of requirements. A World Bank report (2024) highlights that in almost 60% of banks in advanced and emerging economies, climate-related lending is less than 5% of their portfolios, and over a quarter offer no climate financing at all. The adoption of green and sustainable finance taxonomies, crucial for guiding investment, is also significantly lower in advanced and emerging economies (around 10%) compared to advanced economies (around 76%).

Data availability is a major constraint in both economies, hindering effective climate risk assessment, the development of tailored financial products, and the ability to attract international climate finance. Capacity constraints, including limited technical expertise in financial institutions and regulatory bodies and insufficient resources for implementing complex new regulations and risk management practices, are also prevalent.

Fossil-fuel-dependent economies, whether developed or emerging, face particularly significant transition risks as the world moves towards decarbonization. Their institutional responses and strategies must navigate the complex trade-offs between maintaining economic stability, diversifying their economies, and meeting climate commitments.

International financial institutions, such as the International Monetary Fund (IMF) through its Resilience and Sustainability Trust (RST), and the World Bank Group, play a critical role in delivering financial and technical support to both developed and emerging economies. Their efforts aim to facilitate climate action and strengthen the resilience of financial sectors (World Bank Group & International Monetary Fund, 2024). Nevertheless, the scale of support required to address these challenges remains substantial.

The global push for standardised climate risk integration, while essential for overall financial stability and achieving climate goals, thus carries the risk of deepening existing inequalities. Suppose international standards and regulations are imposed without adequate consideration for the unique circumstances of developed and emerging economies—including data scarcity, resource limitations, and pressing development priorities—and without substantial, targeted financial and technical support. In that case, these economies may struggle to comply (Claessens et al., 2023). This may hinder their green transition, lead to carbon leakages if investments shift away from jurisdictions perceived as having weaker standards or higher risks, or result in regulations poorly suited to local needs, potentially undermining both climate objectives and development progress. Bridging this gap is crucial for the success of the global climate agenda.

While international efforts have pushed for greater integration of climate considerations into financial systems, it remains unclear to what extent such measures have effectively contributed to financial stability. Moreover, the potential stabilising role of renewable energy deployment and other mitigation strategies remains underexplored. For this reason, our study focuses on the period from 2020 to 2023, aiming to assess whether climate change risks can be reliably incorporated into financial stability assessments and whether mitigation efforts—especially renewable energy investment—have measurable effects on financial system resilience.

Accordingly, we propose the following research hypotheses:

- *H1: Climate-related risks are significantly associated with declines in financial stability in advanced economies.*
- *H2: Implementing renewable energy strategies has a positive and statistically significant effect on financial stability in advanced economies between 2020 and 2023.*

3 Data and Methodology

For our analysis, we have chosen countries deemed as advanced economies by the International Monetary Fund (IMF) due to their institutional capacity, regulatory maturity, and early adoption of climate-related financial frameworks, especially as these countries have been at the forefront of integrating climate risk into financial supervision. By focusing on advanced economies, the study benefits from relatively high-quality data availability, regulatory consistency, and a greater likelihood of observing the early impacts of mitigation strategies such as renewable energy deployment on financial stability. The countries subject to analysis are: Andorra, Australia, Austria, Belgium, Canada, Croatia, Cyprus, Czech Republic, Denmark, Estonia, Finland, France, Germany, Greece, Hong Kong SAR, Iceland, Ireland, Israel, Italy, Japan, Korea, Latvia, Lithuania, Luxembourg, Macao SAR, Malta, The Netherlands, New Zealand, Norway, Portugal, Puerto Rico, San Marino, Singapore, Slovak Republic, Slovenia, Spain, Sweden, Switzerland, Taiwan Province of China, United Kingdom, United States.

Furthermore, the 2020–2023 timeframe represents a critical period during which climate-related financial policy and regulation rapidly evolved in response to increasing global awareness of climate risk. Table 2 below presents the indicators considered in our analysis:

The selected dataset comprises a comprehensive set of economic, environmental, and institutional indicators that are particularly relevant for assessing the financial stability implications of climate-related risks in advanced economies. Key macroeconomic indicators such as the GDP deflator (GDPD), output gap (OG), inflation (INF), unemployment rate (UEMP), and general government net debt (GGND) are standard measures used by central banks and financial authorities to monitor systemic vulnerabilities. GDP based on purchasing power parity (GDP PPP) provide additional insight into labour market dynamics and global economic integration. Environmental indicators, including renewable energy output (RE), greenhouse gas emissions (GHG), and CO_2 embodied in exports (CO2GE), allow for the assessment of climate transition risks and mitigation outcomes. Figure 1 encompasses the rationale for choosing the indicators depicted above for analysis purposes.

By focusing on advanced economies—where data completeness and governance quality are higher—the study gains empirical precision in investigating how climate-related risks and renewable energy policies affect financial stability. For this reason, as a preliminary step of our analysis, we have used the following regression formula:

$$OG_{it} = \alpha + \beta_1 RE_{it} + \beta_2 GHG_{it} + \beta_3 CO2GE_{it} + \mu it + \varepsilon it \quad (1)$$

Furthermore, in order to explore the pathways through which climate-related indicators might influence financial stability, we have chosen to perform a mediation analysis, as it is essential to understand not just if these factors matter, but how they affect financial outcomes. Mediation analysis helps us uncover these indirect effects, as Baron and Kenny (1986) stated, providing a policy-relevant picture of climate actions' role in strengthening financial stability. The formula used is as follows:

Table 2 Description of the variables employed in the analysis

Abbreviation	Indicator	Description	Unit of Measure	Source
GDPD	Gross Domestic Product deflator	Measures changes in the prices of all new, domestically produced, final goods and services in an economy.	Index, 200 = 100	IMF—World Economic Outlook
OG	Output gap	Difference between actual and potential economic output as a percentage of potential output.	% of Potential GDP	IMF—World Economic Outlook
GDP PPP	Gross Domestic Product based on purchasing-power-parity (PPP)	Compares countries' economic output, accounting for cost of living and inflation differences.	%	IMF—World Economic Outlook
INF	Inflation	Measures the average rate of increase in consumer prices over a period.	%	IMF—World Economic Outlook
GGND	General government net debt	Represents the total gross debt of the government minus financial assets.	%	IMF—World Economic Outlook
UEMP	Unemployment rate	Measures the percentage of the labour force that is unemployed.	%	IMF—World Economic Outlook
RE	Renewable Energy	Amount of energy produced from renewable sources such as wind, solar, and fossil fuels.	Gigawatt-hours (GWh)	IMF—Climate Change Indicators
CO_2 GE	CO_2 Emissions Embodied in Gross Exports	Emissions are released during the production of goods and services that are exported.	Millions of metric tons	IMF—Climate Change Indicators
GHG	Annual greenhouse gas (GHG) air emissions account	Total annual emissions of greenhouse gases expressed in CO_2 equivalent.	Millions of metric tons	IMF—Climate Change Indicators

Source: Authors' compilation

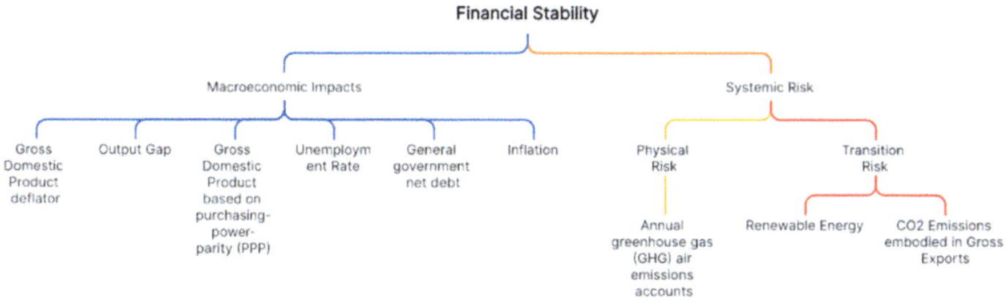

Fig. 1 Conceptual model linking environmental and economic indicators to financial stability. (Source: Authors' compilation in Boardmix)

$$INF_{it} = \alpha_1 + \alpha_{1.1} * RE_{it} + \alpha_{1.2} * CO2GE_{it} + \alpha_{1.3} * GHG_{it} + \varepsilon_{1it} \quad (2)$$

$$GGND_{it} = \alpha_2 + \alpha_{2.1} * RE_{it} + \alpha_{2.2} * CO2GE_{it} + \alpha_{2.3} * GHG_{it} + \varepsilon_{2it} \quad (3)$$

$$UEMP = \alpha_3 + \alpha_{3.1} * RE_{it} + \alpha_{3.2} * CO2GE_{it} + \alpha_{3.3} * GHG_{it} + \varepsilon_{3it} \quad (4)$$

$$\begin{aligned} GDPD_{it} = \alpha_4 + c'_1 * RE_{it} + c'_2 * CO2GE_{it} + c'_3 * GHG_{it} + \beta_1 * INF_{it} \\ + \beta_2 * GGND_{it} + \beta_3 UEMP_{it} + \varepsilon_{4it} \end{aligned} \quad (5)$$

Building upon the robust methodological foundation, further research delves into the granular impacts of climate-related risks and renewable energy policies across specific financial sub-sectors and individual economies, moving beyond an aggregate view to capture unique national contexts and their interplay with global climate trends. Integrating dynamic modelling and forward-looking scenario analysis allows for the simulation of various climate pathways and their evolving effects on financial stability, moving beyond historical correlations to anticipate future vulnerabilities. Furthermore, exploring moderation effects—where specific institutional or policy settings amplify or dampen these relationships—alongside potential feedback loops between financial instability and climate action, provides a more holistic understanding.

4 Results and Discussion

The linear regression analysis scrutinised the relationship between climate-related variables—namely Renewable Energy (RE), Greenhouse Gas Emissions (GHG), and CO_2 Emissions Embodied in Gross Exports (CO2GE)—and macroeconomic imbalances, in line with the Output Gap (OG), in advanced economies between 2020 and 2023.

The model summary (Table 3 and Figs. 2 and 3) indicates a notable improvement in explanatory power when climate-related variables are included (Model M_1), with an R-squared value of 0.686. This suggests that the three predictors account for approximately 68.6% of the variance in the output gap. The adjusted R-squared, which accounts for degrees of freedom, remains relatively robust at 0.552, implying the model retains considerable explanatory strength even after correcting for sample size and model complexity.

Furthermore, the model is statistically significant as a whole, as indicated by the F-statistic of 5.107 and the associated p-value of 0.035. The coefficient estimates provide further insight into the individual contributions of each predictor:

- GHG emissions exhibit a positive and statistically significant association with the output gap ($\beta = 91.462$, $p = 0.012$), with a standardised beta coefficient exceeding 1, denoting a particularly strong effect. This suggests that higher levels of greenhouse gas emissions may correspond with greater deviations of actual

Table 3 Results of the regression analysis

Model Summary—OG						
Model	**R**	**R^2**	**Adjusted R^2**	**RMSE**		
M_0	–	–	–	2.97		
M_1	0.83	0.69	0.55	1.99		
ANOVA						
Model		**Sum of Squares**	**df**	**Mean Square**	**F**	**p**
M_1	Regression	60.50	3.00	20.17	5.11	0.04
	Residual	27.64	7.00	3.95		
Coefficients						
Model		**Unstandardized**	**Standard Error**	**Standardized**	**t**	**p**
M_0	(Intercept)	(0.37)	0.90		(0.42)	0.69
M_1	(Intercept)	0.99	1.01		0.99	0.36
	RE	36.90	70.80	0.15	0.52	0.62
	GHG	91.46	27.07	1.02	3.38	0.01
	CO2GE	6.10	2.88	0.63	2.12	0.07

Note. M_1 includes RE, GHG, and CO2GE
Note. The intercept model is omitted, as no meaningful information can be shown
Source: Authors' computation in JASP Software

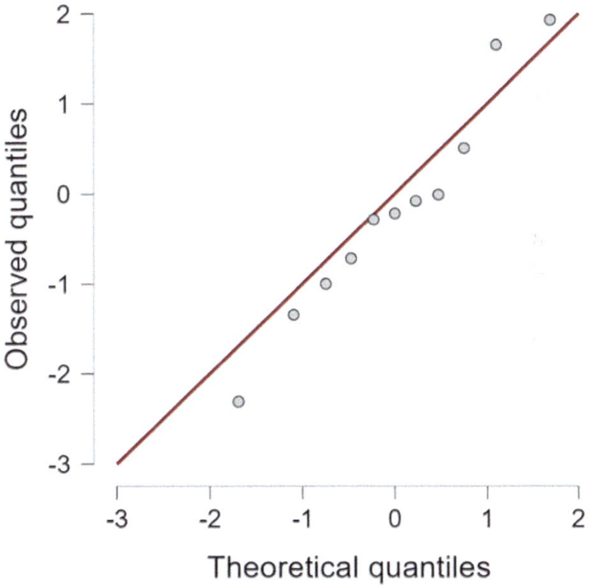

Fig. 2 Q-Q plot standardised residuals. (Source: Authors' computation in JASP Software)

output from potential output, potentially reflecting transition frictions or structural inefficiencies linked to environmental degradation.

- CO2GE also demonstrates a positive and marginally significant relationship with the output gap ($\beta = 6.100$, $p = 0.072$), suggesting that carbon emissions embed-

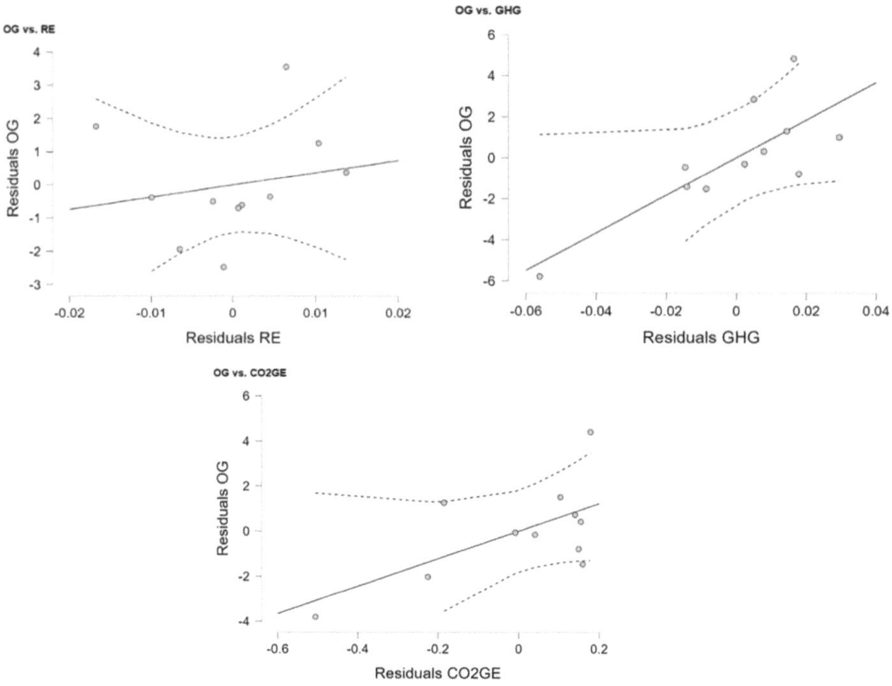

Fig. 3 Partial regression plots. (Source: Authors' computation in JASP Software)

ded in trade flows may influence domestic macroeconomic stability, possibly through competitiveness or external vulnerability channels.

- In contrast, despite a relatively large point estimate, Renewable Energy (RE) does not show a statistically significant relationship with the output gap ($\beta = 36.896$, $p = 0.618$). The large standard error suggests high variability and potential measurement or specification issues.

A mediation analysis was conducted following the regression analysis to explore the potential pathways through which these relationships operate. While regression models identified a clear relationship between climate change and macroeconomic imbalances, there is limited insight into how or through what channels one variable influences another. For this reason, mediation analysis addresses this limitation by explicitly modelling potential intermediate variables—such as inflation (INF), general government net debt (GGND), and unemployment (UEMP)—that may transmit or modify the effects of environmental indicators on economic performance.

The results are presented in Table 4 and Fig. 4 below reveal statistically significant adverse direct effects of RE ($\beta = -0.135$, $p = 0.026$), CO2GE ($\beta = -0.018$, $p < 0.001$), and GHG ($\beta = -0.038$, $p = 0.012$) on GDPD. This suggests that increases in renewable energy production and greenhouse gas emissions, including those embedded in trade, can be associated with a decrease in the GDP deflator's general price level of domestic goods and services.

Table 4 Results of the mediation analysis

Direct effects							95% Confidence Interval	
			Estimate	Std. Error	z-value	p	Lower	Upper
RE	→	GDPD	−0.135	0.06	−2.232	0.026	−0.253	−0.016
CO2GE	→	GDPD	−0.018	0.002	−9.882	<0.001	−0.022	−0.015
GHG	→	GDPD	−0.038	0.015	−2.502	0.012	−0.068	−0.008

Indirect effects									95% Confidence Interval	
					Estimate	Std. Error	z-value	p	Lower	Upper
RE	→	INF	→	GDPD	0.082	0.05	1.64	0.101	−0.016	0.18
RE	→	GGND	→	GDPD	0.091	0.051	1.805	0.071	−0.008	0.19
RE	→	UEMP	→	GDPD	−0.043	0.036	−1.19	0.234	−0.113	0.028
CO2GE	→	INF	→	GDPD	−0.002	0.002	−1.161	0.246	−0.006	0.002
CO2GE	→	GGND	→	GDPD	1.827×10^{-4}	0.002	0.12	0.904	−0.003	0.003
CO2GE	→	UEMP	→	GDPD	−0.002	0.001	−1.188	0.235	−0.005	0.001
GHG	→	INF	→	GDPD	7.915×10^{-4}	0.016	0.048	0.962	−0.031	0.033
GHG	→	GGND	→	GDPD	−0.015	0.015	−0.997	0.319	−0.045	0.015
GHG	→	UEMP	→	GDPD	0.005	0.012	0.437	0.662	−0.019	0.029

Total effects							95% Confidence Interval	
			Estimate	Std. Error	z-value	p	Lower	Upper
RE	→	GDPD	−0.004	0.056	−0.079	0.937	−0.113	0.105
CO2GE	→	GDPD	−0.022	0.002	−9.808	<0.001	−0.027	−0.018
GHG	→	GDPD	−0.047	0.021	−2.22	0.026	−0.089	−0.006
Total indirect effects							**95% Confidence Interval**	
			Estimate	**Std. Error**	**z-value**	**p**	**Lower**	**Upper**
RE	→	GDPD	0.13	0.063	2.08	0.037	0.008	0.253
CO2GE	→	GDPD	−0.004	0.002	−1.877	0.06	−0.008	1.642×10^{-4}
GHG	→	GDPD	−0.009	0.017	−0.554	0.58	−0.042	0.023
Residual covariances							**95% Confidence Interval**	
			Estimate	**Std. Error**	**z-value**	**p**	**Lower**	**Upper**
INF	↔	GGND	0.956	0.705	1.356	0.175	−0.425	2.336
INF	↔	UEMP	0.073	0.158	0.463	0.644	−0.236	0.382
GGND	↔	UEMP	−1.244	1.264	−0.985	0.325	−3.722	1.233
Path coefficients							**95% Confidence Interval**	

(continued)

Table 4 (continued)

Total effects							95% Confidence Interval	
			Estimate	Std. Error	z-value	p	Lower	Upper
			Estimate	**Std. Error**	**z-value**	**p**	**Lower**	**Upper**
INF	→	GDPD	0.23	0.072	3.212	0.001	0.09	0.37
GGND	→	GDPD	−0.026	0.01	−2.669	0.008	−0.045	−0.007
UEMP	→	GDPD	−0.09	0.036	−2.515	0.012	−0.161	−0.02
RE	→	GDPD	−0.135	0.06	−2.232	0.026	−0.253	−0.016
CO2GE	→	GDPD	−0.018	0.002	−9.882	<0.001	−0.022	−0.015
GHG	→	GDPD	−0.038	0.015	−2.502	0.012	−0.068	−0.008
RE	→	INF	0.357	0.187	1.907	0.057	−0.01	0.723
CO2GE	→	INF	−0.009	0.008	−1.245	0.213	−0.024	0.005
GHG	→	INF	0.003	0.072	0.048	0.962	−0.137	0.144
RE	→	GGND	−3.546	1.447	−2.45	0.014	−6.382	−0.709
CO2GE	→	GGND	−0.007	0.059	−0.121	0.904	−0.123	0.108
GHG	→	GGND	0.595	0.553	1.075	0.282	−0.489	1.679
RE	→	UEMP	0.474	0.351	1.351	0.177	−0.214	1.163
CO2GE	→	UEMP	0.019	0.014	1.348	0.178	−0.009	0.047
GHG	→	UEMP	−0.06	0.134	−0.444	0.657	−0.323	0.204

Note. Delta method standard errors, normal theory confidence intervals, ML estimator
Source: Authors' computation in JASP Software

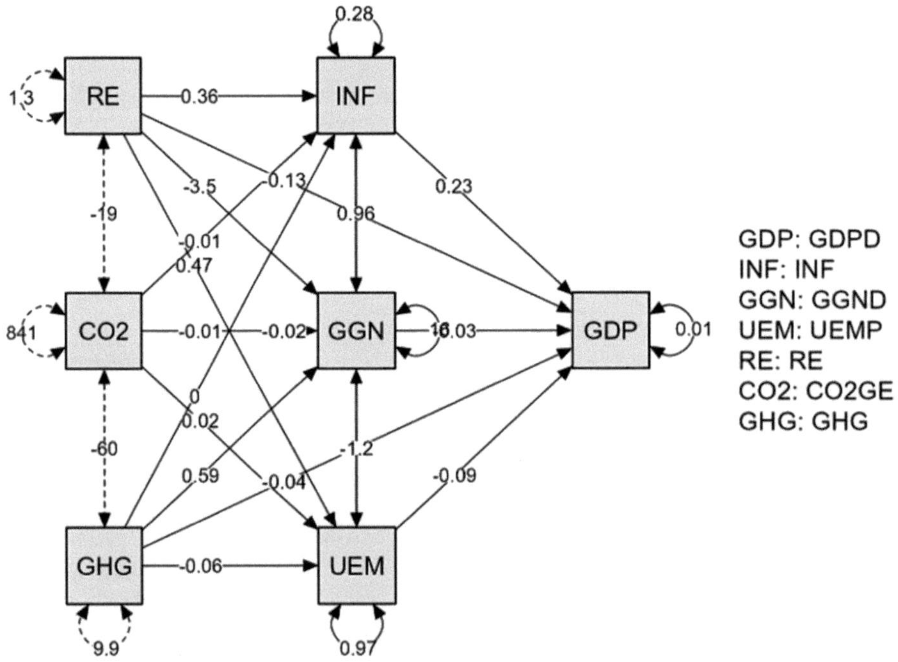

Fig. 4 Mediation path plot. (Source: Authors' computation in JASP Software)

Moreover, while most individual indirect effects through INF, GGND, and UEMP were statistically non-significant, RE showed a positive and statistically significant total indirect effect ($\beta = 0.130$, $p = 0.037$). This implies that although RE reduces GDPD directly, its impact is partly offset by positive mediating pathways, particularly through inflation and fiscal channels. CO2GE had a marginally significant negative total indirect effect ($\beta = -0.004$, $p = 0.060$), while GHG's indirect effects were not significant.

The total effect of RE on GDPD was not statistically significant ($\beta = -0.004$, $p = 0.937$), reflecting a net balancing of negative direct and positive indirect effects. CO2GE ($\beta = -0.022$, $p < 0.001$) and GHG ($\beta = -0.047$, $p = 0.026$) maintained significant negative total effects on GDPD, indicating a consistent negative relationship between emissions and the economy's price level performance. Among mediators, inflation had a considerable positive impact on GDPD ($\beta = 0.230$, $p = 0.001$), while GGND ($\beta = -0.026$, $p = 0.008$) and UEMP ($\beta = -0.090$, $p = 0.012$) had significant negative effects. These results suggest that moderate inflation may stimulate nominal economic output, whereas fiscal debt burdens and labour market slack exert downward pressure on overall price levels.

These findings provide important insights into the complex interplay between climate-related indicators and macroeconomic performance. The GDP deflator, a broad measure of price-level changes across the economy, responded differently to renewable energy investments and carbon emissions. First, the negative direct effect of renewable energy on GDPD likely reflects short-term transitional costs, such as capital intensity, infrastructure shifts, and reduced reliance on fossil fuels. However, the significant positive indirect effects—particularly via inflation and public debt reduction—suggest that renewable energy adoption may have a favourable systemic impact on economic performance over time. This highlights a potential economic rebound from early renewable energy investments, aligning with long-term sustainability goals.

In contrast, CO_2 emissions in exports and total GHG emissions consistently exhibited negative direct and total effects on GDPD. These results reinforce the economic costs of carbon-intensive growth models. Emissions may contribute to domestic and international regulatory burdens, market volatility, and environmental degradation, suppressing output and price stability. The lack of meaningful mediation suggests these impacts are largely structural and direct.

Among mediators, inflation emerged as a double-edged variable. While commonly viewed as detrimental, moderate inflation is strongly correlated with GDPD. This may reflect demand-driven inflation tied to green investments and fiscal expansion. In contrast, government debt and unemployment were detrimental to GDP performance, underscoring the need for sound fiscal management and labour market resilience during economic transitions.

These findings indicate that climate action does not necessarily conflict with macroeconomic stability. Although transitioning to a greener economy may present short-term challenges and trade-offs, it also offers long-term economic benefits. Such advantages include greater price stability, enhanced fiscal sustainability, and innovation-driven productivity gains. In summary, climate change constitutes a systemic risk that could cause extensive disruption across financial institutions, markets, and the broader economy. The physical consequences of global warming and the economic transformations required for a net-zero transition generate a complex array of interconnected risks that may propagate through the financial system. Therefore, the effective incorporation of climate-related risks into financial assessments is essential for mitigating potential losses, strengthening financial resilience, and proactively seizing the significant opportunities presented by the shift toward a sustainable, low-carbon economy.

Enhanced transparency is a critical enabler of effective climate risk management. The global movement towards mandatory, consistent, and comparable climate-related financial disclosures, spearheaded by the TCFD and now evolving under the ISSB's IFRS S2 standard, is a vital step in providing markets with the information needed to accurately price climate risks and opportunities, thereby facilitating more efficient capital allocation. However, the effectiveness of these disclosure frameworks depends on factors extending beyond the standards alone. Rigorous verification and assurance processes are crucial to guarantee the accuracy, reliability, and credibility of the information disclosed. Without independent assurance, the risk of "greenwashing" or misleading disclosures remains high, undermining investor confidence and regulatory utility. Furthermore, the capacity of users—investors, lenders, insurers, and regulators-to interpret this complex information and integrate it meaningfully into their decision-making processes is equally crucial.

5 Conclusions

Beyond disclosure, regulatory and supervisory instruments must continue to evolve. This includes refining climate scenario analysis and stress testing to better capture the nuances of climate risk and develop appropriate micro- and macroprudential tools. Ensuring international interoperability between different jurisdictional regimes and disclosure standards is key to avoiding market fragmentation and reducing the compliance burden on global firms. While global developments show significant progress, instances of resistance or rollback in certain jurisdictions highlight the ongoing tensions and the need for sustained international coordination and political commitment.

Following the empirical evidence, macroeconomic implications of climate-related variables—such as renewable energy use and greenhouse gas emissions—are both direct and indirect, channelled through inflation, debt, and unemployment dynamics, highlighting the importance of financial and regulatory tools that can address climate and macroeconomic risks jointly, rather than in isolation.

The transition to a sustainable global economy requires a monumental redirection of financial flows, particularly towards developed and emerging economies, which face the dual challenge of significant development needs and heightened vulnerability to climate change. Estimates of the climate finance these economies require for mitigation and adaptation run into trillions of dollars annually, far exceeding current commitments and flows. Therefore, financial policies in developed and emerging economies must be carefully designed to mobilise this necessary capital while being tailored to national contexts, considering developmental priorities, existing capacity constraints, and the imperative of financial inclusion. A global approach that transposes regulations from developed economies is unlikely to be effective and could have adverse consequences. Moreover, as the results indicate, renewable energy adoption may initially exert downward pressure on price levels but can also generate positive indirect effects via inflation reduction and debt consolidation. These findings support the argument that well-designed climate policies can align with macro-fiscal objectives with sequencing and support mechanisms if implemented carefully.

International cooperation is indispensable. This includes providing substantial financial assistance (grants, concessional loans, and risk-sharing mechanisms) and targeted technical support by multilateral development banks like the World Bank, international financial institutions like the IMF (e.g., through its Resilience and Sustainability Trust), and developed countries. Organisations like the OECD (2024) are vital in providing data, analysis, and policy guidance to support advanced and emerging economies in aligning their financial systems with climate goals. For advanced and emerging economies, aligning financial policy with sustainability goals is not just about attracting green investment; it is about building resilient economies that can withstand increasing physical climate shocks while ensuring that the transition to a low-carbon future is and equitable, and does not leave vulnerable populations behind or derail progress on broader development objectives. Importantly, our analysis also reveals that CO_2 emissions embedded in exports exert strong adverse direct effects on price dynamics, independent of intermediary factors. This implies that decarbonization of export sectors may stabilise inflation, supporting the case for greening trade and production systems.

The extensive nature of climate-related financial risks necessitates a fundamental integration of climate considerations into the strategic planning processes of all key actors. For financial institutions, this means embedding climate risk into their core governance structures, overall business strategy, enterprise-wide risk management frameworks, and remuneration policies. This involves moving beyond a purely compliance-driven approach to one where climate risks and opportunities actively shape lending decisions, investment strategies, product development, and client engagement.

For non-financial corporations, strategic adaptation requires a thorough assessment of how climate change will impact their operations, supply chains, and markets and the development of robust strategies to enhance resilience and align business models with transition pathways. This includes setting credible, science-based emission reduction targets and investing in decarbonization and adaptation measures. Recent research by Wang et al. (2023) provides new evidence on the complex relationship between energy security and CO_2 emissions, highlighting the need for dynamic and context-specific strategies to manage emissions effectively while ensuring energy reliability. In line with our findings, non-financial firms operating in carbon-intensive sectors should also prepare for macroeconomic feedback loops, where emission-related policies may influence consumer prices, labour markets, and government debt levels. Strategic foresight in these areas will be essential for long-term competitiveness.

Moreover, governments have a critical role in creating an enabling policy environment that facilitates and accelerates the transition. This includes establishing transparent and predictable long-term climate policies (such as carbon pricing mechanisms where feasible), providing targeted support for green innovation and investment, investing in climate-resilient infrastructure, and ensuring that social safety nets are in place to manage the distributional consequences of the transition (Varadi et al., 2025). Developing and communicating national and sectoral transition plans can provide necessary signals to the private sector.

Effective strategic planning for climate change must transcend a purely risk-mitigation mindset. While managing the downside risks is essential, transitioning to a sustainable economy presents significant opportunities for innovation, new market creation, resource efficiency gains, and long-term value generation. Institutions and governments that proactively identify and strategically pursue these climate-related opportunities will be better positioned to thrive in the evolving global economic landscape.

Acknowledgments This work was supported by a grant from the Romanian Ministry of Research, Innovation and Digitalization, the project with the title *"Economics and Policy Options for Climate Change Risk and Global Environmental Governance"* (CF 193/28.11.2022, Funding Contract no. 760078/23.05.2023), within Romania's National Recovery and Resilience Plan (PNRR)—Pillar III, Component C9, Investment I8 (PNRR/2022/C9/MCID/I8)—Development of a program to attract highly specialised human resources from abroad in research, development, and innovation activities.

Competing Interests The authors have no conflicts of interest to declare relevant to this chapter's content.

References

Aglietta, M., & Espagne, É. (2016). *Climate and finance systemic risks, more than an analogy? The climate fragility hypothesis* (CEPII working paper no. 2016-10). Centre d'Études Prospectives et d'Informations Internationales. Retrieved from https://www.cepii.fr/PDF_PUB/wp/2016/wp2016-10.pdf

Baron, R. M., & Kenny, D. A. (1986). The moderator–mediator variable distinction in social psychological research: Conceptual, strategic, and statistical considerations. *Journal of Personality and Social Psychology, 51*(6), 1173–1182. https://doi.org/10.1037/0022-3514.51.6.1173

Carney, M. (2015). *Breaking the tragedy of the horizon – Climate change and financial stability* [Speech transcript]. Bank of England. Retrieved from https://www.bankofengland.co.uk/speech/2015/breaking-the-tragedy-of-the-horizon-climate-change-and-financial-stability/bankofengland.co.uk

Claessens, S., Mehrotra, A., & Shim, I. (2023). Macroprudential policies and global finance. In *Oxford research encyclopedia of economics and finance*. Oxford University Press. https://doi.org/10.1093/acrefore/9780190625979.013.907

European Central Bank. (2021). *Climate-related risk and financial stability*. European Central Bank. Retrieved from https://www.ecb.europa.eu/pub/pdf/other/ecb.climateriskfinancialstability202107~87822fae81.en.pdf

Financial Stability Board. (2022). *Supervisory and regulatory approaches to climate-related risks: Final report*. Financial Stability Board. Retrieved from https://www.fsb.org/uploads/P131022-1.pdf

Financial Stability Board. (2025). *Assessment of climate-related vulnerabilities: Analytical framework and toolkit* (Occasional paper). Financial Stability Board. Retrieved from https://www.fsb.org/2025/01/assessment-of-climate-related-vulnerabilities-analytical-framework-and-toolkit/

Lobonţ, O. R., Varadi, A. E., Vatavu, S., & Doran, N. M. (2025). Bridges or barriers? Unpacking the institutional drivers of business climate adaptation in the EU. *Sustainability, 17*(11), 4865. https://doi.org/10.3390/su17114865

Network for Greening the Financial System. (2019). *A call for action: Climate change as a source of financial risk* (First comprehensive report). Network for Greening the Financial System. Retrieved from https://www.ngfs.net/en/publications-and-statistics/publications/first-comprehensive-report-call-action

Network for Greening the Financial System. (2024a). *Guide on climate-related disclosure for central banks* (2nd ed.). Network for Greening the Financial System. Retrieved from https://www.ngfs.net/system/files/import/ngfs/medias/documents/ngfs_guide_on_climate-related_disclosure_for_central_banks_-_second_edition.pdf

Network for Greening the Financial System. (2024b). *Synthesis report on the greening of the financial system: Insights for financial actors in advanced and emerging economies* (Occasional paper). Network for Greening the Financial System. Retrieved from https://www.ngfs.net/en/publications-and-statistics/publications/synthesis-report-greening-financial-system-insights-financial-actors-advanced-and-emerging-economies

Organisation for Economic Cooperation and Development. (2024). *OECD review on aligning finance with climate goals*. Organisation for Economic Cooperation and Development. Retrieved from https://www.oecd.org/en/publications/oecd-review-on-aligning-finance-with-climate-goals_b9b7ce49-en.html

Task Force on Climate-related Financial Disclosures. (2021). *Implementing the recommendations of the Task Force on Climate-related Financial Disclosures*. TCFD. (Updated October 2021). Retrieved from https://www.fsb-tcfd.org/

Varadi, A. E., Lobonţ, O. R., & Vatavu, S. (2025). EU climate countdown: Economic risks, adaptation and smart policies. *International Journal of Climate Change Strategies and Management, 17*(1), 353–375. https://doi.org/10.1108/IJCCSM-07-2024-0121

Wang, K. H., Zhao, Y. X., Su, Y. H., & Lobonţ, O. R. (2023). Energy security and CO_2 emissions: New evidence from time varying and quantile varying aspects. *Energy, 273*, 127164. https://doi.org/10.1016/j.energy.2023.127164

World Bank. (2024). *Financial sector at a crossroads in emerging markets: Climate risks and financing gaps demand action* [Press release]. World Bank. Retrieved from https://www.worldbank.org/en/news/press-release/2024/08/29/financial-sector-at-a-crossroads-in-emerging-markets-climate-risks-and-financing-gaps-demand-action

World Bank Group & International Monetary Fund. (2024). *World Bank Group and IMF deepen joint effort to scale up climate action* [Press release]. World Bank. Retrieved from https://www.worldbank.org/en/news/press-release/2024/05/31/world-bank-group-and-imf-deepen-joint-effort-to-scale-up-climate-action

Wu, L., Liu, D., & Lin, T. (2023). The impact of climate change on financial stability. *Sustainability, 15*(15), 11744. https://doi.org/10.3390/su151511744

Climate-Driven Displacement and the Necessity of Green Finance

Omer Serkan Gulal and Ebru Topcu

1 Introduction

Climate change has become a profound and accelerating driver of human displacement, generating population movements through both sudden-onset events such as floods, cyclones, and wildfires and slow-onset environmental processes including desertification, salinisation, and rising sea levels. While the impacts of rapid-onset disasters are often immediate and highly visible, the effects of slow-onset processes tend to unfold gradually over time, complicating efforts to distinguish climate-related triggers from other socio-environmental or economic pressures. Displacement resulting from long-term climate stressors such as drought or sea-level rise are generally categorised as climate-induced migration. In contrast, movements caused by geophysical events unrelated to climate change, such as earthquakes or volcanic eruptions, are typically classified under the broader term of environmental migration (Mayer, 2016).

Recent global data underscore the growing scale of this phenomenon. In 2022 alone, disasters triggered an estimated 32.6 million internal displacements, with approximately 98% of these events attributed to weather-related hazards such as storms, floods, and droughts (UNHCR, n.d.). The trend continued in 2023, when an additional 26.4 million people were displaced, 7.7 million of whom remained displaced at year's end (Puşcaş & Guadagno, 2024). Despite heightened concern over cross-border migration, most climate-induced displacements remain internal, confined within national borders (UNHCR, n.d.).

O. S. Gulal
Nevsehir Haci Bektas Veli University, Nevsehir, Türkiye
e-mail: serkangulal@nevsehir.edu.tr

E. Topcu (✉)
Alanya Alaaddin Keykubat University, Alanya, Antalya, Türkiye
e-mail: ebru.topcu@alanya.edu.tr

C.-W. Su et al. (eds.), *Green Finance and the Challenges of Climate Change*, Sustainable Finance, https://doi.org/10.1007/978-3-032-07293-1_6

This growing burden of displacement is not distributed evenly. Populations in regions with significant development constraints, including much of Sub-Saharan Africa, South and Southeast Asia, Central and South America, the Arctic, and Small Island Developing States (SIDS), face heightened vulnerability. Such vulnerability is typically exacerbated by poverty, limited access to essential services and infrastructure, governance deficiencies, and high dependence on climate-sensitive livelihoods like small-scale agriculture, pastoralism, and coastal fisheries. According to the Intergovernmental Panel on Climate Change, mortality from weather-related disasters between 2010 and 2020 was 15 times higher in regions with high vulnerability than in those with very low vulnerability (IPCC, 2022).

The Pacific Island state of Fiji is directly exposed to some of the most severe impacts of climate change. Rising sea levels and increasingly intense extreme weather events threaten infrastructure, livelihoods, and the safety of entire communities, leading to climate-induced displacement. In this context, Fiji urgently requires enhanced climate finance both to mitigate future risks and to strengthen its capacity to adapt to the challenges that are already unfolding (WRI, 2023). In response, Fiji enacted the Climate Change Act in 2021 (Fiji Ministry of Foreign Affairs, 2021).

In South Asia, Bangladesh faces comparable risks on a much larger scale. With 26 of its 64 districts currently experiencing climate-related migration, the government warns that sea-level rise could displace as many as 20 million people by 2050 (Barua et al., 2024). The cumulative human and economic costs of such displacement are already straining governance systems and social cohesion.

Addressing this growing crisis requires legal and policy innovations and substantial and targeted financial investment. Green finance, defined as the mobilisation of financial resources to support environmentally sustainable development, has been increasingly emphasised as a tool to enhance climate resilience and reduce displacement risk (UNEP, 2022; Khemiri, 2024; Campiglio, 2016). Although the urgency of climate-induced mobility is increasingly recognised, financial responses remain fragmented and inadequate. A 2018 review by the Task Force on Displacement revealed that only a small portion of Green Climate Fund-supported projects addressed issues related to migration or displacement. This gap persists, with most interventions depending heavily on humanitarian and development funding outside the formal climate finance framework (Pușcaș & Guadagno, 2024).

While adaptation finance flows to developing countries rose from USD 22 billion in 2021 to USD 28 billion in 2022, the largest single-year increase since the Paris Agreement, this still represents only a modest step toward the scale required to address growing adaptation needs. Meeting the Glasgow Climate Pact's call to double adaptation finance by 2025 would close just a fraction of the current adaptation financing gap. Furthermore, climate finance has not sufficiently targeted the specific vulnerabilities faced by displaced populations and those at imminent risk of displacement (UNEP, 2024).

This financing shortfall is particularly problematic given that those most affected by climate-induced displacement are often among the world's poorest and most marginalised. According to the World Bank (2019), these include populations reliant on rain-fed agriculture, subsistence fishing, and other resource-dependent livelihoods, particularly in fragile coastal and island ecosystems. As climate impacts intensify and the frequency of extreme weather events increases, the need to incorporate human mobility considerations into climate adaptation and finance mechanisms becomes not only a matter of effectiveness but one of justice and long-term sustainability.

This chapter offers a comprehensive inquiry into the nexus of climate change, displacement, and green finance. The first part lays the conceptual groundwork, examining how both sudden and gradual climate-related events influence human mobility and highlighting the definitional complexities surrounding climate-induced migration. The second part explores the continuum between voluntary and forced migration, focusing on structural vulnerabilities. The third part introduces the tools and actors within the green finance ecosystem, assessing their potential in supporting climate adaptation. The fourth part bridges the gap between displacement and financial responses through selected case studies, including those of Fiji and Bangladesh. The final part addresses policy and governance dimensions, emphasising the need for integrated, equitable, and forward-looking financial mechanisms to respond to climate-induced mobility effectively.

2 Conceptual Background: Climate Change, Displacement, and Vulnerability

2.1 Climate Hazards and Displacement Pathways

Environmental factors influence migration processes through mechanisms such as the availability and reliability of ecosystem services and exposure to environmental hazards. These driving forces do not act in isolation; they interact with broader economic, political, social, and demographic contexts, shaping both individual migration choices and larger population movements. Environmental change may render a region uninhabitable directly, but it also has indirect impacts by disrupting livelihoods, leading to economic instability, which in turn can fuel political unrest and intensify conflicts over scarce resources (Black et al., 2011). Throughout history, human environments have evolved under the influence of both natural processes and human activities. However, current human-induced climate change is expected to bring about far more rapid and profound transformations on a global scale over the coming decades. As a result, climate-driven migration and displacement are increasingly understood as multifaceted phenomena, shaped by interlinked environmental, economic, social, and political factors (Mayer, 2016; Wilkinson et al., 2016).

Another prevalent perspective promoted in recent approaches is to view mobility as a risk, placing it primarily as a response to climate change and extreme weather (Wilkinson et al., 2016). Environmental changes can drive migration in two ways: first, by directly threatening survival (push factors); second, by undermining basic living standards and jeopardising livelihoods, especially in sectors such as agriculture that rely on natural resources (Mayer, 2016). However, while this perspective focuses on those who reach the migration areas, it ignores the experiences encountered on the way or those who never get there or choose not to go. Within this framework, risks are divided into two types: intensive risks, which are sudden, high-impact events like hurricanes, and extensive risks, which involve frequent or persistent lower-level stresses such as salinisation and sea-level rise (Wilkinson et al., 2016). Thus, while this assessment keeps pushing towards migration, it's important to note that there are pull factors once the potential migration happens. In fact, the extent to which action is taken for climate adaptations to shape migration outcomes rests heavily on the magnitude of pull factors located in these destination areas (Jacobson et al., 2019).

The categories of environmental risk lead to different migration responses. Sudden and slow onset hazards such as tornadoes, earthquakes, and floods unfold with little warning and often precipitate distress-driven movement, compelling affected individuals to evacuate immediately or seek refuge in the aftermath, and they influence migration in different ways. Sudden events such as tornadoes, earthquakes, and floods occur within short time frames and are typically associated with distress migration, whereby individuals flee preemptively or are evacuated in the aftermath. In contrast, slow-onset environmental changes such as drought or land degradation develop gradually, affect larger geographical areas, and tend to trigger migration at later stages of the process. (McLeman, 2009). Although rapid-onset disasters often result in short-term displacement, repeated occurrences or severe destruction may prompt individuals to relocate permanently or move to more distant areas (Huang, 2023).

Slow-onset environmental changes can escalate into disasters when compounded by sudden-onset events. For example, sea-level rise may lead to abrupt coastal flooding, desertification may trigger wildfires, and increasing temperatures may result in extreme heat waves. When such rapid-onset hazards exceed a community's capacity to cope, the risk of displacement significantly increases. In this context, slow-onset processes act as underlying drivers of displacement by heightening vulnerability to sudden shocks (IDMC, 2018). Decisions to migrate are influenced by vulnerability and adaptive capacity, which are shaped by socio-demographic factors such as gender, age, and ethnicity at the societal, household, and individual levels. Environmental migration tends to be predominantly internal, typically following rural or rural–urban trajectories, while international movements are often directed toward the nearest border (Laczko & Aghazarm, 2009).

While climate change does not directly cause displacement, it contributes to environmental degradation and exacerbates existing vulnerabilities, thereby making continued residence in affected areas increasingly unsustainable. These disruptions manifest through intensified natural disasters, prolonged droughts that threaten food and water security, rising sea levels that render coastal zones uninhabitable, and heightened competition over diminishing resources (Laczko & Aghazarm, 2009). However, climatic hazards alone do not determine human mobility. Decisions to migrate are fundamentally shaped by individuals' and communities' levels of vulnerability and adaptive capacity. These factors influence whether people are able to relocate or remain in place when confronted with climate-related threats. In some instances, such hazards necessitate movement as a coping strategy and simultaneously constrain the ability to move, particularly among the most vulnerable populations (Opitz Stapleton et al., 2017).

2.2 *Forced vs. Voluntary Environmental Migration*

Since the 1980s, environmental refugees, especially climate-induced migration, have been a contentious issue across international organisations, national governments, and local movements (Bose & Lunstrum, 2012).

The act of migration is more accurately understood as occurring along a continuum. At the far right of this spectrum are individuals who possess no control over their displacement, characterised as "involuntary" migrants. As one moves leftward along the continuum, the degree of agency increases, encompassing individuals who retain varying levels of decision-making power in the migration process. At the far left are voluntary migrants, who exercise full autonomy over each aspect of their relocation. Conceptualising migration in this way offers a nuanced alternative to rigid legal definitions of refugee status and accommodates various constraints that may influence individual migration decisions (Bates, 2002) (Table 1).

The terms "climate refugee" and "environmental refugee" are increasingly used in public and academic discourse to describe individuals displaced by climate-related events. However, such terminology lacks formal recognition under international law. People affected by environmental degradation or climate change typically do not meet the criteria outlined in the 1951 Refugee Convention and are only partially addressed by the non-binding Guiding Principles on Internal Displacement, which apply solely within domestic borders. Ongoing debates over the appropriate classification of these individuals further complicate the absence of a universally accepted legal framework. Critics argue that terms like "climate refugee" imply a simplistic, mono-causal explanation that overlooks the complex and interconnected drivers of migration and displacement. They also warn that expanding the refugee definition could weaken existing legal protections. In contrast, proponents of legal recognition contend that climate-induced displacement occurs beyond the control of those affected and thus warrants formal international protection (Shamsuddoha, 2015; Brzoska & Fröhlich, 2016).

Even in its most urgent, forced form, migration is rarely caused solely by environmental "push" factors like rising sea levels. Apart from sudden climate crises

Table 1 Range of control in migration decision-making amid environmental change

Involuntary	Compelled	Voluntary
Environmental Refugee	Environmental Emigrant	Migrant

Source: Bates, 2002

such as floods or hurricanes, which can compel people to flee for their lives, migration usually involves some pull, whether environmental, social, or economic, fueled by the hope of a better future elsewhere (Brown, 2008). In academic discussions, it's often noted that economic migrants respond to a mixture of both push and pull influences. In contrast, "climate migrants" are generally those with no choice but to escape, reacting almost exclusively to immediate threats. In times of crisis, their options for where to go are typically minimal, resulting in disrupted, multi-stage migration journeys that form a distinct profile of repeated displacement (Ahsan et al., 2014). Terms like "environmental refugee" or "climate refugee" have been used to emphasise the involuntary nature of such movements and elicit greater public empathy. In contrast, the term "migrant" often implies a voluntary relocation in pursuit of improved living conditions and thus may fail to capture the coercive nature of climate-related displacement (Brown, 2008). However, these terms remain contested in academic and policy debates. Critics argue they risk oversimplifying the complex and multifactorial causes of displacement and may undermine established international refugee protections. Nevertheless, advocates maintain that climate-displaced individuals, compelled by forces beyond their control, deserve formal recognition and protection under international law (Brzoska & Fröhlich, 2016).

Although many individuals displaced by climate change remain within national borders, an increasing number are forced to cross international boundaries. This trend is projected to intensify as climate impacts worsen. Both rapid-onset events and slow-onset processes, such as land loss from sea-level rise, contribute to displacement. Despite the growing scale of this phenomenon, international and national protection mechanisms remain inadequate (Petz & Ginnetti, 2013).

Migration driven by climate impacts exists along a spectrum. Some people migrate proactively to adapt; others are forcibly displaced. Movements may be temporary or permanent, internal or cross-border. In situations where in-place adaptation is no longer viable, planned relocation becomes necessary. Such relocation often involves the movement of entire communities and requires government coordination, as seen in flood-prone regions, drought-affected rural zones, and low-lying island states (Ferris, 2012).

2.3 Vulnerable Groups and Regions

- **Vulnerable Populations (Trapped Communities and Gendered Risks)**: Climate-induced migration encompasses a spectrum of mobility types, from voluntary migration to forced displacement and planned relocation. The underlying drivers are complex and often overlapping. Effective policy responses must be holistic, context-specific, and grounded in human rights. Migration can be a viable adaptation strategy when supported by appropriate legal protections and inclusive development frameworks. Without such support, climate-induced migration risks deepening existing inequities and evolving into a broader crisis of protection and governance (Zickgraf, 2019; The White House, 2021).

- Overlapping and complex factors, including environmental, economic, and social pressures, drive these movements. A remarkably underexplored dimension of this phenomenon is the plight of "trapped populations", individuals who are unable to migrate due to poverty, disability, or other social constraints. These populations, often located in high-risk zones, face acute exposure to climate hazards without the capacity to adapt or relocate, revealing severe disparities in adaptive capacity (Ayeb-Karlsson et al., 2018).

- Women and girls, particularly in low-income and rural settings, are disproportionately affected by climate change due to their overrepresentation among the global poor and reliance on climate-sensitive resources for daily survival. Despite their vital role in strengthening climate resilience at the household and community levels, they remain among the most vulnerable to its adverse impacts. This vulnerability is driven by structural inequalities, including limited access to financial services, restricted economic participation, and sociocultural norms undermining their decision-making power. In rural areas, their responsibilities for securing food, water, and fuel are intensified by climate-induced phenomena such as droughts and erratic rainfall, increasing physical burdens and exposure to gender-based violence. These compounded challenges not only deepen existing gender disparities but also hinder women's capacity to adapt to climate-related risks effectively (Zetterli, 2023; UN, n.d.).
- **Vulnerable Countries and Regional Disparities:** Least Developed Countries (LDCs) comprise the most climate-vulnerable population of countries in the world. The population dynamics, low human development index, threshold development capacity, and developmental challenges render nations like this more vulnerable to climate change since their human capacity or ability to adapt and recover to climate stressors is essentially limited, more so, as they depend on climate-sensitive sectors for development. The global climate change rate, rising sea levels, natural disasters, and drought can alarm these populations operating in sensitive and vulnerable conditions (UNFCCC, 2023; UNDP, 2015).

- Yet even though climate vulnerability is global, LDCs are more vulnerable for an intersection of reasons. First, geographical characteristics make them more susceptible to climate activity. For example, small island LDCs are prone to flooding and harsher hurricanes/tropical storms; African LDCs have a greater incidence of extended periods of drought, with West and Central African LDCs being the most vulnerable. Second, LDCs do not have the financial capacity which means there's less opportunity/less incentive to spend on adaptations or recuperation from large-scale climatic events. Third, the reliance on exporting primary natural resource products by the LDCs, with minuscule affiliation to regional value chains, increases the likelihood of negative spillovers from external partner nations' environmental policy (UNCTAD, 2022).

- LDCS needs to adapt to climate change. LDCs suffer disproportionate effects of a number of persisting structural challenges outside of limited access to climate financing, most of which are for adaptation and climate resiliency. This shows that climate financing is a small fraction of total climate financing, and the small scale of such economies, geographical disadvantages, remoteness from international markets, and compounded vulnerabilities to climate shocks (UNCTAD, 2022).

- Finally, LDCs, Landlocked Developing Countries (LLDCs), and Small Island Developing States contribute the least to greenhouse gas emissions yet feel the most impact from climate change. Their structural and geographical vulnerabilities decrease their potential for adaptation; thus, these populations are more likely to be climate-displaced and induced to migration (IOM & UN-OHRLLS, 2019).

3 Understanding Green Finance: Scope, Instruments, and Institutional Players

Green finance has emerged as one of the most important tools for addressing sustainability policies, especially with the pressing effects of climate change and environmental degradation. Green finance is defined as public or private financial flows projected to provide environmental benefits and contribute to environmental sustainability, including climate change mitigation, adaptation, biodiversity conservation, and resource efficiency (OECD, 2017). Various financial instruments, such as green bonds, climate risk insurance, sustainable loans, green venture capital, and carbon credits, are integral components of the green finance framework and are designed to promote projects with beneficial environmental consequences (Flammer, 2021). Accordingly, the scale of green finance has significantly expanded due to recent institutional reforms, the investors' activity and most notably, the evolving mandates of the global financial institutions (such as the World Bank, the International Monetary Fund, the European Investment Bank, among others). Furthermore, Bolton et al. (2020) emphasise that central banks and financial supervisors are beginning to integrate climate-related financial risks into their regulatory frameworks and macro prudential policies as part of their financial stability mandates. These trends indicate a market-driven interest in green finance instruments, state-level measures, and a regulatory collaborative stance in facilitating green finance goals (Böhm et al., 2012). Thus, to understand green finance and its potential, the actions of various players in this system, such as banks, international investors, and financial institutions, should be acknowledged.

The growing relevance of green finance is driven by the inadequacy of traditional financial systems in the face of long-term environmental risks and market failures stemming from climate change. In this context, Sustainable Finance 2.0 envisages an approach in which financial institutions include negative environmental and social externalities in their decision-making stages. There is a possibility that these externalities can be priced or, conversely, negatively affect the reputation of a financial institution through mechanisms such as carbon taxes. Internalising such externalities will help reduce the risks related to the viability of financial investments. Although these risks, which will be encountered especially in long-term investment instruments such as stocks, bonds or sustainable loans, are higher, it will be possible for financial institutions and firms to re-establish public trust thanks to the internalization of externalities, which serves as the exact opposite of reputation risk (Schoenmaker & Schramade, 2019). All these systemic challenges have led to the reconsideration of the capital distribution process in line with sustainability goals. Accordingly, green finance has become a concept for channelling funds towards low-carbon infrastructure and environmentally beneficial projects. It is a mechanism that internalises environmental risks and incorporates them in decision-making processes (Campiglio, 2016). In addition, frameworks such as the Paris Agreement (2015) and UNEP FI (2018) emphasise the need to align financial systems with the SDGs and encourage the redirection of private sector capital towards a direction compatible with ecological limits and social equality.

- **Scope:** Green finance is linked to various financial instruments and services that support investment and financing projects with sustainable goals. This is not just limited to financing eco-friendly investments and systems, sustainable development or green loans and green bonds (Bei & Wang, 2023; Zhao & Xing, 2024). Looking at what green finance represents is not restricted to individual financial instruments; it encompasses a variety of mechanisms that support projects and policy developments that fuel a green economy.
- Green bonds are considered a type of financial activity where all the proceeds are used exclusively for financial investments that are green-associated projects related to renewable energies or sustainable infrastructures (ICMA, 2021). This definition suggests green finance's holistic nature, which involves shifting towards a greener economy. The European Parliament goes on to explain this by differentiating green finance from climate finance. Climate finance is intended only for climate change mitigation or adaptation; however, green finance has a wider commitment, with environmental goals including pollution reduction and biodiversity protection. Thus, a more inclusive framework provides for diverse financing needs for sustainable development (EPRS, 2021). Green finance contains many types of financial instruments, such as green bonds and loans. These specific instruments direct funds to projects that are pro-environment. For example, green bonds can create renewable energy infrastructures and sustainable transportation. The proceeds are allocated only for environmentally sustainable projects, indicating that the funding raised directly impacts green objectives (Flammer, 2021).
- Green finance is not exclusive to the private sector. Development banks and governments are critical to allocating funds to green projects. Usually, the latter will attract financial benefits like tax and subsidy incentives to invest in sustainability-driven projects (Taghizadeh-Hesary & Yoshino, 2020). Moreover, regulations and policies that encourage and foster green finance are also evolving, increasing transparency and responsibility in fund allocation. A common feature of these

initiatives is using several financial activities and instruments known as green finance, which promote environmental sustainability. By embedding environmental factors into financial investment decisions, green finance is crucial for a smooth transition to a low-carbon and sustainable economy.

- **Instruments:** In response to the worsening climate crisis and the growing call for sustainable development, green finance instruments have emerged as one of the most essential means of channelling investment towards climate-resilient activities. These instruments are part of an ecosystem which steers financing opportunities to support various actors in climate change mitigation, adaptation, and greater ecological sustainability of the real economy (Babic, 2024; Koczar et al., 2020). Such opportunities are more important than ever, given the estimated need for trillions of dollars annually in green investments to address the financing gap needed to achieve global climate change agreements and the Sustainable Development Goals (SDGs) (Gilchrist et al., 2021; Babic, 2024).

- The green finance framework comprises various instruments with different purposes, actors, and potential impacts. The most established instrument is green bonds, the most mature and widely adopted form. Green bonds are debt securities whose proceeds are used exclusively to finance eligible green projects (Bhutta et al., 2022; Gilchrist et al., 2021). These projects span the spectrum from renewable energy generation, energy efficiency upgrades, low-carbon transportation, climate resiliency efforts, biodiversity, and climate change mitigation (ICMA, 2021). As of 2019, cumulative green bond issuances have reached USD 257 billion, accelerating market growth globally (Wu et al., 2023). Research indicates that the issuance of green bonds creates positive impacts for issuers as they experience a reduction in cost of capital, improved corporate image, and better engagement with stakeholders despite concerns over legitimacy, greenwashing, and credibility of impact reporting persist (Gilchrist et al., 2021). Green equity investment is another critical element of green finance, which differs from debt-type securities as an equity investment. Unlike green bonds, green equity investments give a certain degree of control to the investor firm, increasing the investment's longevity. Equity investment goes green by default (by investing in green projects), through shareholder engagement or divestment. Therefore, large asset managers and ETF funds facilitate directly transferring massive funds into green assets. One of the most significant funds in this space is the Norwegian Sovereign Wealth Fund GPF-G, which owns and controls about 1.5% of the global corporate equities and is a prominent case in this regard (Babic, 2024). Green loans are essentially debt for sustainable projects. They are governed by loan agreements that set out the use of proceeds for environmental objectives as well as ongoing reporting, offering more structural rigour compared to conventional loans. Thus, syndicated green loans enable financing large-scale, high-impact environmental projects, becoming a cornerstone of the expanding green finance toolkit (Aleszczyk et al., 2022). A relatively recent innovation within the green finance ecosystem is the sustainability-linked loan (SLL). Sustainability-linked loans (SLLs) are debt products that change loan pricing—specifically, the interest rate spread—based on the borrower's performance against pre-agreed sustainability targets, which are agreed upon at the beginning of the loan term. SLLs differ from a typical green loan in that the loan amount does not have to be applied exclusively to green or sustainable-related projects; instead, they condition loan pricing on measurable improvements in environmental, social, or ESG rating. For example, an SLL contract with a U.S.-based real estate investment trust offers a better spread if it can show a reduction in its carbon footprint; this demonstrates that lenders reward decarbonization performance through contractually embedded incentives (Gonzalez et al., 2025). Green minibonds are a relatively new tool that allows access to capital markets for project financing in and by small and medium-sized enterprises (SMEs). The minibond concept emerged for small and medium-sized enterprises (SMEs) as an alternative to bank credit in mostly bank-dependent countries such as Italy. Green minibonds are a financing opportunity that allows SMEs to use the proceeds without straying from the intended green project purpose. Thus, minibonds offer financing opportunities to firms that, before this access, were shut out from large capital markets and now can play their part in a green transition. (Cerqueti et al., 2023).

- The green finance framework extends beyond bonds and loans to include public-sector and market-led instruments designed to accelerate decarbonization. These include green investment funds that refer to funds which pursue green investment opportunities, green insurance products that reflect insurance policies leading to climate mitigation and transition to a green economy and green venture capital that funds early-stage climate innovations such as renewable energy and wind power technologies (Chi et al., 2023; Dhayal et al., 2023; Klapkiv & Ülgen, 2023; Randjelovic et al., 2003). Together, these tools channel capital across diverse segments of financial markets, public and private, thereby creating vital links between investment opportunities and real-world sustainability outcomes.
- Green finance has evolved from a singular notion to a multifaceted framework that reallocates capital to sustainable purposes and incentivises financial decisions through an environmentally responsible approach. Whether in the form of green bonds, green equity, or sustainability-linked loans, the wide range of instruments facilitates market failures and vulnerabilities to climate change in a complementary fashion. As the urgency of environmental issues escalates, enhancing the institutional and regulatory framework of green finance will be crucial to ensure its effectiveness, reliability, and global impact.

4 Mapping the Nexus: Climate Displacement and Green Finance

4.1 Economic, Social, and Political Costs of Climate Displacement

To understand the complex costs of climate displacement, it is crucial first to analyse the extent of population shifts induced by climate-related events and subsequently evaluate the wider economic damage caused by these environmental disruptions. Together, these figures lay the groundwork for understanding the cumulative economic, social, and political burdens associated with climate-induced mobility. Therefore, Fig. 1 first presents data on climate-related internal displacement, followed by Fig. 2, which outlines the economic losses caused by climate-induced disasters[1] to illustrate both the scale and economic consequences of climate-related events.

Figure 1 illustrates disaster-induced internal displacement figures (in millions) across 163 countries and regions between 2008 and 2024, based on data from the Internal Displacement Monitoring Centre (IDMC). Earthquake-related

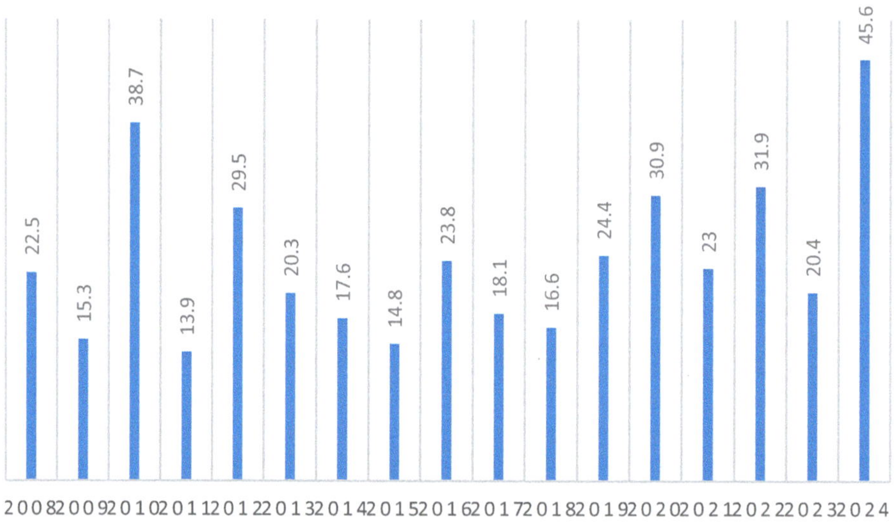

Fig. 1 Disaster-induced internal displacements. (Source: IDMC (n.d.))

[1]Figure 1 includes only internal displacement data, as such movements are more consistently recorded at the national level. In contrast, Figure 2 includes on economic losses reflect the broader impact of climate-related disasters, regardless of whether the affected populations moved or remained in place. This distinction arises from differences in data availability and reporting practices.

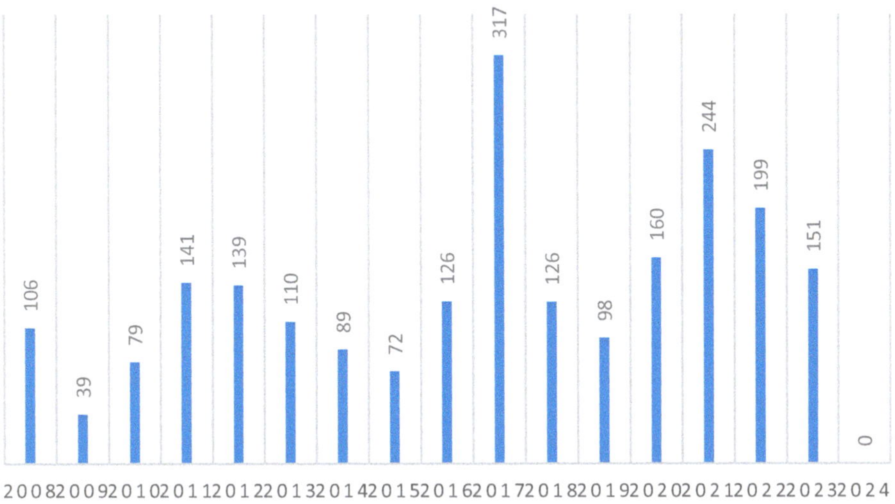

Fig. 2 Global damage costs from natural disasters. (Source: EM-DAT, CRED/UCLouvain (2024)—with major processing by Our World in Data). (Note that while the total global damage costs from natural disasters in 2024—excluding those from earthquakes and volcanic activities—are reported as approximately 426 million dollars (based on data recorded up to April 2024), Munich Re (2025) reported the overall loss as 320 billion dollars. All amounts are reported in units of billion USD.)

displacements were excluded, as they are not directly linked to climate processes. The data reveal a fluctuating but generally increasing trend over the 17 years. Certain years stand out with notable peaks. In 2010, 38.7 million people were displaced, marking the highest level of the early period and reflecting the concentration of disaster impacts in specific contexts. Although fluctuations continued in subsequent years, displacements reached 30.9 million in 2020 and 31.9 million in 2022, both of which represent elevated levels. The highest figure of the entire period was recorded in 2024, with 45.6 million people displaced.

Figure 2, illustrates global economic losses from natural disasters between 2008 and 2024. The year 2017 represents a historical peak with losses amounting to USD 317 billion. Prior to this, costs fluctuated; following a sharp decline in 2009, losses rose again during 2010–2012, declined between 2013 and 2015, and increased notably in 2016. After 2017, losses decreased in 2018–2019 but rose once more from 2020 onward, reaching USD 244 billion in 2021 and USD 199 billion in 2022. In 2023, losses stood at USD 151 billion, a figure below the peak years but still considerably higher than most pre-2017 levels. This pattern likely reflects the increasing severity and frequency of extreme natural disasters and rising exposure of assets and infrastructure to climate-related risks.

According to UNHCR (2024), an increasing number of individuals are being forced to flee persecution, violence, and human rights violations associated with the adverse effects of climate change and disasters, thereby necessitating international protection. Climate change and environmental degradation have become significant drivers of displacement, particularly when compounded by social, economic, and political stressors. Conflict and disaster intersect and reinforce one another in many contexts as primary triggers of forced mobility. The adverse impacts of climate change interact with, and often exacerbate, existing pressures such as violence, insecurity, fragile governance structures, and weak rule of law (UNHCR, 2024). Climate-related hazards have the potential to heighten security risks, particularly when such events result in significant disruptions to livelihoods and undermine food and water security (Opitz Stapleton et al., 2017).

- **Economic Impacts: Household Welfare and Recovery Mechanisms:** From a household perspective, natural disasters constitute a multidimensional shock to overall welfare. The primary impacts can be classified into three categories: threats to physical well-being, loss or damage to assets, and disruptions to income. These adverse effects frequently occur simultaneously and are often exacerbated by displacement from familiar surroundings, the weakening of community support networks, and the interruption or complete absence of public services. The convergence of these factors amplifies the negative consequences for household welfare (Charvériat, 2000).

- The economic impact of a disaster largely depends on whether the damaged assets are replaced. If not replaced, the resulting loss can lead to a decline in consumption and investment, thereby reducing quality of life and economic productivity. Additional financing is required when assets are replaced through in-kind assistance or commercial purchases. This financing typically comes from intertemporal transfers (past savings or future borrowing) or interpersonal transfers (resource redistribution within society). Key financial recovery mechanisms include tax reductions or deferrals, unemployment benefits, low or zero-interest loans, non-repayable grants, insurance payouts, use of personal savings, and the sale of tangible assets. In addition, in some cases, a household member, often the primary income earner, may migrate to a safer region with better housing and employment opportunities as part of the broader recovery strategy (Lindell & Prater, 2003).
- **Social Impacts: Vulnerabilities and Inequalities:** The Nansen Initiative further underscores how disaster displacement deepens pre-existing vulnerabilities, especially among marginalised social groups. Disaster-induced displacement occurs on a large scale, often resulting in profound and far-reaching consequences for individuals and their communities. It generates a wide array of protection concerns and poses significant obstacles to the sustainable development efforts of many states. These challenges are further intensified by the tendency of disasters to amplify existing vulnerabilities within affected populations. Among those most severely impacted are individuals who are sick or injured, unaccompanied or orphaned children, female-headed households, persons with disabilities, older adults, migrants, and members of indigenous communities, all of whom face heightened risks in the aftermath of disaster events (The Nansen Initiative, 2015).

- Such vulnerability is not incidental but deeply rooted in structural inequalities that increase long-term impoverishment. These include limited income diversification, pervasive gender and income disparities, and restricted access to essential assets such as markets, financial capital, insurance, land, education, and information. These underlying drivers reflect deficiencies in the five key capitals of the sustainable livelihoods framework: physical, natural, human, financial, and social capital. In contrast, wealthier populations are generally better equipped to recover from disaster-related losses and, in some cases, may even benefit from disaster economies further deepening existing inequalities (Shepherd et al., 2013).
- **Urbanisation and Infrastructure Strain:** One of the most visible expressions of climate-induced vulnerability is the growing rural-to-urban migration trend, especially in Least Developed Countries. As shifts in the natural environment and demographic pressures intensify, more people are relocating to urban areas for safety and opportunity. Such movements will considerably strain urban infrastructure, environmental systems, and labour markets. In the absence of adequate planning and governance, migrants may become confined to informal settlements or urban slums characterized by vulnerability to flooding, inadequate sanitation, and heightened health risks. These conditions further entrench patterns of inequality, insecurity, and unsustainable development (IOM & UN-OHRLLS, 2019).
- **Political Implications: Governance and Conflict:** Climate change can exacerbate existing social inequalities and contribute to the outbreak or intensification of conflict, especially in regions already experiencing political fragility or resource stress. Environmental degradation may heighten intercommunal tensions and overburden weak governance systems, particularly in peri-urban areas with large migrant populations (The White House, 2021).

4.2 Financing Gaps and Green Finance

Green finance's primary aim is to mobilise financial resources to address the challenges posed by climate change and promote long-term environmental sustainability. It seeks to align the financial sector with sustainability goals by encouraging institutions and investors to adopt comprehensive environmental and social risk management frameworks, incorporate sustainability considerations into investment decision-making processes, and develop innovative financial instruments and services that facilitate the transition toward a more sustainable economy (Khemiri, 2024).

Resilience plays a critical role in determining the extent to which individuals, households, communities, and nations are able to withstand and recover from the effects of both sudden-onset and slow-onset natural hazards, as well as the broader impacts of climate change. Strengthening resilience through disaster risk reduction initiatives, infrastructure development, integrated urban planning, climate adaptation strategies, land reform, and other sustainable development interventions can significantly enhance the capacity of vulnerable populations to remain in their places of origin. These measures not only mitigate exposure to environmental hazards but also serve as a means to substantially reduce the incidence of displacement resulting from disasters (The Nansen Initiative, 2015).

In light of these challenges, various stakeholders have advocated for policy reforms targeting financial institutions, governments, and other relevant actors. To ensure equitable access to climate finance, particularly for developing countries, it is recommended that climate funds adopt more flexible and inclusive frameworks that reflect the institutional and financial limitations many nations face. This includes better serving the needs of SIDS, many of which are ineligible for traditional development aid yet face disproportionate costs from climate impacts (Jensen & Roniger, 2023).

4.3 Case Studies

Fiji: The Pacific Island nation of Fiji became the first emerging market sovereign to issue a green bond, successfully mobilising 100 million Fijian dollars (equivalent to approximately 50 million USD) to finance climate change mitigation and adaptation initiatives. Before the bond issuance, Fiji sought technical support from the International Finance Corporation and the World Bank. With guidance from the World Bank Group, Fiji structured its Green Bond Framework following the internationally recognised Green Bond Principles. The transparent and structured approach adopted by Fiji has since served as a model for other countries, offering a practical framework for sovereign green bond issuance (IFC, 2018).

Seven major projects were funded in 2017–2018 using FSGB proceeds, including (Republic of Fiji, 2018):

- Rural Water Supply Programme—Providing safe water access to over 42,000 people.
- Rainwater Harvesting Systems—Installed over 8000 tanks in drought-prone areas.
- Cyclone-Resilient School Reconstruction—Rebuilt 101 schools damaged by Cyclone Winston.
- Emergency Road Works—Restored roads and built coastal protections.
- Solar Home Systems—Delivered clean and safe energy to over 13,000 rural residents.
- REDD+ Programme (Reducing Emissions from Deforestation and Forest Degradation—Forest conservation, reforestation, and carbon credit generation.
- Naboro Landfill Expansion—Modernised waste management with climate co-benefits.

Fiji's climate finance stems from various sources, such as multilateral development banks, international donors, and bilateral support. Fiji's main sources of climate finance are external donors and the national budget. External donors contribute approximately 79% of all climate finance allocations, with significant contributions from multilateral partners, bilateral donors and non-governmental organisations, including the Global Environmental Facility. On the other hand, the national budget contributes 21% of allocated climate finance, demonstrating the government's commitment to addressing climate change domestically (Mara & Mombauer, 2025).

To facilitate the issuance of its sovereign green bond, Fiji established a Green Bond Framework through a transparent and well-structured process, ensuring strong alignment with the Green Bond Principles set forth by the International Capital Market Association (ICMA). This framework reflected the country's broader commitment to advancing robust climate action. Fiji has positioned itself as a proactive advocate for ambitious international climate policies, a stance supported by its comprehensive domestic agenda. Introducing the National Climate Change Policy in 2012 marked a critical step in institutionalising climate considerations within national development planning. The policy provides strategic direction for implementing adaptation and mitigation measures, reinforcing Fiji's dedication to addressing the multifaceted challenges of climate change (NDC Partnership, 2019).

- **Bangladesh:** For decades, Bangladesh has occupied a frontline position in the global climate crisis, facing recurrent exposure to extreme weather events such as heatwaves, tropical cyclones, floods, and prolonged droughts (Huq et al., 2024). Bangladesh is a global leader in climate change adaptation and resilience. Notably, in 2005, it became one of the first least developed countries to formulate a National Adaptation Programme of Action, marking a significant step toward institutionalising climate adaptation efforts at the national level (Ministry of Environment and Forest, 2005). Bangladesh, one of the countries most exposed to the impacts of climate change, is undertaking a significant transition toward a greener and more inclusive financial system, with the Bangladesh Bank playing a pivotal role. Since initiating its green finance policy in 2009, BB introduced the first green banking guidelines in 2011, requiring all banks and financial institutions 2014 to allocate at least 5% of their funded loans to green finance. By 2015, 10% of Corporate Social Investment budgets were mandated to address climate risk. Bangladeshi banks and financial institutions invest over USD 1.1 billion annually in green finance products, complemented by a new climate resilience policy to support vulnerable communities (Hassan, 2024).

In parallel, the government adopted the Climate Fiscal Framework in 2014 to align the national public finance system with the effective use of domestic and international climate finance. This framework is an enabling mechanism to enhance national ownership of climate finance, ensure alignment with national climate policies, promote coordination among government entities, NGOs, and the private sector, improve results-based management, and increase accountability. Collectively, these measures aim to advance climate-resilient development in Bangladesh (Finance Division, Ministry of Finance, 2023).

5 Policy Gaps, Governance, and Adaptation Strategies

Financial institutions and their regulatory authorities are pivotal in managing financial stability risks linked to climate change and mobilising climate finance. This responsibility includes strengthening the resilience of financial systems to physical risks such as extreme weather events and transition risks arising from the global shift to a low-carbon economy. Furthermore, a supportive regulatory framework is essential to incentivise private-sector investment and promote a sustainable financial ecosystem (Radzewicz-Bak et al., 2024). However, access to climate finance remains limited for local actors. Existing funding mechanisms often entail high transaction costs, rigid eligibility criteria, and short-term investment cycles, posing particular challenges for private finance initiatives. These structural impediments restrict the ability of local communities to pursue tailored, long-term adaptation investments (UNEP FI, 2025).

Climate-related financial risks encompass financial institutions' vulnerabilities due to both the direct physical consequences of climate change and the broader transition to a sustainable economy. These include the economic fallout from extreme weather and the devaluation of carbon-intensive assets. In parallel, environmental risks refer to financial exposures associated with activities that contribute to or are harmed by environmental degradation, posing systemic threats to financial stability (NGFS, 2020).

The Paris Agreement has been a foundational milestone in shaping global governance frameworks addressing human mobility in the context of climate change. Since its adoption in 2015, the Agreement has informed a growing number of international processes that incorporate mobility-related issues into adaptation and resilience planning, elevating the significance of climate-induced displacement within the climate action discourse (IOM, 2018).

As climate change accelerates, countries are increasingly required to undertake strategic cost-benefit evaluations to determine which regions or assets merit protection. These decisions, ranging from the construction of sea defences to the implementation of managed retreat strategies, are shaped by the availability of resources and the foresight exercised by policymakers (Brown, 2008). At the same time, receiving regions that host displaced populations must adopt inclusive planning approaches. Rapid demographic shifts can overburden housing, infrastructure, and public services, risking social unrest. Nonetheless, early integration of displaced populations into local development strategies can enhance cohesion and prevent long-term tensions (Gemenne et al., 2021).

In the immediate aftermath of large-scale displacement, municipalities often bear the brunt of public service delivery, such as water, healthcare, education, and safety. When these systems become overstretched, both host and displaced populations may suffer. Although initial responses tend to be marked by solidarity, sustained pressures can erode support over time, exacerbating pre-existing vulnerabilities in the host context (OECD, 2023).

Despite global commitments, historical emissions and entrenched development trajectories ensure that climate change will have long-lasting impacts for decades, even if countries fully implement their nationally determined contributions under the Paris Agreement (Doelle & Puthucherril, 2023). This underscores the urgency of effective adaptation strategies and equitable access to finance.

At the UN Biodiversity Conference (COP 15) 2009, developed countries pledged to mobilise US$100 billion annually by 2020 to support developing nations' climate response. While this target may be met by 2023, developing countries' financing needs far exceed this benchmark. Due to complex funding procedures and institutional limitations, vulnerable states, including LDCs and SIDS, continue to face barriers in accessing adaptation finance (Jensen & Roniger, 2023).

Although mitigation and adaptation are widely recognised as priorities, financial allocations for adaptation remain disproportionately low. Adaptation finance represents only a small fraction of global climate finance, even as climate shocks become more frequent and severe (Bascunan et al., 2020).

Encouragingly, the institutionalisation of a climate finance ecosystem is advancing. The Task Force on Climate-Related Financial Disclosures (TCFD) is developing standardised reporting frameworks to enhance transparency around climate-related risks. Concurrently, initiatives such as the Green Bond Principles, Social Bond Principles, and Sustainability Bond Guidelines promote credible and transparent investments in climate-aligned projects (Tall et al., 2021).

The scale of required investment is unprecedented. In 2024 alone, climate-induced disasters resulted in economic losses exceeding US$320 billion. Achieving net-zero emissions by 2050 will demand an estimated US$4 trillion annually in global investment. In this context, green finance must evolve, incorporating innovative financial tools, risk mitigation strategies, and resilience-oriented mechanisms to meet the complex demands of a rapidly changing climate and ensure systemic sustainability (Roy, 2025).

To effectively address climate change, displacement, and financial vulnerability, green finance must be reconceptualised as a route to sustainability, social equity and resilience. Easy access to financial resources and mobility-related risks will be critical for just adaptation outcomes. Without such a change, climate financing risks deepening existing inequities rather than alleviating them, exposing vulnerable populations to displacement driven by climate change.

References

Ahsan, R., Kellett, J., & Karuppannan, S. (2014). Climate-induced migration: Lessons from Bangladesh. *The International Journal of Climate Change: Impacts and Responses, 5*(2), 1–15. https://doi.org/10.18848/1835-7156/CGP/v05i02/37204

Aleszczyk, A. A., Loumioti, M., & Serafeim, G. (2022). *The issuance and design of sustainability-linked loans*. Available at SSRN 4287295.

Ayeb-Karlsson, S., Smith, C. D., & Kniveton, D. (2018). A discursive review of the textual use of 'trapped' in environmental migration studies: The conceptual birth and troubled teenage years of trapped populations. *Ambio, 47*, 557–573. https://doi.org/10.1007/s13280-017-1007-6

Babic, M. (2024). Green finance in the global energy transition: Actors, instruments, and politics. *Energy Research & Social Science, 111*, 103482.

Barua, P., Shahjahan, M., Rahman, M. A., & Molla, M. H. (2024). Ensuring the rights of climate-displaced people in Bangladesh. *Forced Migration Review*. Retrieved June 15, 2025, from https://www.fmreview.org/resettlement/barua-shahjahan-rahman-rahman-molla/

Bascuñán, F. L., Molloy, D., & Sauer, B. (2020, July 24). *What are resilience bonds, and how can they protect us against climate crises?* Global Center on Adaptation. Retrieved from https://gca.org/what-are-resilience-bonds-and-how-can-they-protect-us-against-climate-crises/

Bates, D. C. (2002). Environmental refugees? Classifying human migrations caused by environmental change. *Population and Environment, 23*(5), 465–477. http://www.jstor.org/stable/27503806

Bei, J., & Wang, C. (2023). Renewable energy resources and sustainable development goals: Evidence based on green finance, clean energy and environmentally friendly investment. *Resources Policy, 80*, 103194.

Bhutta, U. S., Tariq, A., Farrukh, M., Raza, A., & Iqbal, M. K. (2022). Green bonds for sustainable development: Review of literature on development and impact of green bonds. *Technological Forecasting and Social Change, 175*, 121378.

Black, R., Adger, W. N., Arnell, N. W., Dercon, S., Geddes, A., & Thomas, D. (2011). The effect of environmental change on human migration. *Global Environmental Change, 21*(Suppl 1), S3–S11. https://doi.org/10.1016/j.gloenvcha.2011.10.001

Böhm, S., Misoczky, M. C., & Moog, S. (2012). Greening capitalism? A Marxist critique of carbon markets. *Organization Studies, 33*(11), 1617–1638.

Bolton, P., Després, M., Pereira da Silva, L. A., Samama, F., & Svartzman, R. (2020). *The green swan: Central banking and financial stability in the age of climate change*. BIS.

Bose, P., & Lunstrum, E. (2012). Environmentally induced displacement and forced migration. *Refuge: Canada's Journal on Refugees, 29*(2), 5–10. https://www.jstor.org/stable/48648721

Brown, O. (2008). *Migration and climate change* (IOM migration research series, no. 31). International Organization for Migration.

Brzoska, M., & Fröhlich, C. (2016). Climate change, migration and violent conflict: Vulnerabilities, pathways and adaptation strategies. *Migration and Development, 5*(2), 190–210.

Campiglio, E. (2016). Beyond carbon pricing: The role of banking and monetary policy in financing the transition to a low-carbon economy. *Ecological Economics, 121*, 220–230. https://doi.org/10.1016/j.ecolecon.2015.03.020

Cerqueti, R., Deffains-Crapsky, C., & Storani, S. (2023). Green finance instruments: Exploring minibonds issuance in Italy. *Corporate Social Responsibility and Environmental Management, 30*(4), 1965–1986.

Charvériat, C. (2000). Natural disasters in Latin America and the Caribbean. In *An overview of risk* (Research department working paper, no. 434). Inter-American Development Bank.

Chi, Y., Hu, N., Lu, D., & Yang, Y. (2023). Green investment funds and corporate green innovation: From the logic of social value. *Energy Economics, 119*, 106532.

Dhayal, K. S., Giri, A. K., Esposito, L., & Agrawal, S. (2023). Mapping the significance of green venture capital for sustainable development: A systematic review and future research agenda. *Journal of Cleaner Production, 396*, 136489.

Doelle, M., & Puthucherril, T. G. (2023) Nature-based solutions to sea level rise and other climate change impacts on oceanic and coastal environments: a law and policy perspective. *Nordic Journal of Botany, 1*, e03051. https://doi.org/10.1111/njb.03051

EM-DAT, CRED/UCLouvain (2024). With major processing by Our World in Data. "Global damage costs from natural disasters – EM-DAT" [dataset]. EM-DAT, CRED/UCLouvain, "Natural disasters" [original data]. Retrieved July 3, 2025, from https://archive.ourworldin-

data.org/20250624-125417/grapher/damage-costs-from-natural-disasters.html (archived 24 June 2025).
European Parliamentary Research Service (EPRS). (2021). *Green and sustainable finance* (Briefing – PE 679.081). https://www.europarl.europa.eu/RegData/etudes/BRIE/2021/679081/EPRS_BRI(2021)679081_EN.pdf
Ferris, E. (2012). *Protection and planned relocations in the context of climate change* (Legal and protection policy research series, PPLA/2012/04). United Nations High Commissioner for Refugees.
Fiji Ministry of Foreign Affairs. (2021, September 26). *Fiji legislate 2050 net-zero commitment, challenges industrialised nations to follow suit.* Fiji Ministry of Foreign Affairs. https://www.foreignaffairs.gov.fj/fiji-legislate-2050-net-zero-commitment-challenges-industrialised-nations-to-follow-suit/
Finance Division, Ministry of Finance (2023). *Climate financing for sustainable development: Budget report 2023–24.* Government of the People's Republic of Bangladesh.
Flammer, C. (2021). Corporate green bonds. *Journal of Financial Economics, 142*(2), 499–516.
Gemenne, F., Zickgraf, C., Hut, E., & Castillo Betancourt, T. (2021). *Forced displacement related to the impacts of climate change and disasters.* The Hugo Observatory, University of Liège. Prepared for UNHCR. https://www.unhcr.org/people-forced-to-flee-book/wp-content/uploads/sites/137/2021/10/Franc%CC%A7ois-Gemenne-et-al_Forced-displacement-related-to-the-impacts-of-climate-change-and-disasters.pdf
Gilchrist, D., Yu, J., & Zhong, R. (2021). The limits of green finance: A survey of literature in the context of green bonds and green loans. *Sustainability, 13*(2), 478.
Gonzalez, T. A., Romero, L. C., Karmaziene, E., & Yuan, X. (2025). Green gains: The impact of REITs' environmental performance on sustainability-linked loan interest rates. *Finance Research Letters, 71*, 106415. https://doi.org/10.1016/j.frl.2024.106415
Hassan, M. (2024, January 10). *In Bangladesh, financial inclusion is driving climate resilience.* Alliance for Financial Inclusion. https://www.afi-global.org/opinion/in-bangladesh-financial-inclusion-is-driving-climate-resilience/
Huang, L. (2023, November 16). *Climate migration 101: An explainer.* Migration Policy Institute. https://www.migrationpolicy.org/article/climate-migration-101-explainer
Huq, S., Khan, M., Islam, A. S., & Mirza, A. B. (2024). *Climate change impacts in Bangladesh: What climate change means for a country and its people.* International Centre for Climate Change and Development (ICCCAD.
Internal Displacement Monitoring Centre (IDMC) (2018). *Synthesizing the state of knowledge to better understand displacement related to slow onset events.* Task Force on Displacement, Warsaw International Mechanism for Loss and Damage Associated with Climate Change Impacts. https://unfccc.int/sites/default/files/resource/WIM%20TFD%20I.2%20Output.pdf
Internal Displacement Monitoring Centre (IDMC) (n.d.). *Global internal displacement database – Disasters.* Retrieved July 2, 2025, from https://www.internal-displacement.org/database/displacement-data/
International Capital Market Association (ICMA) (2021). *Green bond principles: Voluntary process guidelines for issuing green bonds.* https://www.icmagroup.org/assets/documents/Sustainable-finance/2022-updates/Green-Bond-Principles-June-2022-060623.pdf
International Finance Corporation (IFC) (2018). *Green bond impact report: Financial year 2018.* https://www.ifc.org/content/dam/ifc/doc/2023/IFC-Green-Report-FY18-FINAL.pdf
International Organization for Migration (IOM) (2018, August). *Mapping human mobility (migration, displacement and planned relocation) and climate change in international processes, policies and legal frameworks.* Analysis report. Implementation of the workplan of the Task Force on Displacement under the Warsaw International Mechanism for Loss and Damage (WIM), United Nations Framework Convention on Climate Change (UNFCCC), Pillar II: Policy – International/Regional, Activity II.2. https://unfccc.int/sites/default/files/resource/WIM%20TFD%20II.2%20Output.pdf

International Organization for Migration (IOM), & United Nations Office of the High Representative for the Least Developed Countries, Landlocked Developing Countries and Small Island Developing States (UN-OHRLLS). (2019). *Climate change and migration in vulnerable countries: A snapshot of least developed countries, landlocked developing countries and small Island developing states*. IOM.

IPCC (2022). Summary for policymakers. In H.-O. Pörtner, D. C. Roberts, E. S. Poloczanska, K. Mintenbeck, M. Tignor, A. Alegría, M. Craig, S. Langsdorf, S. Löschke, V. Möller, & A. Okem (Eds.), *Climate change 2022: Impacts, adaptation and vulnerability. Contribution of Working Group II to the Sixth Assessment Report of the Intergovernmental Panel on Climate Change* (pp. 3–33). Cambridge University Press. https://doi.org/10.1017/9781009325844.001

Jacobson, C., Crevello, S., Chea, C., & others. (2019). When is migration a maladaptive response to climate change? Regional Environmental Change, 19, 101–112. doi: https://doi.org/10.1007/s10113-018-1387-6.

Jensen, L., & Roniger, J. (2023). *International climate finance: Status quo, challenges and policy perspectives*. Members' Research Service, European Parliament.

Khemiri, W. (2024). Green finance and climate adaptation: Instruments, gaps, and governance challenges. *Climate Policy, 24*(1), 14–29. https://doi.org/10.1080/14693062.2023.2269875

Klapkiv, L., & Ülgen, F. (2023). Green products in the insurance market. *Wiadomości Ubezpieczeniowe, 2*, 67–81.

Koczar, J., Vagizova, V., Agliullina, M., & Agliullina, Z. (2020, March). Application of green finance tools in the system of sustainable interaction between the real and financial economy: Opportunities and prospects. In *New Silk Road: Business Cooperation and Prospective of Economic Development (NSRBCPED 2019)* (pp. 604–610). Atlantis Press.

Laczko, F., & Aghazarm, C. (2009). Introduction and overview: Enhancing the knowledge base. In F. Laczko & C. Aghazarm (Eds.), *Migration, environment and climate change: Assessing the evidence* (pp. 9–37). International Organization for Migration. https://publications.iom.int/books/migration-environment-and-climate-change-assessing-evidence

Lindell, M. K., & Prater, C. (2003). Assessing community impacts of natural disasters. *Natural Hazards Review, 4*(4), 176–185. https://doi.org/10.1061/(ASCE)1527-6988(2003)4:4(176)

Mara, G. S. J., & Mombauer, D. (2025). *The nexus of climate change, sovereign debt, and climate finance in Fiji: Building resilience with integrated climate finance and adaptation strategies*. SLYCAN Trust. https://www.slycantrust.org/knowledge-resources/the-nexus-of-climate-change-sovereign-debt-and-climate-finance-in-fiji

Mayer, B. (2016). Chapter one: The concept of climate migration: Advocacy and its prospects. In *The concept of climate migration: Advocacy and its prospects* (pp. 1–37). Edward Elgar Publishing. https://www.elgaronline.com/monochap/9781786431721/11_chapter1.xhtml

McLeman, R. A. (2009). On the origins of environmental migration. *Fordham Environmental Law Review, 20*(2), 403–425. http://www.jstor.org/stable/44175155

Ministry of Environment and Forest (2005). *National Adaptation Programme of Action (NAPA)*. Government of the People's Republic of Bangladesh.

Munich Re. (2025, January 9). *Natural disasters 2024: Climate change is showing its claws* [Media information]. Munich Re. Retrieved July 3, 2025, from https://www.munichre.com/en/company/media-relations/media-information-and-corporate-news/media-information/2025/natural-disaster-figures-2024.html

NDC Partnership (2019). *Pioneering green finance in the developing world: The Fijian sovereign green bond*. Good Practice Database. https://ndcpartnership.org/knowledge-portal/good-practice-database/pioneering-green-finance-developing-world-fijian-sovereign-green-bond

Network for Greening the Financial System (NGFS) (2020, May). *Guide for supervisors: Integrating climate-related and environmental risks into prudential supervision* (NGFS Technical Document). https://www.ngfs.net/sites/default/files/medias/documents/ngfs_guide_for_supervisors.pdf

OECD (2017). *Investing in climate, investing in growth*. https://www.oecd.org/en/publications/investing-in-climate-investing-in-growth_9789264273528-en.html

OECD (2023). *Addressing forced displacement in climate change adaptation: No longer a blind spot* (OECD development policy papers, no. 46). OECD. https://doi.org/10.1787/891ced36-en

Opitz Stapleton, S., Nadin, R., Watson, C., & Kellett, J. (2017). *Climate change, migration and displacement: The need for a risk-informed and coherent approach.* Overseas Development Institute & United Nations Development Programme.

Paris Agreement (2015, December). Paris Agreement. https://unfccc.int/sites/default/files/english_paris_agreement.pdf

Petz, D., & Ginnetti, J. (2013). Neglected displacement: Human mobility in Pacific disaster risk management and climate change adaptation mechanisms. In *Internal Displacement Monitoring Centre (IDMC).* Refugee Council (NRC).

Puşcaş, I. S., & Guadagno, L. (2024, October). *Money changes everything: Leveraging climate finance for human mobility.* Forced Migration Review. Retrieved June 15, 2025, from https://www.fmreview.org/financing-displacement-response/puscas-guadagno/

Radzewicz-Bak, B., Vacher, J., Anderson, G., Gori, F., Harb, M., Korniyenko, Y., Ma, J., Malak, M. T., Nampewo, D., & Sakha, S. (2024). *Preparing financial sectors for a green future: Managing risks and securing sustainable finance* (Departmental paper no. 2024/002). International Monetary Fund. https://doi.org/10.5089/9798400255250.087

Randjelovic, J., O'Rourke, A. R., & Orsato, R. J. (2003). The emergence of green venture capital. *Business Strategy and the Environment, 12*(4), 240–253. https://doi.org/10.1002/bse.361

Republic of Fiji (2018). *Fiji sovereign green bond impact report 2018* (Final ed.). Ministry of Economy, Government of the Republic of Fiji. https://fijiclimatechangeportal.gov.fj/wp-content/uploads/2021/12/Fiji-Sovereign-Green-Bond-Impact-Report-2018-FINAL.pdf

Roy, A. (2025, March 17). *Green finance in a changing climate: Adapting to an uncertain future.* Observer Research Foundation. https://www.orfonline.org/expert-speak/green-finance-in-a-changing-climate-adapting-to-an-uncertain-future

Schoenmaker, D., & Schramade, W. (2019). *Principles of sustainable finance.* Oxford University Press.

Shamsuddoha, M. (2015). *Displacement and migration from climate hotspots: Role of the international community in supporting resilience and protection.* Centre for Participatory Research and Development (CPRD).

Shepherd, A., Mitchell, T., Lewis, K., Lenhardt, A., Jones, L., Scott, L., & Muir-Wood, R. (2013). *The geography of poverty, disasters and climate extremes in 2030.* Overseas Development Institute. https://cdn.odi.org/media/documents/8633.pdf

Taghizadeh-Hesary, F., & Yoshino, N. (2020). Sustainable solutions for green financing and investment in renewable energy projects. *Energies, 13*(4), 788.

Tall, A., Lynagh, S., Blanco Vecchi, C., Bardouille, P., Montoya Pino, F., Shabahat, E., Stenek, V., Stewart, F., Power, S., Paladines, C., Neves, P., & Kerr, L. (2021). *Enabling private investment in climate adaptation & resilience: Current status, barriers to investment and blueprint for action.* World Bank Group; GFDRR. https://documents1.worldbank.org/curated/en/566041614722486484/pdf/Enabling-Private-Investment-in-Climate-Adaptation-and-Resilience-Current-Status-Barriers-to-Investment-and-Blueprint-for-Action.pdf

The Nansen Initiative (2015). *Agenda for the protection of cross-border displaced persons in the context of disasters and climate change* (Vol. 1). https://disasterdisplacement.org/wp-content/uploads/2014/08/EN_Protection_Agenda_Volume_I_-low_res.pdf

The White House (2021, October). *Report on the impact of climate change on migration.*

UNCTAD (2022). *The Least Developed Countries Report 2022: The low-carbon transition and its daunting implications for structural transformation.* United Nations Conference on Trade and Development.

United Nations (UN) (n.d.). *Women and climate change.* United Nations. https://www.un.org/en/climatechange/science/climate-issues/women

United Nations Development Programme (UNDP)(2015). *Responding to climate change in least developed countries.* UNDP.

United Nations Environment Programme (UNEP) (2022). *Adaptation Gap Report 2022: Too little, too slow – Climate adaptation failure puts world at risk*. UNEP. https://www.unep.org/resources/adaptation-gap-report-2022

United Nations Environment Programme (UNEP) (2024). *Adaptation Gap Report 2024: Come hell and high water – As fires and floods hit the poor hardest, it is time for the world to step up adaptation actions*. UNEP. https://www.unep.org/resources/adaptation-gap-report-2024

United Nations Environment Programme (UNEP) Finance Initiative (FI)(2018). *Rethinking impact to finance the SDGs*. https://www.unepfi.org/impact/rethinking-impact/

United Nations Environment Programme (UNEP) Finance Initiative (FI) (2025, May 8). *Finance that works for the frontlines: Elevating local voices in climate adaptation*. https://www.unepfi.org/themes/climate-change/finance-that-works-for-the-frontlines-elevating-local-voices-in-climate-adaptation/

United Nations Framework Convention on Climate Change (UNFCCC) (2023). *Boosting the capacity of least developed countries to address climate change*. UNFCCC.

United Nations High Commissioner for Refugees (UNHCR) (2024). *Focus area strategic plan for climate action 2024–2030*. The UN Refugee Agency. https://www.unhcr.org/media/focus-area-strategic-plan-climate-action-2024-2030

United Nations High Commissioner for Refugees (UNHCR) (n.d.). *Climate change and displacement: The myths and the facts*. UNHCR. https://www.unhcr.org/news/stories/climate-change-and-displacement-myths-and-facts

Wilkinson, E., Kirbyshire, A., Mayhew, L., Batra, P., & Milan, A. (2016). *Climate-induced migration and displacement: Closing the policy gap*. Overseas Development Institute (ODI). ODI Briefing. https://odi.org/en/publications/climate-induced-migration-and-displacement-closing-the-policy-gap/

World Bank (2019). *The World Bank Group's action plan on climate change adaptation and resilience: Managing risks for a more resilient future*. World Bank.

World Resources Institute (WRI) (2023). *Fiji adopts national climate finance strategy*. https://www.wri.org/outcomes/fiji-adopts-national-climate-finance-strategy

Wu, J., Li, J. P., & Su, C. W. (2023). Green bonds and US shale gas prices: Evidence from a novel time-varying causality. *Heliyon, 9*(9), e20027.

Zetterli, P. (2023). *Climate adaptation, resilience, and financial inclusion: A new agenda* (Focus Note). CGAP. https://www.cgap.org/research/publication/climate-adaptation-resilience-and-financial-inclusion-new-agenda

Zickgraf, C. (2019). Keeping people in place: Political factors of (im)mobility and climate change. *Social Sciences, 8*(8), 228. https://doi.org/10.3390/socsci8080228

Zhao, Z., & Xing, Z. (2024). *Corporate management, green finance, and sustainability. Humanities and Social Sciences Communications, 11*, 1–8. https://doi.org/10.1057/s41599-023-02490-4

Climate Change and Corporate Finance: Risk Management and Strategic Response

Alexandra-Mădălina Ţăran, Graţiela Georgiana Noja, Mirela Cristea, Alina Ionaşcu, and Bianca-Raluca Bădiţoiu

1 Introduction

Climate change is one of our time's most complex and pressing challenges, with systemic impacts on all human, natural and economic life dimensions. Beyond the visible effects on ecosystems—such as melting ice caps, rising sea levels or the intensification of extreme weather events—these changes generate profound transformations within modern economies, influencing the structure of financial markets, investor behaviour and corporate strategies. In such a global dynamic, the role

A.-M. Ţăran (✉)
Faculty of Economics and Business Administration, Department of Finance, Information Systems and Business Modelling, West University of Timisoara, Timisoara, Romania
e-mail: alexandra.taran@e-uvt.ro

G. G. Noja
Department of Marketing, International Business and Economics, Faculty of Economics and Business Administration, West University of Timişoara, Timişoara, Romania
e-mail: gratiela.noja@e-uvt.ro

M. Cristea
Faculty of Economics and Business Administration, Department of Finance, Banking and Economic Analysis, University of Craiova, Craiova, Romania
e-mail: simona.cristea@edu.ucv.ro

A. Ionaşcu
Doctoral School of Economics and Business Administration, Faculty of Economics and Business Administration, East-European Center for Research in Economics and Business, West University of Timisoara, Timisoara, Romania
e-mail: alina.ionascu@e-uvt.ro

B.-R. Bădiţoiu
Faculty of Economics and Business Administration, Department of Management and Entrepreneurship, West University of Timisoara, Timisoara, Romania
e-mail: bianca.baditoiu@e-uvt.ro

C.-W. Su et al. (eds.), *Green Finance and the Challenges of Climate Change*, Sustainable Finance, https://doi.org/10.1007/978-3-032-07293-1_7

of companies in the transition towards sustainability becomes crucial, and their ability to integrate climate risks into business strategies and adopt a coherent strategic response is essential for ensuring long-term competitiveness and resilience (Ashrafee & Abdullah-Al, 2022). In recent years, the specialised literature has devoted increasing attention to the intersection between climate change and corporate finance, primarily through corporate social responsibility (CSR), sustainable governance, technological innovations, and green investments (Stankevičienė et al., 2025; Țăran et al., 2025). However, many unknowns exist regarding how environmental, social and governance (ESG) performance influences companies' financial performance and how climate strategies are reflected in economic decisions (Rumansyah & Nainggolan, 2025). In this context, the present study aims to bring an original and methodologically sound contribution to understanding these mechanisms through an empirical analysis applied to a representative sample of European companies in the energy sector. This research aims to analyse how climate change influences corporate financial dynamics, primarily through the lenses of risk management and strategic decision-making.

The paper aims to explore in depth the relationships between climate risks, the strategic behaviour of companies and their financial performance, in an integrated framework that capitalises on both traditional economic performance indicators and those specific to the ESG field. The focus is on identifying the direct and indirect influences of climate strategies on companies' market value, profitability and risk perceived by investors, given the specificity of the energy sector—one of the most important emitters of greenhouse gases, but also a key actor in the transition to a green economy. The study, therefore, examines how different factors—from board structure to emission reduction policies and investments in green technologies—contribute to shaping an effective strategic response to climate challenges.

The paper leverages a rigorous quantitative approach, based on advanced econometric models—robust regression and Gaussian graphical models—applied to an updated 2024 dataset from the LSEG Data & Analytics database. The sample includes over 300 European energy companies, covering a variety of sub-sectors and countries, which allows for a relevant benchmarking and significant generalisation of the results. The selected indicators include classic financial variables and composite and pillar-based ESG scores, allowing for a multidimensional analysis of how sustainability is integrated into business practice. The study opens with a review of the relevant literature, summarising theoretical and empirical perspectives on the link between ESG performance and financial performance and the role of public perception and climate policies in this process. Subsequently, the research methodology is detailed, with emphasis on the structure of the database and the justification for the choice of econometric models used. The results are integrated, highlighting the statistical relationships between the variables and the strategic implications for company management and public policies. The conclusions synthesise the main findings, highlighting the importance of aligning financial and sustainability objectives and formulating concrete recommendations for future research and institutional intervention directions.

Therefore, this article aims to analyse relevant empirical data and contribute to substantiating a new framework of thought regarding how climate change should be treated as a determinant of economic performance. In the context of the transition to a green economy and increasing pressures from investors, authorities and civil society, integrating climate risks into corporate decisions is no longer an option but a strategic necessity. This study therefore offers a comprehensive perspective on this issue, highlighting the role that governance, innovation and human capital can play in ensuring the long-term sustainability of companies in the European energy sector.

The structure of the paper is as follows: Section 2 develops the theoretical framework, in which we synthesise the most relevant studies. Section 3 details the data used and the methodology applied. Section 4 presents the results obtained and analyses their implications. Section 5 formulates the main conclusions. Additional information and empirical evidence are included in the appendix.

2 Literature Review

Numerous scientific studies highlight that climate change is widely associated with rising global temperatures, extreme weather events, sea level rise, biodiversity loss, and economic disruptions. It also influences financial markets, corporate risk assessments, and policy decisions, making it a critical factor in economic and business research. Climate change is a crucial global challenge affecting ecosystems and societies and shaping environmental policies and corporate financial performance, as firms must adapt to regulatory pressures, investor expectations, and climate-related risks. Climate risk management encompasses various strategies designed to tackle both the risks and opportunities linked to climate change. This approach focuses on mitigation and adaptation efforts, ensuring that environmental factors are incorporated into corporate decision-making, daily operations, and long-term strategic planning (Aibar-Guzmán et al., 2024). Driven by the increasing frequency of extreme weather events and the rapid progression of global warming, a growing number of studies are examining climate change and its associated risks (Dai & Zhang, 2023; Ma et al., 2024).

The existing literature research exploring the possible economic effects of climate change is extensive and expanding rapidly, increasingly examining its implications for both environmental sustainability (with regard to ESG factors) and financial performance. Further analysis is needed to draw definitive conclusions regarding the previously mentioned facts. Nevertheless, many authors have discussed climate strategies and their broader impacts within the specialised literature. Thus, Porter and van der Linde (1995) see environmental performance as a chance for companies to become more competitive. They argue that following certain environmental rules can help reduce costs. This aligns with Molina-Azorín et al. (2009), who suggest that from a managerial perspective, companies that adopt proactive environmental strategies can achieve greater market legitimacy and social recognition, potentially leading to higher sales, premium pricing, and improved economic performance.

Additionally, such strategies enable firms to differentiate their products from competitors (Molina-Azorín et al., 2009). Several empirical studies based on accounting data suggest a positive link between financial and environmental performance (King & Lenox, 2002; Melnyk et al., 2003; and others). They indicate that environmental initiatives often bring more benefits than costs, helping firms attract talent, reduce expenses, improve efficiency, and strengthen stakeholder relations. Highlighting the importance of CSR even during economic crises, Gallego-Álvarez et al. (2014) concluded that companies that prioritise CSR during economic downturns tend to perform better, as continued investment in sustainable projects strengthens stakeholder relationships and ultimately leads to greater economic benefits. Alternative researchers examine the impact of climate change on firm performance and find out that firms situated in climate-sensitive coastal regions and affluent provinces face comparatively greater adverse effects on their financial returns (Liu et al., 2024).

Other existing empirical research highlights a positive link between climate risk management and corporate financial performance (Ziegler et al., 2011; Megeid, 2024). Ziegler et al. (2011) found that energy firms in the United States that disclose their climate change strategies tend to experience stronger stock performance. Similarly, Megeid (2024) analysed 25 publicly traded Egyptian companies and found that there is a significant positive relationship between the financial performance, financial reporting, and risk management of industrial organisations and their climate change disclosures.

Different authors investigate the impact of climate change risks on corporate credit ratings globally and reveal that firms in climate-vulnerable countries tend to receive lower credit ratings. This negative relationship stems from weaker firm fundamentals, including higher default risk and cash flow volatility caused by climate-related uncertainties (Nguyen, 2025). According to Pankratz et al. (2023), investors appear to misjudge the economic consequences of rising temperatures, as evidenced by more negative surprises in analyst forecasts and earnings announcement returns. A one standard deviation increase in hot days is associated with a 0.3% decline in revenues and a 1.3% reduction in operating income (Pankratz et al., 2023). Similarly, several studies have reported no significant positive relationship between environmental and financial performance (Gilley et al., 2000; Link & Naveh, 2006; and others). Some scholars argue that any potential connection between the two is nonexistent or too intricate to determine (Margolis & Walsh, 2003).

Although interest in ESG is growing, there is still little research on what influences a company's ESG performance, and the results so far are mixed, with some studies also highlighting the challenge of climate change denial as a significant barrier to progress in this area. While examining the impact of climate change on the ESG performance of A-share listed companies in China from 2011 to 2021, Li et al. (2024) reveal that climate change significantly weakens ESG outcomes, though reducing resource misallocation can help mitigate these effects. Their study shows that ESG performance improves in resource-based cities, suggesting a departure from the resource curse. Firms with greater public and analyst attention and larger and more mature companies show stronger resilience. The study emphasises the importance of improved corporate governance and green innovation to promote sustainable development in the face of climate challenges (Li et al., 2024).

Moreover, some authors (Afzali et al., 2025) address regional climate change denial while studying the linkage between the denial and corporate environmental responsibility in the U.S. Using county-level climate perception surveys, the findings reveal that firms in areas with higher denial exhibit weaker environmental performance, commit more violations, and impose greater societal costs (Afzali et al., 2025). In their 2024 systematic review, Mendy et al. examine 65 peer-reviewed studies focused on strategies to combat climate science denial. They highlighted a variety of approaches, and their findings suggest that tailoring interventions to the specific type of denial is essential, as effectiveness depends on the audience and context. The authors stress the importance of customised strategies and outline key considerations for practitioners, such as identifying the form of denial, clarifying intervention goals, and understanding the communicator's influence.

Moreover, we conducted a bibliometric analysis using the co-occurrence of keywords method, based on a sample of 324 relevant and frequently cited publications retrieved from the Scopus database, following the same methodology as Pirtea et al. (2024). The study covers the period 2014–2025. It is built around the set of keywords employed in the literature review, specifically associating the concepts of "climate change", "environmental performance", and "financial performance". The co-occurrence analysis was conducted using VOSviewer, considering all types of keywords and including only those appearing at least four times across the dataset (minimum keyword occurrence = 4), to highlight the field's most prominent and persistent terms.

Figure 1 illustrates the network of the most relevant keywords and their interconnections. The size of each node reflects the relative frequency, and the thickness of the links represents the strength of co-occurrence between terms. The visualisation reveals three distinct thematic clusters marked by a different colour.

The first cluster (red) is centred around environmental management, accompanied by keywords such as ecological sustainability, emission control, environmental assessment, greenhouse gases, and carbon footprint. This cluster highlights the operational and evaluative dimensions of environmental impact management at the firm level. The prominence of terms related to emissions and environmental responsibility suggests a strong focus on how companies monitor, report, and seek to reduce their environmental footprint. The cluster reflects the growing importance of incorporating structured ecological assessment and sustainability practices into corporate operations and the increasing role of regulatory and strategic tools designed to support emission reduction and long-term environmental goals.

The second cluster (green) revolves around environmental policy, carbon emissions, and ESG criteria, closely associated with terms like stakeholder, ecological economics, corporate governance, and carbon management. This thematic group indicates an institutional and policy-oriented approach to sustainability, capturing how external governance structures and ESG integration influence corporate behaviour and accountability. The relationships between these terms show that firms'

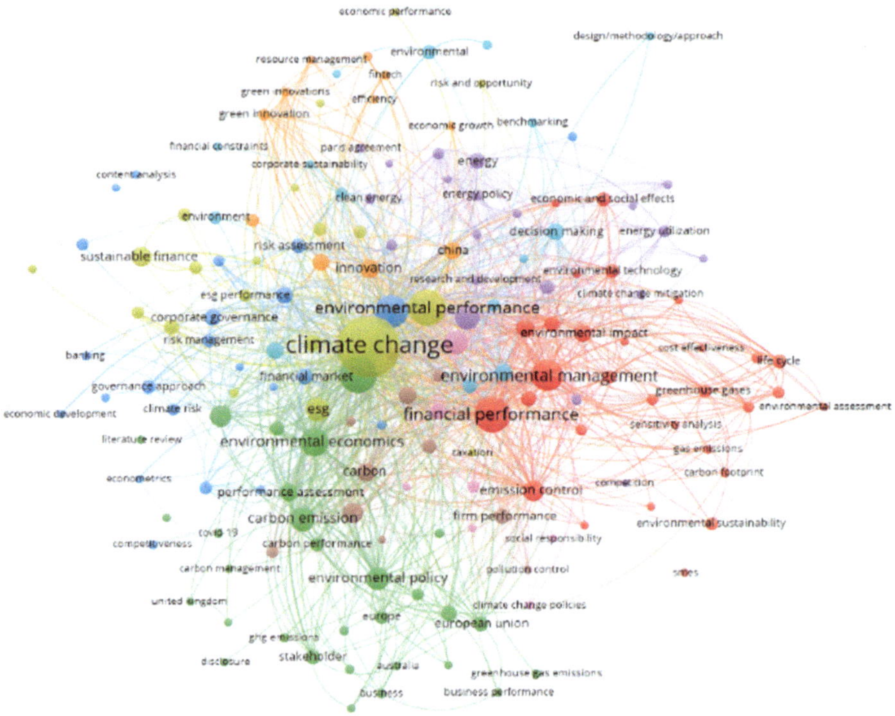

Fig. 1 Co-occurrence network of all keywords within integrated reporting publications. (Source: Own processing using VOSviewer software)

performance is increasingly evaluated in financial terms and through their environmental policy alignment and responsiveness to stakeholder expectations (Wang et al., 2022). This cluster positions ESG as a bridge between corporate transparency, policy compliance, and strategic ecological planning.

The third cluster (blue) is structured around the high-frequency keywords climate change, financial performance, and climate risk, with meaningful connections to climate finance, carbon, and market. It emphasises the economic and financial dimensions of climate impacts, suggesting that how companies address climate risk directly influences market valuation and long-term profitability. The presence of terms like climate finance indicates growing interest in investment mechanisms tailored to manage or capitalise on climate-related risks and opportunities. This cluster, therefore, reflects the deepening intersection between environmental vulnerability and financial strategy.

The fourth cluster (purple) includes keywords such as energy policy, innovation, climate change mitigation, decision making, and technology. This map area reflects the role of technological development, innovation systems, and strategic decision-making in climate adaptation and mitigation. It suggests that successful environmental strategies depend not only on compliance or risk mitigation but also on innovation capacity and the ability to integrate sustainability into core business decisions. The link to energy utilisation and policy also indicates that this cluster captures both micro-level firm strategies and macro-level policy frameworks.

Lastly, the fifth cluster (orange) contains keywords such as economic growth, green innovation, resource management, and corporate sustainability. This cluster explores the interdependence between sustainability-oriented innovation, natural resource efficiency, and the pursuit of economic performance. It underscores the idea that firms engaging in green innovation can generate competitive advantage, especially when sustainability efforts are aligned with economic development objectives. The emphasis on corporate sustainability also reflects the strategic nature of balancing financial goals with environmental and social considerations.

Overall, the network structure and cluster composition confirm the multidimensional nature of current climate-related research, integrating environmental policy, firm-level governance, financial performance, innovation, and sustainability into a complex, interrelated framework.

3 Data and Methodology

In this study, we employed a comprehensive cross-sectional dataset extracted from the LSEG (London Stock Exchange Group) database to assess European companies' financial and sustainability performance in the energy and natural resources sectors for a single fiscal year. The dataset integrates traditional financial metrics and environmental, social, and governance (ESG) indicators, enabling a holistic analysis of firm-level behaviour and strategic orientation.

The analysis includes a representative sample of 95 companies, whose names are listed in the dataset under the variable Company Name. These firms span various subsectors such as oil and gas exploration and production, renewable energy, oilfield services, utilities, and green technologies and are headquartered in different European countries. A few notable examples include BP PLC, Equinor ASA, TotalEnergies SE, Shell PLC, Repsol SA, Siemens Energy AG, and Vestas Wind Systems A/S.

To conduct a robust analysis, we selected a set of 34 key indicators derived from the LSEG database. These indicators cover a wide array of dimensions, including:

- Financial fundamentals (e.g., market capitalisation, total revenue, gross profit);
- Corporate governance (e.g., board diversity, attendance, structure);
- Environmental impact and strategy (e.g., energy use, CO_2 emissions, green buildings);
- Social responsibility (e.g., working conditions, training policies, CSR committees);
- ESG composite scores and thematic pillars.

Each indicator is accompanied by an acronym (used in tables and models) and a detailed definition, which are presented in the next section. This approach allows for consistent measurement and comparability across companies and periods, while enabling a nuanced evaluation of ESG integration and its relationship with firm value and performance. The analysis focuses explicitly on 34 indicators:

- Market Capitalisation (USD) MK_C
  Represents the total market value of a company's outstanding shares, calculated as share price multiplied by the number of shares. Expressed in U.S. dollars, it reflects company size and investor valuation.
- Price to Book Value per Share (Daily Time Series Ratio) PTBV_S
  A financial ratio comparing the market price of a company's stock to its book value per share. It indicates how much investors are willing to pay for each dollar of net assets.
- Beta (5-Year) BETA
  Measures a stock's volatility compared to the overall market over 5 years. A beta above 1 means higher volatility; below 1 indicates lower volatility.
- Total Energy Use (FY0) ENERGY
  The total amount of energy consumed by the company during the most recent fiscal year, regardless of source.
- ESG Combined Score (FY0)—ESG_COMB
  An aggregate score summarising a company's environmental, social, and governance (ESG) performance. Higher scores indicate stronger sustainability practices.
- Environmental Materials Sourcing Score (FY0)—ENV_MAT_SRC
  Evaluates the company's efforts to procure raw materials responsibly and sustainably, minimising environmental harm.
- Green Buildings Score (FY0) GRN_BLD
  Reflects the extent to which the company invests in, owns, or operates buildings certified as environmentally friendly (e.g., LEED, BREEAM).
- Resource Reduction Policy Score (FY0) RES_RED_POL
  Measures the presence and quality of formal policies to reduce resource usage, such as water, materials, and energy.
- Total Energy Use to Revenue (USD Score, FY0) EN_USE_REV
  Indicates energy efficiency by comparing total energy use to total revenue. Lower ratios suggest more efficient energy use per unit of income.
- Total CO_2 Equivalent Emissions to Revenue (USD Score, FY0)—CO2_REV
  Evaluates carbon intensity by comparing total greenhouse gas emissions (in CO_2 equivalents) to revenue. Useful for benchmarking environmental impact per dollar earned.
- Environmental Products Score (FY0)—ENV_PROD
  Measures the share of revenue generated from environmentally beneficial products or services, such as renewable energy or recycling solutions.
- Training and Development Policy Score (FY0)—TRAIN_DEV
  Assess whether the company has formal programs and policies for employee education, upskilling, and career development.
- CSR Sustainability Committee Score (FY0)—CSR_COMM
  Indicates whether the company has a dedicated committee overseeing corporate social responsibility or sustainability matters.

- Total Revenue (FY0, USD)—TOT_REV
  Total income generated from business operations in the latest fiscal year, before expenses are deducted.
- Normalized EBIT (FY0, USD)—NORM_EBIT
  Earnings Before Interest and Taxes, adjusted for non-recurring items, to better reflect core operating performance.
- Normalized EBITDA (FY0, USD)—NORM_EBITDA
  Similar to EBIT, it also excludes depreciation and amortisation. Provides a clearer view of operational profitability and cash flow potential.
- Gross Profit (FY0, USD)—GROSS_PROF
  Revenue minus the cost of goods sold (COGS). Represents the company's profit from core production or delivery service before other expenses.
- Social Pillar Score (FY0)—SOC_PILL
  A composite score summarising the company's performance in social aspects like employee relations, human rights, and community involvement.
- Governance Pillar Score (FY0)—GOV_PILL
  Reflects the quality of the company's governance practices, such as board structure, executive compensation, and shareholder rights.
- Environmental Pillar Score (FY0)—ENV_PILL
  Summarises the company's overall performance on environmental issues, including emissions, resource use, and waste management.
- CSR Strategy Score (FY0)—CSR_STRAT
  Measures the degree to which corporate social responsibility is embedded in the company's strategic planning and operations.
- Environmental Management Team Score (FY0)—ENV_MGMT
  Indicates whether the company has a dedicated and qualified team managing environmental responsibilities and initiatives.
- Biodiversity Impact Reduction Score (FY0)—BIODIV_RED
  Assess the company's actions and policies to minimise adverse effects on biodiversity and ecosystems.
- Flexible Working Hours Score (FY0)—FLEX_HOURS
  Measures the availability and support of flexible work arrangements such as remote work, part-time options, or adjustable schedules.
- Wages & Working Conditions Controversies Score (FY0) –WAGE_CONTRO
  Tracks any controversies or reported incidents related to employee compensation, labour practices, or workplace conditions.
- Corporate Responsibility Awards Score (FY0)—CORP_AWARD
  Indicates whether the company has received third-party recognition or awards for its corporate responsibility efforts.
- Board Meeting Attendance Average Score (FY0)—BRD_ATT_AVG
  Represents the average attendance rate of board members at official meetings, reflecting engagement and oversight.
- Board Structure Policy Score (FY0)—BRD_STRUCT
  Assesses the existence and quality of formal board governance policies, including composition, committees, and term limits.

- Board Attendance Score (FY0)—BRD_ATT
  Measures the actual attendance performance of individual board members or the board.
- Board Gender Diversity, Percent Score (FY0)—BRD_GEN_DIV
  The percentage of seats on the board of directors held by women. A higher percentage indicates greater gender diversity and inclusiveness at the governance level.
- Board Background and Skills Score (FY0)—BRD_SKILLS
  Evaluates whether board members possess a diverse and relevant mix of professional backgrounds and expertise (e.g., finance, law, sustainability).
- Total Assets, Reported (FY0, USD)—TOT_ASSETS
  Represents the total value of the company's assets (current and non-current) as reported in the latest fiscal year.
- Country of Headquarters—COUNTRY
  The primary country where the company's main corporate office or operational control is located.
- TRBC Industry Name—TRBC_IND
  The industry classification assigned by the Thomson Reuters Business Classification (TRBC) system is used to group companies by sector and industry.

Descriptive statistical (Table 1) analysis of the 34 ESG and financial indicators reveals a significant degree of variability among the companies analysed regarding sustainable performance and basic financial characteristics. Regarding financial indicators, the average values of market capitalisation (MK_C), revenues (REV) and profits (EBIT, EBITDA, GROSS_PROF) indicate that the sample is mainly composed of medium to large-sized companies, with a relatively stable financial performance. At the same time, the average beta coefficient of 1.29 suggests a volatility slightly above that of the market, while the price-to-book value indicator (PTBV_S) presents a strongly positive asymmetric distribution, signalling the existence of companies with very high market valuations concerning their book assets.

On the ESG dimension, the average combined score (ESG_COMB) is approximately 51, indicating a moderate integration of sustainability criteria. However, the breakdown by sub-components shows considerable disparities. Environmental indicators, such as resource reduction policy (RES_RED_POL), total energy consumption with revenue (EN_USE_REV) or the score for green products (ENV_PROD), show wide variations and asymmetric distributions, indicating that while some companies implement advanced environmental measures, others are absent from this sphere. This is also evident from the green buildings (GRN_BLD) or biodiversity (BIODIV_RED) scores, where many companies obtain very low or zero scores.

Regarding the social dimension, vocational training policies (TRAIN_DEV) and work flexibility (FLEX_HOURS) are implemented unevenly, although they are at an acceptable level on average. The governance component highlights, on average, balanced scores regarding board structure, gender diversity, and meeting attendance. However, some scores, such as those relating to formal governance policies (BOARD_STR) or board members' skills (BOARD_SK), present an extremely unbalanced distribution, indicating a lack of standardised practices among all firms.

Table 1 Descriptive statistics

	Mean	Std. Deviation	Skewness	Std. Error of Skewness	Kurtosis	Std. Error of Kurtosis	Minimum	Maximum
PTBV_S	1.388	1.782	4.853	0.263	31.319	0.520	−1.600	14.000
BETA	1.295	0.784	0.486	0.263	0.053	0.520	−0.510	3.250
ESG_COMB	50.909	15.236	−0.165	0.261	−0.040	0.517	11.050	84.180
ENV_MAT_SRC	44.490	43.314	−0.056	0.261	−2.025	0.517	0.000	92.380
GRN_BLD	15.102	34.228	1.845	0.261	1.443	0.517	0.000	94.630
RES_RED_POL	67.312	10.367	−6.206	0.263	38.132	0.520	1.830	73.160
EN_USE_REV	52.519	26.395	−0.274	0.263	−0.864	0.520	0.900	99.060
CO2_REV	56.310	28.444	−0.336	0.267	−0.908	0.529	1.720	99.600
ENV_PROD	38.338	39.848	0.134	0.261	−1.923	0.517	0.000	86.360
TRAIN_DEV	56.349	8.904	−6.165	0.261	37.923	0.517	0.000	61.480
CSR_COMM	53.598	24.388	−1.691	0.261	1.151	0.517	0.000	77.330
SOC_PILL	60.517	21.113	−0.563	0.261	−0.285	0.517	7.760	95.850
GOV_PILL	57.403	21.449	−0.148	0.261	−1.020	0.517	12.600	92.240
ENV_PILL	54.576	21.609	−0.016	0.261	−0.562	0.517	1.620	96.020
CSR_STRAT	58.765	27.027	−0.267	0.263	−0.994	0.520	1.920	97.090
ENV_MGMT	44.935	33.512	−0.602	0.261	−1.642	0.517	0.000	75.610
BIODIV_RED	46.749	35.034	−0.569	0.261	−1.626	0.517	0.000	85.880
FLEX_H	41.994	39.924	−0.105	0.261	−2.016	0.517	0.000	85.480
WAGE_C	45.283	21.960	−1.615	0.261	0.638	0.517	0.000	57.720
CORP_AWARD	35.237	40.416	0.272	0.261	−1.964	0.517	0.000	88.720
BOARD_AT_AVG	49.898	27.968	−0.075	0.291	−1.310	0.574	2.220	93.400
BOARD_STR	50.971	5.853	−7.991	0.261	70.492	0.517	0.000	57.140
BOARD_ATT	40.642	35.114	−0.204	0.261	−1.785	0.517	0.000	92.860
BOARD_GEN_DIV	45.515	27.315	0.181	0.261	−1.067	0.517	1.850	99.690

(continued)

Table 1 (continued)

	Mean	Std. Deviation	Skewness	Std. Error of Skewness	Kurtosis	Std. Error of Kurtosis	Minimum	Maximum
BOARD_SK	50.081	0.748	9.220	0.261	85.000	0.517	50.000	56.900
MK_C	21.502	2.251	−0.518	0.261	0.940	0.517	13.545	26.156
ENERGY	15.514	3.137	−0.500	0.261	−0.117	0.517	6.995	21.510
REV	21.738	2.298	−0.110	0.263	−0.122	0.520	15.191	26.481
EBIT	20.334	2.122	−0.018	0.287	−0.435	0.566	15.543	24.373
EBITDA	20.766	1.962	0.188	0.281	−0.429	0.555	16.310	24.852
GROSS_PROF	20.727	2.311	0.017	0.272	−0.061	0.538	14.603	25.278
ASSETS	22.420	1.994	0.246	0.261	−0.605	0.517	17.967	26.730

Source: Own processing using STATA18 software

In line with the primary objective of the research, the empirical analysis was structured in two complementary methodological directions, using advanced econometric procedures. First, robust regression models were applied, including Huber and biweight iterations, to ensure the stability of the estimates in the presence of outliers and non-normal distributions, which are evident features in the structure of the analysed indicators. These models allow the identification of significant relationships between the financial performance of companies and the integration of ESG criteria, without being distorted by extreme observations. Second, the network analysis used Gaussian graphical models (GGM), highlighting interdependence and influence structures between the studied variables. The GGM estimates were carried out using two distinct methods: EBICglasso (Extended Bayesian Information Criteria Graphical Lasso), which optimises the selection of sparse networks, and the partial correlation method (PCOR), which captures the conditional relationships between the indicators. This approach highlights the central nodes and strong links in the ESG network, providing insight into the dimensions that systemically influence the sustainable behaviour of companies. The entire analysis is conducted on a cross-sectional data set, corresponding to a single fiscal year, reflecting European companies' financial and non-financial performance in the energy and natural resources sectors.

Robust regression models with Huber and biweight iterations were selected for their ability to reduce the influence of extreme observations, providing more reliable estimates than classical OLS regressions in the context of heterogeneous data. Huber iteration gradually controls the impact of outliers through an adaptive loss function. At the same time, the biweight method adjusts the estimates more aggressively in the presence of significant deviations from the mean. By applying these techniques, we aimed to highlight relationships between ESG and financial variables, without compromising the robustness of the conclusions due to abnormal values. In addition, network analysis using Gaussian Graphical Models (GGM) allows the examination of conditional relationships between variables, going beyond raw correlations. Using the EBICglasso method, we obtained sparse networks, optimised by an extended Bayesian criterion, favouring the selection of the most relevant connections. The partial correlation method (PCOR) was applied complementarily to identify direct dependencies between indicators, excluding collateral influences.

Thus, GGM facilitated the identification of direct, indirect and overall connections between ESG indicators, highlighting, for example, structural relationships between governance policies and environmental scores or between social performance and the level of internal controls, thus outlining nuclei of influence in the architecture of corporate sustainability.

4 Results and Discussion

4.1 Results of Robust Regression Models

The results of RREG (Table 2) comprise six regression models, according to each dependent variable, namely: model (1) for Market Capitalization (MK_C); model (2) for Earnings Before Interest and Taxes (EBIT); model (3) Earnings Before Interest, Taxes, Depreciation, and Amortization (EBITDA); model (4) Price-to-Book Value Ratio (PTBV_S); model (5) market risk (BETA); model (6) Gross Profit (GROSS_PROF).

The coefficient of determination (R^2) values for each robust regression model (Table 2) indicate strong explanatory power of the independent ESG and control variables in the case of models (1), (2), (3), and (6), with very high R^2 values (over 0.95), while for the models (4) and (5), R^2 perform significantly lower (0.382 and 0.553), indicating weaker explanatory power, suggesting that PTBV_S and BETA are influenced by other factors not captured in the current model specification.

Direct influences of the ESG-related credentials on the market indicator, namely market capitalisation (MK_C)—model (1)—of the companies from the energy sector reveal the following: higher ESG scores (ESG_COMB) are associated with

Table 2 Results of the Robust regression models (RREG) for the companies from the energy sector

	(1)	(2)	(3)	(4)	(5)	(6)
	MK_C	EBIT	EBITDA	PTBV_S	BETA	GROSS_ PROF
ESG_COMB	−0.0368** (0.0111)	−0.00263 (0.0119)	0.00193 (0.00739)	0.00191 (0.0155)	0.00416 (0.0118)	−0.00893 (0.0116)
ENV_MAT_ SRC	0.00279 (0.00242)	0.000620 (0.00223)	−0.000462 (0.00138)	0.00854* (0.00322)	−0.00505* (0.00246)	0.000849 (0.00228)
GRN_BLD	−0.00639* (0.00252)	−0.00251 (0.00237)	−0.00700*** (0.00144)	−0.00220 (0.00328)	−0.00226 (0.00256)	0.000655 (0.00231)
RES_RED_ POL	0.175* (0.0791)	0.0128 (0.0101)	0.00243 (0.00591)	0.0136 (0.0146)	0.0233 (0.0806)	0.0800 (0.0990)
EN_USE_REV	0.00643 (0.00624)	0.0184* (0.00751)	0.0100* (0.00457)	−0.0141 (0.00884)	−0.00541 (0.00639)	0.0243*** (0.00656)
CO2_REV	−0.00500 (0.00408)	−0.00488 (0.00414)	−0.00734** (0.00239)	0.00378 (0.00535)	−0.00282 (0.00415)	−0.00696 (0.00391)
ENV_PROD	−0.00157 (0.00289)	−0.00224 (0.00311)	0.00134 (0.00180)	−0.00440 (0.00391)	0.00546 (0.00294)	−0.00127 (0.00314)
TRAIN_DEV	−0.0321* (0.0135)	0.0527 (0.0919)	−0.00328 (0.0549)	−0.0905 (0.119)	0.0246 (0.0137)	−0.0164 (0.0949)
CSR_COMM	−0.00167 (0.00532)	−0.0000278 (0.00536)	0.00291 (0.00322)	−0.00394 (0.00670)	0.0120* (0.00547)	−0.0123* (0.00594)
SOC_PILL	0.0145 (0.00872)	0.00454 (0.00787)	−0.00793 (0.00471)	0.00984 (0.0111)	−0.0216* (0.00927)	0.0114 (0.00767)

(continued)

Table 2 (continued)

	(1)	(2)	(3)	(4)	(5)	(6)
	MK_C	EBIT	EBITDA	PTBV_S	BETA	GROSS_ PROF
GOV_PILL	0.00939 (0.00585)	−0.000345 (0.00570)	0.00481 (0.00353)	−0.000408 (0.00769)	0.0114 (0.00603)	0.00846 (0.00605)
ENV_PILL	0.0343** (0.00998)	−0.0136 (0.0101)	−0.00181 (0.00624)	−0.00208 (0.0140)	0.00836 (0.0102)	−0.00972 (0.0105)
CSR_STRAT	−0.000819 (0.00474)	0.0000112 (0.00484)	−0.00440 (0.00264)	0.00421 (0.00621)	−0.00704 (0.00482)	0.00311 (0.00430)
ENV_MGMT	0.00416 (0.00268)	−0.00333 (0.00264)	−0.00221 (0.00159)	−0.00300 (0.00369)	0.00316 (0.00278)	−0.000897 (0.00269)
BIODIV_RED	−0.0142*** (0.00312)	−0.000183 (0.00276)	0.00148 (0.00170)	0.000492 (0.00407)	0.00587 (0.00320)	−0.00266 (0.00279)
FLEX_H	−0.00685* (0.00261)	0.00807* (0.00307)	0.00383* (0.00184)	0.00172 (0.00360)	0.00371 (0.00269)	−0.00190 (0.00257)
WAGE_C	0.0159** (0.00490)	0.00383 (0.00487)	−0.00349 (0.00300)	−0.00147 (0.00658)	−0.00829 (0.00545)	−0.0000111 (0.00491)
CORP_ AWARD	−0.00157 (0.00243)	−0.00338 (0.00254)	0.00201 (0.00151)	−0.00494 (0.00315)	−0.000375 (0.00249)	0.000669 (0.00248)
BOARD_AT_ AVG	−0.0113*** (0.00286)	−0.00245 (0.00296)	−0.00372* (0.00180)	0.00222 (0.00374)	−0.00506 (0.00291)	0.00210 (0.00267)
BOARD_STR	−0.00373 (0.0115)	0.0169 (0.0105)	0.0101 (0.00643)	−0.00258 (0.0150)	0.000944 (0.0117)	0.131* (0.0506)
BOARD_ATT	−0.00231 (0.00275)	−0.00174 (0.00255)	0.000173 (0.00157)	−0.00395 (0.00365)	0.00403 (0.00285)	−0.00348 (0.00258)
BOARD_GEN_ DIV	−0.00799* (0.00336)	−0.00462 (0.00350)	−0.00917*** (0.00214)	−0.00263 (0.00487)	0.00420 (0.00342)	−0.00485 (0.00364)
ENERGY	−0.112 (0.0760)	0.320** (0.102)	0.0934 (0.0624)	−0.128 (0.109)	−0.133 (0.0773)	0.293*** (0.0770)
ASSETS	1.094*** (0.102)	0.745*** (0.134)	0.839*** (0.0780)	0.155 (0.150)	−0.0209 (0.105)	0.700*** (0.120)
_cons	−11.10* (5.043)	−6.073 (4.577)	0.724 (2.743)	4.250 (6.227)	0.606 (5.132)	−10.77 (7.385)
R^2	0.955	0.966	0.984	0.382	0.553	0.971

Note: "Standard errors in parentheses; * $p < 0.05$, ** $p < 0.01$, *** $p < 0.001$". Source: authors' research

lower market capitalization, potentially reflecting the cost burden or investor skepticism; environmental pillar score (ENV_PILL) indicates a positive relationship between environmental performance and market capitalization (0.0343**); Resource Reduction Policies (RES_RED_POL) suggests such policies are rewarded by the market (positive and statistically significant coefficient, 0.175*); there are an inverse association of the biodiversity impact reduction (BIODIV_RED) with market value (negative and highly important, −0.0142***); training and development of the employees (TRAIN_DEV) could reflect higher short-term costs, induced by inverse association with market capitalization (negative and significant, −0.0321*); positive influence of the wages working condition controversies (WAGE_C) with market capitalization could be explained by the fact that controversies around wages or working conditions may arise from efforts to reduce labor costs, which can improve profit margins in the short term; flexible working hours (FLEX_H) reflects that increased flexibility is associated with slightly lower capitalization, the market values being behind in valuing these benefits for employees; negative implications of the board meeting attendance (BOARD_AT_AVG) on market capitalization may reveal that boards that meet too often or become overly involved may slow down decision-making, low flexibility and reduce innovation; gender diversity in the board (BOARD_GEN_DIV) induced negative implications on market capitalization.

Moreover, the other used market dimensions, Price to Book Value per Share (PTBV_S)—model (4), stock's volatility compared to the overall market over 5 years Beta (BETA)—model (5), were favorable influenced by the CSR Sustainability committee (CSR_COMM), and unfavorable by the social pillar score (SOC_PILL), while Environmental Materials Sourcing Score (ENV_MAT_SRC) induced mixed implications, favorable in case of Price to Book Value per Share (PTBV_S), and unfavorable for the BETA value.

As regards profitability, revealed by EBIT—model (2), and EBITDA—model (3), and gross profit (GROSS_PROF)—model (6), this was positive influenced by the following ESG indicators: total energy use to revenues (EN_USE_REV) that indicated strong returns from energy efficiency; flexible working hours (FLEX_H), which induced higher profitability (measured by EBIT and EBITDA) of the companies from energy sectors; total use of energy (ENERGY) that reflected substantial positive implications on profitability of the energy companies. Opposite, negative impacts of the ESG-related indicators on profitability were induce by the following measures: green buildings (GRN_BLD), CO_2 emissions revenue (CO2_REV); board meeting attendance (BOARD_AT_AVG); gender diversity in the board (BOARD_GEN_DIV); and CSR Sustainability committee score (CSR_COMM), which imply dedicated measures to improve profitability of these companies.

As regards assets value (ASSETS), as a control variable, this dimension suggests that firm size drives profitability for companies in the energy sector.

Furthermore, the analysis highlighted that several ESG indicators significantly influenced the financial performance of companies in the energy sectors, although the direction of this impact varies. The environmental pillar (ENV_PILL) and energy efficiency relative to revenue (EN_USE_REV) were positively associated with financial performance. In contrast, factors such as biodiversity reduction (BIODIV_RED), green buildings (GRN_BLD), and gender board diversity (BOARD_GEN_DIV) had a negative impact. Board characteristics—including structure, composition, and diversity—also significantly influence financial performance. Firm size (measured by assets) and energy sector affiliation were consistent and positive factors of profitability and market value, indicating a correlation between operational capacity and financial resilience.

As a summary, we highlight the following:

- Several ESG indicators significantly affected the financial performance of the companies from energy sectors, but the direction of the impact varies: positive implications for the environmental pillar (ENV_PILL) and energy use to revenue (EN_USE_REV); and negative in case of Biodiversity Reduction (BIODIV_RED), green buildings (GRN_BLD) or board gender diversity (BOARD_GEN_DIV);
- Board characteristics—attendance, structure, diversity- also influence financial outcomes.
- Firm size (ASSETS) and sector (ENERGY) have consistently positive and significant effects on profitability and value.

4.2 *Results of GGM Models*

The results of the GGM (Gaussian graphical model) are shown in Figs. 2 and 3, which highlight strong connections between financial variables. In particular, MK_C (market capitalisation) is highly correlated with variables such as REV, EBIT, EBITDA, and BETA, emphasising the internal consistency of financial indicators. Moreover, we have identified strong interlinkages between MK_C and BETA, suggesting a direct dependence between the firm's market value and systemic risk (volatility of the stock relative to the market).

Regarding the relationship between ESG credentials and financial dimensions, we observed that ENV_PILL (Environmental Pillar Score), ENV_PROD (Environmental Products Score), and CO2_REV (Total CO2 Equivalent Emissions to Revenue) exhibit weak connections with financial indicators. The weaker connections between environmental indicators such as ENV_PILL (Environmental Pillar Score), ENV_PROD (Environmental Products Score), and CO2_REV (Total CO2 Equivalent Emissions to Revenue) and financial indicators can be attributed to the indirect, delayed, and often difficult-to-quantify nature of the impact of green activities on economic performance. Investments in sustainable products or in reducing emissions typically involve high initial costs, with the associated financial benefits—such as increased operational efficiency, consumer loyalty, or access to green finance—manifesting over the long term and being influenced by contextual factors such as industry, national regulations, and the quality of corporate governance. Consequently, the weak relationships observed in the GGM model reflect the complexity of the mechanisms through which environmental sustainability influences financial performance and the fact that other structural and strategic variables often mediate these effects.

GRN_BLD (Green Buildings Score) and BIODIV_RED (Biodiversity Impact Reduction Score) are weakly linked to financial indicators, with some connections being negative, indicating potential costs or adverse effects on financial performance. Governance and human resource factors, such as BOARD_SK (Board Backgorund and Skills Score), BOARD_GEN_DIV (Board Gender Diversity), BOARD_STRUCT (Board Structure Policy Score), WAGE_CONTRO (Wages &

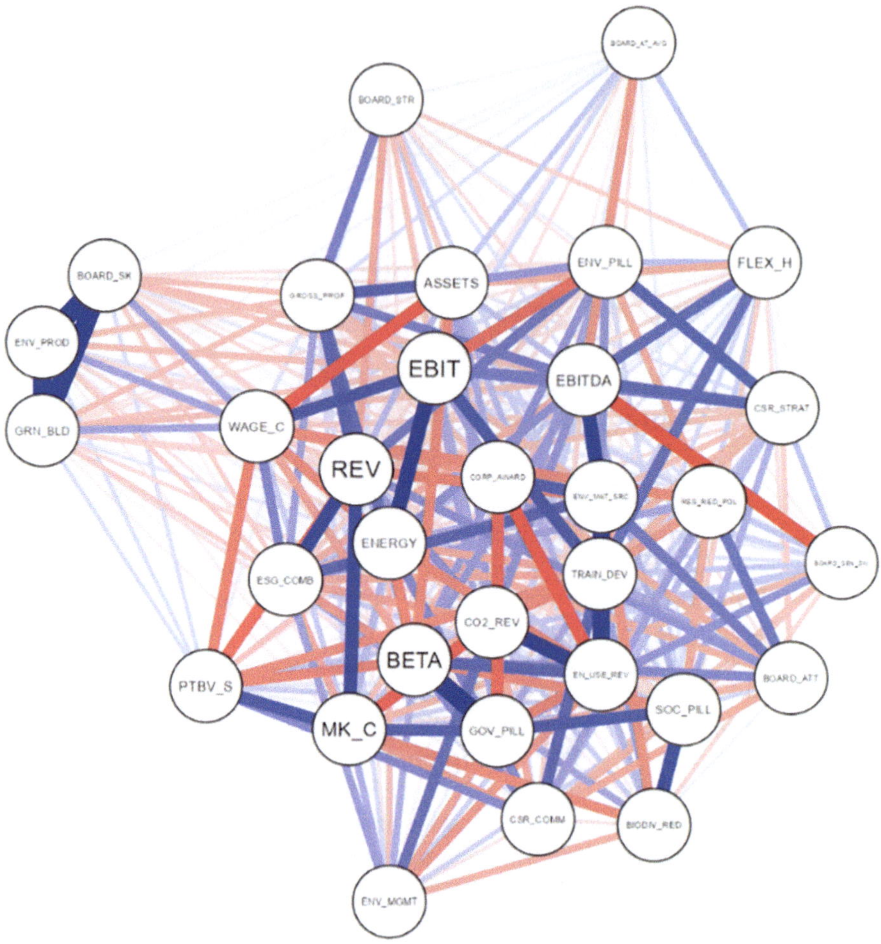

Fig. 2 Network structure of Gaussian Graphical Model (GGM) using force-directed layout. (Source: Authors' research in RStudio)

Working Conditions Controversies Score), and FLEX_HOURS (Flexible Working Hours Score), demonstrate numerous interconnections both among themselves and with other ESG indicators. Furthermore, BOARD_SK (Board Background and Skills Score) is associated with MK_C (Market Capitalisation), suggesting that board skills impact market capitalisation. Thus, a board of directors with relevant experience, strategic knowledge and expertise in key areas contributes to informed decision-making, reduces management risks and increases investor confidence (a higher market valuation, reflected in an increase in stock market capitalisation). In addition, a competent board can more effectively support the adaptation processes

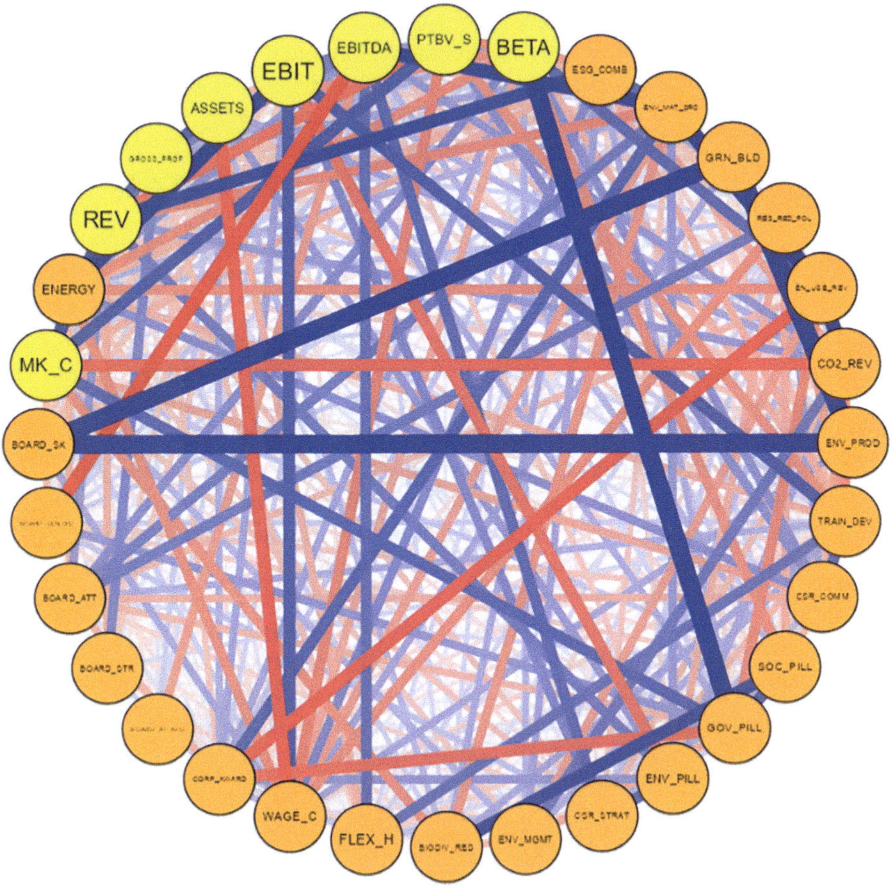

Fig. 3 Circular visualisation of the Gaussian Graphical Model (GGM) depicting variable interconnections. (Source: Authors' research in RStudio)

to market challenges, including those related to sustainability and climate change, strengthening the company's strategic position and attracting long-term capital.

5 Conclusions

The results of the bibliometric analysis highlight that the three clusters illustrate the complexity and interdependence of the central themes (climate change & corporate finance) in the corporate social responsibility (CSR) field. The first cluster emphasises the essential role of management in integrating the principles of governance, responsible investment, and cost control—fundamental aspects for the effective implementation of CSR. The second cluster highlights the profound connection between social responsibility, environmental performance, and sustainability, indicating that a commitment to reducing carbon emissions and enhancing overall performance is vital for sustainable corporate strategies. Finally, the third cluster underscores the relationship between climate change and financial performance, emphasising the need for organisations to adapt their practices to address climate risks and capitalise on the economic opportunities associated with the transition to sustainability. Furthermore, these results provide an integrated perspective on how economic, social, and environmental elements converge within the framework of corporate responsibility practices.

Moreover, the results indicate that integrating risk management into sustainability strategies is not merely a tool for compliance with environmental regulations but also a means of ensuring financial resilience and long-term competitiveness. The main results reveal how climate change influences corporate finance through various dynamics, from risk management to strategic adaptation, highlighting the central role of management in implementing corporate social responsibility (CSR), governance, and cost control; all these factors are fundamental for an effective response to climate risks and for strengthening organisational resilience. Furthermore, the results underscore the strong link between social responsibility, sustainability, and environmental performance, demonstrating that reducing carbon emissions and aligning with climate objectives have become essential for maintaining corporate performance. Additionally, based on our findings, we observed that integrating climate risks into financial decisions is crucial for long-term economic sustainability. The findings emphasise the need for a strategic framework that enables companies to comply with environmental regulations and capitalise on the financial opportunities associated with the transition to a green economy.

The empirical data presented unconventional statistical features (asymmetry, extreme values), which justified using robust regressions to ensure the accuracy and consistency of ESG–financial performance relationship estimates. The descriptive results highlighted a deeply uneven ESG integration at the level of the companies analysed, which suggests a need for standardisation of reporting practices and caution in interpreting the impact of ESG on financial performance. The analysis reveals areas of excellence and absence in applying ESG principles, especially concerning environmental policies, which underlines the need for targeted and differentiated interventions on specific ESG components.

The robust regression model (RREG) results highlight complex relationships between organisational factors, financial performance, and environmental sustainability of energy sector companies, in the context of the transition to sustainable practices. Developed human resources and attractive research systems positively impacted environmental sustainability, but negatively influenced the level of investments, while information technologies supported investments but adversely affected the environmental component. Regarding ESG indicators, the environmental pillar (ENV_PILL) and energy efficiency relative to revenues (EN_USE_REV) positively impacted financial performance, unlike factors such as biodiversity reduction (BIODIV_RED), green buildings (GRN_BLD), or gender diversity on the board of directors (BOARD_GEN_DIV), which were negatively correlated. Simultaneously, the structural characteristics of the board of directors, company size and membership in the energy sector were found to have a significant and consistently positive effect on profitability and firm value. These findings highlight the importance of a strategic balance between environmental, social, and governance objectives and companies' financial priorities in the context of the challenges posed by climate change.

Nevertheless, our results highlight the novelty of our study, suggesting the need to i) balance investment policies with environmental objectives. It also suggests that specific sustainability-oriented initiatives—such as strengthening human capital and research—can temporarily reduce firms' direct investments. Therefore, public policies should create support mechanisms (grants, green credits, tax deductibility) that compensate for these effects and encourage firms to integrate environmental objectives without compromising economic performance; ii) regulate digital technologies within sustainability. Given that digitalisation is associated with increased investment but can negatively affect environmental sustainability, it is necessary to develop policies that promote "green technologies" and the energy-efficient use of IT infrastructure (e.g. low-emission cloud computing, energy-efficient data centres); iii) clarify and standardise the ESG framework. The uneven results of different ESG indicators on financial performance suggest the need for a standardised and better-calibrated ESG reporting framework. European policies should aim to harmonise indicators to reduce uncertainties about the impact of these factors on the value of companies.

The GGM model highlights that within the companies analysed, there are complex and conditional relationships among ESG variables, governance, HR, and financial performance. Financial indicators demonstrate strong links between these elements, which confirms the coherence of economic data; however, the connections between ESG and financial indicators are weaker and more heterogeneous. This suggests that the impact of ESG on financial performance is nuanced, potentially indirect, and significantly influenced by the internal context of the company (e.g. board structure, innovation, sector). The model thus emphasises the necessity for multidimensional analyses and the integration of ESG factors into companies' strategic and financial assessment frameworks.

Corporate finance must consider both accurate risk management and proactive strategic responses when evaluating climate change as a crucial determinant that continues to pose various complex economic and financial challenges. Furthermore, our main conclusions offer fresh insights for companies regarding the intricate development of a strategic response and risk management, considering the multiple effects resulting from the repercussions of climate change.

Broadly, our study also includes several recommendations. Firstly, integrating sustainability into corporate financial strategy means that companies should treat sustainability not merely as a compliance objective but as an integral part of their economic strategy, incorporating environmental and governance risks and opportunities into decision-making models. Secondly, investing in human capital can have a positive long-term financial impact. Although short-term investments in human resources may temporarily divert funds from other projects, they contribute to innovation, adaptability, and sustainable performance in the long run. It is recommended that frameworks be developed to assess the social and environmental return on investments in HR and research. Furthermore, diversifying the boards of directors with a strategic approach is essential. Given the mixed results regarding gender diversity on boards, it is advised that internal company policies and external regulations should aim not only at formal representation but also at integrating diversity into decision-making and strategic processes to maximise benefits on financial performance. Ultimately, we should encourage large companies to act as transition leaders. As firm size is associated with increased profitability, large companies can be motivated through "lead-by-example" policies to adopt advanced sustainability models and support their integration into wider value chains.

A future research direction aims to deepen and expand the methodological framework by identifying additional relevant indicators that can enhance the accuracy and robustness of the analyses. In this regard, it is essential to develop more precise measurement criteria and to establish new analytical procedures capable of capturing the nuances of the relationship between climate change and corporate finance. By continuously updating the database and adapting the dimensions investigated, our research will dynamically integrate new areas of interest, particularly those related to environmental performance and technological innovation, to more accurately reflect the complexity of the current economic and climate landscape. Concurrently, another vital direction for the project's development involves broadening the analysis internationally, encompassing a larger number of companies and a longer time frame. This approach will facilitate the examination of structural and regional variations in firms' responses to climate risks, contributing to the generalisation of conclusions regarding the relationship between climate change and corporate finance within the framework of strategic risk management and response. Through this expansion, the study will provide a deeper and more robust statistical perspective on the interconnection of climate policies, ESG strategies, and financial decisions within the global economic context.

Our main limitations stem from the limited availability of relevant data concerning certain analysed variables. We are also aware of various other factors that could potentially influence our results, but are not included in the analysis. Future research will focus on incorporating ESG credentials, considering their bidirectional implications for firms and the impact of corporate governance quality on the financial performance of the companies considered.

Acknowledgments This work was supported by a grant from the Romanian Ministry of Research, Innovation and Digitalization, the project with the title "*Economics and Policy Options for Climate Change Risk and Global Environmental Governance*" (CF 193/28.11.2022, Funding Contract no. 760078/23.05.2023), within Romania's National Recovery and Resilience Plan (PNRR)—Pillar

III, Component C9, Investment I8 (PNRR/2022/C9/MCID/I8)—Development of a program to attract highly specialised human resources from abroad in research, development, and innovation activities.

Competing Interests The authors have no conflicts of interest to declare relevant to this chapter's content.

Ethics Approval No ethics approval was required.

Appendix 1

Summary of Network		
Number of nodes	Number of non-zero edges	Sparsity
32	496/496	0.000

Note. Data was binarised using the median

Centrality measures per variable				
	Network			
Variable	Betweenness	Closeness	Strength	Expected influence
PTBV_S	0.075	0.255	−0.642	−1.304
BETA	1.040	1.227	1.299	−0.504
ESG_COMB	−0.269	0.885	0.955	−0.061
ENV_MAT_SRC	1.661	1.541	0.872	0.475
GRN_BLD	−0.752	−1.688	−0.749	−0.345
RES_RED_POL	−0.476	−0.413	−0.096	0.256
EN_USE_REV	0.144	0.669	0.586	−0.036
CO2_REV	0.144	0.729	0.270	−1.065
ENV_PROD	−0.752	−1.688	−0.749	−0.345
TRAIN_DEV	0.144	0.618	1.221	−0.604
CSR_COMM	−0.752	−0.543	−0.368	−0.422
SOC_PILL	−0.062	−0.437	0.141	0.300
GOV_PILL	0.282	0.427	0.598	0.730
ENV_PILL	−0.545	0.031	−0.222	0.884
CSR_STRAT	−0.407	−0.140	−0.414	−0.516
ENV_MGMT	−0.752	−1.093	−1.345	0.052
BIODIV_RED	−0.614	−0.612	−0.721	0.205
FLEX_H	−0.614	−0.530	−1.065	0.184
WAGE_C	1.799	0.835	0.461	−2.110
CORP_AWARD	−0.683	0.186	0.428	0.462
BOARD_AT_AVG	−0.752	−1.935	−2.480	−1.244
BOARD_STR	−0.752	−1.739	−1.895	−0.836
BOARD_ATT	−0.407	−0.396	−0.692	−0.861

Centrality measures per variable				
	Network			
Variable	Betweenness	Closeness	Strength	Expected influence
BOARD_GEN_DIV	−0.752	−0.366	−1.088	−0.661
BOARD_SK	−0.752	−1.688	−0.749	−0.345
MK_C	0.006	0.417	0.889	0.341
ENERGY	−0.200	1.015	1.093	−0.884
REV	0.971	0.766	1.113	1.284
GROSS_PROF	0.006	0.532	0.414	2.515
ASSETS	−0.200	0.526	0.175	0.648
EBIT	0.213	1.090	0.939	1.490
EBITDA	4.004	1.521	1.824	2.317

Clustering measures per variable				
	Network			
Variable	Barrat[a]	Onnela	WS[a]	Zhang
PTBV_S	0.000	−0.777	0.000	−0.028
BETA	0.000	0.807	0.000	−0.171
CSR_COMM	0.000	−0.252	0.000	0.866
SOC_PILL	0.000	0.460	0.000	−1.011
GOV_PILL	0.000	0.457	0.000	−0.272
ENV_PILL	0.000	−0.273	0.000	−0.325
CSR_STRAT	0.000	−0.447	0.000	0.642
ENV_MGMT	0.000	−1.586	0.000	2.594
BIODIV_RED	0.000	−0.544	0.000	−0.104
FLEX_H	0.000	−0.799	0.000	−0.334
WAGE_C	0.000	0.105	0.000	−0.410
CORP_AWARD	0.000	0.777	0.000	0.149
ESG_COMB	0.000	1.105	0.000	−0.033
BOARD_AT_AVG	0.000	−2.399	0.000	0.976
BOARD_STR	0.000	−1.655	0.000	0.406
BOARD_ATT	0.000	−0.796	0.000	0.738
BOARD_GEN_DIV	0.000	−0.912	0.000	1.141
BOARD_SK	0.000	−0.942	0.000	1.426
MK_C	0.000	1.025	0.000	−0.828
ENERGY	0.000	1.284	0.000	0.104
ENV_MAT_SRC	0.000	0.567	0.000	0.073
REV	0.000	1.138	0.000	−1.533

Clustering measures per variable				
	Network			
Variable	Barrat[a]	Onnela	WS[a]	Zhang
EBIT	0.000	0.940	0.000	−0.517
EBITDA	0.000	1.878	0.000	−1.941
GROSS_PROF	0.000	0.594	0.000	−0.583
ASSETS	0.000	0.273	0.000	−1.092
GRN_BLD	0.000	−0.942	0.000	1.426
RES_RED_POL	0.000	0.212	0.000	−1.415
EN_USE_REV	0.000	0.431	0.000	0.178
CO2_REV	0.000	−0.084	0.000	−0.205
ENV_PROD	0.000	−0.942	0.000	1.426
TRAIN_DEV	0.000	1.298	0.000	−1.342

[a]Coefficient could not be standardised because the variance is too small

Appendix 2

Centrality Plot

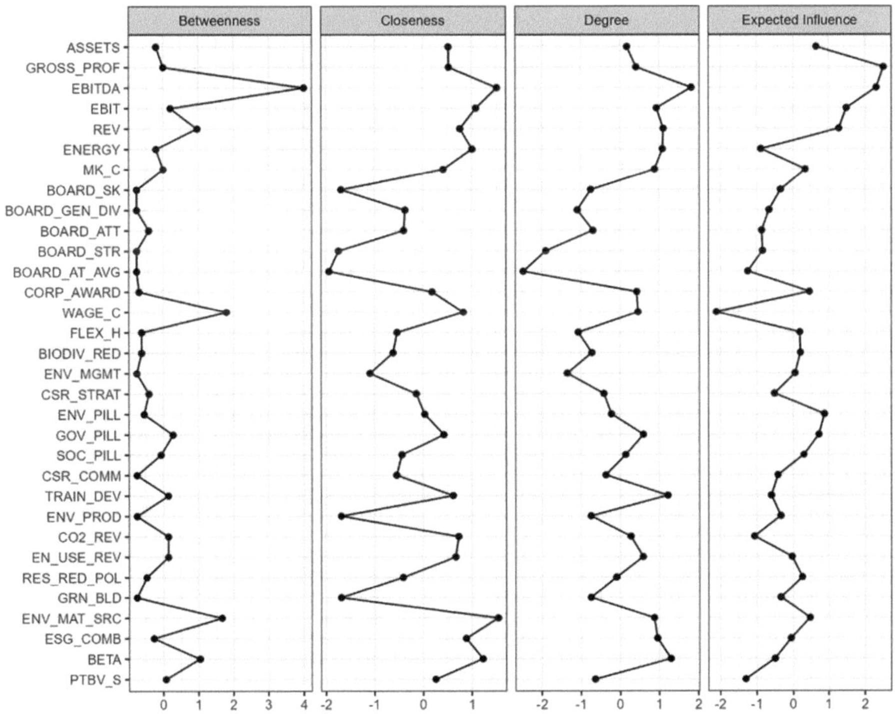

Appendix 3

Clustering Plot

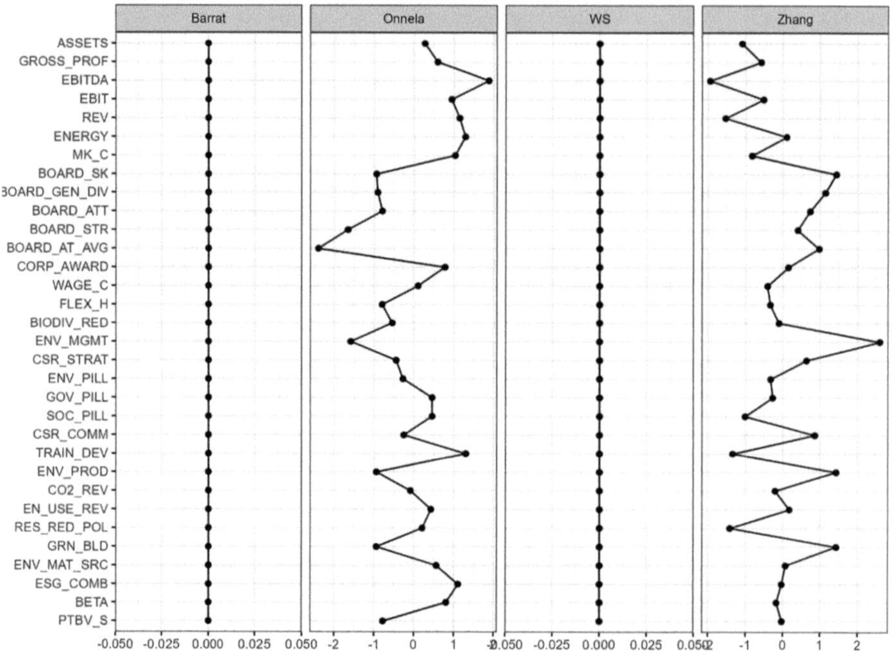

References

Afzali, M., Colak, G., & Vähämaa, S. (2025). Climate change denial and corporate environmental responsibility. *Journal of Business Ethics, 196*, 31–59.

Aibar-Guzmán, B., Raimo, N., Vitolla, F., & García-Sánchez, I.-M. (2024). Corporate governance and financial performance: Reframing their relationship in the context of climate change. *Corporate Social Responsibility and Environmental Management, 31*, 1493–1509.

Ashrafee, T. H., & Abdullah-Al, M. (2022). Does corporate social responsibility help mitigate firm-level climate change risk? *Finance Research Letters, 47*, 102791. https://doi.org/10.1016/j.frl.2022.102791

Dai, Z., & Zhang, X. (2023). Climate policy uncertainty and risks taken by the bank: Evidence from China. *International Review of Financial Analysis, 87*, 102579.

Gallego-Álvarez, I., García-Sánchez, I., & Da Silva Vieira, C. (2014). Climate change and financial performance in times of crisis. *Business Strategy and the Environment, 23*, 361–374.

Gilley, K., Worrell, D., & El-Jelly, A. (2000). Corporate environmental initiatives and anticipated firm performance: The differential effects of process-driven versus product-driven greening initiatives. *Journal of Management, 26*(1), 199–216.

King, A., & Lenox, M. (2002). Exploring the locus of profitable pollution reduction. *Management Science, 48*, 289–299.

Li, C., Tang, W., Liang, F., & Wang, Z. (2024). The impact of climate change on corporate ESG performance: The role of resource misallocation in enterprises. *Journal of Cleaner Production, 445*, 141263.

Link, S., & Naveh, E. (2006). Standardization and discretion: Does the environmental standard ISO 14001 lead to performance benefits? *IEEE Transactions on Engineering Management, 53*, 508–519.

Liu, L., Beirne, J., Azhgaliyeva, D., & Rahut, D. (2024). Climate change and corporate financial performance. *Journal of Risk and Financial Management, 17*(7), 267.

Ma, Q., Zhou, Y., & Wang, J. (2024). The impact of climate change on credit risk of rural financial institutions: A threshold effect based on agricultural insurance. *The North American Journal of Economics and Finance, 71*, 102086.

Margolis, J. D., & Walsh, J. P. (2003). Misery loves companies: Rethinking social initiatives by business. *Administrative Science Quarterly, 48*, 268–305.

Megeid, N. S. A. (2024). The impact of climate risk disclosure on financial performance, financial reporting and risk management: Evidence from Egypt. *Future Business Journal, 10*(1), 21.

Melnyk, S., Sroufe, R., & Calantone, R. (2003). Assessing the impact of environmental management systems on corporate and environmental performance. *Journal of Operations Management, 21*(3), 329–351.

Molina-Azorín, J., Claver-Cortés, E., López-Gamero, M., & Tari, J. (2009). Green management and financial performance: A literature review. *Management Decision, 47*(7), 1080–1100.

Nguyen, Q. K. (2025). Green finance, climate risk and financial stability: Evidence from ASEAN+4 countries. *Environmental and Sustainability Indicators, 28*, 100922. https://doi.org/10.1016/j.indic.2025.100922

Pankratz, N., Bauer, R., & Derwall, J. (2023). Climate change, firm performance, and investor surprises. *Management Science, 69*(12), 7352–7398.

Pirtea, M.-G., Noja, G. G., Moldovan, N.-C., Grecu, I.-M., & Ţăran, A.-M. (2024). Managing climate change risks and environmental challenges towards sustainable development within the European Union. In S. Grima, I. Romānova, G. G. Noja, & T. Dorożyński (Eds.), *Economic development and resilience by EU member states* (Contemporary studies in economic and financial analysis) (Vol. 115, pp. 159–176). Emerald Publishing Limited. https://doi.org/10.1108/S1569-375920240000115010

Porter, M. E., & Van der Linde, C. (1995). Green and competitive: Ending the stalemate. *Harvard Business Review, 73*(5), 120–134.

Rumansyah, A. M. A., & Nainggolan, Y. A. (2025). The influenced of ESG environmental performance on financial performance: A study of Indonesian publicly listed companies. *Journal of Accounting, Finance and Management, 5*(6), 1894–1903. https://doi.org/10.38035/jafm.v5i6

Stankevičienė, J., Noja, G. G., Pirtea, M.-G., Ţăran, A.-M., Gînguţă, A., & Giovanis, A. (2025). Technological innovations for sustainable business success: Enhancing processes and performance. *Business Process Management Journal*. https://doi.org/10.1108/BPMJ-07-2024-0667

Ţăran, A.-M., Gînguţă, A., Nachescu, M.-L., Pirtea, M.-G., Tarczynska-Luniewska, M., & Noja, G. G. (2025). Renewable energy, climate change, and sustainable development nexus in an era of globalisation. The role of global governance. *Transformations in Business & Economics, 24*(1), 218.

Wang, Y., Wu, Z., & Zhang, G. (2022). Firms and climate change: A review of carbon risk in corporate finance. *Carbon Neutrality, 1*(1), 6.

Ziegler, A., Busch, T., & Hoffmann, V. H. (2011). Disclosed corporate responses to climate change and stock performance: An international empirical analysis. *Energy Economics, 33*, 1283–1294.

Regulatory Landscape in Green Finance: Global Standards and Local Challenges

Anastasia Doraş Lisnic, Diana Balan, and Ana-Cristina Nicolescu

1 Introduction

In the wake of the global climate crisis and the increasing demand for financial systems to support the green transition, the relevance of green finance regulations is on the rise. These regulations not only facilitate the efficient allocation of capital to sustainable projects but also set clear criteria for transparency, accountability, and the prevention of greenwashing. However, the regulatory landscape is marked by significant diversity in terms of instruments, institutional actors, and the maturity of the regulatory framework, especially in emerging economies. In this context, the purpose of this chapter is to provide a rigorous analysis of how global standards are defined, transposed, and applied, as well as the regional challenges that influence their effective implementation.

The purpose of this chapter is to conduct a critical analysis of the regulatory framework in the field of green finance, with a focus on the global convergence trend in the definition and application of standards, as well as on the regional particularities and challenges that influence their implementation. Using a methodological approach based on semantic and institutional network analysis, the chapter seeks to identify dominant regulatory models, differences between developed and emerging economies, and factors influencing the coherence, effectiveness, and

A. Doraş Lisnic (✉) · D. Balan
Doctoral School of Economics and Business Administration, Faculty of Economics and Business Administration, West University of Timişoara, Timişoara, Timiş County, Romania
e-mail: anastasia.doras01@e-uvt.ro; diana.balan01@e-uvt.ro

A.-C. Nicolescu
Faculty of Economics and Business Administration, Department of Finance, Business Information Systems and Modelling, West University of Timişoara,
Timişoara, Timiş County, Romania
e-mail: cristina.nicolescu@e-uvt.ro

C.-W. Su et al. (eds.), *Green Finance and the Challenges of Climate Change*, Sustainable Finance, https://doi.org/10.1007/978-3-032-07293-1_8

adaptability of national sustainable finance policies. Green finance, now at the forefront of international economic agendas, offers hope in the face of the growing need to combat climate change and advance sustainable development. It necessitates a comprehensive understanding of its regulatory environment (Andreeva et al., 2018). Green finance, recognised as an effective tool for reducing carbon emissions and promoting environmental sustainability, has expanded rapidly and garnered significant interest worldwide (Liu & Zhu, 2024). Diverse actors and differing environmental priorities across various economies have contributed to the growth of green finance activities, resulting in a complex web of regulatory standards (Nedopil et al., 2021). To meet international climate goals on time, the private sector must make a significant contribution to the global shift toward a low-carbon economy (Haas & Popov, 2019).

For green finance to become an effective tool in the transition to a sustainable economy, a coherent regulatory framework and reliable institutional mechanisms must be established. Practically, it involves clear taxonomies, transparent reporting policies, and rigorous verification procedures to provide confidence to investors and strengthen the viability of projects with a positive environmental impact. These tools will shape a functional system for classifying activities as "green", ensure that economic actors are accountable for their declared practices, and create robust structures for validating compliance with sustainability criteria. At the same time, the role of digital innovation in providing transparency and efficiency in these processes cannot be overstated. Harmonising standards and regulations at the international level is crucial for facilitating cross-border investment and preventing opportunistic behaviour that arises from differences between national regulatory systems.

The originality of this chapter lies in its dual approach—semantic and institutional—applied to an extensive corpus of national green finance regulations, using advanced analytical tools such as InfraNodus. Unlike previous studies that analyse regulations from a descriptive or legal-normative perspective only, this research introduces a structural and relational dimension to policy discourse, identifying not only the formal content but also the networks of influence and governance logic that shape sustainability standards.

This research adds value by providing a comparative mapping between developed and emerging economies, highlighting not only semantic divergences but also institutional imbalances that affect the effective implementation of international standards. This is the first time that an analysis of this kind has been applied to such a large and diverse sample, offering a critical perspective on how global convergence can coexist with local fragmentation. In addition, the results provide a methodological framework that can be replicated in other areas of public policy with a transnational dimension.

This chapter examines the tension between global regulatory convergence and the diversity of local implementation. Methodologically, the chapter employs network analysis using InfraNodus, a visual and analytical tool designed to capture complex semantic and institutional interconnections within policy documents. Based on over 100 national frameworks from 32 economies, the analysis identifies key regulatory clusters—such as sustainable finance, green guidelines, and climate innovation—and reveals how different terms and issuing institutions combine into dominant narratives and governance models. This approach facilitates a deeper understanding of the formal content of regulations and the underlying structure of influence and discourse that shape green finance policy. This chapter maps the regulatory landscape of green finance across multiple dimensions—normative, semantic, and institutional—to provide a nuanced understanding of the global standards shaping sustainable finance and the local challenges that continue to influence its adoption and effectiveness.

Particular attention is paid to asymmetry, fragmentation, and coherence trends in the institutional and semantic architecture of regulatory frameworks in developed and emerging economies. The chapter also examines the enabling elements, such as institutional trust, digital innovation, and policy coherence, as well as the obstacles to effective regulation, including greenwashing, regulatory complexity, and capacity constraints.

This chapter is structured into three main sections, preceded by an introduction. The first part delves into the literature, highlighting recent theoretical and empirical contributions to the regulation of green finance. The second section outlines the methodology employed and presents the results obtained through the analysis of semantic and institutional networks, providing a comparative mapping of the regulatory frameworks in developed and emerging economies. The final part synthesises the study's conclusions, summarising the main findings and highlighting both the challenges and opportunities associated with the regulatory landscape of sustainable finance. It underscores the crucial and urgent need for global harmonisation of standards and their adaptation to national institutional specificities to effectively transform green policies from mere rhetoric into powerful instruments for the green transition.

2 Green Finance Regulation: Global Convergence and Local Discontinuities

Green finance is vital in supporting sustainable development and addressing environmental issues. Its growing importance is supported by several international initiatives and strategic frameworks that are shaping financial systems towards environmental and social objectives. Among the most influential are the Sustainable Development Goals (SDGs) and the Paris Agreement, which have significantly contributed to public policy reforms and the redirection of investment flows towards climate change mitigation and adaptation measures. At the same time, financial instruments such as green bonds, green loans, and impact investing have become essential channels for financing sustainable projects. These are complemented by the development of regulatory frameworks, including green taxonomies and ESG (environmental, social, and governance) standards, designed to bring transparency and consistency to the assessment of environmentally impactful investments. At the regional and national levels, initiatives such as China's Belt and Road Initiative and green financing supported by Awqaf in countries like Indonesia, Malaysia, and Turkey reflect the adaptation of global principles to local cultural and institutional contexts. In addition, the integration of financial technologies (FinTech) and clean technologies (CleanTech) is helping to expand and streamline green financing instruments, highlighting the role of digital innovation in the green transition.

However, the widespread adoption of green financing still faces many challenges. These include the fragmentation of the regulatory framework, uneven institutional capacities, and the risks of greenwashing. Overcoming these obstacles requires not only closer international cooperation but also a deeper understanding of how global standards are translated into local practices.

Against this backdrop, this review summarises the main recent academic contributions on the regulatory, institutional, and operational dimensions of green finance. Based on a systematic analysis of studies indexed in the Web of Science database for the period 2017–2025, it examines how global standards are developed and implemented, the roles of international and national actors, and the policy architectures shaping the green transition in both developed and emerging economies. Particular attention is paid to regulatory convergence mechanisms, institutional fragmentation, and the political and economic conditions that influence the effective adoption of sustainable financial practices.

The framework for regulating green finance is harmonising definitions, improving transparency, and encouraging market growth. As Zain et al. (2024) argue, developing green finance regulations, including taxonomies, product standards, disclosure rules, and labelling mechanisms, is essential to manage climate and Environmental, Social, and Governance (ESG) risks. Furthermore, international agreements and frameworks, presented by Baştürk (2024) and Jenei et al. (2024), such as the Paris Agreement (2015), the European Green Deal (2019), the Green Bond Principles (GBP) issued by the International Capital Markets Association (ICMA), alongside with initiatives at European Union (EU) level (EU Taxonomy, the Corporate Sustainability Reporting Directive (CSRD), the Sustainable Finance Disclosure Regulation (SFDR) and the Union Directive on Corporate Sustainability Due Diligence (CSDD)), provide a strategic foundation for mobilising capital toward green transition.

According to Bhandary et al. (2021), institutions such as the United Nations (UN), the International Monetary Fund (IMF), the World Bank, and the G20 exert political and economic pressure on member states to adopt green finance principles. Park (2019) and Ikevuje et al. (2024) emphasise the importance of integrating ESG factors into sustainable development mechanisms, particularly through financial instruments such as green bonds and green credits (Hafner et al., 2020), which serve as crucial tools for financing environmentally beneficial projects.

The European Commission aims to make Europe the first climate-neutral continent by 2050. This ambition is reflected in its adoption, in July 2021, of the European Green Deal (EGD), which seeks to reallocate financial flows towards sustainable projects and impose mandatory sustainability reporting standards. Tagliaferri (2022) notes that this initiative is expected to yield a cleaner environment, more affordable energy, smarter transportation, new job opportunities, and an improved quality of life. The EU's regulatory frameworks and the UN's Sustainable Development Goals (SDGs) both strive to mobilise resources and investment towards initiatives that have socially and environmentally beneficial outcomes. These frameworks share the objectives of addressing global issues, promoting sustainable development, and integrating environmental, social, and governance (ESG) considerations into financial decision-making. Jenei et al. (2024) analysed the alignment between the SDGs and EU regulations, highlighting their mutual objectives, which are schematised in Fig. 1.

A systematic review was conducted to ensure a comprehensive understanding of global standards and local challenges in the green finance regulatory framework. We conducted a more detailed review using a systematic analysis, emphasising the most revolutionary and recent scientific studies. The review applied Boolean search operators and wildcard characters to the Web of Science—Core Collection (WOS) database, we used the following keywords: ("regulat*" or "compliance" or "polic*") and ("green finance" or "sustainable finance" or "environmental finance" or "eco-finance") and ("landscape" or "framework" or "environment" or "context") and ("investment" or "fund*" or "capital" or "finance") and ("risk" or "assessment" or "challenge*" or "management" or "evaluation"). The review covered only articles in English published between 2017 and 2025. Selection criteria included relevance to the regulatory dimension of green finance, considered through the title and abstract (Fig. 2).

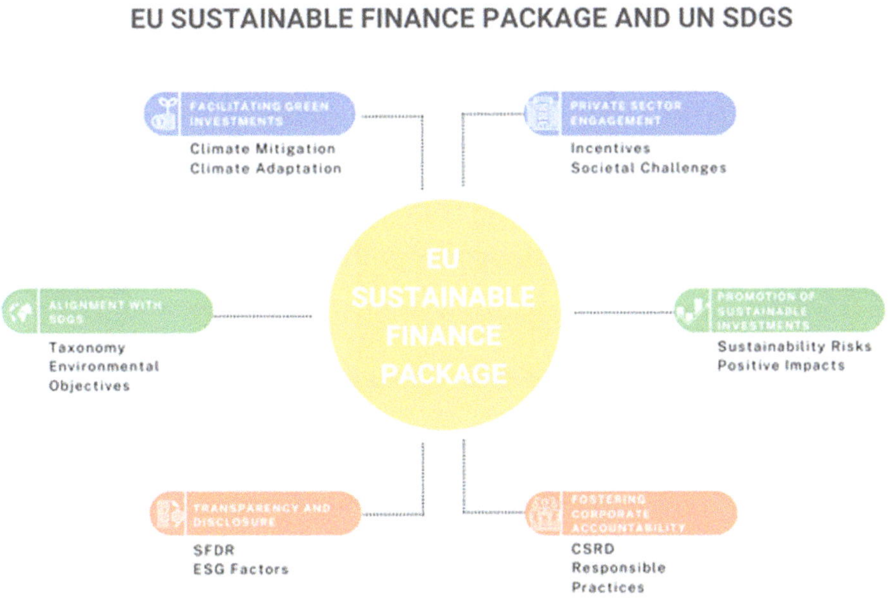

Fig. 1 The interconnections between the SDGs and the European Union regulatory frameworks. (Source: Authors' processing based on Jenei et al. (2024), via Canva)

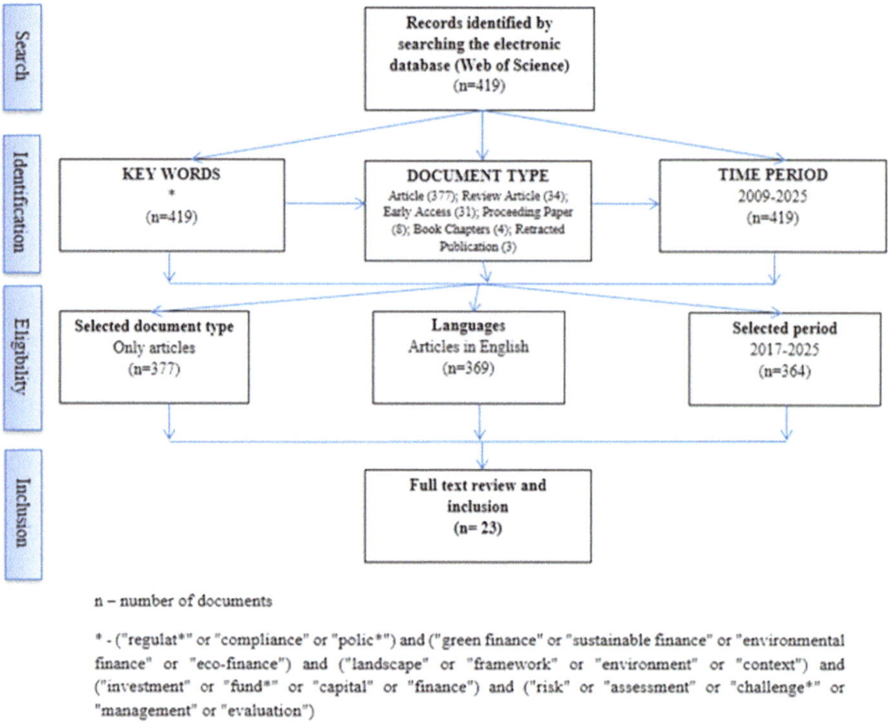

Fig. 2 The systematic selection strategy of the studies highlights the regulatory framework in green finance, including global standards and local challenges. (Source: Authors' processing)

Table 1 reports relevant articles on the regulatory framework in green finance.

Nupieri (2024) highlights the EU's evolving sustainable finance regulations since 2018, as part of the European Commission's efforts to align the financial system with global sustainability goals. Moreover, Ahlström and Sjåfjell (2023) and Bengo et al. (2022) argue that the EU's SFDR (2019/2088) and the Taxonomy Regulation (2020/852) establish transparent reporting standards and a classification system for sustainable economic activities. Furthermore, Streimikiene et al. (2024) and Chiu (2022) emphasise that these efforts aim to prevent greenwashing and promote convergence in the global ESG framework, while Kemfert and Schmalz (2019) highlight the role of the Paris Agreement in triggering the establishment of these international standards. Moreover, Hu et al. (2022) and Zhao et al. (2022) argue that environmental regulation can incentivise innovation and sustainable development if supported by a coherent and predictable framework. Still, as Nedopil et al. (2021) emphasised, inconsistency in international standards continues to create uncertainty and affect the attraction of green capital.

At the international level, organisations such as the Task Force on Climate-related Financial Disclosures (TCFD) and the International Platform for Sustainable Finance (IPSF) are key actors trying to harmonise regulatory efforts. Still,

Table 1 Relevant articles on the research about the regulatory framework in green finance: global standards and local challenges

Title	Authors	Journal	Year
Bridging funding gaps for climate and sustainable development: Pitfalls, progress, and potential of private finance	Clark, R., Reed, J., and Sunderland, T.	*Land Use Policy*	2018
Sustainable finance: political challenges of development and implementation of framework conditions	Kemfert, C., and Schmalz, S.	*Green Finance*	2019
EU financial regulations and social impact measurement practices: A comprehensive framework on finance for sustainable development	Bengo, I., Boni, L., and Sancino, A.	*Corporate Social Responsibility And Environmental Management*	2022
Evolution of green finance and its enablers: A bibliometric analysis	Bhatnagar, S., and Sharma, D.	*Renewable & Sustainable Energy Reviews*	2022
A Study On Progress And Challenges Of Green Finance In Madurai District	Iswaryalakshmi, P.	*Journal of Pharmaceutical Negative Results*	2022
Do climate technology, financialisation, and sustainable finance impede environmental challenges? Evidence from G10 economies	Li, Q.T., Sharif, A., Razzaq, A., and Yu, Y.Y.	*Technological Forecasting and Social Change*	2022
Development of renewable energy resources by green finance, volatility and risk: Empirical evidence from China	Lin, C.Y., Chau, K.Y., Tran, T.K., Sadiq, M., Van, L., and Phan, T.T.H.	*Renewable Energy*	2022
Nexus between green financing, economic risk, political risk and environment: evidence from China	Ning, Q.Q., Guo, S.L., and Chang, X.C.	*Economic Research-Ekonomska Istrazivanja*	2022
The involvement of national development banks in promoting sustainable finance	Nyikos, G., and Kondor, Z.	*Deturope-The Central European Journal Of Regional Development And Tourism*	2022
Does green finance counteract the climate change mitigation: Asymmetric effect of renewable energy investment and R&D	Zhang, D., Mohsin, M., and Taghizadeh-Hesary, F.	*Energy Economics*	2022
Why policy coherence in the European Union matters for global sustainability	Ahlström, H., and Sjåfjell, B.	*Environmental Policy and Governance*	2023

(continued)

Table 1 (continued)

Title	Authors	Journal	Year
Role of green finance in greening the economy: conceptual approach	Alekneviciene, V., and Bendoraityte, A.	*Central European Business Review*	2023
Studying the green economic growth with clean energy and green finance: The role of financial policy	Liu, X.Y., Zhao, T., and Li, R.	*Renewable Energy*	2023
The role of green finance in mitigating environmental degradation: Empirical evidence and policy implications from complex economies	Numan, U., Ma, B.J., Sadiq, M., Bedru, H.D., and Jiang, C.	*Journal of Cleaner Production*	2023
Integrating Climate Change Risks and Sustainability Goals into Saudi Arabia's Financial Regulation: Pathways to Green Finance	Alhejaili, M.O.	*Sustainability*	2024
Can green finance policy improve public health?	Feng, T.C., and Zhou, Q.	*Global Public Health*	2024
Green horizons: Enabling the energy transition through climate change policies	Guesmi, K., Mohammed, K.S., and Tiwari, S.	*International Review of Economics & Finance*	2024
Quantile modeling for environmental risk: SAARC's journey with green finance, policies, and regulations	Kaewsaeng-on, R., and Mehmood, S.	*Journal of Cleaner Production*	2024
Green finance policy, ESG rating, and cost of debt—Evidence from China	Li, W.H., Hu, H.B., and Hong, Z.K.	*International Review of Financial Analysis*	2024
Laying Stress on Ideas: Actors, Connections, and Meanings of Sustainable Finance in Italy	Nupieri, T.	*Societamutamentopolitica-Rivista Italiana di Sociologia*	2024
The role of sustainable finance in the context of the European green course	Streimikiene, D., Mikalauskas, I., Leckiene, V., Pisula, T., and Mikalauskiene, A.	*Economics & Sociology*	2024
The role of green finance and governance effectiveness in the impact of renewable energy investment on CO_2 emissions in BRICS economies	Yadav, A., Gyamfi, BA., Asongu, S.A., and Behera, D.K.	*Journal of Environmental Management*	2024
Building an International Regulatory and Legal Framework for Green Digital Finance	Winarto, W.W.A., and Parwoto	*Revista Juridica Portucalense*	2025

Source: Authors' processing

according to Winarto and Parwoto (2025), these initiatives are not sufficiently adapted for new digital technologies such as blockchain, Artificial Intelligence, big data, and the Internet of Things. Moreover, as noted by Numan et al. (2023) and Ning et al. (2022), political instability, accelerated urbanisation, and institutional fragmentation pose substantial implementation barriers, including the absence of a uniform global reporting system, which limits the comparability and effectiveness of transnational green policies. However, Zetzsche et al. (2022) caution that overlaps between international ESG standards and the complexity of the EU taxonomy may hinder the integration of green capital and market convergence.

Recognising public-private partnerships is also crucial. According to Clark et al. (2018), unlocking private capital is vital for achieving international goals, such as the Sustainable Development Goals (SDGs) and the Paris Agreement. At the institutional level, global financial institutions such as the World Bank and the Green Climate Fund play central roles in enabling the transition to public-private partnerships and low-emission economies (Li et al., 2022; Iswaryalakshmi, 2022).

While global regulations provide strategic direction, numerous challenges often limit national and local implementation. For instance, China has strengthened its position in green finance by piloting policies and integrating sustainability criteria into central bank macroprudential assessments (Feng & Zhou, 2024; Li et al., 2024). However, Lin et al. (2022) argue that green finance faces a market framework that cannot fully consolidate support for green projects across the public, private, and non-profit sectors. Mungai et al. (2022) demonstrate that the benefits of the green transition are uneven, with limited benefits in countries characterised by political instability, weak financial markets, and inadequate infrastructure for monitoring and reporting. Furthermore, Streimikiene et al. (2024) highlight the differences between EU countries in implementing the European Green Deal, particularly in the Baltic States, where the lack of a practical framework hinders progress.

Saudi Arabia serves as a poignant example of the dissonance between strategic ambitions and operational limitations. Although committed to Vision 2030, the country faces a lack of standardised metrics for assessing the impact of green investments (Alhejaili, 2024). Similarly, Yadav et al. (2024) demonstrate that weak governance and inconsistent policies in BRICS countries hinder the effectiveness of renewable energy investments, despite these countries playing a crucial role in the global transition and being responsible for a significant share of global emissions.

Local institutional quality also affects policy outcomes. Integrated financial and non-financial communication (Avram et al., 2019), strong internal governance structures (Pirtea & Nicolescu, 2013), gender-balanced leadership (Gravili et al., 2019), and robust ESG performance metrics (Lobonț et al., 2022) are essential elements for building institutional trust and enhancing the effectiveness of green finance frameworks.

Local governments also play a critical role in supporting the green transition. To ensure financial transparency, Bhatnagar and Sharma (2022) advocate for the creation of local monitoring bodies. Similarly, Zhang et al. (2022) argue that the absence of clear regulations and inadequate incentives hinders the scaling of green finance in its initial phases. Abdel-Aziem and Soliman (2023), along with Liu and

Wu (2023), draw attention to local barriers, including low administrative capacity, a lack of technical expertise, and legislative instability, which hinder the practical transposition of international regulations.

National institutions must, therefore, prioritise capacity building and policy coherence. Zain et al. (2024) emphasise that strengthening regulatory capabilities is key to effective alignment with the international framework. Meanwhile, Dima et al. (2009) highlight that short-term fiscal policies and rigid public spending can undermine long-term financial sustainability in emerging economies such as Romania.

In addition to institutional problems, there are also challenges at the structural level. Greenwashing is an additional challenge, aggravated by the lack of a uniform categorisation of the concept of 'green'. A lack of coherence favours greenwashing (Buttigieg & Pulis, 2024) and renders ESG reporting merely a form of marketing (Tanguy, 2024). Aleknevičienė and Bendoraitytė (2023) warn that, in the absence of clear labelling standards, green marketing can be used misleadingly, complicating and hindering the mobilisation of private capital. Liu et al. (2023) also emphasise the need for coordinated financial policies that favour shifting towards a green economy through strict incentives and regulation.

Finally, according to Nyikos and Kondor (2022), the traditional finance paradigm, which separates financial returns from social and environmental considerations, remains a barrier in many countries. Guesmi et al. (2024) note that the energy transition, which is essential for achieving the SDGs, requires significant local adjustments and adequate financial support. The effectiveness of green finance, as suggested by Kaewsaeng-on and Mehmood (2024) and Bhatnagar and Sharma (2022), depends on harmonisation between regulatory policies, monitoring mechanisms, and the involvement of local actors. Simplified standards, digital innovation, and a more integrated social and governance dimension into regulation, including the expansion of ESG taxonomies, are recommended by Zetzsche et al. (2022), Hu et al. (2022), and Ikevuje et al. (2024), while Chiu (2020) warns against regulatory complexity, which may provoke political resistance, claiming the need for a more transparent and efficient framework.

Based on the literature analysis, it becomes clear that green finance regulations are shaped by a complex network of international and local factors, in which the convergence of global standards faces institutional fragmentation and implementation challenges. This reality requires not only a conceptual understanding of the phenomenon, but also a structured exploration of how policies are formulated, articulated and implemented in different economic contexts. In this regard, the following section outlines the methodological approach, a crucial element of our research, employed to map the regulatory landscape in the field of green finance. It highlights the structure of the semantic and institutional networks that define this field at both the global and national levels. The chosen methodology aims to provide a systematic and comparative representation of existing strategies, identifying dominant patterns, points of coherence, and areas of discontinuity that affect the effectiveness of regulations in the transition to a sustainable economy.

3 Methodological and Analytical Insights: A Network-Based Exploration of Green Finance Regulations

3.1 Network Analysis Methodology and Policy Data Overview

The analysis of green finance regulations requires a comprehensive approach to capture the formal structure of existing policies and their inter-institutional and semantic relationships. To accomplish this, our study used network analysis to comprehensively examine the relationships between financial regulation and climate policy at both global and national levels.

The study is based on an extensive database of over 100 national plans from 32 economies and country groupings, sourced from The Sustainable Finance Regulations Platform. The dataset provides information about National Strategy and Frameworks for Climate Transition and Sustainable Finance, providing a detailed breakdown by country, policy name, policy category, year, subject, issuer, status, compliance, and object scope.

This research's methodology centres around network analysis, conducted using InfraNodus, a sophisticated tool for visualising and analysing complex networks within large datasets. The proposed method utilises a text network analysis algorithm to model a text as a network, identifying the most influential words based on their co-occurrence patterns. Subsequently, a graph community detection algorithm is applied to discern distinct topical clusters that represent the text's primary themes and elucidate the interrelationships among these themes (Paranyushkin, 2019).

3.2 Cross-National Mapping of Regulations for Green Finance

The first part of the analysis examines the commonalities and notable differences between green finance policies across the 32 economies, offering a comparative perspective on their structures and content. The study is conducted along two main lines: policy names and types of issuing institutions.

Nomenclature and Thematic Focus of National Strategy and Frameworks

The nomenclature in naming regulatory instruments for green finance provides essential clues regarding states' regulatory orientation and strategic priorities. The analysis of the 103 national strategies reveals a coherent conceptual architecture centred on anchor terms (Fig. 3).

The semantic centrality of the term "finance", both in terms of network connectivity (degree = 46) and frequency of occurrence (45 mentions), confirms the dominant position of the financial dimension in the current sustainability discourse. This indicates a global recognition of financial markets, investment instruments, and the

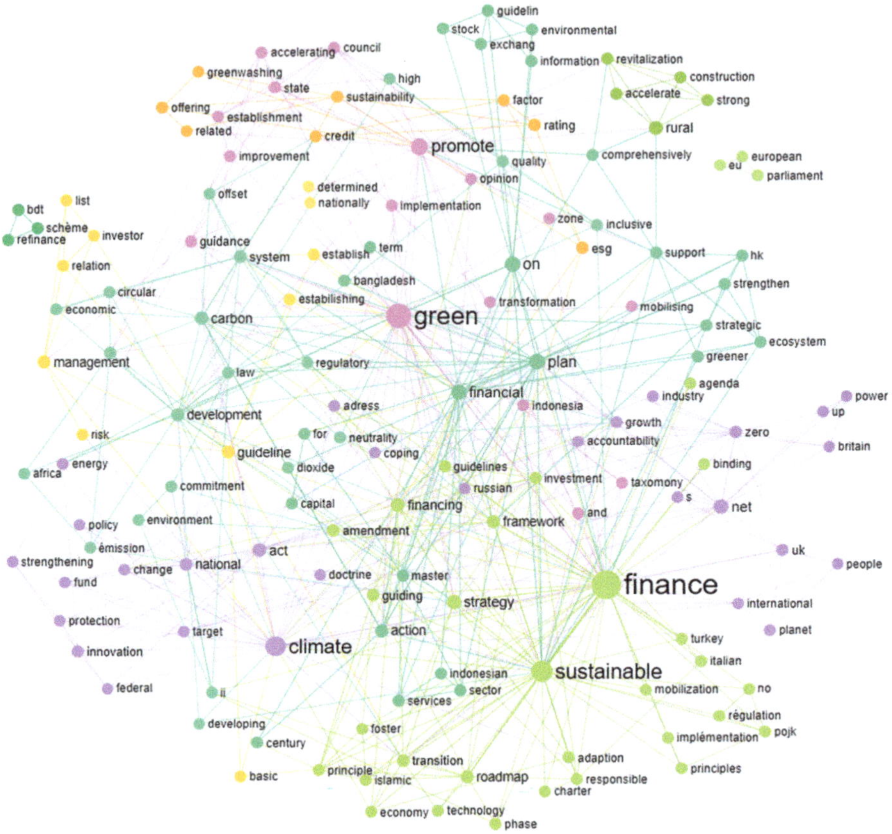

Fig. 3 Network of policy names for 103 national strategies and frameworks for climate transition and sustainable finance. (Source: Author's processing based on available data using InfraNodus)

role of financial institutions in regulating sustainability. The term "sustainable", with an almost equivalent centrality (degree = 41), signals a clear move from simple environmental policies to a systemic integration of sustainability principles into economic architecture. This association between "finance" and "sustainable" reflects the process of normalising ESG principles in national financial regulations.

The terms "green" (degree = 37, frequency = 45) and "climate" (degree = 35, frequency = 38) occupy central positions in the semantic network, indicating their pivotal role in articulating the discourse on the ecological transition. Although less frequent than "finance" or "sustainable", their centrality signals the deep interconnection with other strategic concepts in the regulatory architecture. This positioning reflects a firm anchoring of policies in climate and ecological objectives, even if implementation is often indirect or mediated through broader terms, such as "plan" or "development". Also, the moderate frequencies suggest a tendency towards institutionalisation, but not yet a terminological maturity equivalent to that observed in terms such as "sustainable finance". The terms "plan" (degree = 28, frequency = 13), "strategy" (degree = 18, frequency = 18), "development" (degree = 21, frequency = 5) and "promote" (degree = 23, frequency = 5) indicate a processual and programmatic dimension of policies. They reflect that, in many cases, the regulatory frameworks analysed are formulated as forward-looking initiatives, in the planning or promotion stages, rather than in advanced phases of implementation and control.

The network analysis also provides a synthetic illustration of how green finance policies are conceptualised globally, distributing the official names of national policies into four distinct clusters (Table 2).

The most influential cluster, Sustainable Finance, with 40% influence and 30 nodes, reflects a pronounced focus on integrating sustainability into the financial system. The thematic concentration, sustained by terms such as "strategy", "investment", "guidelines", "adaptation", and "implementation", reveals a maturation of the regulatory discourse, in which financing the ecological transition is approached not only as an objective but as a structuring process.

On the other hand, the Green Guidance cluster, with an influence of 22%, reflects an earlier stage of policy development, highlighted by terms such as "promote", "guidance", "establishment", and "accelerating". This suggests that in many countries, policies are still in the phase of formulating general directions. The Climate Action cluster (17%) highlights the operational dimension of the transition, as illustrated by terms such as "net zero", "innovation", "energy", "growth", "accountability", and "coping". The last cluster, Financial Strategy (12%), is more technical and functional, including terms such as "exchange", "capital", "services", and "ecosystem", and reflects how market instruments are mobilised to support sustainability.

Table 2 Keyword clusters of policy names

Topical cluster	Influence	Total nodes	Category	Keywords
1.	40%	30	Sustainable Finance	Finance, sustainable, strategy, financing, roadmap, framework, transition, amendment, guiding, investment, principle, adaptation, binding, foster, economy, guidelines, agenda, mobilisation, technology, responsible, charter, phase, implementation, principles, regulation
2.	22%	16	Green Guidance	Green, promote, guidance, establishment, improvement, council, opinion, state, implementation, accelerating, zone, transformation, taxonomy, mobilising.
3.	17%	29	Climate Action	Climate, net, act, national, innovation, zero, international, energy, growth, target, strengthening, planet, industry, address, accountability, policy, fund, protection, federal, change, doctrine, coping, people, power
4.	12%	25	Financial Strategy	Plan, financial, action, exchange, guideline, master, support, comprehensive, environmental, strength, services, greener, ecosystem, strategic, information, stock, term, capital, century, sector

Source: Author's processing

Institutional Issuers

Identifying and analysing the institutional issuers of green finance policies is crucial in understanding the governance structure behind regulations. Institutional actors formulate and implement policies, shaping the discourse and strategic direction of the green transition.

This section examines the main categories of issuers, including governments, financial authorities, environmental ministries, and central banks, to highlight the specific institutional architecture of green finance regulation in different economic contexts.

The network analysis of issuing institutions (Fig. 4) highlights a complex balance between financial and governmental actors in the architecture of green finance regulations. The terms "financial" (degree = 28), "authority" (degree = 24), and "bank" (degree = 17) indicate a significant centrality of financial actors, suggesting that institutions such as central banks, financial supervisory authorities, and regulatory commissions play a pivotal role in the development and implementation of

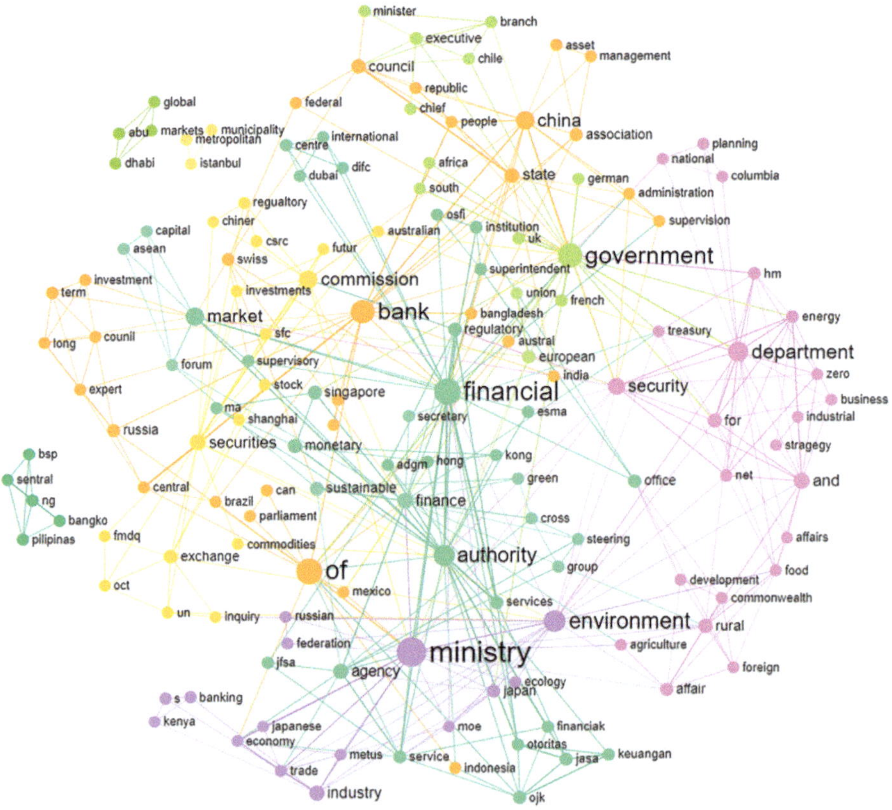

Fig. 4 Network of issuers for 103 national strategies and frameworks for climate transition and sustainable finance. (Source: Author's processing based on available data using InfraNodus)

policies. Although the frequency of the term "bank" is higher (16) compared to "financial" (13), the higher semantic centrality of the latter suggests a more transversal and interconnected role in the network. Thus, a picture emerges of a regulation with a financial-technocratic nature, in which financial expertise is a central pillar in operationalising sustainability objectives.

On the other hand, government institutions are also prominent, as indicated by the values for "government" (degree = 26, frequency = 29) and "ministry" (degree = 24, frequency = 11). The term "government" is the most frequently encountered, indicating a broad political involvement and possibly an institutional vagueness where responsibilities are distributed among several entities. The terms "environment" (degree = 23, frequency = 7) and "department" (degree = 17, frequency = 7) suggest an active participation of thematic structures, but with a relatively reduced influence in the network. Also interesting is the relatively discreet role of "security" (degree = 18, frequency = 5), which can be associated with regulating capital markets and emerging issues related to climate security. Overall, the semantic network suggests a multi-actor governance framework in green finance, where financial authorities and government structures intersect, albeit not always unified or integrated.

The institutional classification resulting from the network analysis highlights three distinct thematic clusters, which reflect the governance configuration in the regulation of green finance: Financial Institutions, Government Strategy, and Ministry Affairs (Table 3).

The most influential cluster, Financial Institutions (36% influence, 10 nodes), groups terms such as "financial", "institution", "superintendent", and "international". This cluster highlights the epistemic and operational centrality of financial actors, particularly supervisory authorities, central banks, and international regulatory platforms. The high semantic density surrounding these terms suggests a transfer of regulatory authority to financial entities with advanced analytical capabilities and global reach.

Table 3 Keyword clusters of issuers

Topical cluster	Influence	Total nodes	Category	Keywords
1.	36%	10	Financial Institutions	financial, corporate, institution, OSFI, superintendent, DIFC, centre, international, institutions
2.	24%	29	Government Strategy	government, department, energy, corporation, net, security, industrial, German, French, Colombia, planning, Australia, UK, municipality, metropolitan, federal, Africa, national, Philippines, zero, strategy, business, treasury
3.	12%	33	Ministry Affairs	environment, ministry, rural, agency, office, finance, affairs, service, Japan, industry, sustainable, Russian, commonwealth, development, foreign, steering, group, cross, agriculture, ecology, Swiss, federation, Mexico, JFSA, MOE, METUS, trade, economy, Japanese, forum, financial, food

The Government Strategy cluster (24% influence, 29 nodes) indicates a vital role of central and decentralised government structures in articulating green transition policies, but with a greater terminological dispersion. Terms such as "planning", "treasury", "federal", and "municipality" denote a heterogeneous institutional architecture, in which climate strategies are integrated into economic or territorial policies, rather than being confined to specialised sectoral regulations.

In contrast, Ministry Affairs (12% influence, 33 nodes) reflects a broader spectrum of institutions—from ministries of environment and industry to development agencies or agriculture—illustrating a complex inter-ministerial governance model, but with potentially low regulatory efficiency. The low influence of this cluster suggests that, although numerous, these entities are not positioned as decision-making centres in the network, but rather as vectors of implementation. Thus, the analysis confirms the existence of an asymmetric dynamic in the institutional architecture of green finance, in which financial expertise prevails over political-administrative coordination.

3.3 *Regulatory Coherence in Developed Countries*

The network analysis of regulations in developed economies highlights a network dominated by concepts aligned with the global sustainability agenda (Fig. 5). The terms "finance" (degree = 25, frequency = 19), "climate" (degree = 26, frequency = 11), "sustainable" (degree = 22, frequency = 13), and "strategy" (degree = 19, frequency = 14) signal a high thematic coherence and a precise alignment between climate policy objectives and financial infrastructure. The terms "finance" and "strategy", with close values in both degree of centrality and frequency, highlight that regulations in these economies are not fragmentary, but part of deliberate and planned strategies, in which the financial component is functionally integrated into the logic of sustainability. The frequent term "sustainable" indicates a solid institutional commitment to ESG principles, and its coupling with "climate" reflects a systemic approach to the green transition.

At the same time, the terms "green", "net", "action", and "act" denote a clear operational orientation: policies are not limited to programmatic rhetoric, but involve concrete actions aimed at objectives such as net zero and emission reduction. Although the terms "net" (degree = 15, frequency = 5), "action" (degree = 12, frequency = 6), and "act" (degree = 14, frequency = 6) have a lower centrality, their relative frequency suggests that they are integrated into the institutional discourse, but in supporting positions, not as primary vectors. Overall, the semantic profile of developed economies reflects an integrated, stable, and functional approach, in which sustainability is incorporated into strategic financial planning, and the climate transition is treated as a governed and monitored process, not as a mere rhetorical imperative.

The thematic classification of policies in developed economies, obtained through semantic network analysis, reveals a clear conceptual structuring around three

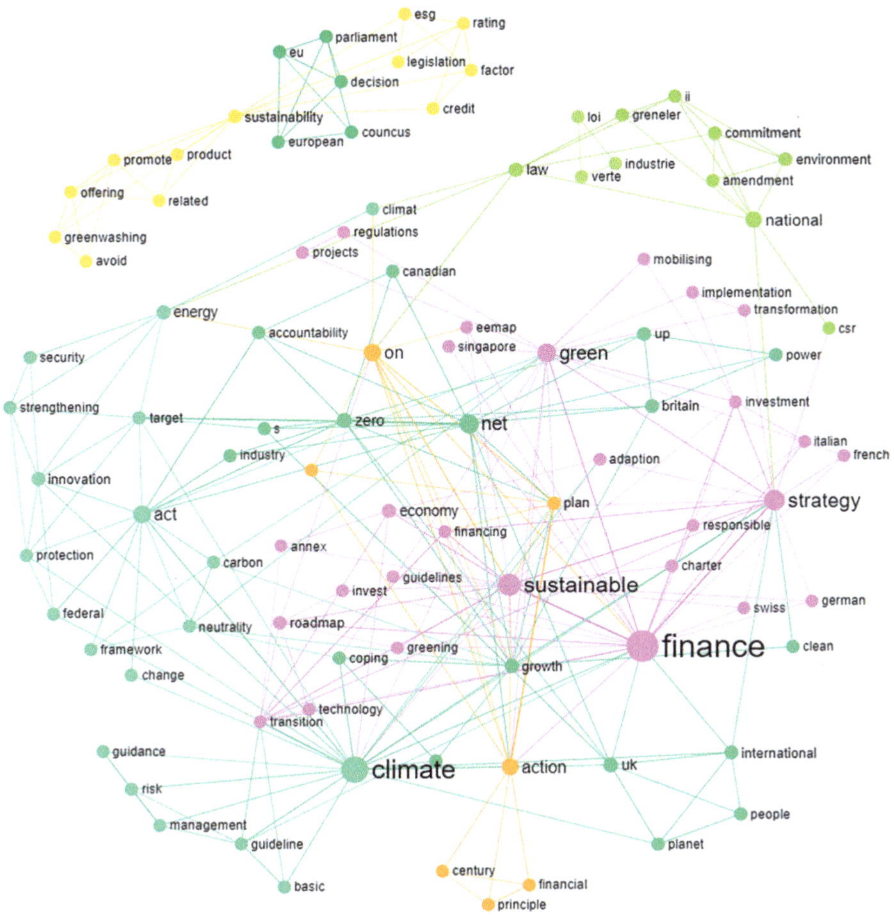

Fig. 5 Network of policy names in developed countries. (Source: Author's processing based on available data using InfraNodus)

major clusters: Sustainable Finance (50%), Climate Innovation (22%), and Net Growth (12%) (Table 4). The distribution of influence between clusters suggests that, in developed economies, green finance regulation is marked by a dominance of the financial dimension integrated with sustainability, complemented by technological innovation in support of the transition and an approach oriented towards responsible growth and economic resilience.

The semantic structure of regulations in developed economies is dominated by the Sustainable Finance cluster (50% influence, 28 nodes), highlighting the financial centrality of the sustainability discourse and a high level of institutional maturity. Terms such as "finance", "sustainable", "strategy", "green", "transition", "guidelines", "implementation", or "investment" reflect a well-consolidated policy architecture that is strategic and oriented towards integrating ESG objectives into the economic system. It is also worth noting the presence of geographical terms

Table 4 Keyword clusters of policy names in developed countries

Topical cluster	Influence	Total nodes	Category	Keywords
1.	50%	28	Sustainable Finance	finance, sustainable, strategy, green, economy, roadmap, transition, German, French, annex, adaptation, financing, invest, greening, regulations, projects, Swiss, guidelines, Singapore, implementation, transformation, technology, Italian, responsible, charter, investment, mobilising
2.	22%	19	Climate Innovation	climate, act, energy, innovation, target, strengthening, management, neutrality, carbon, change, security, protection, federal, risk, guidance, framework, guideline, basic
3.	12%	17	Net Growth	net, zero, Britain, international, growth, coping, planet, industry, accountability, Canadian, clean, crisis, people, power

Source: Author's processing

(German, French, Italian, Singapore, Swiss), which suggest a high degree of customisation of policies depending on the national context while maintaining a coherent and transferable normative structure. Thus, regulations in this cluster are not only descriptive or aspirational but are based on concrete plans, emerging technologies, and clearly defined market instruments.

The Climate Innovation cluster (22%) reflects the technological and operational dimensions of regulations, combining terms such as "innovation", "management", "target", "risk", "neutrality", and "framework". This group emphasises the pivotal role of technological innovation, risk management, and inter-institutional coordination in facilitating the green transition. In parallel, the Net Growth cluster (12%) suggests a discursive articulation around the net-zero paradigm, accompanied by terms such as "coping", "accountability", "crisis", and "clean". This semantic core appears to capture the adaptive and social dimensions of regulations, in which sustainability is interpreted as a response to challenges to structural and democratic legitimacy. Together, these clusters indicate a coherent and layered network in which the financial dimension is the core, innovation functions as an operational vector, and social and equity concerns support the political narrative of sustainability.

3.4 Policy Fragmentation in Emerging Countries

The semantic structure of green finance policies in emerging economies (Fig. 6) is marked by a pronounced presence of the terms "sustainable" (degree = 36, frequency = 22), "finance" (degree = 35, frequency = 22) and "green" (degree = 32, frequency = 11), indicating a discursive alignment with the global terminology, but not necessarily an institutional or normative coherence. This semantic triad reflects

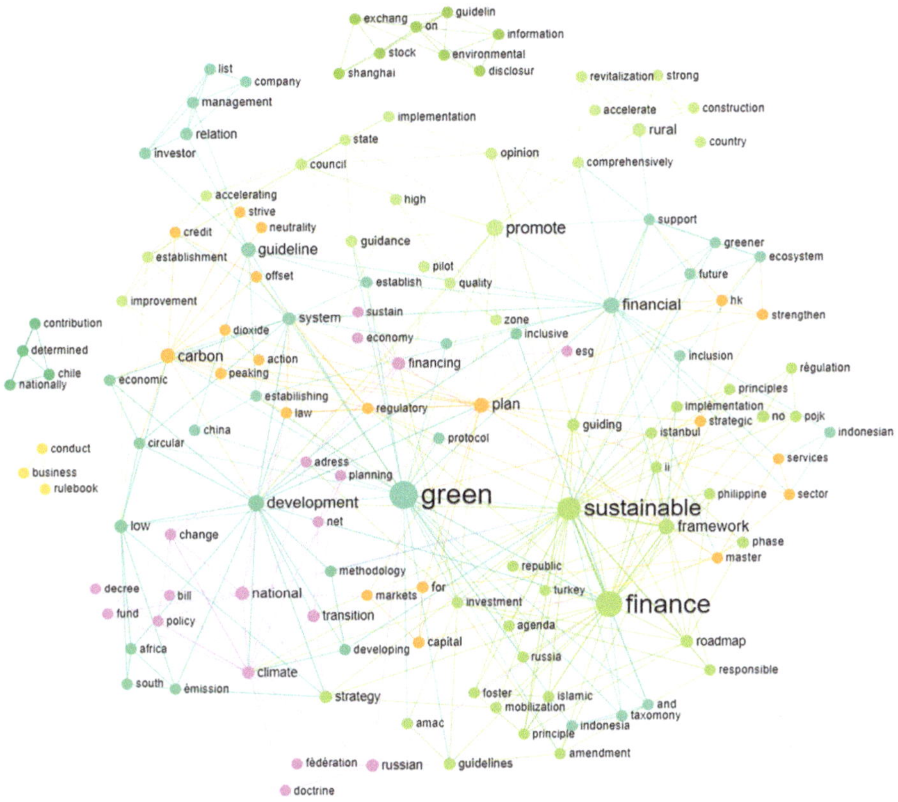

Fig. 6 Network of policy names in emerging countries. (Source: Author's processing based on available data using InfraNodus)

an internalisation of the sustainability vocabulary rather than an effective operational integration into the national financial architecture. Compared to developed economies, the terminological network in emerging countries appears denser but less structured, suggesting the existence of an influential discursive framework that has not been sufficiently transformed into functional public policies. The terms "development" (degree = 22, frequency = 5), "promote" (degree = 20, frequency = 4), and "plan" (degree = 19, frequency = 5) betray an approach oriented towards attracting external resources, top-down strategic planning, and promoting sustainability as a general objective, rather than as a normative practice.

Additionally, the presence of the terms "carbon", "climate", and "framework" in the network, with moderate centrality and frequency values, suggests an increased interest in aligning with international requirements for decarbonisation and climate reporting. However, the role of these terms in the network appears rather reactive, indicating a dependence on external pressure and standards imposed by international bodies or development partners. Thus, the semantic network of emerging economies reflects a superficial assimilation of the green finance paradigm, with a discourse that mimics the coherence of developed regulations, but without solid institutional support, robust compliance mechanisms, or well-defined financial-technical instruments.

Semantic clustering of regulatory discourse in emerging economies reveals four broad thematic categories, reflecting distinct structural priorities and different levels of institutional maturity in green finance (Table 5).

The thematic classification of green finance policies in emerging economies highlights a heterogeneous discursive network, dominated by the Sustainable Finance cluster (37% influence, 28 nodes), in which terms such as "framework", "strategy", "guidelines", "investment", "regulation", and "implementation" predominate. This lexical distribution suggests a declarative concern for constructing regulatory frameworks and attracting investments, but a consolidated institutional structure does not accompany it. The presence of terms such as "Islamic", "Russia", "Philippines", or "Turkey" denotes a marked regional and cultural diversity, in which regulatory models are adapted to national legal and economic realities, but without an apparent convergence towards robust international standards.

At the same time, the Green Guidelines cluster (influence = 31%, 23 nodes) indicates an emphasis on guides, protocols, and terms such as "support", "ecosystem", "establish", and "taxonomy", which reflects an intermediate stage of regulation, focused on guidelines and principles, not on formal constraints. The Rural Revitalisation cluster (9%) introduces a distinct dimension, specific to emerging regions, where climate policies are correlated with the need for rural development,

Table 5 Keyword clusters of policy names in emerging countries

Topical cluster	Influence	Total nodes	Category	Keywords
1.	37%	28	Sustainable Finance	finance, sustainable, framework, strategy, roadmap, guidelines, guiding, investment, principle, Islamic, foster, Istanbul, Turkey, republic, Russia, agenda, Philippines, mobilisation, amendment, responsible, phase, implementation, principles, regulation
2.	31%	23	Green Guidelines	green, financial, guideline, system, relation, investor, management, list, support, future, greener, ecosystem, protocol, company, establish, clause, establishing, China, inclusion, taxonomy, Indonesia
3.	9%	20	Rural Revitalisation	promote, rural, opinion, guidance, zone, improvement, council, establishment, state, high, comprehensively, accelerating, quality, implementation, country, strong, construction, accelerate, revitalisation, pilot
4.	8%	18	Climate Policy	national, financing, transition, climate, Russian, change, doctrine, net, planning, address, policy, fund, decree, bill, economy, sustain, federation, ESG

Source: Author's processing

infrastructure, and social cohesion. Terms such as "guidance", "construction", "quality", or "pilot" suggest partially integrating climate objectives into national development agendas. Finally, the climate policy cluster (8%) has the lowest influence. It concentrates terms such as "doctrine", "fund", "decree", and "ESG", signalling an incipient effort for formal regulation and adaptation to international trends. These four clusters reflect a fragmented approach, emphasising disparate initiatives, in which green finance regulations coexist with economic development and political stability objectives, without generating a coherent and systemic architecture.

4 Conclusion

This chapter confirms the structural tension between international standard convergence and local discontinuities in implementing the green finance regulatory framework. The analysis of over 100 national frameworks from 32 economies highlights significant differences in the coherence, complexity, and regulatory efficiency of green finance regulations between developing and emerging economies.

At the international level, standards are characterised by a densely connected semantic structure, in which terms such as "sustainable", "green", "climate", and "finance" occupy central positions. However, terminological maturity is uneven, which supports the hypothesis that the global framework's high complexity may create obstacles in the transposition process at the national level (Zetzsche et al., 2022).

In developed economies, the regulatory architecture is coherent and strategic, highlighted by the predominance of the "Sustainable Finance" cluster and the integration of the terms "strategy", "transition", "implementation" and "guidelines", in line with the conclusions of Bengo et al. (2022) and Ahlström and Sjåfjell (2023).

In contrast, emerging economies present a fragmented and institutionally incoherent regulatory framework, illustrated by the relatively balanced but weakly connected distribution of four clusters: "Sustainable Finance", "Green Guidelines", "Rural Revitalisation", and "Climate Policy". This semantic profile suggests a declarative orientation, lacking a clear operational anchor, which confirms the conclusions of Mungai et al. (2022) and Ning et al. (2022).

The high risk of greenwashing is also noted, amplified by the lack of clear labelling and classification standards. The frequent use of generic terms such as "green", "guideline", or "promote", in the absence of references to verification and compliance mechanisms weakens the credibility of sustainability policies and investor confidence.

Finally, although most regulations are not fully adapted to new digital technologies, developed economies are integrating them into the language and structure of policies. This partially confirms Winarto and Parwoto's (2025) criticisms of the gap between technological innovation and regulation.

The analysis of institutional networks reveals a clear dominance of financial actors—central banks, supervisory authorities, and international institutions—in defining the regulatory agenda. The "Financial Institutions" cluster is the most influential (36% influence), which confirms the findings of Zain et al. (2024) and Nyikos and Kondor (2022). At the same time, the relatively peripheral role of ministries of environment or agriculture signals a deficit of inter-ministerial integration, affecting the coordination of policies and their concrete implementation.

In conclusion, regulatory coherence in green finance is deeply conditioned by the institutional context, the level of digitalisation, and the capacity to translate global convergence into functional policies. Without administrative reforms and effective inter-institutional governance, many emerging economies risk turning green regulations into a mere rhetorical exercise, with little impact on the transition to a sustainable economy.

To increase the efficiency of regulations in green finance, an integrated approach that combines the harmonisation of international standards with their adaptation to national institutional specificities is essential. First, it is necessary to strengthen administrative and technical capacity in emerging economies through training, digitalisation, and institutional support to ensure the practical application of the global framework in locally relevant policies. Second, governments should develop clear labelling, auditing, and verification mechanisms to prevent greenwashing and increase investor confidence. Integrating new digital technologies—such as blockchain, artificial intelligence, and big data—into reporting and monitoring processes can increase transparency and efficiency of implementation. A better coordinated multi-stakeholder governance, including financial institutions and ministries of environment, agriculture, energy, and infrastructure, is also recommended for cross-cutting and sustainable regulation. Finally, the customisation of national regulations based on flexible global standards should be encouraged, so that policies reflect the socio-economic realities of each country, without compromising global goals regarding the ecological transition.

The implications for public policy are multiple and concern both the international and national levels. First, to avoid green regulations becoming mere rhetorical exercises, it is essential to establish international audit and interoperability mechanisms between taxonomies, allowing for a comparable assessment of sustainable performance. Global actors should support standardisation initiatives that include emerging economies in normative co-creation processes, avoiding top-down approaches.

At the national level, governments should incorporate concrete measures to enhance institutional capacity in their green transition strategies, including training public officials in ESG reporting, integrating digital technologies for enhanced traceability, and establishing public-private partnerships for green financial innovation. It is advisable to introduce interministerial and intersectoral dialogue platforms that integrate the perspectives of the environment, finance, infrastructure and agriculture, reflecting the fundamental interdependencies of the ecological transition.

Ultimately, to enhance the effectiveness of green policies, a differentiated approach is necessary, combining flexible international guidelines with local adaptations that are tailored to the economic, cultural, and political contexts. Policies must reflect both the global imperatives of emissions reduction and local realities to stimulate green investment and avoid implementation failures.

Acknowledgments This work was supported by a grant from the Romanian Ministry of Research, Innovation and Digitalization, the project with the title *"Economics and Policy Options for Climate Change Risk and Global Environmental Governance"* (CF 193/28.11.2022, Funding Contract no. 760078/23.05.2023), within Romania's National Recovery and Resilience Plan (PNRR)—Pillar III, Component C9, Investment I8 (PNRR/2022/C9/MCID/I8)—Development of a program to attract highly specialised human resources from abroad in research, development, and innovation activities.

Competing Interests The authors declare no conflicts of interest relevant to the content of this chapter.

References

Abdel-aziem, A. H., & Soliman, T. H. M. (2023). The regulatory landscape of green finance: An improved approach for market development. *Financial Technology and Innovation, 2*(1), 26–34. https://doi.org/10.54216/FinTech-I.020103

Ahlström, H., & Sjåfjell, B. (2023). Why policy coherence in the European Union matters for global sustainability. *Environmental Policy and Governance, 33*(3), 272–287. https://doi.org/10.1002/eet.2029

Aleknevičiene, V., & Bendoraityte, A. (2023). Role of green finance in greening the economy: Conceptual approach. *Central European Business Review, 12*(2), 105–130. https://doi.org/10.18267/j.cebr.317

Alhejaili, M. O. (2024). Integrating climate change risks and sustainability goals into Saudi Arabia's financial regulation: Pathways to green finance. *Sustainability, 16*(10). https://doi.org/10.3390/su16104159

Andreeva, O. V., Vovchenko, N. G., Ivanova, O. B., & Kostoglodova, E. D. (2018). Green finance: Trends and financial regulation prospects. *Contemporary Studies in Economic and Financial Analysis, 100*, 9–17. https://doi.org/10.1108/S1569-375920180000100003

Avram, A., Nicolescu, A. C., Avram, C. D., & Dan, R. L. (2019). Financial communication in the context of corporate social responsibility growth. *Amfiteatru Economic, 21*(52), 623–638. https://doi.org/10.24818/EA/2019/52/623

Baştürk, M. F. (2024). Does green finance reduce carbon emissions? Global evidence based on system generalized method of moments. *Sustainability, 16*(18), 8210. https://doi.org/10.3390/su16188210

Bengo, I., Boni, L., & Sancino, A. (2022). EU financial regulations and social impact measurement practices: A comprehensive framework on finance for sustainable development. *Corporate Social Responsibility and Environmental Management, 39*(4), 809–819. https://doi.org/10.1002/csr.2235

Bhandary, R. R., Gallagher, K. S., & Zhang, F. (2021). Climate finance policy in practice: A review of the evidence. *Climate Policy, 21*(4), 529–545. https://doi.org/10.1080/14693062.2020.1871313

Bhatnagar, S., & Sharma, D. (2022). Evolution of green finance and its enablers: A bibliometric analysis. *Renewable & Sustainable Energy Reviews, 162*. https://doi.org/10.1016/j.rser.2022.112405

Buttigieg, C. P., & Pulis, S. (2024). Strategic initiatives to address greenwashing. *ERA Forum, 25*, 327–337. https://doi.org/10.1007/s12027-024-00815-7

Chiu, I. (2020). Building a single market for sustainable finance in the EU- idealism, policy and mixed messages. *SSRN Electronic Journal*. https://doi.org/10.2139/ssrn.3631946
Chiu, I. H. Y. (2022). The EU sustainable finance agenda: Developing governance for double materiality in sustainability metrics. *European Business Organization Law Review, 23*, 87–123. https://doi.org/10.1007/s40804-021-00229-9
Clark, R., Reed, J., & Sunderland, T. (2018). Bridging funding gaps for climate and sustainable development: Pitfalls, progress and potential of private finance. *Land Use Policy, 71*, 335–346. https://doi.org/10.1016/j.landusepol.2017.12.013
Dima, B., Lobonţ, O., & Nicolescu, C. (2009). The fiscal revenues and public expenditures: Is their evolution sustainable? The Romanian case. *Annales Universitatis Apulensis Series Oeconomica, 11*(1), 416–425.
Feng, T. C., & Zhou, Q. (2024). Can green finance policy improve public health? *Global Public Health, 19*(1). https://doi.org/10.1080/17441692.2024.2406480
Gravili, G., Avram, A., & Nicolescu, A. C. (2019, April). Gender equality and firm financial performance: The case of Central and Eastern Europe financial and IT&C sectors. In *Proceedings of the ICGR 2019 2nd International Conference on Gender Research*, Rome, pp. 11–12.
Guesmi, K., Mohammed, K. S., & Tiwari, S. (2024). Green horizons: Enabling the energy transition through climate change policies. *International Review of Economics & Finance, 94*. https://doi.org/10.1016/j.iref.2024.103409
Haas, R., & Popov, A. (2019). Finance and decarbonisation: Why equity markets do it better. *Research Bulletin, 64*. https://www.ecb.europa.eu/press/research-publications/resbull/2019/html/ecb.rb191127~79fa1d3b70.en.pdf
Hafner, S., Jones, A., Anger-Kraavi, A., & Pohl, J. (2020). Closing the green finance gap – A systems perspective. *Environmental Innovation and Societal Transitions, 34*, 26–60. https://doi.org/10.1016/j.eist.2019.11.007
Hu, Y., Dai, X., & Zhao, L. (2022). Digital finance, environmental regulation, and green technology innovation: An empirical study of 278 cities in China. *Sustainability, 14*(14), 8652. https://doi.org/10.3390/su14148652
Ikevuje, A. H., Anaba, D. C., & Iheanyichukwu, U. T. (2024). Exploring sustainable finance mechanisms for green energy transition: A comprehensive review and analysis. *Finance & Accounting Research Journal, 6*(7), 1224–1247. https://doi.org/10.51594/farj.v6i7.1314
Iswaryalakshmi, P. (2022). A study on progress and challenges of green finance in Madurai District. *Journal of Pharmaceutical Negative Results, 13*, 1048–1054. https://doi.org/10.47750/pnr.2022.13.S08.132
Jenei, S., Tóth, A., Afadzinu, K. S., & Kálmán, B. G. (2024). EU sustainable finance framework. *Journal of Infrastructure, Policy and Development, 8*(15), 9485. https://doi.org/10.24294/jipd9485
Kaewsaeng-on, R., & Mehmood, S. (2024). Quantile modeling for environmental risk: SAARC's journey with green finance, policies, and regulations. *Journal of Cleaner Production, 434*. https://doi.org/10.1016/j.jclepro.2023.140234
Kemfert, C., & Schmalz, S. (2019). Sustainable finance: Political challenges of development and implementation of framework conditions. *Green Finance, 1*(3), 237–248. https://doi.org/10.3934/GF.2019.3.237
Li, Q. T., Sharif, A., Razzaq, A., & Yu, Y. Y. (2022). Do climate technology, financialization, and sustainable finance impede environmental challenges? Evidence from G10 economies. *Technological Forecasting and Social Change, 185*. https://doi.org/10.1016/j.techfore.2022.122095
Li, W.H., Hu, H.B., and Hong, Z.K. (2024). Green finance policy, ESG rating, and cost of debt – Evidence from China. International Review of Financial Analysis, 92. doi: https://doi.org/10.1016/j.irfa.2023.103051.
Lin, C. Y., Chau, K. Y., Tran, T. K., Sadiq, M., Van, L., & Phan, T. T. H. (2022). Development of renewable energy resources by green finance, volatility and risk: Empirical evidence from China. *Renewable Energy, 201*, 821–831. https://doi.org/10.1016/j.renene.2022.10.086

Liu, C., & Wu, S. S. (2023). Green finance, sustainability disclosure and economic implications. *Fulbright Review of Economics and Policy, 3*(1), 1–24. https://doi.org/10.1108/FREP-03-2022-0021

Liu, W., & Zhu, P. (2024). The impact of green finance on the intensity and efficiency of carbon emissions: The moderating effect of the digital economy. *Frontiers in Environmental Science, 12*, 1362932. https://doi.org/10.3389/fenvs.2024.1362932

Liu, X. Y., Zhao, T., & Li, R. (2023). Studying the green economic growth with clean energy and green finance: The role of financial policy. *Renewable Energy, 215*. https://doi.org/10.1016/j.renene.2023.118971

Lobonţ, O. R., Nicolescu, A. C., Costea, F., Li, Z. Z., Ţăran, A. M., & Davidescu, A. (2022). A panel threshold model to capture the nonlinear nexus between public policy and entrepreneurial activities in EU countries. *Mathematics, 10*(8), 1265. https://doi.org/10.3390/math10081265

Mungai, E. M., Ndiritu, S. W., & Da Silva, I. (2022). Unlocking climate finance potential and policy barriers – A case of renewable energy and energy efficiency in Sub-Saharan Africa. *Resources, Environment and Sustainability, 7*, 100043. https://doi.org/10.1016/j.resenv.2021.100043

Nedopil, C., Dordi, T., & Weber, O. (2021). The nature of global green finance standards – Evolution, differences, and three models. *Sustainability, 13*(7), 3723. https://doi.org/10.3390/su13073723

Ning, Q. Q., Guo, S. L., & Chang, X. C. (2022). Nexus between green financing, economic risk, political risk and environment: Evidence from China. *Economic Research-Ekonomska Istrazivanja, 35*(1), 4195–4219. https://doi.org/10.1080/1331677X.2021.2012710

Numan, U., Ma, B. J., Sadiq, M., Bedru, H. D., & Jiang, C. (2023). The role of green finance in mitigating environmental degradation: Empirical evidence and policy implications from complex economies. *Journal of Cleaner Production, 400*. https://doi.org/10.1016/j.jclepro.2023.136693

Nupieri, T. (2024). Laying stress on ideas: Actors, connections, and meanings of sustainable finance in Italy. *Societamutamentopolitica-Rivista Italiana di Sociologia, 15*(30), 53–63. https://doi.org/10.36253/smp-15351

Nyikos, G., & Kondor, Z. (2022). The involvement of national development banks in promoting sustainable finance. *Deturope-The Central European Journal Of Regional Development And Tourism, 14*(1), 147–163.

Paranyushkin, D. (2019). InfraNodus: Generating insight using text network analysis. In *The World Wide Web Conference*, pp. 3584–3589. https://doi.org/10.1145/3308558.3314123

Park, S. (2019). Green bonds and beyond: Debt financing as a sustainability driver, Chapter 42. In B. Sjåfjell & C. M. Bruner (Eds.), *Cambridge handbook of corporate law, corporate governance and sustainability*. Cambridge University Press. (University of Oslo Faculty of Law Research Paper No. 2019–46, University of Connecticut School of Business Research Paper No. 19–12). Available at SSRN: https://ssrn.com/abstract=3383561

Pirtea, M. & Nicolescu, A. C. (2013). Corporate governance codes of best practice of top Romanian banks. *Annals of the University of Oradea, Economic Science Series, 22*(2), 390–397.

Streimikiene, D., Mikalauskas, I., Leckiene, V., Pisula, T., & Mikalauskiene, A. (2024). The role of sustainable finance in the context of the European Green course. *Economics & Sociology, 17*(2). https://doi.org/10.14254/2071-789X.2024/17-2/3

Tagliaferri, M. (2022). EU's Green deal turns ESG programs into a "must have". https://trendspotting.sidley.com/trends/greendeal.

Tanguy, C. (2024). Greenwashing in ESG reporting: Risks, consequences, and strategies for transparency. *Deepki*, 12.4.2024. https://www.deepki.com/blog/greenwashing-esg-reporting/

Winarto, W. W. A., & Parwoto. (2025). Building an international regulatory and legal framework for green digital finance. *Revista Juridica Portucalense, 37*. https://doi.org/10.34625/issn.2183-2705(37)2025.ic-9

Yadav, A., Gyamfi, BA., Asongu, S.A., and Behera, D.K. (2024). The role of green finance and governance effectiveness in the impact of renewable energy investment on CO_2 emissions in BRICS economies. Journal of Environmental Management, 358. doi: https://doi.org/10.1016/j.jenvman.2024.120906.

Zain, N. R. M., Ali, E. R. A. E., Tounchibine, O., & Alias, R. M. R. (2024). SDGs and VBI: A META Analysis from Islamic Finance Perspective. In *Islamic Sustainable Finance* (pp. 41–50). Routledge. https://doi.org/10.4324/9781003395447

Zetzsche, D. A., Bodellini, M., & Consiglio, R. (2022). The EU sustainable finance framework in light of international standards. *Journal of International Economic Law, 25*(4), 659–679. https://doi.org/10.2139/ssrn.3984511

Zhang, D., Mohsin, M., & Taghizadeh-Hesary, F. (2022). Does green finance counteract the climate change mitigation: Asymmetric effect of renewable energy investment and R&D. *Energy Economics, 113*. https://doi.org/10.1016/j.eneco.2022.106183

Zhao, X., Mahendru, M., Ma, X., Rao, Y., & Shang, Y. (2022). Impacts of environmental regulations on green economic growth in China: New guidelines regarding renewable energy and energy efficiency. *Renewable Energy, 187*, 728–742. https://doi.org/10.1016/j.renene.2022.01.076

Green Finance as a Driver for Sustainable Development: The Role of Emerging Economies

Nicoleta Mihaela Doran, Badareu Gabriela, Doran Marius Dalian, and Mihiț Lavinia Daniela

1 Introduction

The global urgency to transition toward sustainable development has intensified in recent years, driven by escalating environmental degradation, climate-related risks, and the recognition of planetary boundaries. As the Intergovernmental Panel on Climate Change (IPCC, 2022) underscored, limiting global warming to 1.5 °C requires systemic energy, infrastructure, and financial systems transformations. Against this backdrop, green finance has emerged as a policy instrument and a fundamental enabler of structural economic change.

Green finance encompasses a spectrum of financial products, institutional arrangements, and regulatory frameworks aimed at aligning capital flows with environmental objectives (OECD, 2017; UNEP, 2021). It includes mechanisms such as green bonds, green credit, environmental risk disclosure, and sustainability-linked loans designed to mobilise public and private investments in low-carbon and climate-resilient projects. According to the Global Landscape of Climate Finance (CPI, 2023), green finance flows exceeded USD 850 billion annually in the post-pandemic recovery period—yet remain well below the estimated USD 4.3 trillion needed per year to meet climate goals by 2030.

In emerging economies, the role of green finance is particularly salient. These countries face a dual imperative: sustaining economic growth while mitigating and adapting to environmental challenges. However, institutional weaknesses, shallow capital markets, and limited fiscal space often hinder the deployment of green

N. M. Doran (✉) · B. Gabriela
Faculty of Economics and Business Administration, University of Craiova, Craiova, Romania
e-mail: nicoleta.doran@edu.ucv.ro; gabriela.badareu@edu.ucv.ro

D. M. Dalian · M. L. Daniela
Doctoral School in Economic Sciences "Eugeniu Carada", Faculty of Economics and Business Administration, University of Craiova, Craiova, Romania

C.-W. Su et al. (eds.), *Green Finance and the Challenges of Climate Change*, Sustainable Finance, https://doi.org/10.1007/978-3-032-07293-1_9

capital at scale (G20 Green Finance Study Group, 2016; World Bank, 2022). As Lee and Lee (2022) demonstrate in the case of China, green finance significantly enhances green total factor productivity (GTFP), with effects amplified in regions exhibiting high pollution intensity and weaker environmental governance. Based on a panel of 30 Chinese provinces (2006–2018), their findings provide robust empirical support for the productivity-enhancing effects of green financial deepening.

Similarly, Yu et al. (2021) highlight that green finance can mitigate firm-level financing constraints to support green innovation, especially among private enterprises. Their evidence suggests that green credit allocation is uneven, disproportionately favouring state-owned enterprises, despite private firms exhibiting greater innovation potential. This underscores the need for improved transparency and targeted regulatory intervention to ensure equitable access to green financial instruments.

Furthermore, in the Belt and Road Initiative (BRI) context, Zhang et al. (2019) show that public expenditure on R&D and human capital development can foster green economic growth and energy efficiency when complemented by sustainable green finance. Their study, using GMM and DEA methodologies, reveals heterogeneity across BRI countries, suggesting that income levels and institutional quality mediate policy effectiveness.

The European Union's Sustainable Finance Strategy (EC, 2021) and the Taxonomy Regulation further reinforce the normative framework for green finance by standardising definitions and reducing information asymmetries. Meanwhile, the UNEP FI Principles for Responsible Banking and the Network for Greening the Financial System (NGFS) have expanded the role of financial institutions in internalising climate-related financial risks and redirecting capital toward sustainable activities.

Despite this progress, significant challenges remain. Market fragmentation, lack of harmonised standards, and the dominance of short-term investment horizons continue to impede the scalability of green finance in emerging contexts. Bridging these gaps requires coherent policy architectures, capacity building, and de-risking instruments to crowd in private capital. It also demands aligning monetary, fiscal, and industrial policy tools with green finance objectives.

In light of these dynamics, this paper investigates the extent to which green finance drives sustainable development in emerging economies. By synthesising recent empirical findings with institutional and policy developments, it aims to shed light on how green finance contributes to environmental performance, innovation, and inclusive growth.

The primary objective of this study is to explore the intellectual landscape and thematic evolution of green finance research with a specific focus on the role played by emerging economies. While previous studies have investigated the impact of green finance on productivity, innovation, or environmental performance, few have systematically analysed how knowledge on green finance has been generated, disseminated, and institutionalised in the global academic space, particularly from the perspective of developing regions. By applying bibliometric techniques to a curated dataset extracted from the Web of Science Core Collection, this research aims to identify the most influential contributions, thematic clusters, collaboration patterns, and knowledge flows that define the field of green finance. Special emphasis is placed on capturing the scientific output and academic influence of scholars and institutions based in emerging economies to understand their contribution to the evolution of sustainability-oriented financial thinking.

The novelty of this study lies in its integrative bibliometric approach, which combines quantitative mapping with a critical interpretation of thematic trends to uncover underexplored dynamics within green finance literature. Unlike existing bibliometric analyses that often adopt a global or Eurocentric lens, this research shifts the analytical focus toward the academic voice and positioning of emerging economies—regions that are simultaneously vulnerable to climate risks and essential for the success of the global green transition. Furthermore, by bridging green finance with bibliometric intelligence, this study offers a novel framework for evaluating the diffusion of environmental finance knowledge and identifying strategic gaps in research collaboration, policy alignment, and thematic integration. The findings are expected to inform academic debate and policy development by highlighting the areas where green finance can be further leveraged to support sustainable development pathways in less-advanced but fast-growing economies.

The structure of the study is organised as follows: Section 1 outlines the conceptual foundations of green finance and situates it within the broader discourse on sustainable development, with a particular emphasis on its relevance for emerging economies. Section 2 details the research design and bibliometric methodology employed to map the field's scientific production and intellectual structure. Section 3 presents the key findings, including the most influential authors, journals, keywords, thematic clusters, and collaboration networks, focusing on the contribution and positioning of emerging countries. Finally, Section 4 concludes by summarising the main insights, highlighting the study's academic and policy relevance, and offering directions for future research on sustainable finance in developing regions.

2 Methodology and Data

Bibliometric analysis is widely acknowledged as a powerful methodological tool for evaluating academic knowledge's development, structure, and dissemination across scientific domains (Ellegaard & Wallin, 2015; Donthu et al., 2021). Through the quantitative examination of citation patterns, co-authorship networks, and thematic clusters, bibliometrics enables researchers to map the intellectual landscape of a given field, identify key actors, and track the emergence of research frontiers. While initially more prevalent in the natural and information sciences, bibliometric methods have gained increasing relevance in fields such as environmental economics and green finance, particularly due to the growing availability of high-quality bibliographic databases and visualisation software (Aria & Cuccurullo, 2017; Verma & Gustafsson, 2020).

This study employs a focused bibliometric approach to explore the scientific production on green finance in emerging economies. It aims to identify the most influential contributions in the field and analyse their thematic orientation, methodological design, and policy relevance. Rather than surveying the entire literature, we deliberately concentrate on a sample of 13 highly cited articles, selected as the most frequently referenced publications within the Web of Science Core Collection as of 2025. This purposive sampling strategy is consistent with previous bibliometric studies that aim to capture the intellectual structure of a field through its most impactful documents (Linnenluecke et al., 2017; Badareu et al., 2024).

To construct the sample, a search was conducted in the Web of Science Core Collection, using the keyword "green finance", refined by applying geographic filters relevant to emerging economies (based on World Bank classifications and EU enlargement status). Total citation count sorted the results, and the top 13 articles were selected for analysis. All selected documents were peer-reviewed articles published between 2010 and 2025 in English-language journals indexed in WoS categories such as Environmental Studies, Business Finance, and Economics.

The selected articles were processed using VOSviewer (version 1.6.18), a bibliometric visualisation tool that enables the construction of maps based on citation data, co-authorship, and keyword co-occurrence. Although our sample size did not allow for the creation of large-scale statistical clusters, it was suitable for generating network maps of co-authorship and keyword links, particularly when viewed at the micro level. In the VOSviewer visualisations, nodes represent individual authors or terms, the node size corresponds to frequency (e.g., citation count or keyword appearance), and the edges between them illustrate the strength of co-occurrence or collaboration.

In parallel with network mapping, we conducted an in-depth qualitative content analysis of each article. This dual methodology—quantitative visualisation and qualitative analysis—enabled us to triangulate findings and offer both breadth and depth in our literature evaluation. By focusing on the most cited studies, we ensured that the analysis captured the publications that have exerted the most significant academic influence in shaping the field.

Several preliminary observations emerged from our review. First, the dominant themes among the most cited articles include the role of monetary policy and private investment in green finance, the institutional barriers to financing green transitions, the role of local governments, and the financial sector's responsibility in funding renewable energy and climate resilience. Second, although the articles span multiple countries—including Hungary, Poland, and Romania—they share a common analytical concern with the structural limitations faced by emerging economies in aligning financial systems with environmental objectives.

Importantly, our methodology also allowed us to identify conceptual and methodological diversity within the sample. While some studies employed panel data econometrics to evaluate policy impacts, others used case studies or surveys to explore the behaviour of banks, municipalities, or investors. This variety reflects the interdisciplinary nature of green finance and underscores the value of a multimethod approach to bibliometric synthesis.

Finally, focusing on a small but high-impact corpus of publications allows for a detailed and critical engagement with the literature, beyond the descriptive statistics that dominate larger-scale reviews. It also facilitates the development of actionable insights for future research and policymaking by spotlighting what has been studied most intensively and what remains underexplored, such as the integration of fintech in green finance or the role of microfinance in ecological transition in low-income settings.

3 Results

3.1 Temporal Dynamics of Green Finance Research in Emerging EU Economies

Understanding the temporal evolution of scholarly contributions is critical for assessing the maturity and progression of an emerging research field such as green finance, particularly within less economically advanced regions of the European Union. Figure 1 illustrates the annual distribution of scientific publications retrieved from the Web of Science Core Collection from 2020 to 2025, based on a focused bibliometric query using the keywords "green finance" and "emerging economies". The results were filtered to include only articles affiliated with EU member states classified as emerging markets, by standard economic criteria established by institutions such as the IMF and Eurostat.

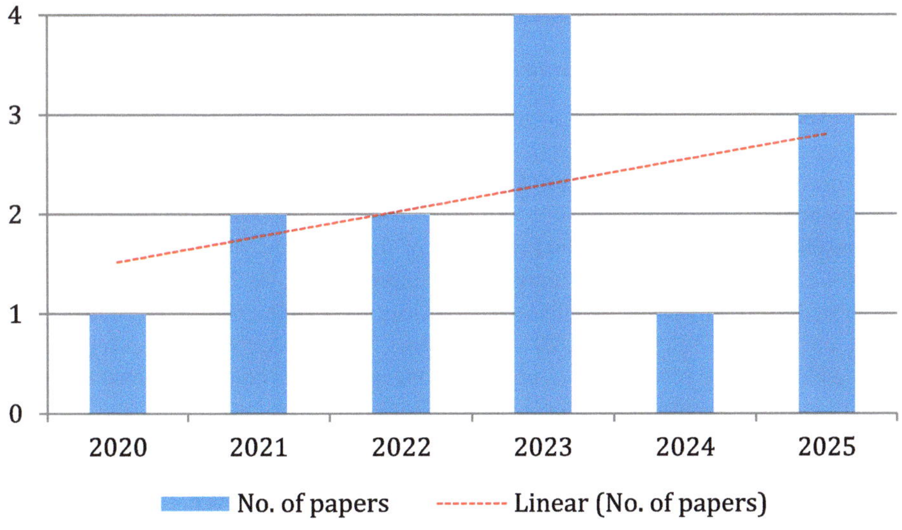

Fig. 1 Evolution of scientific publications on green finance. (Source: Web of Science, 2025)

The data reveal a non-linear but progressively upward trajectory in the number of scientific contributions from these countries. Starting from a single publication in 2020, the research volume doubled in 2021 and remained at the same level in 2022. A notable surge was observed in 2023, when the number of publications peaked at four—a development that likely correlates with implementing major EU-level policy instruments promoting green investment, such as the European Green Deal and the Recovery and Resilience Facility. These mechanisms have amplified the relevance of green finance both in national policy agendas and academic discourse.

A temporary contraction occurred in 2024, possibly due to publication lag, decreased research funding, or institutional constraints within some emerging economies. Nonetheless, the partial rebound observed in 2025 suggests restoring scholarly interest and possibly consolidating academic capacities in the domain.

Notably, the linear trendline superimposed on the bar chart confirms a moderately positive long-term trajectory, signalling that the field is gaining intellectual momentum, albeit from a relatively modest baseline. This observation underscores the gradual—but uneven—integration of green finance into the research ecosystems of emerging EU countries. While these findings point to increased awareness and engagement with sustainability-oriented financial mechanisms, they also reflect persistent disparities in research infrastructure, international visibility, and institutional prioritisation across member states.

In sum, the temporal dynamics revealed by the data suggest that green finance has entered the academic agenda of emerging EU economies. Still, its diffusion remains vulnerable to systemic and structural factors. Supporting the sustained growth of this field will likely require targeted investment in research capacity, enhanced cross-border collaboration, and the alignment of national research priorities with EU sustainability frameworks.

3.2 Keyword Co-occurrence Network in Green Finance Research

The co-occurrence analysis of keywords (with a minimum threshold of 2 occurrences)

present an in-depth conceptual mapping of the scientific discourse surrounding green finance in the context of emerging economies. Figure 2 depicts the network generated through VOSviewer, based on bibliographic data extracted from the Web of Science Core Collection (2020–2025). Thus, the co-occurrence map represents how green finance research is structured, conceptualised, and interconnected specifically within the intellectual ecosystem of emerging economies.

In bibliometric terms, co-occurrence analysis is an established method for uncovering the cognitive architecture of a research domain. Keywords are treated as semantic signifiers that index the conceptual content of a study. When two keywords appear together in multiple documents, they signal thematic proximity or shared

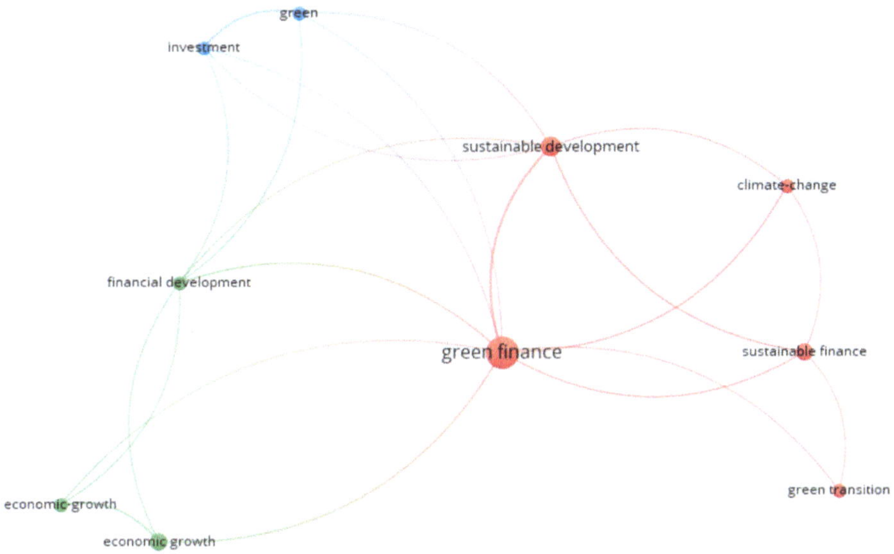

Fig. 2 Keyword co-occurrence network in green finance research. (Source: Own Data Processing Using VOSviewer, 2025)

research concerns. VOSviewer operationalises this by creating a two-dimensional spatial layout in which each node represents a keyword, the node size reflects the frequency of occurrence, and the links between nodes reflect total link strength (TLS), or the cumulative intensity of co-occurrence. Clusters are algorithmically generated based on modularity and association strength, grouping terms that co-occur more frequently with each other than with terms outside the cluster.

In the current network, "green finance" appears as the most dominant and central keyword, with 10 occurrences and a total link strength of 16. This highlights its frequency and role as an integrative hub in the thematic landscape. Given the search string used, this is expected. Still, its central position also affirms the degree to which green finance is embedded in discussions across multiple adjacent domains within emerging market contexts.

Notably, "green finance" is strongly linked to "sustainable development" (4 occurrences; TLS: 10), "sustainable finance" (3 occurrences; TLS: 6), and "climate change," indicating that the dominant strand of research is normative, situated at the intersection of environmental governance, developmental objectives, and financial innovation. This finding resonates with the global policy discourse that positions green finance as a catalytic instrument for achieving the United Nations Sustainable Development Goals (SDGs), particularly in economies undergoing industrial transition or structural reform.

One of the most thematically significant clusters centres on the triad "green finance"—"economic growth"—"financial development." This configuration indicates that researchers from emerging economies are increasingly concerned with the dual imperative of reconciling economic expansion with environmental responsibility. The term "economic growth" appears with 3 occurrences, while the variant "economic-growth" is listed separately with 2 occurrences—an artefact of author inconsistency in hyphenation, but one which jointly underlines the centrality of growth-oriented development in these regions. "Financial development" (2 occurrences; TLS: 7) further strengthens this interpretation, suggesting that much of the academic work explores how financial systems can be leveraged to support green transitions without undermining macroeconomic stability.

Interestingly, terms such as "green bonds," "investment," and "green transition" also occur within the network, albeit with lower frequencies, indicating a growing—though still nascent—interest in market-based instruments and financial engineering mechanisms. Their presence, however, is critical, as it signals a shift from purely conceptual discussions toward more applied, policy-relevant and investment-oriented approaches.

The keyword "green" (2 occurrences; TLS: 5) appears as an isolated but linking term. Although its generic nature might dilute semantic specificity, it functions as a semantic bridge between more defined concepts like "green bonds" and "sustainable development." Its usage may reflect diverse framing strategies adopted by scholars across different countries and disciplines, pointing to the interdisciplinary nature of green finance research in emerging markets, where environmental science, economics, and finance converge.

Another important observation relates to the relative absence of some expected keywords, such as "carbon markets," "green taxonomy," or "ESG regulation," which are more common in studies from developed economies. This suggests that while emerging countries engage with green finance discourse, their engagement's depth and institutional complexity remain early. The literature focuses more on macro-level themes (growth, development, sustainability) than regulatory or technical aspects. This aligns with many emerging economies that are still developing their legal and policy frameworks for sustainable finance.

Moreover, the overall density of the network is relatively low, and the average TLS values are modest, indicating that the field is still fragmented and in a consolidation phase. This observation has methodological implications: the relatively small size of the network suggests a narrow pool of indexed literature and a need for broader coverage, while the low co-occurrence intensity points to limited conceptual convergence across authors. These characteristics are typical of emerging research fields, particularly in geographic contexts where academic infrastructure, funding, and international visibility may be unevenly distributed.

Nevertheless, the presence of structured clusters and the emergence of key integrative nodes (e.g., "green finance," "sustainable development," "financial development") provide strong evidence that a core vocabulary and conceptual backbone are forming. This indicates a promising trajectory for future research consolidation. With increased policy focus at the European and global level—through instruments such as the EU Green Deal, the Global Gateway, and the Sustainable Finance Disclosure Regulation—scholarly production from emerging economies is likely to grow in volume and complexity.

3.3 *Mapping Scholarly Collaboration in Green Finance: A Co-authorship Network Analysis in Emerging Economies*

Scientific collaboration plays a pivotal role in shaping the intellectual landscape of an emerging research field. In the context of green finance, which intersects environmental sustainability, economic development, and financial innovation, the structure of academic co-authorship networks can offer deep insights into the maturity, connectivity, and diffusion of ideas across scholarly communities. Figure 3 presents the co-authorship network generated through VOSviewer, based on a bibliometric corpus restricted to publications originating from emerging economies between 2020 and 2025. The visualisation captures the relationships among authors who have contributed to the green finance discourse in these settings, offering a valuable lens to assess the development of collaborative research practices in a domain of growing strategic relevance.

Each node in the network represents an individual author, with node size indicating publication frequency and link thickness denoting the total link strength

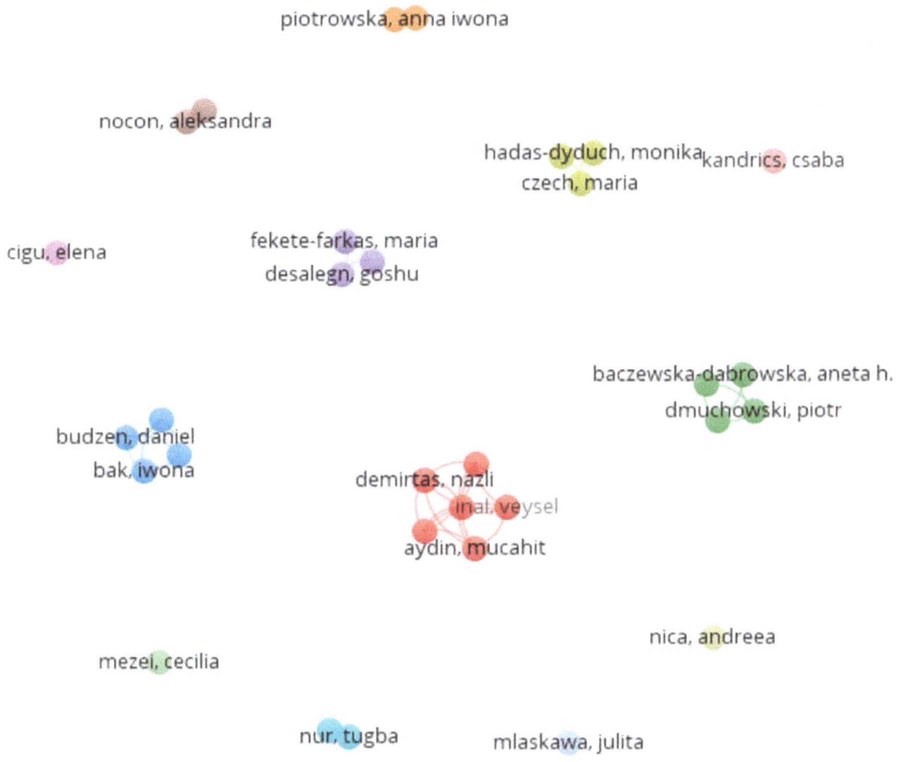

Fig. 3 Co-authorship network in green finance research in emerging economies. (Source: Own Data Processing Using VOSviewer, 2025)

(TLS)—a cumulative measure of collaborative intensity. Despite the rising number of publications on green finance in recent years, the network reveals a relatively low density of connections and limited clustering. This suggests that the field remains fragmented, with a predominance of isolated efforts and minimal cross-border or interdisciplinary collaboration, especially when compared to more established domains in environmental economics or sustainable finance in developed economies.

One of the most prominent clusters in the network consists of a group of Turkish researchers, including Demirtas, Nazli, Aydin, Mucahit, and Inal, Veysel (Sogut et al., 2025), who each exhibit strong mutual linkages (TLS = 5). This cluster reflects a localised and likely institutionalised form of collaboration, indicative of sustained co-authorship relationships within a specific academic environment. Their repeated contributions, though modest in citation impact (2 citations each), highlight their role in maintaining continuity and thematic cohesion in green finance research in Turkey—an emerging economy with increasingly robust sustainability ambitions.

A second notable collaboration is observed between Baczewska-Dąbrowska, Aneta H. and Dmuchowski, Piotr (Dmuchowski et al., 2021), researchers affiliated with institutions in Poland. While their total link strength is lower (TLS = 3), both scholars have recorded significant academic visibility, with 30 citations each. Despite limited collaborative intensity, their relatively high impact points to the potential influence of landmark publications or positioning in high-impact journals. Such cases demonstrate that impactful scholarship can emerge within and outside densely connected co-authorship networks.

Further instances of individual excellence are evident in the positions of authors such as Fekete-Farkas, Maria, Desalegn, Goshu, and Tangl, Anita (Desalegn, Fekete-Farkas & Tangl, 2022), each of whom has accrued 23 citations while maintaining limited co-authorship links. These authors, positioned more peripherally in the network, may operate within broader international or interdisciplinary teams without recurring collaboration patterns or publish solo-authored works with substantial analytical depth. Their presence underscores the potential for intellectual influence beyond tightly knit academic circles, although the absence of strong network integration could limit long-term knowledge diffusion and capacity building.

Similarly, Cigu, Elena (Cigu, 2020), Kandrics, Csaba (Kandrics), Czech, Maria and Hadas-Dyduch, Monika (Czech, Hadas-Dyduch & Puszer, 2023), Budzen, Daniel and Bak, Iwona (Bak et al., 2023), and Nica, Andreea (Nica, 2023) represent smaller, nationally anchored nodes that contribute to the diversification of the field through context-specific analyses of green finance in Central and Eastern Europe. Mezei, Cecilia (Mezei, 2024), Piotrowska, Anna Iwona (Piotrowska & Piotrowski, 2025), and Mlaskawa, Julita (Mlaskawa, 2022) appear as isolated yet visible contributors, reflecting the relatively early and dispersed stage of network consolidation across the region.

A notable characteristic of the network is its national clustering and absence of strong transnational research linkages. Most collaborative relationships are confined within individual countries, and there is limited evidence of cross-country co-authorships among emerging EU or non-EU economies. This reflects the broader structural challenges facing research communities in these regions, including language barriers, limited access to international funding, and underdeveloped mechanisms for institutional cooperation. Given green finance's global and policy-driven nature, such constraints present significant obstacles to academic consolidation and practical policy relevance.

The overall structure of the co-authorship network confirms that green finance remains a nascent yet promising area of research in emerging economies. The modest levels of collaboration and the presence of highly cited but unconnected authors suggest that while interest in the field grows, institutional and systemic mechanisms for fostering sustainable scientific partnerships remain underdeveloped. Research efforts appear to be predominantly individual or localised, with limited strategic coordination at national or regional levels.

To accelerate the development of the field, policy and funding bodies in emerging economies should consider promoting collaborative models through targeted initiatives, such as cross-border calls for proposals, joint doctoral programs, and co-authored policy-research platforms. Equally important is promoting interdisciplinary research clusters that integrate finance, climate science, environmental economics, and public policy under the green finance umbrella.

In conclusion, the co-authorship network presented in Fig. 3 offers a diagnostic snapshot of the current state of scholarly collaboration in green finance within emerging economies. While the structure is marked by fragmentation and low connectivity, there are encouraging signs of localised research capacity, high-impact individual contributions, and potential for expansion through strategic coordination. As the global sustainability agenda increasingly centres green finance as a critical lever for transformation, strengthening collaborative research ecosystems in emerging markets becomes both a scholarly imperative and a policy priority.

3.4 Authors Engagement in Green Finance Research Within Emerging Economies

Table 1 synthesises selected emerging economies' quantitative contribution to green finance academic production, based on bibliometric data retrieved from the Web of Science Core Collection. The five countries identified—Poland, Hungary, Türkiye, Romania, and Azerbaijan—represent a geographically and institutionally diverse

Table 1 Country-level engagement in green finance research in emerging economies (2020–2025)

Country	No. of documents	Total citations	Total Link Strength (TLS)
Poland	6	52	0
Hungary	3	25	0
Türkiye	2	2	1
Romania	2	2	0
Azerbaijan	1	2	1

Source: Own Data Processing Using VOSviewer, 2025

subset of emerging markets, each at different stages of alignment with the European Union's green transition agenda. The indicators considered include the number of documents, citation counts, and total link strength (TLS), the latter serving as a proxy for international collaboration.

Poland emerges as the most active contributor, with six documents and a total of 52 citations. Despite having no recorded TLS (TLS = 0), which suggests a lack of indexed international co-authorships, the relatively high citation count indicates a degree of academic visibility and impact. This may reflect either the quality of the journals in which Polish authors have published or the intrinsic policy relevance of the research outputs. The absence of cross-border co-authorship, however, points to a research environment that is productive but still inward-facing, with limited integration into broader European or global networks.

Hungary also exhibits a noteworthy contribution, with three documents and 25 citations, yet records no total link strength. As in Poland's case, this suggests that the Hungarian scholarly community has produced green finance research of recognized value, but that these contributions have emerged in relative isolation from transnational collaborations. The presence of highly cited authors such as Fekete-Farkas, Maria further supports this interpretation. Hungary thus appears to be developing intellectually significant but structurally fragmented research efforts in this domain.

Türkiye and Romania each contribute two publications, yet show a markedly lower citation impact (2 citations per country). While Türkiye registers a minimal TLS value of 1—indicating at least one international co-authorship link—Romania shows no evidence of external collaboration. These figures highlight limited academic integration and low visibility, which may reflect structural barriers (funding, language, access to high-impact journals) and the relatively recent institutional adoption of green finance themes in national research agendas. Given Türkiye's geographical position and growing engagement with EU green policy frameworks, forming stronger collaborative ties could significantly enhance its scholarly influence.

Lastly, Azerbaijan contributes a single publication, with two citations and TLS = 1. While modest in absolute terms, this data point signals an emerging academic interest in green finance within the South Caucasus region and the potential for future engagement. Even at this early stage, cross-border collaboration is a positive indicator for Azerbaijan's integration into wider research networks.

Taken together, these results point to several critical insights. First, the overall volume of green finance publications in emerging European and adjacent economies remains low, reflecting the relatively recent emergence of this thematic area in regional academic discourses. Second, although some countries (notably Poland and Hungary) demonstrate notable impact in citations, the lack of transnational co-authorships suggests that these national research efforts are not yet fully embedded in the collaborative structures that characterise more mature scientific domains.

This has significant implications. Green finance, by its very nature, is transdisciplinary and transnational, touching upon global value chains, EU regulatory regimes, international financial markets, and shared environmental challenges. The absence of robust international co-authorship networks limits academic visibility. It may hinder the practical policy uptake of research findings, particularly in contexts where cross-border coordination and alignment with EU directives are critical.

To overcome these limitations, targeted policy interventions and academic incentives are needed to foster collaboration across emerging economies. Initiatives such as EU-funded Horizon Europe consortia, COST Actions, or regional platforms (e.g., the Visegrád Group, Black Sea Economic Cooperation) could be leveraged to stimulate joint research programs, co-authored publications, and cross-institutional doctoral training. Equally important are promoting English-language publication strategies, open access dissemination, and active participation in international conferences.

In conclusion, while the data reflect early-stage engagement with green finance research across emerging economies, they also highlight a latent capacity for growth. Countries like Poland and Hungary have demonstrated substantive intellectual output, while Türkiye and Azerbaijan signal nascent but promising collaborative potential. Strengthening institutional linkages and promoting research integration will transform these isolated contributions into a cohesive, impactful, and policy-relevant research ecosystem in green finance across the emerging European landscape.

3.5 Authors Engagement in Green Finance Research Within Emerging Economies

The visual representation in Fig. 4 encapsulates the intellectual influence and temporal dynamics of key scholarly contributions in green finance across emerging economies between 2020 and 2025. The co-citation network, generated through VOSviewer, reflects the absolute volume of citations associated with each publication and its structural positioning within the broader research landscape.

From a general perspective, the spatial layout of the nodes indicates a relatively fragmented but evolving citation ecosystem, where a few core works exhibit a dominant citation footprint, while others remain peripheral despite their potential relevance. The size and density of the nodes—notably for authors like Dmuchowski (2021), Desalegn (2022), and Bak (2023)—signal higher levels of scholarly attention and influence, suggesting these publications serve as reference points for subsequent studies in the green finance domain.

The chronological spread of cited works is also telling. With visible representation from every year between 2020 and 2025, the figure underscores a sustained scholarly interest in the intersection between finance and sustainability in emerging markets. However, earlier publications such as Cigu (2020) or Dmuchowski (2021) occupy central positions due to longevity and their methodological rigour or conceptual contributions that resonated with subsequent authors.

Furthermore, the authors' diverse geographic affiliations—ranging from Poland, Romania, and Hungary to Türkiye and Azerbaijan—reflect the regional diversification of green finance discourse, fueled by policy alignment with EU sustainability

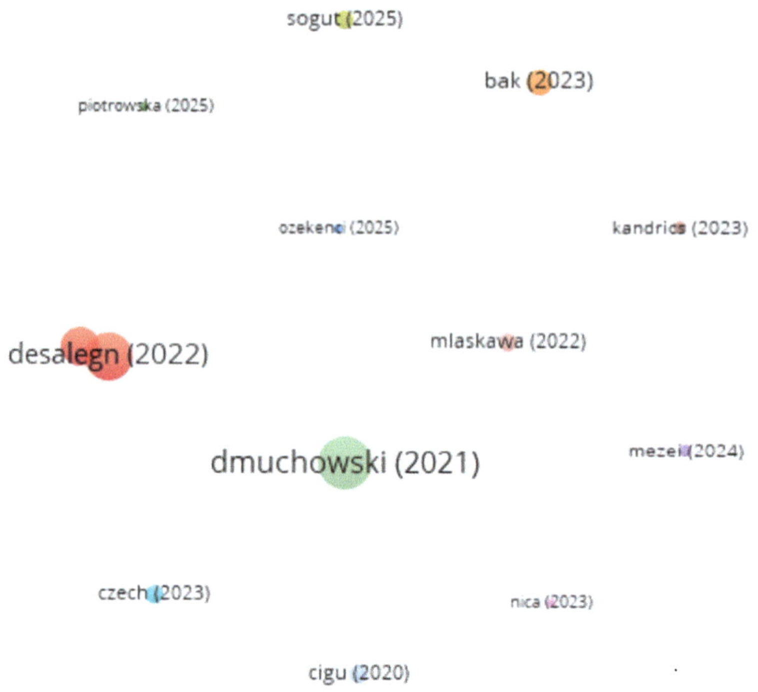

Fig. 4 Most cited publications in green finance in emerging economies. (Source: Own Data Processing Using VOSviewer, 2025)

frameworks and international financial commitments (e.g., the Paris Agreement or the European Green Deal). The emergence of newer contributors in 2023–2025 suggests an ongoing generational transition and academic rejuvenation in this research niche.

Another noteworthy observation is the limited co-citation clustering, which indicates that while the field is growing, the theoretical or methodological convergence across authors is still in its nascent phase. This dispersion of citation pathways suggests multiple analytical approaches are being applied, ranging from econometric modelling and bibliometric analysis to case study evaluations and policy impact assessments.

As shown by the bibliometric analysis, the article by Dmuchowski et al. (2021) titled "Green economy – growth and maintenance of the conditions of green growth at the level of Polish local authorities", published in the Journal of Cleaner Production, stands out with the highest citation impact (30 citations in the Web of Science Core Collection). This study offers a comprehensive assessment of the structural and financial constraints local governments in Poland face as they attempt to implement sustainable development strategies amidst pressing environmental challenges. The authors emphasise that Poland's reliance on coal as a primary energy source has resulted in some of Europe's highest urban air pollution levels, which makes the need for climate adaptation and mitigation investments at the local level all the more urgent. The paper's strength lies in its critical analysis of the financial mismatches and institutional gaps that hinder the transition toward green growth at the municipal level. It identifies key barriers, such as the limited engagement of the private sector in local financing, regulatory and fiscal rigidities, and the lack of risk-mitigation tools like currency hedging for local authorities. Furthermore, the authors highlight the growing investment gap between the climate goals and the actual capacity of local governments to finance green infrastructure. The study's relevance is amplified by its emphasis on the need for new policy instruments, including guarantee schemes, innovative financial mechanisms, and potentially new institutions tailored to support the green transition at the local scale. Notably, the article contributes both theoretically and practically to the green finance discourse. It bridges the gap between sustainable finance and public policy by demonstrating how aligning local investment priorities with national climate goals remains inadequate without enabling financial architecture. The policy recommendations, such as increasing access to sustainable finance for municipalities and establishing central coordination structures, resonate strongly with the broader EU Green Deal objectives. As such, this article serves as a reference point for policymakers and researchers interested in the nexus between local governance, environmental finance, and sustainable development.

The second most cited paper in the dataset is authored by Desalegn, Fekete-Farkas, and Tangl (2022). It is entitled "The Effect of Monetary Policy and Private Investment on Green Finance: Evidence from Hungary", published in the Journal of Risk and Financial Management. With 23 citations in the Web of Science Core Collection, this article provides a robust empirical investigation into how monetary dynamics and private capital flows influence the evolution of green finance in an emerging economy context. The authors adopt a quantitative research design, relying on quarterly time series data from 2013 to 2020, and apply advanced econometric techniques such as Johansen cointegration and vector error correction models to capture long-run and short-run relationships among variables. The study finds that interest rates significantly negatively influence green finance. At the same time, private and inward foreign direct investments (FDI) have strong positive effects in both the short and long term. On the contrary, outward FDI and greenhouse gas emissions negatively correlate with green financial flows. Importantly, the role of domestic private investment emerges as the most influential determinant in the variance decomposition of green financing, suggesting that internal capital dynamics may be more critical than external factors in Hungary's green transition. The study also uncovers that most Hungarian credit institutions prioritise corporate over sustainability strategies, underscoring a potential mismatch between institutional behaviour and national environmental goals. This article contributes to the literature by highlighting how macroeconomic tools—particularly interest rate manipulation—can catalyse or hinder sustainable investment. The findings hold substantial policy implications, especially for central banks and financial regulators in emerging economies, who are encouraged to blend neoliberal and reformist approaches to foster green finance. Moreover, the paper supports the argument that the banking sector plays a pivotal role in operationalising the Sustainable Development Goals (SDGs) by aligning credit policies with environmental objectives. In the broader scope of green finance scholarship, this work exemplifies the intricate interplay between fiscal instruments, institutional behaviour, and sustainable economic transformation in post-transition economies like Hungary.

The third most cited article in the dataset is "Responsible Lending Policy of Green Investments in the Energy Sector in Poland" by Pyka and Nocon (2021), published in the journal *Energies*. With 13 citations, the study offers an insightful perspective on the role of commercial banks in supporting green transformation through lending practices, focusing particularly on Poland's energy sector, one of the most carbon-intensive areas in Eastern Europe. The paper's primary objective is to explore the determinants of changes in bank lending associated with green energy investments and to assess whether banks are effectively transitioning toward responsible and sustainable finance frameworks. The research design combines quantitative and qualitative methods, comprising a two-stage empirical investigation. The first stage involves a questionnaire-based survey distributed to key decision-makers in Poland's largest commercial banks to assess their acceptance of sustainable financing principles and practices. The second stage builds upon case studies and secondary quantitative data analysis to validate observed trends and lending behaviours. The findings reveal banks' growing but uneven commitment toward sustainable finance, with clear sectoral tendencies to green the loan portfolio in favour of environmentally responsible projects. However, the study also notes that traditional, ecologically harmful projects dominate bank lending, especially in the energy market. A key contribution of the study lies in identifying structural and behavioural inertia within the Polish banking sector. While the awareness of environmental responsibility is increasing, this alone is insufficient to drive comprehensive change. The authors argue that a responsible lending policy that integrates environmental criteria into loan evaluation and approval requires complex business decisions, regulatory alignment, and long-term strategic vision. They emphasise the need for a more profound institutional transformation, where sustainability is embedded in the core business model of financial institutions rather than treated as a peripheral concern. Overall, this article provides a valuable empirical addition to the emerging literature on green finance in post-socialist economies. It highlights the potential and limitations of financial institutions in accelerating the green transition, particularly in high-impact sectors like energy. From a policy perspective, the paper underlines the urgency of reinforcing regulatory incentives and developing targeted instruments to steer the financial system toward more sustainable investment pathways. It positions the Polish experience as a relevant case study for other emerging European economies seeking to harmonise energy security, economic growth, and environmental sustainability.

4 Conclusions

This bibliometric study offers a comprehensive overview of green finance research's evolution, structure, and intellectual patterns in emerging economies. By systematically analysing a curated sample of 13 peer-reviewed publications, the study mapped the co-authorship networks, most influential contributors, and thematic concentrations, providing valuable insights into how this research domain has developed concerning sustainability and economic development imperatives. The findings confirm that green finance in emerging economies is a dynamic yet fragmented field, characterised by multidisciplinarity and influenced by regional policy frameworks and global environmental agendas.

One of the most significant contributions of this study lies in its ability to highlight the structural barriers and knowledge gaps that persist across the literature. Despite a growing interest in the subject, much of the green finance research remains concentrated in a limited number of countries—especially within the Central and Eastern European region—and often reflects localised experiences rather than transnational best practices. The thematic fragmentation—evident through co-occurrence analysis—suggests that while concepts like green bonds, sustainable banking, and climate-resilient investment are frequently discussed, they are rarely examined in a truly integrated or comparative manner.

The analysis also revealed a lack of robust methodological convergence across studies, with most employing qualitative assessments (e.g., case studies, institutional reviews) or descriptive quantitative techniques (e.g., usage statistics, regression models) that offer limited generalisability. Only a few articles employed advanced econometric or time-series tools to model causal relationships or to predict policy effects. Moreover, the observed co-authorship network shows low levels of cross-border academic collaboration, suggesting that knowledge diffusion remains bounded by institutional or linguistic barriers, limiting the emergence of a unified research front.

Another important conclusion is that most studies are policy-driven and pragmatic in scope, reflecting the urgent need to address the financing gap for sustainable development in emerging economies. However, this policy orientation does not always translate into operational solutions, and many papers fall short of proposing actionable frameworks or tools that governments, financial institutions, or private investors can adopt. This gap between theoretical contributions and policy application points to a critical area for future research and academic innovation.

Acknowledgments This work was supported by a grant from the Romanian Ministry of Research, Innovation and Digitalization, the project with the title *"Economics and Policy Options for Climate Change Risk and Global Environmental Governance"* (CF 193/28.11.2022, Funding Contract no. 760078/23.05.2023), within Romania's National Recovery and Resilience Plan (PNRR)—Pillar III, Component C9, Investment I8 (PNRR/2022/C9/MCID/I8)—Development of a program to attract highly specialised human resources from abroad in research, development, and innovation activities.

Competing Interests The authors have no conflicts of interest to declare relevant to this chapter's content.

References

Aria, M., & Cuccurullo, C. (2017). Bibliometrix: An R-tool for comprehensive science mapping analysis. *Journal of Informetrics, 11*(4), 959–975. https://doi.org/10.1016/j.joi.2017.08.007

Badareu, G., Doran, M. D., Firu, M. A., Olaru, S. M., & Doran, N. M. (2024). Financial development and climate change: A detailed bibliometric investigation. *Economies, 12*(288).

Bak, I., Budzen, D., Kryk, B., & Sobczyk, A. (2023). Financial involvement of local government units in achieving environmental objectives of sustainable development in Poland. *Economics and Environment, 86*(3). https://doi.org/10.34659/eis.2023.86.3.631

Battiston, S., Mandel, A., Monasterolo, I., Schütze, F., & Visentin, G. (2017). A climate stress-test of the financial system. *Nature Climate Change, 7*(4), 283–288. https://doi.org/10.1038/nclimate3255

Campiglio, E. (2016). Beyond carbon pricing: The role of banking and monetary policy in financing the transition to a low-carbon economy. *Ecological Economics, 121*, 220–230. https://doi.org/10.1016/j.ecolecon.2015.03.020

Cigu, E. (2020). The role of green finance: An overview. In *European Finance, Business and Regulation (EUFIRE 2020).*

Climate Policy Initiative. (2023). *Global Landscape of Climate Finance 2023*. Climate Policy Initiative. https://www.climatepolicyinitiative.org/

Czech, M., Hadas-Dyduch, M., & Puszer, B. (2023). Effectiveness of green bonds in selected CEE countries: Analysis of similarities. *Risks, 11*(12). https://doi.org/10.3390/risks11120214

D'Orazio, P. (2022). Green macro-financial policies in emerging markets: Signals and patterns of recent developments. *Environmental Economics and Policy Studies, 24*(1), 1–23. https://doi.org/10.1007/s10018-021-00301-z

Desalegn, G., Fekete-Farkas, M., & Tangl, A. (2022). The effect of monetary policy and private investment on green finance: Evidence from Hungary. *Journal of Risk and Financial Management, 15*(3). https://doi.org/10.3390/jrfm15030117

Dmuchowski, P., Dmuchowski, W., Baczewska-Dąbrowska, A. H., & Gworek, B. (2021). Green economy – Growth and maintenance of the conditions of green growth at the level of Polish local authorities. *Journal of Cleaner Production, 301*. https://doi.org/10.1016/j.jclepro.2021.126975

Donthu, N., Kumar, S., Mukherjee, D., Pandey, N., & Lim, W. N. (2021). How to conduct a bibliometric analysis: An overview and guidelines. *Journal of Business Research, 133*, 285–296. https://doi.org/10.1016/j.jbusres.2021.04.070

Ellegaard, O., & Wallin, J. A. (2015). The bibliometric analysis of scholarly production: How great is the impact? *Scientometrics, 105*(3), 1809–1831. https://doi.org/10.1007/s11192-015-1645-z

European Commission. (2018). *Final report of the high-level expert group on sustainable finance*. Publications Office of the European Union. https://ec.europa.eu/info/publications/180131-sustainable-finance-report_en

European Commission (EC). (2021). *Strategy for Financing the Transition to a Sustainable Economy*. European Commission. https://finance.ec.europa.eu/publications/strategy-financing-transition-sustainable-economy_en

G20 Green Finance Study Group. (2016). *G20 Green Finance Synthesis Report*. G20 Green Finance Study Group. https://g20sfwg.org/

Flammer, C. (2021). Corporate green bonds. *Journal of Financial Economics, 142*(2), 499–516. https://doi.org/10.1016/j.jfineco.2021.05.014

Intergovernmental Panel on Climate Change (IPCC). (2022). *Climate Change 2022: Mitigation of Climate Change. Contribution of Working Group III to the Sixth Assessment Report of the Intergovernmental Panel on Climate Change*. Cambridge University Press. https://www.ipcc.ch/report/ar6-wg3/

Kandrics, C. (2023). Financing a sustainable economy in Hungary, opportunities and challenges: Decarbonisation, green transition, sustainable finance, central bank. *Public Finance Quarterly – Hungary, 69*(1). https://doi.org/10.35551/PFQ_2023_1_2

Lee, C.-C., & Lee, C.-C. (2022). Green finance and green total factor productivity: Evidence from China's provincial data. *Energy Economics, 109*, 105958. https://doi.org/10.1016/j.eneco.2022.105958
Linnenluecke, M. K., Chen, X., Ling, X., Smith, T., & Zhu, Y. (2017). Research in finance: A review of influential publications and a research agenda. *Pacific-Basin Finance Journal, 43*, 188–199. https://doi.org/10.1016/j.pacfin.2017.04.005
MacNeil, I., & Kokoric, A. (2021). Sustainable finance and the EU green bond standard. *Law and Financial Markets Review, 15*(1), 12–24. https://doi.org/10.1080/17521440.2021.1881896
Mezei, C. (2024). Developing a green financing framework for local municipalities in Hungary. *Tér és Társadalom, 38*(3). https://doi.org/10.17649/TET.38.3.3585
Mlaskawa, J. (2022). Does taxonomy in banking support sustainable development of Polish companies? Entrepreneurship and sustainability. *Issues, 10*(1). https://doi.org/10.9770/jesi.2022.10.1(14)
Monasterolo, I. (2020). Climate-related financial risk: An emerging field in climate finance. *Climate Policy, 20*(6), 752–764. https://doi.org/10.1080/14693062.2019.1704678
Network for Greening the Financial System. (2019). A call for action: Climate change as a source of financial risk. https://www.ngfs.net/en/call-action
Network for Greening the Financial System (NGFS). (2021). *NGFS Progress Report on Bridging Data Gaps*. NGFS. https://www.ngfs.net/
Nica, A. (2023). The banking system – Limping toward a sustainable future. Is green finance a thing in Romania as of March 2023? *Europolity – Continuity and Change in European Governance, 17*(2), 209–255.
Organisation for Economic Co-operation and Development. (2017). *Investing in climate, investing in growth*. OECD Publishing. https://doi.org/10.1787/9789264273526-en
Ozekenci, S. Y., & Nur, T. (2025). Determining the role of green logistics and green finance in achieving carbon neutrality. *LogForum, 21*(1). https://doi.org/10.17270/J.LOG.001147
Piotrowska, A. I., & Piotrowski, D. (2025). Green FinTech: A consumer awareness study. *Sustainability, 17*(8). https://doi.org/10.3390/su17083701
Pyka, I., & Nocon, A. (2021). Responsible lending policy of green investments in the energy sector in Poland. *Energies, 14*(21). https://doi.org/10.3390/en14217298
Ravikumar, S., Agrahari, A., & Singh, S. N. (2015). Mapping the intellectual structure of Scientometrics: A co-word analysis of the Journal Scientometrics (2005–2010). *Scientometrics, 102*(1), 929–955. https://doi.org/10.1007/s11192-014-1445-7
Sogut, Y., Demirtas, N., Teksan, E., Torusdag, M., Inal, V., & Aydin, M. (2025). Building a sustainable environment in EU countries: The critical role of infrastructure investments and institutional quality. *International Journal of Sustainable Development and World Ecology, 32*(1). https://doi.org/10.1080/13504509.2024.2410763
Taghizadeh-Hesary, F., & Yoshino, N. (2019). *The way to induce private participation in green finance and investment*. ADB Institute working paper series. https://www.adb.org/publications/way-induce-private-participation-green-finance-and-investment
United Nations Environment Programme (UNEP). (2021). *State of Finance for Nature: Time to Act – Doubling Investments by 2025 and Tripling by 2030*. UNEP. https://www.unep.org/resources/state-finance-nature
United Nations Environment Programme Finance Initiative (UNEP FI). (2019). *Principles for Responsible Banking: Shaping the Future of Banking*. UNEP FI. https://www.unepfi.org/banking/bankingprinciples/
Verma, S., & Gustafsson, A. (2020). Investigating the emerging COVID-19 research trends in the field of business and management: A bibliometric analysis approach. *Journal of Business Research, 118*, 253–261. https://doi.org/10.1016/j.jbusres.2020.06.057
Volz, U., Böhnke, J., Knierim, L., Richert, K., & Roehrl, R. A. (2015). *Financing the green transformation*. UNEP inquiry working paper. https://wedocs.unep.org/handle/20.500.11822/20624
World Bank. (2022). *Mobilizing Private Finance for Climate Action in Emerging Markets*. World Bank Publications. https://www.worldbank.org/

Yu, Z., Khan, S. U., & Zhang, R. (2021). Green credit policy and firm-level green innovation: Evidence from China. *Journal of Cleaner Production, 278*, 123403. https://doi.org/10.1016/j.jclepro.2020.123403

Zhang, Y., Zhou, X., & Zhou, D. (2019). Green finance and sustainable development in Belt and Road Initiative countries: The role of R&D and human capital. *Environmental Science and Pollution Research, 26*(36), 37136–37152 https://doi.org/10.1007/s11356-019-07177-6

Leveraging Technological Innovations and FinTech Solutions for Sustainable Green Finance

Cristina Criste, Hadasa Ligia Pricopiuc, and Oana-Ramona Lobonț

1 Introduction

In the context of intensifying climate change, resource depletion and promoting social inequalities, this study examines the role of Fintech and related technological innovations in sustainability and carbon reduction in Europe, highlighting their contribution to achieving financial sustainability and environmental protection goals. To support the transition to a green economy, the importance of technological innovation with a focus on financial technology (FinTech) was highlighted to change the way financial resources are mobilised and directed towards sustainable projects (Iqbal et al., 2025; Mirza et al., 2023). FinTech has become a fundamental component of the financial sector, making a significant contribution to the development of green finance (Muganyi et al., 2021). In addition, FinTech is involved in digital payments and other technological innovations, such as AI-based algorithms (Panetta et al., 2023). By incorporating environmental, social and governance (ESG) metrics, digital innovation is reshaping the core services and eco-friendly bond transactions (Mertzanis et al., 2025). Information technologies such as blockchain, big data and bitcoin have a beneficial effect on companies' production factors and promote the adoption of green finance (Xu et al., 2023; Yáñez-Valdés & Guerrero, 2024). In China and 58 other developed countries, FinTech has been shown to promote technological advancement and environmental sustainability. It reduces carbon

C. Criste · H. L. Pricopiuc (✉)
Faculty of Economics and Business Administration, Doctoral School of Economics and Business Administration, West University of Timişoara, Timişoara, Romania
e-mail: cristina.criste@e-uvt.ro; hadasa.gag01@e-uvt.ro

O.-R. Lobonț
Faculty of Economics and Business Administration, Department of Finance, Business Information Systems and Modelling, West University of Timişoara, Timişoara, Timiş County, Romania
e-mail: oana.lobont@e-uvt.ro

C.-W. Su et al. (eds.), *Green Finance and the Challenges of Climate Change*, Sustainable Finance, https://doi.org/10.1007/978-3-032-07293-1_10

emissions and energy consumption, fostering the sustainable growth of the real economy and alleviating environmental damage (Cheng et al., 2023; Kashif et al., 2024; Qin et al., 2024a). The path to digital finance success relies on technological innovations and economic behaviours that influence digital finance (Dong et al., 2025). The same impact can be observed among OECD countries, highlighting how FinTech and green innovations can reduce carbon emissions through the efficient use of natural and economic resources (He et al., 2024). The uptake of FinTech can promote sustainable economic development (Mirza et al., 2023). Through its green attributes, FinTech can contribute to Europe's Green Deal, a pact that aims for Europe to achieve climate neutrality by 2050 and emphasizes the importance of technology in achieving this goal (European Parliament, 2018). To support the European Green Deal, a set of "Fit for 55" proposals have been developed to reduce emissions by up to 55% by 2030 (European Council, 2025). In order to stimulate sustainable financing, the EU has defined the general conditions that must be met for an economic activity to be sustainable, known as the EU taxonomy. It redefines how sustainability is measured and integrated into investors' financial decision-making processes (European Commission, n.d.).

This paper contributes to the existing literature on this topic by providing additional information. Previous studies have investigated the impact of FinTech environmental issues as well as the connection between FinTech and economic factors (Zia et al., 2024; Awais et al., 2023). Although studies have shown that FinTech has a significant environmental impact, a survey showed that respondents did not perceive a positive relationship between the two (Piotrowska & Piotrowski, 2025). Interest in the topic of how FinTech impacts carbon emissions is on the rise, yet few studies on this topic have been conducted in Europe (Le & Pham, 2024; Obobisa & Ahakwa, 2024). The rapid evolution of FinTech in Europe is supported by a strong digital infrastructure and regulatory frameworks such as PSD2 (Revised Payment Services Directive) (European Central Bank, 2018) and open banking initiatives. Therefore, this study focuses on understanding how FinTech innovations can contribute to reducing carbon emissions, a matter of significant importance. Empirical results confirm that technological innovation has a dual impact on FinTech development and the environment: it contributes positively to green finance and energy transition (Sreenu & Mishra, 2023; Wang et al., 2024), but can generate negative effects by intensifying economic activities and increasing emissions.

The data are analyzed using robust econometric methods, such as 2SLS and GMM, frequently applied in recent literature on topics such as: FinTech development and emission reduction (Tao et al., 2022), the impact of technology and the financial sector on public health (Jing et al., 2022) and the contribution of the media to green energy, energy and renewable energy (Wu et al., 2023). In furtherance of these research directions, our study aims to contribute to the understanding of green Europe, applying robust methods and focusing on the role of FinTech and technological innovation. Thus, we align ourselves with the findings of Hernández de Rojas et al. (2025), who highlights the role of regional competitiveness and shared vision in promoting economic cohesion and growth, as well as those of Lyulyov and Pimonenko (2025), who draw attention to the impact of external shocks and ecological constraints on EU convergence. In this context, European policies targeting digitalisation, sustainability and resilience become essential for an effective transition.

This research makes several significant contributions. The innovative nature of our research in comparison with previous studies is underpinned both by the way in which the empirical analysis is carried out, in particular the techniques used, and by the way in which the research framework was set up. This paper offers a novel investigation into the direct and indirect effects of FinTech development on carbon emissions across European countries, an area that remains insufficiently explored, despite the region's strong commitment to sustainability policies. The study adopts an innovative approach by examining these complex dynamics using two robust econometric techniques: Generalised Method of Moments (GMM) and Two-Stage Least Squares (2SLS) regression. These methods enable us to demonstrate the pattern and direction of influence of FinTech development on all the considered factors. Technological innovation indicators, environmental degradation and economic expansion were chosen for analysing the impact of financial technologies and digital transformation. The study focuses on Europe because several interrelated factors make the region a relevant and strategic case for investigating the relationship between FinTech, technological innovation and environmental sustainability. Firstly, Europe is a global leader in environmental regulation and climate policy. The developed initiatives that support this claim include the European Green Deal (European Parliament, 2018), the Fit for 55 package (European Council, 2025), and the EU Taxonomy for Sustainable Activities (European Commission, n.d.). Europe supports the fintech ecosystem and its rapid evolution through open and digital banking infrastructures, including PSD2 (European Central Bank, 2018). The rich context for analysis is provided by the variation in the impacts of financial technologies depending on the economic structures, environmental performance and energy profiles of different European countries (Llorca & Rodriguez-Alvarez, 2024). This heterogeneity enables a nuanced understanding of how FinTech affects the environment in different ways, depending on national context, institutional quality, and levels of technological readiness (Mertzanis, 2023). Moreover, the study results suggest a stable and dynamic long-term relationship between FinTech development, technological innovation and environmental degradation. These findings are consistent with the model developed by previous studies that explored the innovation-fragility perspective. For instance, FinTech technological innovations are shown to increase pollution in the short term but decrease it in the long term. However, the impact of FinTech depends on economic and environmental circumstances, which makes it difficult (Jie et al., 2024). Another study has shown that high carbon emissions from FinTech usage and resource availability deteriorate the quality of the environment (Tu, 2024). Further research indicates that the impact of FinTech on social-environmental performance is small yet statistically significant (Mertzanis, 2023). For us, this is shown by the negative effects of FinTech on the environment, even though there is the possibility for efficiency improvements. Furthermore, the findings of this study are particularly valuable for understanding how common strategies in the area of digital finance and climate action can be coordinated effectively, given the EU's focus on cohesion policy and cross-border financial integration.

This paper is organised into distinct sections. The first section outlines the research framework, emphasising the importance of FinTech and technological innovations in the environment. Section 2 reviews relevant studies on the topic, highlighting advances in understanding FinTech technologies and their impact on climate issues. The growing interest in FinTech, technological innovations, and carbon emissions demonstrates the importance of this subject. Section 3 presents the methodology employed: 2SLS and GMM. This section outlines the underlying research hypotheses. Section 4 presents the data sources used in the research, including the relevant datasets underlying the analysis and information on descriptive statistics. Section 5 systematically presents the results, often using tables and figures to illustrate the impact of FinTech on CO_2 emissions and the relationship between FinTech, R&D and emissions. The final section summarises the study's conclusions, synthesising the key insights derived from the analysis. This synthesis reiterates the importance of the findings and discusses their broader implications for policymakers.

2 Literature Review

In recent years, the question of whether technological innovation catalyses sustainable green finance has attracted considerable attention. Considering global economic policy uncertainty, we can address this question by examining the impact of technological innovation on emissions. According to Huang (2025), contemporary research in FinTech and sustainability has demonstrated that these two areas should not be studied in isolation due to their profound interconnections. Almaqtari (2024) states that FinTech maintains the connection between technology and sustainability, and that technological innovation enhances FinTech performance. Tu (2024) concludes that, although FinTech and innovative technologies have considerable sustainability potential, existing research is limited and requires further exploration.

Liu et al. (2023) emphasise the positive correlation between FinTech development and green innovations within enterprises, highlighting FinTech's contribution to the transition to a green economy at a micro level. Du et al. (2022b) also support the positive relationship between FinTech and green innovation. As carbon emissions are a key contributor to climate change, Song et al. (2023) reveal that digital finance reduces CO_2 emissions in Chinese cities, particularly in the east, by promoting technological innovation. Similarly, Du et al. (2022a) show that the rapid development of FinTech contributes significantly to ESG performance, which is essential for achieving China's "dual carbon goal" and sustainable development of the whole society. It has also been confirmed that Fintech can alleviate firms' financial constraints. Wang et al. (2023) concluded that in China, energy security has the most significant influence on CO_2 emissions. Still, government green expenditures do not produce a noticeable long-term impact on CO_2 emissions, regardless of the pollution level. Using evidence from Australia, Anwar et al. (2024) found that green finance supported by FinTech solutions contributes to carbon reduction by combining financial and technological innovations. Combining the two increases efficiency and reduces waste in financial processes. Li and Zhou's (2024) study of 281 Chinese cities demonstrates that developing the digital economy significantly reduces urban carbon emissions, particularly in regions with high market liberalisation. However, Ahmad et al. (2024) say that although favouring green energy, globalisation and economic growth simultaneously amplify the ecological footprint, negatively impacting sustainability. Mavlutova et al. (2025) argue that green digital investments reduce emissions and support renewable energy. However, the EU's lack of digital literacy and standardisation limits the real impact of these investments on sustainability.

In contrast, Li et al. (2024) found that FinTech and financial markets increased carbon emissions in Asian economies, whereas green innovations and access to financial institutions supported sustainable green finance. Jia et al. (2024) argue that FinTech can reduce carbon emissions by promoting green investment and transparency. However, it can also support polluting industries, thereby undermining the balance between economic growth and environmental protection. Ahmad et al. (2024) reveal that FinTech accelerates the green transition and reduces CO_2 emissions in 25 European Union countries.

Technological innovations, natural resources, FinTech, climate and economic performance are all interconnected, and further investigation is required to determine whether negative environmental influences can be transformed into positive ones, particularly in Europe (Ahmad et al., 2024; Samour et al., 2024).

According to Kowalewski and Pisany (2023), FinTech development is positively influenced by the quality of research and the level of collaboration between academia and industry. Huang et al. (2025) found that advances in FinTech stimulate investment in R&D in innovative sectors, thereby fostering regional economic output growth. In terms of bibliometric analysis, Judijanto et al. (2024) identify a growing interest in sustainable finance and emphasise the key contribution of technologies such as FinTech and artificial intelligence in promoting sustainable finance globally. Puschmann and Khmarskyi (2024), by analysing 74 relevant academic papers, demonstrate that green FinTech is an emerging and under-explored field with significant potential to support sustainable finance by providing technological solutions dedicated to environmental sustainability. Du et al. (2022b) state that green finance policies stimulate technological innovation and improve the performance of green firms, particularly in regions with advanced financial digitalisation and a strong focus on sustainability. Furthermore, Irfan et al. (2022) point out that even though green finance can stimulate technological innovation and financial performance, its effectiveness is limited in areas with low digitalisation.

Samour et al. (2024) emphasise that using renewable energy and promoting financial inclusion can greatly support environmental sustainability. However, technological innovation and economic growth tend to have a negative impact on environmental quality. Xiao et al. (2024) prove that FinTech reduces the ecological footprint by amplifying the positive effects of the circular economy and technological innovation, but only partially offsets the increase in the environmental footprint caused by the exploitation of natural resources. According to Chen and Liu (2024), countries with advanced technology and abundant renewable energy sources exhibit negative material footprints. However, the positive effects of FinTech, GDP and globalisation must also be considered. The findings are further reinforced by Razzaq et al. (2021), who concluded that infrastructure development, GDP, globalisation, and population growth all positively impact the material footprint. Furthermore, Qin et al. (2024b) found that technological development helps mitigate climate risk in a particular area and worldwide. However, its effect may not be observed during an economic recession.

Croutzet and Dabbous (2021) highlight the essential role of FinTech in stimulating the adoption of renewable energy in OECD economies, by facilitating access to finance, reducing investment risks and increasing transparency, while a slight increase in GDP is associated with higher consumption of renewable energy. Teng and Shen (2023) highlight that the impact of FinTech on energy efficiency is positive but varies depending on the level of economic development. In a study dedicated to Western European countries, Oyebanji and Kirikkaleli (2023) show that green electricity and innovation reduce CO_2 emissions, while economic growth has the opposite effect. The authors highlight the importance of green electricity in combating climate change, given its central role in reducing emissions globally.

Yoo and Yi (2022) imply that the use of digital technologies is altering the economic and social framework, impacting economic growth, income inequality, and changes in work patterns and social interactions, thereby affecting the well-being of society. In addition, Demir et al. (2022) emphasise that while FinTech supports the UN goals and the G20 Principles on Digital Financial Inclusion, its benefits are particularly concentrated in high-income countries. Moreover, Arner et al. (2020) argue that although FinTech supports financial inclusion and the UN's SDGs (United Nations Sustainable Development Goals), regulatory risks and digital exclusion persist, as one-third of older people in the EU have never used the internet, and 1 in 5 people have low digital skills.

Various aspects have been examined in previous studies regarding the impact of FinTech technologies, for example carbon emissions (Anwar et al., 2024; Song et al., 2023; Li & Zhou, 2024; Li et al., 2024), environmental innovations (Liu et al., 2023; Du et al., 2022b), technology innovations (Judijanto et al., 2024; Puschmann & Khmarskyi, 2024; Irfan et al., 2022), renewable energy (Croutzet & Dabbous, 2021; Teng & Shen, 2023), material and ecological footprint (Chen & Liu, 2024; Razzaq et al., 2021; Xiao et al., 2024; Ahmad et al., 2024), R&D (Huang et al., 2025; Kowalewski & Pisany, 2023), economic growth (Yoo & Yi, 2022) and financial inclusion (Demir et al., 2022; Arner et al., 2020). Also, we can see illustrated in Fig. 1. what are the negative and positive effects that impact upon sustainable green finance, effects supported by both past and present studies. The reviewed studies show that this topic has not yet been adequately analysed in Europe.

Numerous studies explore the impact of FinTech on environmental sustainability (Macchiavello & Siri, 2020; Xiao et al., 2024; Tu, 2024; Li et al., 2024; Jie et al., 2024; Kakar et al., 2025), but few focus specifically on Europe as a cohesive region

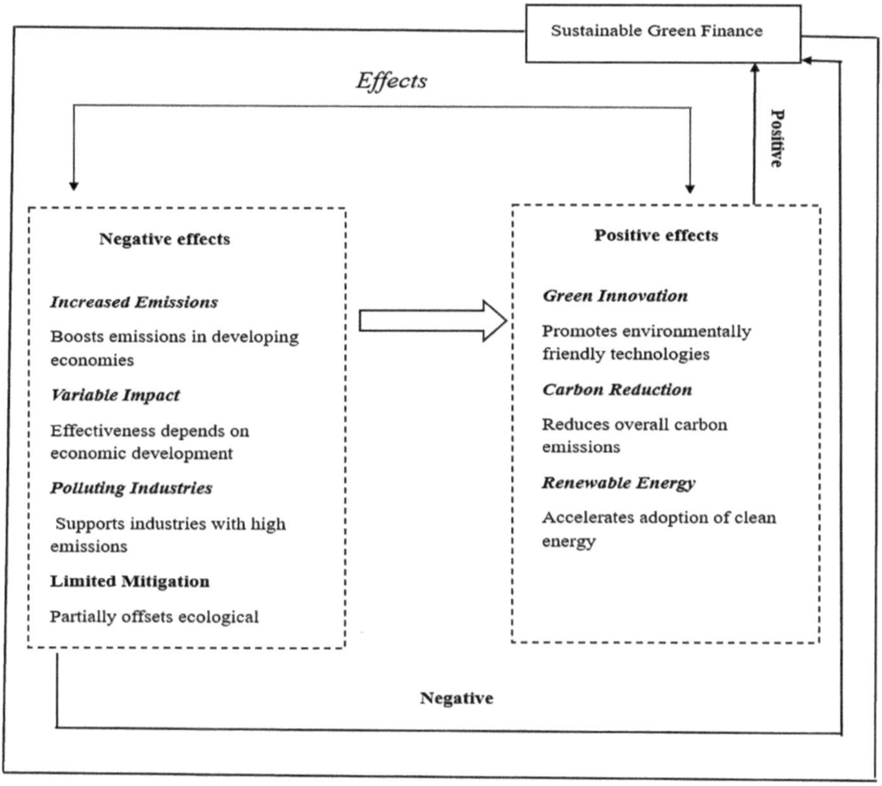

Fig. 1 Understanding the spectrum of digital finance's environmental impact

(Oyebanji & Kırıkkaleli, 2023). The existing literature often examines individual countries or sub-regions, neglecting the systemic interactions (Du et al., 2022a, b; Hazla, 2023; Muhammad et al., 2022) and policy implications across the entire European landscape. For instance, some researchers focus on the Polish financial sector (Hazla, 2023) or on specific Western European countries (Oyebanji & Kırıkkaleli, 2023). Thus, a dedicated study is significant to understand the nuanced relationship between FinTech development, technological innovation and environmental outcomes within the region, given Europe's unique economic and political integration efforts, which are driven by EU directives and frameworks. This is particularly important considering the EU's commitment to sustainability through initiatives like the EU Taxonomy (European Commission, n.d.; Bassen et al., 2025; Davrinche et al., 2025) and the Sustainable Development Goals (Arner et al., 2020; Mavlutova et al., 2025; Choudhary et al., 2025; Jie et al., 2024), which necessitate a coordinated and comprehensive approach to green finance and FinTech policies.

Current research often examines the relationship between FinTech and environmental impact in a fragmented manner. Only a few factors show the relationship, such as CO_2, technological innovation, and renewable energy. However, these factors are either studied in isolation or combination, without considering the bigger picture. Therefore, a more systemic perspective is needed, considering the interplay between FinTech development, CO_2 emissions and material footprint, technological innovations, and green policies (Du et al., 2022b). This holistic approach would provide a more comprehensive understanding of how these factors interact to encourage or hinder sustainable development, enabling the design of effective policy interventions.

Contrary to existing studies, this paper uses 2SLS and GMM methods to investigate the impact of FinTech and technological innovations on carbon emissions. Additionally, the research gap reflected in the literature is the distinction between FinTech and Green FinTech, which underlines the need for an integrated approach to explore the intersection between these fields (Kozar & Wodnicka, 2024). In summary, we can conclude that FinTech and sustainability are key areas to research to understand how to harness financial innovation to achieve environmental goals. This requires a nuanced approach to recognise the opportunities and challenges of FinTech in a context of globalisation and natural resource management. This research contributes to overcoming the outlined limitations by integrating rigorous empirical analysis and utilising a broad framework of relevant economic factors. To examine the complex relationship between FinTech, technological innovation, and environmental outcomes, this study applies two robust econometric methods: 2SLS and GMM (Afzal et al., 2022). These techniques address endogeneity and omitted variable bias, with GMM allowing dynamic panel analysis and strengthening the reliability of policy-relevant findings (Dabbous et al., 2025).

3 Methodology

This study aims to evaluate the impact of FinTech development and technological innovations on environmental quality: CO_2 Emissions and Mat. Footprint, in Europe over the 2010–2022 period, investigating whether FinTech facilitates the transition to lower carbon emissions and reduced greenhouse gases, by employing complex empirical methods to establish a causal link with CO_2 emissions and material footprint, in the context of sustainable green finance and key macroeconomic variables, namely, GDP (Destek & Manga, 2021; Kan et al., 2022; Muhammad et al., 2022). Including gross domestic product, research and development expenditures, and environment-related variables such as CO_2 emissions and carbon footprint as determinants is a strongly supported phenomenon (Kumar et al., 2025; Sikdar & Mukhopadhyay, 2017; Tnani, 2018). Likewise, the negative role played by FinTech development on the environment and energy consumption has been predominantly reported by (Dilek & Furuncu, 2019; Hileman & Rauchs, 2017; Krause & Tolaymat, 2018; Jangid et al., 2025). However, it should be mentioned that opinions regarding this correlation are divided, whereas the positive contribution of FinTech development to the environment has also been strongly supported in the literature (Sreenu & Mishra, 2023; Aboulajras et al., 2025; Muhammad et al., 2022).

Based on those methodologies utilised in prior studies by Ozturk and Ullah (2022), Tao et al. (2022), and Yao et al. (2021), the present research included comparable values and related methodological specifications. To fulfil our general objective and following the methodological procedures that are desired to be applied, we have formulated the following hypotheses for testing:

H1: The expansion of FinTech companies is positively associated with CO_2 emissions, indicating that an increase in the number of FinTech firms contributes to higher environmental degradation.

H2: R&D expenditures moderate the relationship between FinTech development and environmental impact, in that higher levels of R&D expenditures reduce the positive effect of FinTech expansion on CO_2 emissions and material footprint.

Our analysis framework focuses mainly on the presumption that propagates the evaluation of both the positive and negative roles of FinTech within a complex cross-sectional framework. This study aimed to evaluate the role of FinTech development and its contribution to the transition towards Sustainable Green Finance. Therefore, the Empirical model framework can be described as:

$$Emissions = f\left(FinTech\,Development, Control\,Variables\right) \tag{1}$$

This methodology can be conceptualised as a regression model, specifying the emissions proxy as a dependent variable, along with the independent variable proxy and relevant control variables.

$$\mathrm{y} = \beta_0 + \beta_1 x_1 + \cdots + \beta_k x_k + \varepsilon \tag{2}$$

Equation 2 y denotes emissions, representing the average of data at the European level for 37 EU and non-EU member states taken as a whole, regarding CO_2 emissions and carbon footprint. Subsequently, x1 signifies the FinTech Development proxy, while xk is the vector of macroeconomic control variables, which includes Gross Domestic Product. Given the non-linear character observed in the descriptive statistical table, the model specification will be considered in logarithmic form. Within the log-log or log-linear specifications, the estimated coefficients represent elasticities; this can be interpreted as the percentage change in the dependent variable resulting from a percentage change in the independent variable. This specification is also widely accepted in the literature on greenhouse gas emissions. The final model is specified according to the two models represented by Eqs. 3 and 4.

$$\log\left(CO_2\right) = \beta_0 + \beta_1 \log\left(FinTech\right) + \beta_2 \log\left(TI\right) + \beta_3 \log\left(GDP\right) + \varepsilon_i \tag{3}$$

$$\log\left(Mat.footprint\right) = \beta_0 + \beta_1 \log\left(FinTech\right) + \beta_2 \log\left(TI\right) + \beta_3 \log\left(GDP\right) + \varepsilon_i \tag{4}$$

Equations 3 and 4 can be estimated using numerous data methods. These methods include instrumental variable methods. Classical estimation methods, such as FE and RE, may provide biased and inconsistent estimates due to endogeneity issues. Therefore, using the 2SLS approach in basic models is preferable to address the endogeneity problem. GMM approach for Eq. (3) developed by Arellano and Bond (1991), which can eliminate the problem of autocorrelation, heteroscedasticity, and endogenous variables, and GMM is considered to be an ideal approach for empirical studies involving data limitations such as periods and a large number of intersections. Since our group data contains 37 economies taken as a whole, stability is achieved by averaging them over a relatively small period of 13 years (2010–2022), and it is deduced that GMM is an appropriate econometric method. Thus, to avoid problems related to endogeneity, autocorrelation, and heteroscedasticity, this study used 2SLS and GMM approaches.

4 Data

This study evaluates the impact of FinTech (FT) and technological innovations (TI) on environmental quality: CO_2 Emissions (CO_2) and Mat. Footprint (MF), investigating whether FinTech development facilitates a smooth transition towards lower carbon emissions and greenhouse gas levels, in 37 European economies, using the average value of each economy over the 13 years from 2010 to 2022. A series of variables was used to gain a nuanced understanding of these dynamics. For this research, data were collected from various official sources.

The Data were divided into four dimensions to better understand and operationalise the dataset. *I. FinTech Development*, the primary explanatory variable, presents the total number of FinTech firms operating in Europe. The data for this indicator comes from the Statista database and provides comprehensive and up-to-date information on the digital finance sector. *II. Technological Innovation*, technological innovation, is measured in our study by research and development expenditures as a percentage of gross domestic product. This indicator reflects a country's commitment to support and stimulate innovation, and the indicator is taken from the World Bank database. *III. Environmental Emissions*: the environmental results are evaluated using two key indicators: carbon dioxide emissions and material footprint. These values capture the ecological impact of economic and technological activities and are obtained from the Sustainable Development Agenda database; *IV. Sustainable Growth*, reflected in this case by GDP per capita, sourced from the World Bank database. The analysis includes this variable to account for potential income-related effects on both FinTech adoption and environmental outcomes.

Therefore, the analysis used data from official sources such as Statista, the World Bank, and official reports. The dataset was configured based on the primary evidence found in the scientific literature (Table 1).

The indicators retained in our dataset reflect different aspects of the relationship between technological advancement, economic development, and sustainability. As

Table 1 Description of the indicators retained in the analysis

Category	Indicator/metric	Source	Unit/description
FinTech Development	Number of FinTech firms	Statista	Count (firms)
Technological Innovation	R&D expenditure as % of GDP	World Bank	Percentage (%)
Environmental Emissions	CO_2 emissions	Sustainable Development Agenda	Metric tons (total or per capita)
	Material Footprint	Sustainable Development Agenda	Tons per capita
Economic Development	GDP (for normalisation/ reference)	World Bank	(current US$)

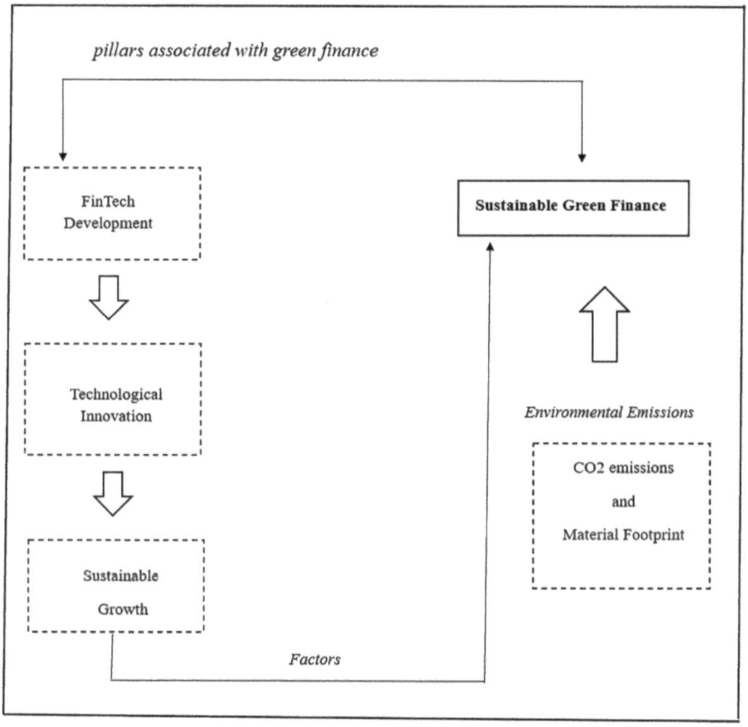

Fig. 2 Key variables and data sources used in the empirical analysis

illustrated in Fig. 2, the indicators were divided into three pillars associated with green finance sustainability: FinTech, Technological Innovations, and Economics.

The Total number of FinTech companies in Europe measures the first and most important FinTech variable. Technological innovations, which represent Adaptation to the environment, are measured by research and development expenditure as a percentage of GDP. This measure provides insights into the level of allocations

Table 2 Descriptive statistics

Variables	Mean	Std. Dev.	Min	Max	Skewness	Kurtosis
TI	302.692	39.61352	245.0000	370.0000	0.190496	1.863608
MF	8.283	0.907721	6.959459	9.670000	0.131111	1.871554
CO_2	2.125	2.754643	−4.763150	7.799853	−0.629256	5.208503
FinTech	20.533	1.050980	19.04243	22.33216	0.395312	2.078602
GDP	4942.385	2576.698	1754.000	8906.000	0.250664	1.591683

within the studied economy and reflects the degree of access to resources and tools. In this case, GDP reflects the value of environmental resources, and the costs associated with environmental degradation. While proxy variables, CO_2 emissions and Mat measure Emissions. Footprint (MF). MF indicates the average amount of raw materials, measured in tons, needed to support per capita consumption, while CO_2 allows us to track emission changes over time.

Table 2 provides detailed information regarding descriptive statistics, variable definitions, and data sources.

The study employs a structural set of variables that captures the multidimensional relationship between financial innovation and environmental sustainability. The selected variables form the model's foundation and support subsequent econometric estimations designed to test the study's basic hypotheses.

5 Results and Discussion

In order to evaluates the impact of FinTech (FT) and technological innovations (TI) on environmental quality: CO_2 Emissions (CO_2) and Mat. Footprint (MF), investigating whether FinTech development facilitates a smooth transition towards lower carbon emissions and greenhouse gas levels, we consider two proxies to measure emissions: CO_2 Emissions (CO_2) and Mat. Footprint (MF). This study considers both Two-Stage Least Squares (2SLS) regressions and the Generalised Method of Moments (GMM) to empirically investigate the relationship between FinTech, technological innovation (proxied by R&D), and CO_2 emissions and Mat.Footprint. Therefore, two distinct models were applied to provide a robust analysis of the impact of FinTech on polluting emissions.

To test and validate our first hypothesis, that an increased number of FinTech companies leads to higher CO_2 emissions, we started by building robust models focused on CO_2 emissions. These models are presented in Table 3, which shows the results for both the 2SLS regression and the GMM regression.

In Table 3, columns 2 and 3 present the results of the 2SLS model for FinTech and the CO_2 nexus, columns 4 and 5 illustrate the results of the GMM model, and column 4 reports the findings of the impact of FinTech on CO_2 for the GMM model. The findings of the 2SLS models demonstrate that FinTech exerts a significant positive impact, while IT exerts a negative effect on CO_2. Therefore, our results confirm

Table 3 Estimates of the model for CO_2 Emission (CO_2) using 2 two-stage least squares model and GMM as a robustness check

	2SLS		GMM	
Variable	Coefficient	T-stat	Coefficient	T-statistic
Constant	48.111***	2.99	39.18589	3.528873
FinTech	0.0023**	2.21	0.00155	2.072599
R&D	−0.1689**	−2.4	0.134362	1.535034
GDP_Growth	0.0422	0.75	−0.08787	−1.79723

T-stat in parentheses. ***$p < 0.01$, **$p < 0.05$, *$p < 0.1$

the hypothesis that the development of the FinTech sector is associated with an increase in CO_2 carbon emissions. Coefficient estimates show that a 1 percent increase in FinTech tends to increase CO_2 emissions by 0.23 percent, while a 1 percent increase in IT leads to a 0.16 percent decrease in CO_2 emissions, thus improving outcomes in the case of polluting emissions in the analysed economies. GDP tends to reduce CO_2 emissions in the GMM regression model; in this model, macroeconomic variables GDP is positively associated with improved environmental performance in European economies. GDP growth leads to a 0.04 percentage point increase in CO_2 emissions. This finding suggests that, according to this model, even with potential improvements in efficiency, technological advancements, and a shift toward cleaner industries, economic expansion is still associated with increased pollution in the context of this model (Akey & Appel, 2020; Hartman & Kwon, 2004; Marinaş et al., 2018; Ouyang et al., 2020).

The GMM regression model confirms the improvement of environmental performance in European economies. Conversely, the 2SLS model indicates that an increase in GDP leads to a 0.04 percentage point increase in CO_2 emissions. FinTech (0.00155) is represented by a positive and significant coefficient (5%); an increase in FinTech usage is associated with an increase in CO_2, with a 1-unit increase in the FinTech indicator leading to a 0.00155-unit increase in CO_2 emissions. Financial technologies can contribute to intensifying economic activities such as consumption, easy investments, and digital expansion, increasing energy consumption and, implicitly, emissions (Elheddad et al., 2020; Muhammad et al., 2022). This fact suggests a lack of integration of sustainability in FinTech development.

In the case of IT, the coefficient is positive but not statistically significant at 0.1344; R&D is not oriented towards green innovations. Investments in technology may not only be aimed at energy efficiency or sustainable solutions. In the case of GDP Growth, the coefficient is negative: −0.0879 and significant; the results suggest that an increase in GDP is associated with a decrease in CO_2 emissions. These results reflect a decoupling between macroeconomic variables GDP and pollution, possibly due to the transition to services, energy efficiency, or environmental regulations.

To test the second hypothesis, according to which the increase in research and development expenditures mitigates the positive relationship between the number of FinTech companies and emissions, as well as the material footprint, we

Table 4 Estimates of the model for Mat. footprint (MF) using a two-stage least squares model and GMM as a robustness check

	2SLS		GMM	
Variable	Coefficient	T-statistic	Coefficient	T-statistic
Constant	39.185888***	3.528873	0.059464	1.479402
FinTech	0.00155*	2.072599	0.081224	25.554009
GDP growth	0.134362	1.535034	−0.000889	−5.853834
R&D	−0.087868	−1.79723	0.059464	1.479402

T-stat in parentheses. ***p 0.01, **p 0.05, *p 0.1

conceptualised a second model, which considers Mat. footprint as a dependent variable. Table 4 presents the coefficient estimates representing the impact of FinTech on Mat. Footprint and the control variables TI and GDP in the case of the 2SLS and GGM models.

The results of the 2SLS model demonstrate that the coefficient associated with FinTech is positive (0.00155) and significant at a level of 10%, indicating that an increase in the use or development of financial technologies is associated with an increase in Mat. Footprint. Although FinTech is often associated with a non-polluting or digital sector, its economic effects can stimulate consumption and production, driving the demand for material resources. Integrating sustainability objectives into financial digitisation policies is necessary to support this endeavour (Afzal et al., 2022; Irajifar et al., 2023). In the case of GDP Growth, the coefficient is positive (0.134). However, the result may reflect an economy in transition, where GDP growth remains resource-dependent. Regarding TI, the coefficient is negative (−0.0879) and significant (t = −1.80), suggesting that investments in research and development contribute to the reduction of Mat. Footprint. Moreover, R&D activities lead to technological innovations that, in turn, streamline resource utilisation. This can lead to significant reductions in resource consumption and pollution. Targeted promotion of R&D towards green technologies and circular economy could amplify these effects (Bednarčíková & Repiská, 2021; Feroz et al., 2021; Quttainah & Ayadi, 2024). The constant is significant at 1%, reflecting a baseline, Mat. Footprint when all other variables are zero, which supports the robustness of the model. Thus, our research hypothesis was partially confirmed (H2): Increased R&D expenditures, along with green finance policies, reduce the impact of FinTech development on CO_2 emissions and material footprint by promoting sustainable technologies and renewable energy alternatives.

Previous studies have reported the effects of financial innovations on Sustainable Green Finance (Dong et al., 2024; Monk & Perkins, 2020; Sreenu & Mishra, 2023; Zhang et al., 2022). In this perspective, the notable innovation-fragility perspective versus innovative growth, presented by (Hsu et al., 2021; Kwong et al., 2023; Lv et al., 2021; Umar & Safi, 2023), argues that financial technology and innovations have either positive or negative implications for economic development. Correspondingly, the negative role of FinTech development on the environment and energy consumption has been mainly reported by (Yuan et al., 2021; Zhang et al., 2022).

Ultimately, although slightly counterintuitive, a negative coefficient between FinTech and emissions could support a more nuanced perspective, indicating that, under certain conditions, FinTech development could contribute to reducing emissions. This hypothesis is based on the idea that FinTech innovations can streamline the allocation of capital towards sustainable projects (Loukoianova et al., 2024), improve emissions monitoring and reporting (Teng & He, 2025), or promote more responsible and environmentally conscious consumption (Piotrowska & Piotrowski, 2025).

As previously demonstrated, our findings indicate a positive association between FinTech development and emissions. This seemingly paradoxical relationship suggests that the expansion of FinTech, although a digital sector, can amplify resource consumption and production; these factors contribute to a greater demand for material resources, and the increased demand can implicitly lead to higher energy consumption and, therefore, increased emissions. However, our results remain robust.

Green finance sustainability is measured through various factors, with natural resources, FinTech, and technological innovation being key considerations (Manta et al., 2025). On the one hand, innovative technology has a dual impact: negative, but also supports the potential to promote sustainability if managed efficiently (Sreenu & Mishra, 2023). Also, the impact of FinTech can vary depending on different economies, registering both positive and negative results. Implying that FinTech has a constructive effect on environmental sustainability, as Europe can reduce its dependence on carbon-emitting resources by using green resources for energy production and can significantly reduce CO_2 emissions (Muhammad et al., 2022; Puschmann et al., 2020; Sreenu & Mishra, 2023). This helps to mitigate adverse impacts by decreasing pollution. At the same time, IT encourages the growth and use of less carbon-emitting innovations, such as energy-efficient innovations and storage systems (Brodny & Tutak, 2020; Carvalho et al., 2010). Such advances help lower the level of carbon pollutants, which have a detrimental effect on air quality. Therefore, it is necessary to encourage using technologies with lower carbon emissions to achieve carbon neutrality goals. The results of the study are similar to those of Albino et al. (2014), Baker et al. (2020), and Ganda (2019).

6 Conclusions and Policy Implications

Our study explored, from an innovative perspective, the impact of FinTech (FT) and technological innovations (TI) on environmental quality, investigating whether FinTech development facilitates a smooth transition towards lower carbon emissions and greenhouse gas levels. Our empirical analysis focused on two different estimation techniques, namely two-stage least squares (2SLS) and generalised methods of moments (GMM), to investigate two distinct models. The main results highlight the effects of digital finance, technological innovations supported by research and development activities, renewable energy consumption reflected in CO_2 emissions and Material Footprint, and economic growth; these findings underscore the need to integrate sustainability objectives into financial digitisation policies to support to support the transition to a low-carbon economy and the achievement of climate neutrality.

The estimates obtained in the two models for digital financial inclusion demonstrate significant effects in the 2SLS and GMM models, demonstrating that digital financial services can contribute to promoting environmental quality. Similarly, the estimates associated with IT present positive and significant effects to some extent in the estimation techniques analysed, suggesting that the use of IT can contribute to improved environmental outcomes and the growth of sustainable green finance, but also presents adverse effects through its potential to amplify energy consumption and production, which leads to negative impacts on environmental quality. These mixed results, both positive and negative, amplify the need to implement measures that support sustainability, such as carbon taxes or increasingly strict regulations regarding pollution control. These results underscore the importance of integrating research and innovation activity into the development of renewable energy to promote technological advancements and cost reductions in the renewable energy sector.

In addition, the estimates related to FinTech are significantly positive in the 2SLS and GMM models, highlighting an unfavourable impact on pollution. On the other hand, the estimate associated with FinTech is significant in both the GMM and 2SLS models; FinTech has a positive and significant effect on Mat. Footprint and on CO_2, which can be explained through a series of factors such as: Increased digital consumption, increased consumption in the chain, which implies greater use of material resources, Increased investment and rapid lending, which will facilitate consumption and production; therefore, more raw materials will be used. Although FinTech has demonstrated its contribution to economic efficiency and financial inclusion, it is attributed an ecological cost, namely amplifying the pressure on the environment if sustainable policies do not accompany it. Moreover, our study indicates that FinTech, GDP, and research and development activities are essential contributors to pollutant emissions based on the results of applying estimation techniques. This research provided a series of key findings and policy implications. The estimated models indicate technological innovations' significant positive and negative influence on emissions, thereby affecting environmental quality.

Based on this finding, it is recommended that nations adopt different strategies to accelerate the adoption of IT-related activities to enhance environmental quality. Nations should give special importance to the strategic implementation of IT solutions, directing these efforts towards industries with significant pollution levels. In addition, governments should focus on promoting the digitalisation of the financial sector, as they have the potential to impact the development of innovative financial instruments that support environmental sustainability.

Therefore, policies that encourage innovation in green technologies and the adoption of sustainable practices within the financial sector can also help minimise the negative impact of FinTech on the environment. Also, consumer awareness is needed by making them aware of the effects of their financial decisions on the environment, information campaigns should be integrated regarding the benefits of green investments and supporting initiatives that promote new sustainable financial practices. Thus, consumers would be more predisposed towards green products and services. At the same time, promotion campaigns should facilitate the awareness of financial services consumers, and digital literacy could improve the efficient use of digital financial services. These campaigns should focus to some extent on the potential benefits of digital financial inclusion and the possible risks of integrating digital technology, while providing consumers with informational support.

Considering the significant impact of the financial sector's digitalisation on the transition to a sustainable, low-carbon economy policymakers should give special importance to expanding digital financial inclusion. This could facilitate the launch of regulatory reforms to stimulate financial innovation and promote the development of digital financial infrastructure by supporting public-private partnerships.

In addition, we must also emphasise the possible negative effects of FinTech in amplifying energy consumption, as this consumption will negatively affect environmental quality. Therefore, policymakers should consider reducing energy consumption by launching initiatives or restrictive measures, such as carbon taxes or emission taxes, along with regulations on pollution control. All of these measures are needed to alleviate the negative environmental impacts. In conclusion, the study's findings underscore the necessity for a comprehensive and integrated policymaking approach. Achieving sustainable environmental quality demands coordinated efforts across various sectors and effective cross-sectoral coordination between different levels of government, the private sector, and civil society.

Embracing a holistic approach enables decision-makers to foster an environment conducive to sustainable innovation in the financial sector, promoting environmental responsibility. Despite the meaningful findings, this study acknowledges its limitations. The data may not encompass all facets of green growth and IT diffusion, potentially overlooking some determinants. The used models inherently carry assumptions and potential issues; for example, the reliability of instrumental variables strongly impacts the outcome of the GMM model. In addition, the European focus limits the generalisation of findings to different social and regulatory frameworks. The cross-sectional design fails to capture dynamic changes over time, and measuring complex variables such as FinTech or IT diffusion may not be accurate. Ultimately, while this study points to relationships among variables, establishing clear causality is still a challenge. Future research could address these limitations and provide various insights into the subject. Further research could explore conceptual differences between developed and developing countries regarding green growth. Another essential element to analyse in future studies should focus on green growth and poverty reduction.

Acknowledgments This work was supported by a grant from the Romanian Ministry of Research, Innovation and Digitalization, the project with the title "Economics and Policy Options for Climate Change Risk and Global Environmental Governance" (CF 193/28.11.2022, Funding Contract no. 760078/23.05.2023), within Romania's National Recovery and Resilience Plan (PNRR)—Pillar III, Component C9, Investment I8 (PNRR/2022/C9/MCID/I8)—Development of a program to attract highly specialised human resources from abroad in research, development, and innovation activities.

Competing Interests The authors have no conflicts of interest to declare relevant to this chapter's content.

Appendix

See Tables 5 and 6.

Table 5 Correlation matrix

	CO_2	FinTech	GDP	Mat.Footprint	TI
CO_2	1.000	−0.723	−0.211	−0.0759	0.745
FinTech	−0.723	1.000	0.205	0.5845	0.992
GDP	−0.211	0.205	1.000	0.3760	0.229
Mat.Footprint	−0.075	0.584	0.376	1.0000	0.539
TI	−0.745	0.992	0.229	0.5397	1.000

Table 6 European countries

No.	ID	Country	No.	ID	Country
1	ALB	Albania	20	LUX	Luxembourg
2	AUT	Austria	21	MKD	Macedonia
3	BEL	Belgium	22	MLT	Malta
4	BIH	Bosnia and Herzegovina	23	MNE	Montenegro
5	BGR	Bulgaria	24	NLD	Netherlands
6	HRV	Croatia	25	NOR	Norway
7	CYP	Cyprus	26	POL	Poland
8	CZE	Czech Republic	27	PRT	Portugal
9	DNK	Denmark	28	ROU	Romania
10	FIN	Finland	29	RUS	Russian Federation
11	FRA	France	30	SRB	Serbia
12	DEU	Germany	31	SVK	Slovakia
13	GRC	Greece	32	SVN	Slovenia
14	HUN	Hungary	33	ESP	Spain
15	ISL	Iceland	34	SWE	Sweden
16	IRL	Ireland	35	CHE	Switzerland
17	ITA	Italy	36	UKR	Ukraine
18	LVA	Latvia	37	GBR	United Kingdom
19	LTU	Lithuania			

References

Aboulajras, A. S. A., Khalifa, W., & Kareem, P. H. (2025). Environmental sustainability in emerging economies: The impact of natural resource rents, energy efficiency, and economic growth via quantile regression analysis. *Sustainability, 17*(8), 3670. https://doi.org/10.3390/su17083670

Afzal, A., Rasoulinezhad, E., & Malik, Z. (2022). Green finance and sustainable development in Europe. *Economic Research-Ekonomska Istraživanja, 35*(1), 5150. https://doi.org/10.1080/1331677x.2021.2024081

Ahmad, M., Pata, U. K., Ahmed, Z., & Zhao, R. (2024). Fintech, natural resources management, green energy transition, and ecological footprint: Empirical insights from EU countries. *Resources Policy, 92*, 104972. https://doi.org/10.1016/j.resourpol.2024.104972

Akey, P., & Appel, I. (2020). The limits of limited liability: Evidence from industrial pollution. *The Journal of Finance, 76(1)*, 5–55. https://doi.org/10.1111/jofi.12978

Albino, V., Ardito, L., Dangelico, R. M., & Petruzzelli, A. M. (2014). Understanding the development trends of low-carbon energy technologies: A patent analysis. *Applied Energy, 135*, 836. https://doi.org/10.1016/j.apenergy.2014.08.012

Almaqtari, F. A. (2024). The moderating role of IT governance on the relationship between FinTech and sustainability performance. *Journal of Open Innovation: Technology, Market, and Complexity, 10*(2), 100267. https://doi.org/10.1016/j.joitmc.2024.100267

Anwar, H., Waheed, R., & Aziz, G. (2024). Importance of FinTech and green finance to achieve the carbon neutrality targets: A study of Australian perspective. *Environmental Research Communications, 6*, 115007. https://doi.org/10.1088/2515-7620/ad853d

Arellano, M., & Bond, S. (1991). Some Tests of Specification for Panel Data: Monte Carlo Evidence and an Application to Employment Equations. *The Review of Economic Studies, 58*(2), 277–297. https://doi.org/10.2307/2297968

Arner, D., Buckley, R., Veidt, R., & Zehzche, D. (2020). Sustainability, FinTech and financial inclusion. *European Business Organization Law Review, 21*, 7–35.

Awais, M., Afzal, A., Firdousi, S., & Hasnaoui, A. (2023). Is fintech the new path to sustainable resource utilisation and economic development? *Resources Policy, 81*, 103309. https://doi.org/10.1016/j.resourpol.2023.103309

Baker, S. E., Stolaroff, J. K., Peridas, G., Pang, S. H., Goldstein, H., Lucci, F. R., Li, W., Slessarev, E., Pett-Ridge, J., Ryerson, F. J., Wagoner, J. L., Kirkendall, W., Aines, R. D., Sanchez, D. L., Cabiyo, B., Baker, J. W., McCoy, S., Uden, S., Runnebaum, R. C., ... McCormick, C. (2020). *Getting to neutral: Options for negative carbon emissions in California*. https://doi.org/10.2172/1597217

Bassen, A., Kordsachia, O., Lopatta, K., & Tan, W. (2025). Revenue alignment with the EU taxonomy regulation in developed markets. *Journal of Banking & Finance, 170*, 107339. https://doi.org/10.1016/j.jbankfin.2024.107339

Bednarčíková, D., & Repiská, R. (2021). Digital transformation in the context of the European Union and the use of digital technologies as a tool for business sustainability. *SHS Web of Conferences, 115*, 1001. https://doi.org/10.1051/shsconf/202111501001

Brodny, J., & Tutak, M. (2020). Analyzing similarities between the European Union countries in terms of the structure and volume of energy production from renewable energy sources. *Energies, 13*(4), 913. https://doi.org/10.3390/en13040913

Carvalho, M. d. G., Bonifacio, M., & Dechamps, P. (2010). Building a low carbon society. *Energy, 36*(4), 1842. https://doi.org/10.1016/j.energy.2010.09.030

Chen, X., & Liu, Z. (2024). Fintech and sustainable resources management: Role of trade openness and globalization in BRICS countries. *Resources Policy, 90*, 104700. https://doi.org/10.1016/j.resourpol.2024.104700

Cheng, X., Yao, D., Qian, Y., Wang, B., & Zhang, D. (2023). How does fintech influence carbon emissions: Evidence from China's prefecture-level cities. *International Review of Financial Analysis, 87*, 102655. https://doi.org/10.1016/j.irfa.2023.102655

Choudhary, P., Ghosh, C., & Thenmozhi, M. (2025). Impact of fintech and financial inclusion on sustainable development goals: Evidence from cross country analysis. *Finance Research Letters, 72*, 106573. https://doi.org/10.1016/j.frl.2024.106573

Croutzet, A., & Dabbous, A. (2021). Do FinTech trigger renewable energy use? Evidence from OECD countries. *Renewable Energy, 179*, 1608–1617. https://doi.org/10.1016/j.renene.2021.07.144

Dabbous, A., Barakat, K. A., Arfi, W. B., & Nammouri, H. (2025). The impact of environmental policy stringency and economic complexity on nations' energy transitions: The mediating role of fintech financing. *Energy Economics, 147*, 108540. https://doi.org/10.1016/j.eneco.2025.108540

Davrinche, G., Filbien, J.-Y., & Vigneron, L. (2025). Do analysts matter for green investment? Evidence from the EU taxonomy. *Economics Letters, 250*, 112277. https://doi.org/10.1016/j.econlet.2025.112277

Demir, A., Pesqu´e-Cela, V., Altunbas, Y., & Murinde, V. (2022). Fintech, financial inclusion and income inequality: A quantile regression approach. *The European Journal of Finance, 28*(1), 86–107. https://doi.org/10.1080/1351847X.2020.1772335

Destek, M. A., & Manga, M. (2021). Technological innovation, financialization, and ecological footprint: Evidence from BEM economies. *Environmental Science and Pollution Research, 28*(17), 21991. https://doi.org/10.1007/s11356-020-11845-2

Dilek, Ş., & Furuncu, Y. (2019). Bitcoin Mining and Its Environmental Effects. *Atatürk Üniversitesi İktisadi ve İdari Bilimler Dergisi, 33(1)*, 91–106.

Dong, H., Zhang, L., & Zheng, H. (2024). Green bonds: Fueling green innovation or just a fad? *Energy Economics, 135*, 107660. https://doi.org/10.1016/j.eneco.2024.107660

Dong, X., Lv, S., & Wu, S. (2025). The green promise of digital finance: Exploring its effects and mechanisms on carbon reduction. *China Economic Review, 90*, 102374. https://doi.org/10.1016/j.chieco.2025.102374

Du, P., Huang, S., Hong, Y., & Wu, W. (2022a). Can FinTech improve corporate environmental, social, and governance performance?—A study based on the dual path of internal financing constraints and external fiscal incentives. *Frontiers in Environmental Science, 10*, 1061454. https://doi.org/10.3389/fenvs.2022.1061454

Du, M., Zhang, R., Chai, S., Li, Q., Sun, R., & Chu, W. (2022b). Can green finance policies stimulate technological innovation and financial performance? Evidence from Chinese listed green enterprises. *Sustainability, 14*, 9287. https://doi.org/10.3390/su14159287

Elheddad, M., Benjasak, C., Deljavan, R., Alharthi, M., & Almabrok, J. M. (2020). The effect of the Fourth Industrial Revolution on the environment: The relationship between electronic finance and pollution in OECD countries. *Technological Forecasting and Social Change, 163*, 120485. https://doi.org/10.1016/j.techfore.2020.120485

European Central Bank. (2018, March). *The revised Payment Services Directive (PSD2) and the transition to stronger payments security*. https://www.ecb.europa.eu/press/intro/mip-online/2018/html/1803_revisedpsd.en.html

European Commission. (n.d.). *EU taxonomy for sustainable activities*. https://finance.ec.europa.eu/sustainable-finance/tools-and-standards/eu-taxonomy-sustainable-activities_en

European Council. (2025, March 17). *Fit for 55*. https://www.consilium.europa.eu/en/policies/fit-for-55/

European Parliament. (2018, March 8). *Reducing carbon emissions: EU targets and policies*. European Parliament. https://www.europarl.europa.eu/topics/en/article/20180305STO99003/reducing-carbon-emissions-eu-targets-and-policies

Feroz, A. K., Zo, H., & Chiravuri, A. (2021). Digital transformation and environmental sustainability: A review and research agenda [Review of digital transformation and environmental sustainability: A review and research agenda]. *Sustainability, 13*(3), 1530. Multidisciplinary Digital Publishing Institute. https://doi.org/10.3390/su13031530

Ganda, F. (2019). The impact of innovation and technology investments on carbon emissions in selected organisation for economic co-operation and development countries. *Journal of Cleaner Production, 217*, 469. https://doi.org/10.1016/j.jclepro.2019.01.235

Hartman, R., & Kwon, O.-S. (2004). Sustainable growth and the environmental Kuznets curve. *Journal of Economic Dynamics and Control, 29*(10), 1701. https://doi.org/10.1016/j.jedc.2004.10.001

Hazla, M. W. (2023, May 19). *Can FinTechs become a catalyst for ESG rules in the Polish financial sector?* Blue Europe. https://www.blue-europe.eu/analysis-en/short-analysis/can-fintechs-become-a-catalyst-for-esg-rules-in-the-polish-financial-sector/?utm

He, B., Jie, W., He, H., Alsubih, M., Arnone, G., & Makhmudov, S. (2024). From resources to resilience: How green innovation, fintech and natural resources shape sustainability in OECD countries. *Resources Policy, 91*, 104856. https://doi.org/10.1016/j.resourpol.2024.104856

Hernández de Rojas, F., Pita, P. R., & Martínez, J. E. P. (2025). The phenomena of spatial spillovers of digitalization and competitiveness inside European regions. *Telecommunications Policy*, 102989. https://doi.org/10.1016/j.telpol.2025.102989

Hileman, G., & Rauchs, M. (2017). 2017 Global Blockchain Benchmarking Study. *Cambridge Centre for Alternative Finance.* https://www.jbs.cam.ac.uk/wp-content/uploads/2020/08/2017-09-27-ccaf-globalbchain.pdf

Hsu, C., Ngo, T. Q., Chien, F., Li, L., & Mohsin, M. (2021). Evaluating green innovation and performance of financial development: Mediating concerns of environmental regulation. *Environmental Science and Pollution Research, 28*(40), 57386. https://doi.org/10.1007/s11356-021-14499-w

Huang, Y. (2025). Fintech as a catalyst for sustainable development: A bibliographic review of drivers, technologies, econometrics and regional insights. *Journal of Accounting Literature.* https://doi.org/10.1108/JAL-02-2025-0094

Huang, X., Dong, J., & Li, X. (2025). Fintech, technological innovation and regional economic growth: Theoretical modeling and empirical evidence. *China Economic Review, 91*, 102397. https://doi.org/10.1016/j.chieco.2025.102397

Iqbal, M. A., Shaheen, W. A., Shabir, S., Ullah, U., et al. (2025). Towards a green economy: Investigating the impact of sustainable finance, green technologies, and environmental policies on environmental degradation. *Journal of Environmental Management, 374*, 124047. https://doi.org/10.1016/j.jenvman.2025.124047

Irajifar, L., Chen, H., Lak, A., Sharifi, A., & Cheshmehzangi, A. (2023). The nexus between digitalization and sustainability: A scientometrics analysis. *Heliyon, 9*(5). https://doi.org/10.1016/j.heliyon.2023.e15172

Irfan, M., Razzaq, A., Sharif, A., & Yang, X. (2022). *Technological Forecasting and Social Change, 182*, 121882. https://doi.org/10.1016/j.techfore.2022.121882

Jangid, H., Bal, D. P., & Rao, N. V. M. (2025). Role of FinTech and technological innovation towards energy, growth, and environment nexus in G20 economies. *Scientific Reports, 15*, 20057. https://doi.org/10.1038/s41598-025-02794-2

Jia, Z., Alharthi, M., Haijun, T., Mehmood, S., & Hanif, I. (2024). Relationship between natural resources, economic growth, and carbon emissions: The role of fintech, information technology and corruption to achieve the targets of COP-27. *Resources Policy, 90*, 104751. https://doi.org/10.1016/j.resourpol.2024.104751

Jie, Y., Rasool, Z., Nassani, A. A., Mattayaphutron, S., & Murad, M. (2024). Sustainable Central Asia: Impact of fintech, natural resources, renewable energy, and financial inclusion to combat environmental degradation and achieving sustainable development goals. *Resources Policy, 95*, 105138. https://doi.org/10.1016/j.resourpol.2024.105138

Jing, R., Ma, Y., Zhang, L., & Hafeez, M. (2022). Does financial technology improve health in Asian economies? *Frontiers in Public Health, 10*, 843379. https://doi.org/10.3389/fpubh.2022.843379

Judijanto, L., Qosim, N., Syamsulbahri, Utami, E. Y., & Afandy, C. (2024). Sustainable finance: A bibliometric analysis of green finance and its role in global markets. *West Science Business and Management, 2*(4), 1206–1215.

Kakar, S. K., Wang, J., Arshed, N., Hien, T. T. L., Akhter, S., & Abdullahi, N. M. (2025). The impact of circular economy, sustainable infrastructure, and green FinTech on biodiversity in Europe: A holistic approach. *Technology in Society, 81*, 102841. https://doi.org/10.1016/j.techsoc.2025.102841

Kan, X., Lv, R., Chen, Y., & Fu, Y. (2022). Research on the influence of inclusive finance on carbon emissions. *Frontiers in Business Economics and Management, 6*(1), 19. https://doi.org/10.54097/fbem.v6i1.2253

Kashif, M., Ullah, A., Ullah, S., & Qian, N. (2024). Towards a greener future: The impact of financial technology (FinTech) and climate finance on ecological sustainability. *Journal of Environmental Management, 370*, 122876. https://doi.org/10.1016/j.jenvman.2024.122876

Kowalewski, O., & Pisany, P. (2023). The rise of fintech: A cross-country perspective. *Technovation, 122*, 102642. https://doi.org/10.1016/j.technovation.2022.102642

Kozar, Ł. J., & Wodnicka, M. (2024). FinTech and green finance: Bibliometric analysis. *Procedia Computer Science, 246*, 2080–2089. https://doi.org/10.1016/j.procs.2024.09.643

Krause, M. J., & Tolaymat, T. (2018). Quantification of energy and carbon costs for mining cryptocurrencies. *Nature Sustainability, 1*, 711–718. https://doi.org/10.1038/s41893-018-0152-7

Kumar, P., Kaur, R., Radulescu, M., Kalaš, B., & Hagiu, A. (2025). Drivers of environmental sustainability, economic growth, and inequality: A study of economic complexity, FDI, and human development role in BRICS+ nations. *Sustainability, 17*(9), 4180. https://doi.org/10.3390/su17094180

Kwong, R., Kwok, M. L. J., & Wong, H. (2023). Green FinTech innovation as a future research direction: A bibliometric analysis on green finance and FinTech. *Sustainability, 15*(20), 14683. https://doi.org/10.3390/su152014683

Le, V. L. T., & Pham, K. D. (2024). The impact of financial inclusion and digitalization on CO_2 emissions: A cross-country empirical analysis. *Sustainability, 16*(23), 10491. https://doi.org/10.3390/su162310491

Li, C., & Zhou, W. (2024). Can digital economy development contribute to urban carbon emission reduction? Empirical evidence from China. *Journal of Environmental Management, 357*, 120680. https://doi.org/10.1016/j.jenvman.2024.120680

Li, Y., Shahid, M. N., Islam, M. U., & Deme, F. (2024). The role of green technological innovation, Fintech, and financial development in environmental sustainability: A study on selected Asian countries. *Journal of Economics, Finance and Accounting Studies, 6*(1), 32–45. https://doi.org/10.32996/jefas

Liu, J., Zhang, Y., & Kuang, J. (2023). Fintech development and green innovation: Evidence from China. *Energy Policy, 183*, 113827. https://doi.org/10.1016/j.enpol.2023.113827

Llorca, M., & Rodriguez-Alvarez, A. (2024). Economic, environmental, and energy equity convergence: Evidence of a multi speed Europe. *Ecological Economics, 219*, 108133. https://doi.org/10.1016/j.ecolecon.2024.108133

Loukoianova, E., Natalucci, F. M., Wang, Y. D., & Kanada, S. (2024). *Fintech Applications for Boosting Climate Finance (Staff Climate Note No. 2024/008)*. International Monetary Fund. https://doi.org/10.5089/9798400293740.066

Lv, C., Shao, C., & Lee, C. (2021). Green technology innovation and financial development: Do environmental regulation and innovation output matter? *Energy Economics, 98*, 105237. https://doi.org/10.1016/j.eneco.2021.105237

Lyulyov, O., & Pimonenko, T. (2025). Green economic growth: Convergence patterns and eco-productivity clusters. *Journal of Open Innovation: Technology, Market, and Complexity, 11*(2), 100567. https://doi.org/10.1016/j.joitmc.2025.100567

Macchiavello, E., & Siri, M. (2020). Sustainable Finance and Fintech: Can Technology Contribute to Achieving Environmental Goals? A Preliminary Assessment of 'Green FinTech'. *European Banking Institute Working Paper Series, 2020*(71). https://doi.org/10.2139/ssrn.3672989

Manta, O., Vasile, V., & Rusu, E. (2025). Banking Transformation Through FinTech and the Integration of Artificial Intelligence in Payments. *FinTech, 4*(2), 13. https://doi.org/10.3390/fintech4020013

Marinaş, M.-C., Dinu, M., Socol, A. G., & Socol, C. (2018). Renewable energy consumption and economic growth. Causality relationship in Central and Eastern European countries. *PLoS One, 13*(10). https://doi.org/10.1371/journal.pone.0202951

Mavlutova, I., Spilbergs, A., Romanova, I., Kuzmina, J., Fomins, A., Verdenhofs, A., & Natrins, A. (2025). The role of green digital investments in promoting sustainable development goals and green energy consumption. *Journal of Open Innovation: Technology, Market, and Complexity, 11*, 100518. https://doi.org/10.1016/j.joitmc.2025.100518

Mertzanis, C. (2023). FinTech finance and social-environmental performance around the world. *Finance Research Letters, 56*, 104107. https://doi.org/10.1016/j.frl.2023.104107

Mertzanis, C., Marashdeh, H., Kampouris, I., & Atayah, O. (2025). Managing environmental finance in the digital era: Evidence from green bonds. *Journal of Environmental Management, 373*, 123434. https://doi.org/10.1016/j.jenvman.2024.123434

Mirza, N., Umar, M., Afzal, A., & Firdousi, S. F. (2023). The role of fintech in promoting green finance, and profitability: Evidence from the banking sector in the euro zone. *Economic Analysis and Policy, 78*, 33–40. https://doi.org/10.1016/j.eap.2023.02.001

Monk, A., & Perkins, R. (2020). What explains the emergence and diffusion of green bonds? *Energy Policy, 145*, 111641. https://doi.org/10.1016/j.enpol.2020.111641

Muganyi, T., Yan, L., & Sun, H.-P. (2021). Green finance, fintech and environmental protection: Evidence from China. *Environmental Science and Ecotechnology, 7*, 100107. https://doi.org/10.1016/j.ese.2021.100107

Muhammad, S., Pan, Y., Magazzino, C., Luo, Y., & Waqas, M. (2022). The fourth industrial revolution and environmental efficiency: The role of fintech industry. *Journal of Cleaner Production, 381*, 135196. https://doi.org/10.1016/j.jclepro.2022.135196

Obobisa, E. S., & Ahakwa, I. (2024). Stimulating the adoption of green technology innovation, clean energy resources, green finance, and environmental taxes: The way to achieve net zero CO_2 emissions in Europe? *Technological Forecasting and Social Change, 205*, 123489. https://doi.org/10.1016/j.techfore.2024.123489

Ouyang, X., Li, Q., & Du, K. (2020). How does environmental regulation promote technological innovations in the industrial sector? Evidence from Chinese provincial panel data. *Energy Policy, 139*, 111310. https://doi.org/10.1016/j.enpol.2020.111310

Oyebanji, M. O., & Kirikkaleli, D. (2023). Green technology, green electricity, and environmental sustainability in Western European countries. *Environmental Science and Pollution Research, 30*(13), 38525–38534. https://doi.org/10.1007/s11356-022-24793-w

Ozturk, I., & Ullah, S. (2022). Does digital financial inclusion matter for economic growth and environmental sustainability in OBRI economies? An empirical analysis. *Resources, Conservation and Recycling, 185*, 106489.

Panetta, I. C., Leo, S., & Foglie, A. D. (2023). The development of digital payments – Past, present, and future – From the literature. *Research in International Business and Finance, 64*, 101855. https://doi.org/10.1016/j.ribaf.2022.101855

Piotrowska, A. I., & Piotrowski, D. (2025). Green FinTech: A consumer awareness study. *Sustainability, 17*(8), 3701. https://doi.org/10.3390/su17083701

Puschmann, T., & Khmarskyi, V. (2024). Green fintech: Developing a research agenda. *Corporate Social Responsibility and Environmental Management, 31*(6), 2823–2837. https://doi.org/10.1002/csr.2675

Puschmann, T., Hoffmann, C. H., & Khmarskyi, V. (2020). How green FinTech can alleviate the impact of climate change—The case of Switzerland. *Sustainability, 12*(24), 10691. https://doi.org/10.3390/su122410691

Qin, L., Aziz, G., Hussan, M. W., Qadeer, A., et al. (2024a). Empirical evidence of fintech and green environment: Using the green finance as a mediating variable. *International Review of Economics & Finance, 89*, 33–49. https://doi.org/10.1016/j.iref.2023.07.056

Qin, M., Zhu, Y., Xie, X., Shao, X., & Lobonţ, O.-R. (2024b). The impact of climate risk on technological progress under the fourth industrial era. *Technological Forecasting and Social Change, 202*, 123325. https://doi.org/10.1016/j.techfore.2024.123325

Quttainah, M. A., & Ayadi, I. (2024). The impact of digital integration on corporate sustainability: Emissions reduction, environmental innovation, and resource efficiency in the European. *Journal of Innovation & Knowledge, 9*(3), 100525. https://doi.org/10.1016/j.jik.2024.100525

Razzaq, A., Ajaz, T., Li, J. C., Irfan, M., & Suksatan, W. (2021). Investigating the asymmetric linkages between infrastructure development, green innovation, and consumption-based material footprint: Novel empirical estimations from highly resource-consuming economies. *Resources Policy, 74*, 102302. https://doi.org/10.1016/j.resourpol.2021.102302

Samour, A., Radmehr, R., Ali, E. B., Shayanmehr, S., Ofori, E. K., Porhajašová, J. I., Babošová, M., Kačániová, M., & Dimnwobi, S. K. (2024). The role of financial inclusion and technological innovation in stimulating environmental sustainability in the European countries: A new perspective based on load capacity factor. *Heliyon, 10*, e39970. https://doi.org/10.1016/j.heliyon.2024.e39970

Sikdar, C., & Mukhopadhyay, K. (2017). The nexus between carbon emission, energy consumption, economic growth and changing economic structure in India: A multivariate cointegration approach. *The Journal of Developing Areas, 52*(4), 67. https://doi.org/10.1353/jda.2018.0052

Song, X., Yao, Y., & Wu, X. (2023). Digital finance, technological innovation, and carbon dioxide emissions. *Economic Analysis and Policy, 80*, 482–494. https://doi.org/10.1016/j.eap.2023.01.004

Sreenu, N., & Mishra, S. (2023). Impact of green finance and fintech on sustainable economic growth: Empirical evidence from India. *Heliyon, 9*(5). https://doi.org/10.1016/j.heliyon.2023.e16301

Tao, R., Su, C.-W., Naqvi, B., & Rizvi, S. K. A. (2022). Can Fintech development pave the way for a transition towards low-carbon economy: A global perspective. *Technological Forecasting and Social Change, 174*, 121278. https://doi.org/10.1016/j.techfore.2021.121278

Teng, M., & Shen, M. (2023). Fintech and energy efficiency: Evidence from OECD countries. *Resources Policy, 82*, 103550. https://doi.org/10.1016/j.resourpol.2023.103550

Teng, Z., Xia, H., & He, Y. (2025). Rewiring Sustainability: How Digital Transformation and Fintech Innovation Reshape Environmental Trajectories in the Industry 4.0 *Era. Systems, 13*(6), 400. https://doi.org/10.3390/systems13060400

Tnani, M. (2018). Relationships between economic growth, CO_2 emissions, and innovation for nations with the highest patent applications. *Environmental Economics, 9*(2), 47. https://doi.org/10.21511/ee.09(2).2018.04

Tu, Y.-T. (2024). Drivers of environmental performance in Asian economies: Do natural resources, green innovation and Fintech really matter? *Resources Policy, 90*, 104832. https://doi.org/10.1016/j.resourpol.2024.104832

Umar, M., & Safi, A. (2023). Do green finance and innovation matter for environmental protection? A case of OECD economies. *Energy Economics, 119*, 106560. https://doi.org/10.1016/j.eneco.2023.106560

Wang, K.-H., Zhao, Y.-X., Su, Y. H., & Lobonţ, O.-R. (2023). Energy security and CO_2 emissions: New evidence from time-varying and quantile-varying aspects. *Energy, 273*, 127164. https://doi.org/10.1016/j.energy.2023.127164

Wang, Y., Han, M., & Liu, K. (2024). Towards sustainable path: Measuring the effectiveness of natural resource utilization, renewable energy resources, fintech, environmental investments, and research & development. *Resources Policy, 93*, 105053. https://doi.org/10.1016/j.resourpol.2024.105053

Wu, S., Wang, Z., Su, C., & Zhang, W. (2023). Digital media and green development path in Asia: Does digital financial inclusion matter? *Sustainability, 15*, 11359. https://doi.org/10.3390/su151411359

Xiao, A., Xu, Z., Skare, M., Xiao, J. L., & Qin, Y. (2024). Unlocking the potential of FinTech: A pathway to sustainable resource management in the EU. *Resources Policy, 98*. https://doi.org/10.1016/j.resourpol.2024.105358

Xu, J., Chen, F., Zhang, W., Liu, Y., & Li, T. (2023). Analysis of the carbon emission reduction effect of Fintech and the transmission channel of green finance. *Finance Research Letters, 56*, 104127. https://doi.org/10.1016/j.frl.2023.104127

Yáñez-Valdés, C., & Guerrero, M. (2024). Equity crowdfunding platforms and sustainable impacts: Encountering investors and technological initiatives for tackling social and environmental challenges. *European Journal of Innovation Management, 27*(7). https://doi.org/10.1108/EJIM-03-2022-0127

Yao, Y., Hu, D., Yang, C., & Tan, Y. (2021). The impact and mechanism of fintech on green total factor productivity. *Green Finance, 3*(2), 198–221.

Yoo, I., & Yi, C.-G. (2022). Economic innovation caused by digital transformation and impact on social systems. *Sustainability, 14*(5), 2600. https://doi.org/10.3390/su14052600

Yuan, G., Ye, Q., & Sun, Y. (2021). Financial innovation, information screening and industries' green innovation — Industry-level evidence from the OECD. *Technological Forecasting and Social Change, 171*, 120998. https://doi.org/10.1016/j.techfore.2021.120998

Zhang, K. Q., Chen, H. H., Tang, L. Z., & Qiao, S. (2022). Green finance, innovation and the energy-environment-climate nexus. *Frontiers in Environmental Science, 10*. https://doi.org/10.3389/fenvs.2022.879681

Zia, Z., Zhong, R., & Akbar, M. W. (2024). Analyzing the impact of fintech industry and green financing on energy poverty in the European countries. *Heliyon, 10*, e27532. https://doi.org/10.1016/j.heliyon.2024.e27532

Financial Institutions and Central Banks as Catalysts for Green Transformation and Climate Risk Mitigation

Manta Alina Georgiana, Bădîrcea Roxana Maria, Gherțescu Claudia, Pîrvu Ramona, Manta Liviu Florin, Berceanu Dorel, and Poenaru Maria Magdalena

1 Introduction and Context of the Research

In the context of market developments leading to deeper social awareness, sustainability trends have significantly influenced business strategies and operations. This paradigm shift has prompted the banking sector to review and adapt its practices to be more socially and environmentally responsible. The importance of sustainability in the banking sector is not limited to financial aspects. Banking institutions have become increasingly aware of their impact on communities, the environment and society at large. They are therefore committed to integrating sustainability principles into all aspects of their business.

The concept of social and environmental sustainability is not new. The term "ESG (environmental, social, governance)" and the idea were first introduced in June 2004 by the UN Global Compact initiative "Who Cares Wins" to emphasise the materiality and interplay between environmental, social and governance issues. A significant growth in the literature confirms that integrating sustainability factors

M. A. Georgiana · B. R. Maria (✉) · P. Ramona · B. Dorel
Faculty of Economics and Business Administration, University of Craiova, Craiova, Romania
e-mail: alina.manta@edu.ucv.ro; roxana.badircea@edu.ucv.ro; dorel.berceanu@edu.ucv.ro

G. Claudia
Doctoral School in Economic Sciences "Eugeniu Carada", Faculty of Economics and Business Administration, University of Craiova, Craiova, Romania
e-mail: ghertescu.claudia.x6e@student.ucv.ro

M. L. Florin
Faculty of Automation, Computers and Electronics, University of Craiova, Craiova, Romania
e-mail: florin.manta@edu.ucv.ro

P. M. Magdalena
Faculty of Horticulture, University of Craiova, Craiova, Romania
e-mail: maria.poenaru@edu.ucv.ro

C.-W. Su et al. (eds.), *Green Finance and the Challenges of Climate Change*, Sustainable Finance, https://doi.org/10.1007/978-3-032-07293-1_11

into portfolio strategy and investment analysis can influence the potential long-term performance of investors (Auer & Schuhmacher, 2016; Nagy et al., 2013, 2015; Novy-Marx, 2013; Vilanova et al., 2009). Other authors like Nizam et al. (2019) consider that investors can identify emerging market opportunities in companies that place the management of sustainability factors at the core of their business and investment managers and financial analysts who can interpret and correlate sustainability factors with a company's prospects can develop a significant competitive advantage (Doran et al., 2025).

In the area of banking sustainability, research conducted by Stauropoulou and Sardianou in 2019 highlights an increase in interest from policymakers. They are exerting their influence on the administrative structure to encourage financial institutions to plan their products and services to meet customer needs and take a proactive approach to address economic, social and environmental issues. The banking sector has a responsibility in terms of its practices that influence the monetary, social and environmental environment, and this needs to be reflected across its entire portfolio of activities (Abuzahra & Nobanee, 2022). To understand the impact of sustainability and climate change mitigation strategies in banking, it is necessary to develop a clear understanding of the concept of sustainable finance.

According to Sommer (2020), sustainable finance involves reconciling sustainability perspectives within the dynamic cycles of financial market actors, financial market strategies and associated institutional and market arrangements to help achieve sound, feasible, adaptive and comprehensive development. And in the context of European Union policies, sustainable finance is defined as finance that supports economic growth while reducing environmental pressures, thereby contributing to the environmental and climate objectives set by the European Ecological Pact. This concept also includes social and governance aspects. Sustainable finance implies transparency about the risks associated with ESG factors that may influence the financial system and adopts measures to mitigate these risks through appropriate governance of financial and corporate actors (European Central Bank, 2020).

Also, according to the same author, Sommer (2020), sustainable finance empowers the financial sector to identify and allocate the capital needed for progress towards a more sustainable economy. In this context, Coleton et al. (2020) explain the impact of banking activities through the direct effects on climate and society using natural resources in its processes and the indirect effects generated by its lending activity and the companies it finances. These advanced interpretations underline the importance of integrating sustainability principles into the banking sector's practices and strategies to fulfil its social and environmental responsibilities and help create a more resilient and sustainable economy in the long term (Manta et al., 2025a).

In addition, the banking sector has a key role to play in promoting sustainable development. Sustainability is currently one of the industry's most significant trends. It can manifest in investors' interest in responsible, sustainable investments, corporate management's focus on corporate social responsibility, or investors' attention to sustainability and environmental impact issues (Nizam et al., 2019; Manta et al., 2024a).

On the other hand, in their efforts to achieve sustainability, many banks use efficiency to reduce operational costs and operate on a limited budget (Meghişan-Toma et al., 2022). This approach helps save resources and creates an environment that promotes long-term success. According to Smith and Nobanee (2020), integrating artificial intelligence into the banking sector is one of the surest ways to build sustainable systems and achieve significant, long-term growth and development (Manta et al., 2024b). This strategy demonstrates that adopting advanced technologies can play an essential role in ensuring the continued viability and success of the banking sector in an ever-changing environment (Manta et al., 2025b).

The gaps in the literature identified focus mainly on a few key issues. First, while there has been a significant increase in interest in the role of the banking sector in tackling climate change, many studies tend to approach the issue in a shallow manner. This unexplored significant interdependencies between environmental, social and economic factors and banking practices. There is also insufficient integration of practical case studies and recent empirical evidence to provide a full picture of how sustainability strategies are implemented and their impact on carbon reduction. On the other hand, another gap in the literature is the lack of comparative studies across different regions or financial systems that could highlight the diversity of ways in which banking policies and practices adapt to sustainability requirements. In addition, contributions from grey literature, such as banks' internal reports or regulatory documents, are not fully integrated into the academic analysis, which limits a practical and up-to-date understanding of these topics.

Therefore, the methodology used in this study is relevant and enhances the novelty of the research by applying a systematic and rigorous bibliometric approach. This methodology allows for collecting and analysing a large volume of data from appropriate academic sources, as well as the objective and quantifiability of identifying and synthesising emerging research trends and directions (Gherțescu et al., 2025). Using keyword co-occurrence analysis, the study identifies dominant themes and reveals the thematic connections between relevant concepts, thus providing a clearer and interconnected perspective on the topic. Therefore, the bibliometric methodology validates the results obtained and enhances novelty by highlighting existing gaps and providing a solid basis for future research. In this way, the study contributes to a deeper and more structured understanding of how the banking sector can play a significant role in climate change mitigation.

The article is structured according to the standards of a scientific article, including an introductory section, a section for literature review, a methodology section, a section for the presentation of the results obtained, the conclusions of the research and a reference list. Due to the general nature of this paper, which serves as a literature review, the literature review is integrated into the research results section. In order to provide an appropriate context, in the first part of the paper, we briefly set out the general framework within which this research is conducted.

2 Literature Review

Climate change has emerged as a systemic risk with far-reaching implications for the financial sector. The Paris Agreement explicitly calls for aligning "finance flows" with low-carbon, climate-resilient development (UNFCCC, 2015). Financial institutions and central banks are increasingly considered pivotal catalysts in driving the green transformation and mitigating climate-related economic risks (Carney, 2015; Campiglio et al., 2018). A growing body of literature highlights that the actions of banks, institutional investors, and regulators can significantly influence the pace of decarbonization by reallocating capital towards sustainable investments and away from carbon-intensive activities (Krogstrup & Oman, 2019; Bolton et al., 2020). This literature review synthesises recent academic research and institutional reports on these themes. It is organized into six major sections: (1) the role of financial institutions in promoting sustainability, (2) ESG integration in finance and implications for stability, (3) the development of green finance instruments such as green bonds and sustainable loans, (4) climate change as a source of banking sector vulnerability, (5) the response of central banks, regulators, and the evolution of climate-related financial disclosure, and (6) empirical evidence on how climate risks and sustainability efforts are affecting financial performance.

2.1 Role of Financial Institutions in Sustainability

Financial institutions are critical in facilitating the transition to a sustainable, low-carbon economy by mobilising and directing capital toward environmentally friendly projects. Banks and asset managers act as intermediaries that can either accelerate or hinder the green transformation, depending on their lending and investment decisions (Schoenmaker & Schramade, 2019). In recent years, many banks have committed to financing renewable energy, energy efficiency, and other climate-friendly initiatives, recognising that financing sustainability is an ethical imperative and a source of long-term value (Contreras et al., 2019; World Bank, 2020). Studies note that private finance is essential to fill the multitrillion-dollar investment gap needed to meet climate goals, as public funding alone is insufficient (OECD, 2017; Krogstrup & Oman, 2019). Financial institutions can influence corporate behaviour and spur innovation in green technologies by allocating credit to low-carbon sectors and engaging with clients on decarbonization strategies (UNEP FI, 2018). For example, institutional investors leveraging shareholder activism and stewardship can push high-emitting firms to adopt cleaner practices or risk divestment (Busch et al., 2016).

Equally, financial institutions serve as risk managers in the transition, pricing and managing climate risks (Manta et al., 2023). Banks are increasingly integrating climate criteria into due diligence—for instance, factoring in the potential future carbon price or physical climate risks when evaluating loan applications (Delis et al.,

2019). This reallocation of capital is reinforcing a "virtuous cycle" where sustainable businesses gain easier access to funds, whereas carbon-intensive industries face higher capital costs or credit restrictions (Ongena et al., 2019). Such trends are evident in initiatives like the Equator Principles adopted by major project-finance banks to ensure projects meet environmental standards, and in the rapidly growing membership of the Principles for Responsible Investment among asset managers (Cosma et al., 2022). However, researchers also caution that without clear standards, there is a risk of greenwashing, where institutions mislabel activities as "green" without substantial climate impact (Boffo & Patalano, 2020). Overall, the literature underscores that the financial sector's proactive participation is indispensable for channelling funds in line with the Paris Agreement objectives and for driving the corporate shift toward sustainability (UNFCCC, 2015; Schoenmaker & Schramade, 2019).

2.2 *ESG Integration and Financial Stability*

Integrating Environmental, Social, and Governance (ESG) factors into financial decision-making has become a mainstream strategy, premised on the view that ESG issues can be material to firms' risk and return—and by extension, to financial stability. In particular, the "E" (environmental) component, which includes climate risk, has gained prominence as a factor that can affect credit risk, market valuations, and operational resilience of financial institutions (FSB, 2020). A broad meta-analysis of over 2000 studies finds that the majority report a non-negative or positive relationship between ESG performance and corporate financial performance, suggesting that considering ESG does not necessarily compromise returns and often correlates with better risk-adjusted outcomes (Friede et al., 2015). In banking, recent studies show that institutions with higher ESG scores tend to have lower idiosyncratic risk and more stable earnings, indicating potential resilience benefits (Goss & Roberts, 2011). By incorporating ESG criteria—for example, assessing a borrower's carbon footprint or governance practices—banks can improve their credit risk assessments and potentially reduce default probabilities in their loan portfolios (Cosma et al., 2022; Gherţescu et al., 2024). Furthermore, strong performance on ESG metrics can signal prudent management and long-term orientation, traits that are associated with lower volatility and tail risks for financial firms (Nollet et al., 2016; Albuquerque et al., 2020).

From a systemic perspective, integrating climate considerations (the environmental pillar of ESG) is increasingly seen as essential for preserving financial stability (FSB, 2020; ECB, 2021a). Regulators point out that if banks and investors systematically underestimate climate-related risks, financial institutions could abruptly face large losses when these risks materialise, potentially destabilising the financial system (Carney, 2015; BCBS, 2021). In response, prudential authorities now expect banks to embed climate and broader ESG factors into risk management. The European Central Bank, for instance, issued a Guide on climate-related and environmental risks

(ECB, 2020) outlining supervisory expectations for banks to incorporate these risks into governance, strategy, and risk management processes. Similarly, the Bank of England's Prudential Regulation Authority (PRA) has stressed that banks and insurers should assess how climate change could affect their balance sheets and has begun climate "stress tests" to evaluate resilience (Bank of England PRA, 2019). These developments reflect a consensus that environmental risks are *financial* risks—a paradigm shift from viewing sustainability as purely an ethical concern (FSB, 2020; TCFD, 2017). Nonetheless, ongoing debate exists on how exactly to integrate ESG into regulatory frameworks. Some policymakers proposed a "green supporting factor" (lower capital requirements for green assets) to incentivise sustainable lending. Still, research warns this could undermine stability if such assets do not actually bear lower risk (D'Orazio & Popoyan, 2019). Others advocate a "brown penalising factor" (higher capital for carbon-intensive exposures), aligning capital buffers with the higher risk of heavy-polluting assets. While consensus is still forming, the trajectory is clear: ESG considerations, especially climate risk, are becoming embedded in the fabric of risk management and prudential supervision to bolster the resilience of the financial system (ECB, 2020; Cosma et al., 2022).

2.3 Green Finance Instruments

The rise of green finance instruments is a tangible manifestation of the financial sector's support for sustainability. Chief among these are green bonds, which are fixed-income securities whose proceeds are earmarked for environmentally beneficial projects (e.g. renewable energy, clean transportation, green buildings). The market has expanded exponentially since the first green bonds were issued in the late 2000s. Global green bond issuance reached a record high in recent years—over USD 500 billion issued in 2021 alone—reflecting strong investor demand for climate-friendly assets (Climate Bonds Initiative, 2022). These bonds have attracted diverse issuers, including sovereigns, development banks, municipalities, and corporations, thereby mainstreaming sustainable finance within capital markets (Ehlers & Packer, 2017). A notable feature observed in multiple studies is the existence of a modest "greenium," where green bonds often carry a slightly lower yield (higher price) compared to otherwise similar conventional bonds, indicating investors' willingness to accept a smaller return in exchange for environmental impact (Zerbib, 2019; Bachelet et al., 2019). For instance, Zerbib (2019) finds a green yield discount on average of a few basis points, suggesting that pro-environmental investor preferences affect bond pricing. This pricing differential, albeit small, is significant as it implies issuers can potentially enjoy a lower cost of capital by going green, creating a market-based incentive for sustainable financing (Zerbib, 2019; Baker et al., 2018; Poenaru et al., 2025). Furthermore, evidence suggests that issuing a green bond can produce positive reputational effects and often coincides with improvements in the issuer's environmental performance post-issuance, indicating that green bonds may contribute to genuine environmental impact beyond mere rebranding (Flammer, 2021).

In addition to green bonds, various other sustainable finance instruments have gained traction. Sustainability-linked bonds and loans are innovative debt instruments whereby the interest rate or coupon is tied to the borrower's achievement of predefined sustainability targets (such as greenhouse gas emission reductions or renewable energy capacity installation). Unlike green bonds, the proceeds of sustainability-linked loans need not be used exclusively for green projects; instead, the instrument incentivises overall ESG performance improvements through pricing mechanisms. Early evidence indicates these loans are effective in aligning borrowers' incentives with sustainability goals, as companies stand to benefit from lower borrowing costs if they hit their ESG targets (Krogstrup & Oman, 2019). Green loans and mortgages have also grown, whereby banks offer favourable terms for projects like green buildings or electric vehicles (Ferrari & Landi, 2024). Moreover, transition bonds have emerged to fund emissions-reduction efforts in "brown" industries that are not yet green but are transitioning (e.g. funding a shift from coal to gas or gas to renewables in energy companies), reflecting the need to finance incremental improvements in hard-to-abate sectors (ICMA, 2020). Beyond debt markets, equity investors are pursuing thematic ESG funds and climate indices, and products like green exchange-traded funds (ETFs) now channel funds into baskets of climate-leading firms (Reboredo & Ugolini, 2020). The proliferation of these instruments illustrates the financial innovation underway to support sustainability objectives. However, researchers highlight challenges such as verifying the environmental impact of these instruments, preventing "ESG-washing," and ensuring that standards (e.g., the Green Bond Principles or EU Green Bond Standard) are rigorous and globally harmonized (Boffo & Patalano, 2020; Ehlers et al., 2020). Overall, green financial instruments are important tools enabling investors to participate in climate financing, and their rapid growth is a positive sign that capital markets are aligning, albeit gradually, with climate mitigation goals (Climate Bonds Initiative, 2022; Flammer, 2021).

2.4 *Climate Risk and Banking Sector Vulnerabilities*

Climate change introduces physical and transition risks that can profoundly affect the banking sector's stability. Physical risks refer to the potential for extreme weather events and longer-term shifts in climate patterns (rising sea levels, chronic droughts, etc.) to erode asset values and borrowers' repayment capacity. Transition risks stem from the economic adjustment toward a low-carbon economy—policy changes, technological disruptions, and shifts in consumer preferences could lead to sudden devaluations of carbon-intensive assets (NGFS, 2019; TCFD, 2017). Banks are exposed to these risks through their loan books, investment portfolios, and collateral values. A growing literature seeks to quantify these exposures and potential losses. For example, credit risk may increase in regions or industries vulnerable to climate change: a severe flood can damage properties and factories, undermining the value of collateral and the ability of borrowers to service mortgages or business loans. Empirical studies have found higher loan default rates following natural disasters, especially when insurance penetration is low, highlighting direct physical risk to bank balance sheets (Collier et al., 2022). On the transition side, policy risk is significant: if governments impose carbon pricing or strict emissions regulations, firms in fossil fuel extraction, power generation, heavy industry, and transportation could see their profits and asset values decline. Researchers have introduced the concept of "stranded assets," particularly fossil fuel reserves and coal-fired power plants that may become economically unviable in a carbon-constrained future (Caldecott, 2018). Banks that have heavily financed such assets face the prospect of credit losses as those business models suffer (Battiston et al., 2017). Delis, Kok, and Ongena (2019) provide evidence that banks started to price transition risk into loan contracts after the Paris Agreement: high-emitting firms began to face higher loan spreads, indicating lenders demanded a risk premium for climate policy exposure. This suggests that banks are starting to differentiate clients by climate riskiness, but it also implies that those banks that do not adjust could underestimate their risk.

The aggregate exposures can be large. An ECB study found that a significant share of European banks' corporate loan portfolios is tilted towards high-emission industries, implying substantial transition risk under scenarios of aggressive climate policy (Alogoskoufis et al., 2021). In a scenario where carbon prices rise sharply, credit losses for those banks could spike and capital adequacy could be strained (Alogoskoufis et al., 2021; Vermeulen et al., 2021). Similarly, the NGFS (Network for Greening the Financial System) scenario analysis (NGFS, 2020) projects that if the transition to a low-carbon economy is delayed and then happens abruptly ("disorderly" scenario), financial institutions could face much higher losses by mid-century compared to an orderly transition, due to both physical damages and stranded assets. Climate stress tests conducted by central banks provide further evidence of vulnerabilities: the Bank of England's 2021 Climate Biennial Exploratory Scenario, for instance, found that under a no-policy-action scenario, British banks and insurers could incur cumulative losses of over £200 billion by 2050 due to credit impairments and market losses (Bank of England, 2022). The European Central Bank's 2022 climate stress test similarly revealed that, in an adverse climate scenario, the average bank's credit provisions could increase markedly, and for certain vulnerable institutions, capital ratios would fall significantly (ECB, 2022). While exploratory, these tests underscore that climate risks are not just hypothetical future concerns—they can crystallise into financial losses that could threaten individual banks and, if widespread, the banking system's stability (BCBS, 2021; FSB, 2020).

A key challenge highlighted in the literature is the mismatch of time horizons: climate risks often manifest over more extended periods than the typical business or financial cycle (Carney, 2015). This "tragedy of the horizon" implies that it may be too late to prevent severe financial consequences by the time climate risks become undeniably clear and near-term. Therefore, researchers argue for proactive risk management today. This includes improving climate-risk disclosure (so that banks and investors can evaluate their exposures), developing forward-looking risk modelling tools (like climate stress tests and scenario analysis), and considering macroprudential measures to mitigate systemic risk buildup (Bolton et al., 2020; Battiston et al., 2017). In summary, the banking sector's vulnerability to climate change is well-documented: without intervention, physical disasters can spike default rates and damage assets, and a late or disorderly transition can trigger a sudden re-pricing of carbon-intensive assets. These findings justify the growing emphasis by regulators and central banks on integrating climate risk into the core financial stability framework (NGFS, 2019; ECB, 2021a).

2.5 Central Banks, Regulation and Climate Disclosure

Central banks and financial regulators have become key agents in addressing climate-related financial risks and promoting a greener financial system. Over the past 5 years, there has been a remarkable shift in central bank discourse and action on climate change (Dikau & Volz, 2021). One landmark development was the creation of the Network for Greening the Financial System (NGFS) in 2017 by eight founding central banks and supervisors under the leadership of Banque de France and others (NGFS, 2019). The NGFS has since grown to include central banks worldwide (over 100 members), and it serves as a collaborative platform to develop best practices. In its first comprehensive report, the NGFS (2019) declared climate change a source of financial risk and urged central banks to integrate climate considerations into their supervision and financial stability monitoring—effectively a call to action that "green" the mandates of central banks where applicable. The NGFS has produced influential guidance, such as climate scenario models for stress-testing (NGFS, 2020) and recommendations on how central bank operations (monetary policy, collateral frameworks, portfolio management) could incorporate climate factors (NGFS, 2021).

Regulatory bodies are also actively responding. The Basel Committee on Banking Supervision (BCBS) released principles in 2022 for the effective management and supervision of climate-related financial risks, which set out how banks and supervisors should govern, identify, measure, and monitor climate risks within the existing Basel risk categories (BCBS, 2022). These principles signal that climate risk should become part of the routine risk management and regulatory review process, even though precise capital requirements for climate exposures are not yet formulated. In the European Union, a sweeping sustainable finance regulatory agenda has been rolled out: the EU Taxonomy (Regulation 2020) defines what qualifies as a green economic activity, the Sustainable Finance Disclosure Regulation (SFDR) (2019) mandates transparency from asset managers and financial advisors on ESG risks in their products, and the Corporate Sustainability Reporting Directive (CSRD) (2022) will require companies (including banks) to disclose detailed sustainability information, including climate risks. These regulations aim to improve information flow and prevent greenwashing, enabling markets and regulators to assess better institutions' climate-related exposures (European Commission, 2018, 2019). Enhanced disclosure is crucial, as incomplete or poor-quality data on climate risks has been a significant obstacle to risk analysis (TCFD, 2017; FSB, 2020). The Task Force on Climate-related Financial Disclosures (TCFD, 2017) recommendations have become the global baseline for climate risk reporting, and many jurisdictions—the UK, EU, New Zealand, among others—are making TCFD-aligned disclosures mandatory for large financial institutions and corporations. By improving disclosure, regulators believe that market discipline will reward firms that manage climate risks well and penalise those that do not, thereby reinforcing stability (TCFD, 2017; Ameli et al., 2021).

Central banks themselves are adjusting their operations. The European Central Bank, following a strategic review in 2021, announced an action plan to incorporate climate change into its monetary policy framework (ECB, 2021b). Measures include integrating climate risk into corporate bond purchase programs (tilting purchases toward greener issuers), refining collateral rules (penalising assets with higher climate risk or favouring sustainable assets as collateral), and developing new climate-adjusted economic models (ECB, 2021b; Lagarde, 2021). The People's Bank of China has implemented a green lending facility that provides funds to banks at lower rates for lending to environmentally beneficial projects, and the Bank of Japan introduced a climate financing facility in 2021 to support green loans and bonds (Bank of Japan, 2021). These actions, while careful to stay within the bounds of mandates (typically focused on price stability and financial stability), represent a *greening* of central banking tools (Dikau & Volz, 2021). In the prudential space, supervisory authorities are increasingly conducting climate risk stress tests and embedding findings into banks' Supervisory Review and Evaluation Process (SREP) (EBA, 2021; ECB, 2022).

An essential aspect of the regulatory push is climate risk disclosure and data. Regulators are aware that "you cannot manage what you cannot measure." Initiatives like the TCFD have led to more firms reporting their carbon footprints, climate scenarios, and risk management strategies. Additionally, new sustainable finance standards are emerging. For instance, the International Sustainability Standards Board (ISSB), formed under the IFRS Foundation in 2021, is developing uniform global disclosure standards for climate and ESG, which could become widely adopted to ensure comparability (IFRS, 2022). The hope is that better disclosure will inform regulators and investors and encourage firms to actively mitigate climate risks and align their business models with climate objectives, knowing they will be held accountable (TCFD, 2017; Ameli et al., 2021). There is, however, an active debate on central bank mandates and legitimacy. Some scholars argue that by taking on climate change, central banks risk overstepping their traditional remit and venturing into fiscal or political territory (Honohan, 2019). Others counter that ignoring climate risks would compromise central banks' core objectives in the long run, as unchecked climate change could destabilise the macroeconomy and financial system; thus, integrating climate is within the scope of ensuring long-term price and financial stability (Bolton et al., 2020; Dikau & Volz, 2021). The consensus in policy circles appears to be that central banks should act as facilitators of the low-carbon transition *within* the bounds of their mandates—for example, by correcting market failures related to climate risk information and by safeguarding financial stability—while governments must lead on climate policy (Krogstrup & Oman,

2019; NGFS, 2019). In sum, through enhanced supervision, regulatory frameworks, scenario analysis, and support for green finance, central banks and regulators are increasingly aligning the financial system with climate imperatives and guarding against the systemic risks posed by climate change.

2.6 Empirical Evidence on Climate Risk and Financial Performance

Empirical research on the intersection of climate risks, sustainability practices, and financial performance is burgeoning, providing nuanced insights into how "green" strategies and climate exposures translate into financial outcomes. One robust finding across many studies is that high sustainability or ESG performance is associated with *comparable or slightly superior* financial performance relative to peers, especially over the long term (Friede et al., 2015). Firms with strong environmental performance often enjoy lower capital costs and higher valuations, as investors perceive them as less exposed to future climate liabilities and more adaptive to regulatory changes (Chava, 2014; Gregory et al., 2014). For example, Chava (2014) found that U.S. companies with significant environmental concerns (e.g., toxic emissions) faced higher interest spreads on bank loans, suggesting lenders demand a premium for firms carrying environmental risks. This aligns with more recent evidence in Europe that, after 2015, banks started incorporating carbon intensity into loan pricing, effectively penalising more carbon-intensive companies with higher loan costs (Delis et al., 2019). In equity markets, climate risk has become a priced factor. Engle et al. (2020) constructed a "Climate Risk Index" from news and found that stocks more sensitive to this index earn higher expected returns, indicating investors require compensation for bearing climate risk. Similarly, empirical asset pricing studies show that companies with higher carbon emissions or lower carbon efficiency have begun to trade at discounts (or face higher required returns) as climate-aware investors adjust their portfolios (Görgen et al., 2020; Imam et al., 2019). Conversely, "green" firms or those with proactive climate strategies often see positive abnormal returns around key climate policy events. Ramelli et al. (2018) demonstrated that in the days around the 2015 Paris Agreement, stock prices of firms in carbon-intensive industries underperformed, whereas renewable energy companies and those with lower emissions outperformed—a market reaction reflecting anticipated winners and losers of tighter climate policy.

In the banking sector, evidence is emerging on how climate risk exposure correlates with bank performance and stability metrics. A recent study by Reboredo and Ugolini (2022) finds that banks with greater exposure to climate-sensitive sectors experienced higher credit risk and volatility, especially following climate-related stress events or policy announcements. Their results suggest that prudent management of climate risks (e.g., diversifying away from fossil-fuel-heavy borrowers) is associated with better financial performance for banks regarding risk-adjusted returns. Likewise, Jung et al. (2018) showed that banks which voluntarily disclose more climate-related information (indicating they are actively assessing climate risks) tend to have lower capital costs and better market valuations, implying that investors reward transparency and preparedness on climate issues. On the bond side, green bonds have generally performed comparably to conventional bonds from a total return perspective, with some studies noting slightly lower yield spreads (consistent with the greenium concept discussed earlier) but no evidence of higher default rates—suggesting that investing in green bonds has not come at the expense of credit quality (Reboredo, 2018; Gianfrate & Peri, 2019). Moreover, portfolio-level studies indicate that integrating climate criteria can improve risk-adjusted performance. For instance, climate-aligned equity indices have had risk/return profiles similar to or better than broad market indices in recent years, partly because they avoided the steep losses in coal and oil stocks and captured gains in clean tech (MSCI, 2021). There is also evidence that ESG-focused funds, including those with climate screens, may offer downside protection in turbulent markets. Lins et al. (2017) found that during the 2008 financial crisis, companies with higher CSR (which includes environmental) scores had significantly higher stock returns, as investors flocked to firms deemed socially responsible, perceiving them as lower-risk in a crisis—a pattern echoed in the early COVID-19 pandemic market downturn (Albuquerque et al., 2020). These findings reinforce the argument that considering sustainability can be aligned with fiduciary duty and financial prudence, rather than opposed to it.

That said, not all empirical evidence is one-sided; some studies find neutral or context-dependent effects. For example, Pástor et al. (2021) present a theoretical and empirical case that when demand for green assets is robust, those assets can become overvalued in the short run (yielding lower expected returns in the future), even as brown assets carry a risk premium for being shunned. They suggest an equilibrium where investors may accept lower returns from green investments in exchange for non-financial utility (a cleaner conscience or lower perceived risk), which could slightly reduce long-term return prospects for green assets relative to brown. However, even this framework acknowledges that in "bad times"—e.g., when carbon taxes or climate shocks hit—green assets should outperform brown due to lower exposure to those shocks (Pástor et al., 2021). Thus, the risk-return tradeoff of sustainable investing can depend on the time horizon and specific risk realised. On balance, the empirical literature to date provides encouraging evidence that integrating climate and sustainability factors is compatible with strong financial performance. Effective climate risk management tends to be associated with improved creditworthiness and market valuations (Jung et al., 2018; Delis et al., 2019), while ignoring climate trends can lead to mispricing and abrupt corrections as seen in market reactions around major climate events (Ramelli et al., 2018). As data quality improves and time passes, we expect even clearer evidence to emerge on how climate change reshapes financial risk-return profiles across asset classes (Kling et al., 2021). The prudent conclusion is that climate risk is material and managing it well can yield a competitive advantage, whereas inaction or complacency may carry significant financial penalties.

2.7 *Consolidated Insights from the Literature*

The literature reviewed demonstrates that financial institutions and central banks are integral to pursuing a greener economy and mitigating climate-related risks. Banks, investors, and other financial intermediaries can act as catalysts for green transformation by reallocating capital toward sustainable enterprises, integrating ESG factors into risk management, and developing innovative financial instruments that support climate objectives. At the same time, the financial system faces non-trivial risks from climate change, both in the form of physical damages and the transition to a low-carbon paradigm, which, if unmanaged, could undermine financial stability. Central banks and regulators have recognised this challenge and are actively reshaping regulatory frameworks and monetary operations to foster transparency, resilience, and alignment with climate goals. The surge in climate stress-testing, mandatory disclosures, and international cooperation (such as through the NGFS) reflects an unprecedented mobilisation of the financial policy community around climate action. Empirical evidence mostly corroborates that embracing sustainability is financially sound: institutions that proactively manage climate risks and seize green financing opportunities tend to perform as well as, if not better than, their less sustainable counterparts, especially over the long run.

This convergence of research findings and policy initiatives suggests a future where the financial sector protects itself from climate-related losses and *drives* decarbonization by influencing the cost and availability of capital. Nevertheless, significant challenges and open questions remain. These include better data and modelling techniques to quantify climate risks, developing internationally consistent taxonomies and reporting standards, and carefully calibrating policies (like green supporting or brown penalising factors) to avoid unintended consequences. Moreover, the balance between central bank independence and support for broader climate policy will likely continue to be debated. In the future, interdisciplinary research—bridging finance, economics, and climate science—is crucial to refine our understanding of how financial institutions can most effectively contribute to climate change mitigation and how policy frameworks can best incentivise this role. In conclusion, the literature to date makes a compelling case that the green transformation of the financial system is underway: financial institutions and central banks, armed with appropriate tools and safeguards, are positioned to be catalysts for a more sustainable economy while safeguarding the system against the escalating risks of climate change (Bolton et al., 2020; NGFS, 2019; Reboredo & Ugolini, 2022). The continued evolution of market practices and regulatory oversight will determine how successfully finance can be harnessed to meet the climate challenge in the critical years ahead.

3 Materials and Methods

Nearly five decades ago, De Solla Price (1965) was among the first to propose applying scientific methods to the study of science (Boyack et al., 2005). Bibliometric methods, such as co-citation analysis and bibliographic linkage, use bibliographic data from publication databases to build structural pictures of scientific fields. These methods bring a degree of objectivity to the assessment of the scientific literature (Garfield, 1979) and can identify informal research networks, known as "invisible colleagues", that exist below the surface but are not formally connected (Crane, 1972; Price, 1965).

Bibliometric methods serve two primary purposes: performance analysis and science mapping (Cobo et al., 2012). Performance analysis aims to assess researchers' and institutions' research and publication performance. Science mapping focuses on revealing the structure and dynamics of scientific domains. This information is helpful for researchers when they want to analyse a particular line of research. Bibliometric methods bring quantitative rigour to an otherwise subjective assessment, providing evidence for theoretical categories derivable from an analysis. This approach to bibliometric analysis has become an essential tool in the study and understanding of the evolution and structure of scientific fields.

The specialised software VOSviewer will allow us to perform a bibliometric analysis. This is a computer program used to create maps based on network data and to visualise and explore these maps. According to the VOSviewer Manual, this program allows the generation of maps based on various types of data, including network adjacency matrix, scientific publications, scientific journals, researchers, research organizations, countries or keywords, based on co-authorship, co-occurrence, citation, bibliographic linkage or co-citation networks extracted from Web of Science, Scopus, PubMed or RIS files. These methodologies allow us to analyse keywords, authors, institutions, countries, etc., helping to identify the most frequently used keywords, the most cited authors and research centres.

Bibliometric analysis can also be used to assess a publication's popularity among scholars and to check the author's reputation (Ball & Dirk, 2005). According to Zupic and Čater (2015), this type of investigation facilitates literature review, guiding the researcher to influential research works or publications and objectively mapping the field of study.

In this research, we conducted a detailed analysis of the co-occurrence of keywords in papers, focusing on keywords appearing below the abstract, setting a minimum threshold of 25 papers in which these keywords are present. We also examined co-citation, setting a minimum threshold of 25 papers receiving citations from the same paper. In addition, we investigated co-authorship from the perspective of institutions and countries, considering the number of co-authors among the most frequent sources, setting a minimum threshold of 25 papers. Using these criteria, we constructed maps and performed analysis using the VOSviewer tool, highlighting connections between publications, researchers, institutions and keywords. The links were rated according to their strength and expressed as positive or negative values to provide a detailed picture of networks and relationships in the field of study.

This complex and detailed approach to bibliometric analysis has allowed us to comprehensively understand the field's scientific networks, collaborations and research themes. The size of the nodes in the analysis map indicates the importance and contribution of each article to the whole sample. Larger nodes suggest a more significant contribution of the article to the overall bibliometric analysis. The items are grouped to form clusters, but it is important to emphasise that these clusters do not necessarily cover all items in the generated map.

4 Results and Discussions

A detailed bibliometric analysis of relevant Web of Science resources, scholarly articles, publications and other sources of information identified significant trends, connections and developments in the use and development of the terms 'climate change' and 'banking'. This analysis aimed to gain a deeper understanding of the implications of sustainability strategy for the banking sector and its evolution over time. It also aimed to identify the most common research topics related to the banking sector's involvement in climate change mitigation. To carry out a relevant analysis, we selected publications in the field of climate change in the banking sector, resulting in a sample of 5596 articles. These results are associated with the terms "climate change" and "banking sector" without imposing any further restrictions. The evolution of the number of publications over time is detailed in the analysis below. The terms 'climate change' and 'banking sector' started to be significantly more used in Web of Science publications since 2005 and show an upward trend in usage, with a total of 670 articles published in 2022 (Fig. 1). This significant increase in interest in sustainability in banking indicates a paradigm shift in the approach and awareness of the role of financial institutions in promoting socially and environmentally responsible practices.

4.1 The Co-occurrence Network of the Keyword of the Author of Publications

The purpose of keyword co-occurrence network analysis is to determine the terms frequently used in this study's various papers. This type of analysis can help researchers understand the topics and issues that have provoked the most interest among researchers in this sector. By identifying keywords that appear together frequently in the papers included in the analysis, it is possible to highlight the dominant themes and relationships between the concepts addressed in the arca of sustainability in the banking sector.

The keyword co-occurrence network map (Fig. 2) highlights six main clusters distinguished by their colours: red, green, blue, purple, lime green and azure. The most frequently occurring word in the purple cluster, the one with the largest node, is "climate-change" (1160), followed by "climate change" (989), "climate" (389), "dynamics" (303), "impact" (295), "management" (250) and "growth" (246). Thus, the key term "climate change" is the most frequently encountered, emphasising the importance and relevance of climate change within the banking sector. Observations indicate an evolution of the concept of climate change in the banking sector over time.

The first *red cluster* highlights the primary interest in the linkages between environmental performance in the banking sector, through keywords such as sustainable finance, risk, sustainability, impact, emissions and policy. These linkages

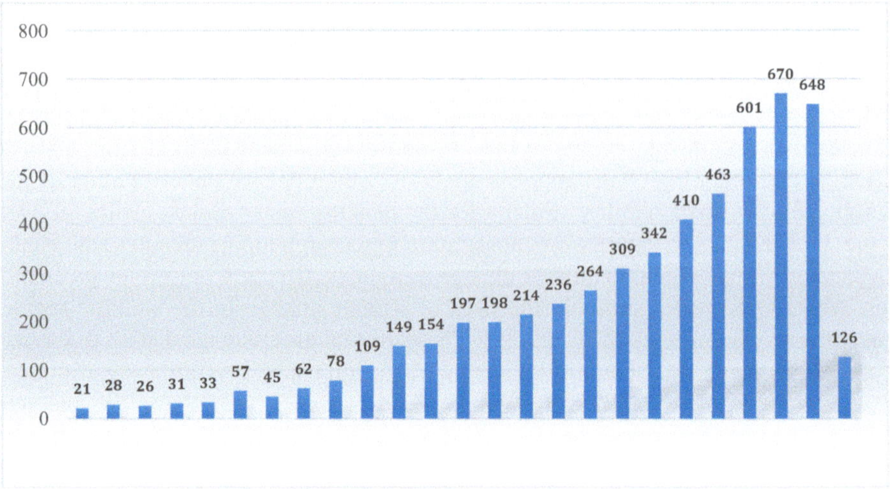

Fig. 1 Sustainability publications in the banking sector. (Source: Processed data from Web of Science Database, 2024)

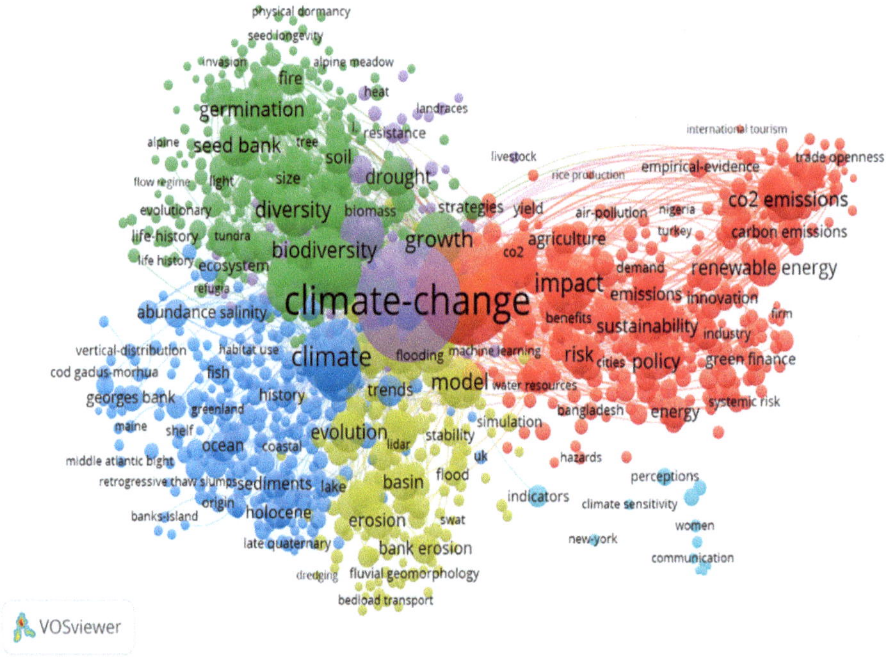

Fig. 2 Co-author network of authors' keywords. (Source: Own processing in VOSviewer)

underscore that green banking is recognised as sustainable, contributing to environmental protection and ensuring long-term economic prosperity (Islam et al., 2020). Interpretation of these results indicates a significant concern among researchers for the banking sector's involvement in addressing climate change issues and promoting sustainability. In the red cluster, there is a frequent mention of the link between the keywords "impact" (295 occurrences) and "CO_2 emissions" (151 occurrences). This suggests that the papers investigate the environmental implications of the transition from classical to more sustainable banking models. Banking institutions appear to be deeply engaged in social and environmental responsibility, highlighting their concern for CO_2 emissions and social and financial efficiency. In this context, studies exploring microfinance approaches (Gutierrez-Nieto et al., 2009; Chahine & Tannir, 2010; Allet & Hudon, 2015; Forcella & Hudon, 2016), environmental awareness initiatives (Miralles-Quiros et al., 2019; Batae et al., 2021), and the development of investment and financing policies that aim to improve the quality of the environment (Wellalage & Kumar, 2021) are included.

The *green cluster* is represented by the following nodes: "dynamics" with 303 occurrences, "growth" with 246 occurrences and "diversity" with 196 occurrences. The "dynamics" node, with 2016 strong links, suggests a focus on changes and developments in sustainability practices and policy in the banking sector. This may include adapting to new requirements and regulations on sustainable finance, integrating ESG (environmental, social and governance) criteria into lending and investment decisions, and changing customer and investor perceptions of sustainability. According to the European Central Bank (2020), in the context of banks offering sustainable lending, the interest rate adjustment process could be linked to the fulfilment by the client of sustainability objectives over a pre-determined period. This practice encourages adopting and implementing sustainability practices and policies, and may include criteria to reduce the risks associated with climate change and environmental impacts.

Considering the *blue cluster* with the central nodes "climate" (389) and "variability" (231), it suggests that there is a strong link between climate change and associated variability, as well as banks' concerns and actions in the area of sustainability. Over the last decade, banks have strengthened their sustainability actions. Some authors believe these initiatives are likely motivated by consumer dissatisfaction after the 2008 financial crisis (Pomering & Dolnicar, 2006). This sustainable behaviour is perceived as restoring or building the bank's reputation (Forcadell & Aracil, 2017; Tran, 2014).

The *lime green cluster* with the main nodes "impacts" (230) and "model" (211) shows links with the following words: evolution, stability, rates. The strong links of over 1200 of the main nodes suggest the idea that there is a concern in sustainable banking for impact analysis, model development and application, long-term evolution and stability, and the relationship between interest rates and financial issues. These issues are key to understanding and promoting a sustainable approach to banking, with significant implications for the environment, society and financial stability.

The *purple cluster* is one of the smallest clusters, but with a significant influence, being represented by the node with the most occurrences (1160) and the strongest links (6623)-"climate-change". Showing links to all clusters, with the most occurrences and strongest links, this cluster indicates that climate change is a central and interconnected issue in discussions on sustainability in banking. Climate issues affect all banking activities, including risk management, investments and assessment of their operations' social and environmental impacts. Thus, the banking sector needs to address and integrate climate change into its strategies to ensure sustainability and resilience when facing these global challenges.

The last *azure cluster* is the smallest cluster but also the least connected, as it presents small nodes only connected to the main "climate-change" node. The presence of small nodes in this cluster indicates less diversity and limited connection to other concepts or aspects in the analysed network. In the context of the banking sector, this interpretation could suggest that certain elements or topics related to sustainability and climate change are less explored or addressed in discussions and analysis in this area. These issues may be considered less relevant or prioritised than other concerns such as financial risks, operational efficiency or innovation in banking products and services.

4.2 Authors' Co-authorship Network

Bibliometric analysis, performed with the VosViewer application, allows one to identify the most cited authors in a field of interest through author citations and author co-citations in publications. This analysis highlights the most influential authors in the field, is a valuable tool in the documentation process, and is used to identify trends in academic research.

Figure 3 shows the co-citation map, where the analysed unit is "authors". By setting a minimum number of citations per author of 25, 218 authors meeting this condition were identified out of a total of 19,803. The larger the node (circle), the more cited the author. Likewise, the thicker the link between the nodes, the stronger the link, and the more often these authors are co-cited. These authors were grouped into 6 clusters: red, green, blue, lime green, purple and azure. Thus, Fig. 3 contains 35 strongly linked authors and forms a network, while the other 183 authors are not interlinked to each other and do not show strong links. Table 1 presents the ranking of the top 10 authors out of the total 218 authors, who have published the most papers and the top 10 authors with the most citations. At the top of the ranking is the author Kirikkaleli, Dervis, with the most published papers (23) and the most citations (1346), but he does not belong to any cluster, as he does not show any links.

The first cluster (red) includes 7 authors, and the most cited authors in this cluster are Poschlod, Peter (833 citations); Thompson, Ken (827 citations); Dixon, Kingsley W. (848 citations), and Walk, Jeffrey I. (710 citations). These authors' high number of citations indicates significant influence in their field of research, suggesting important contributions and recognition among the scientific community.

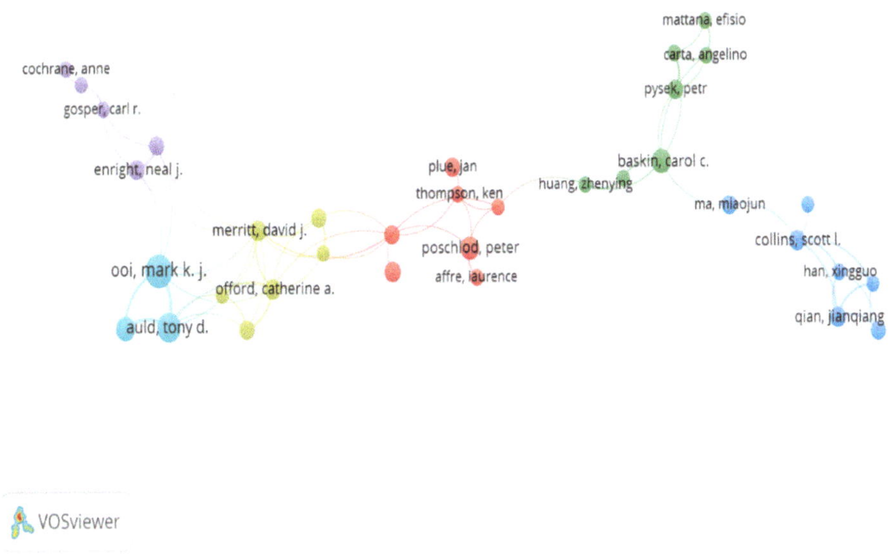

Fig. 3 Authors' co-citation. (Source: Own processing in VOSviewer)

The second cluster (green) comprises 7 authors, with the author Baskin, Carol C. (135 citations) representing the most influential author in this cluster, with a total of 8 papers. Even though the number of citations is lower than that of the first cluster, this author can be considered an essential author in his field.

Clusters three (blue) and four (lime green) have 7 and 6 authors, respectively. The most influential author in these clusters is Suding, Katharine N. (951 citations) for the blue cluster. This author has a significant number of citations, indicating a considerable contribution to the literature in his field.

The last cluster comprises 3 authors and is characterised by the author Ooi, Mark K.J. (772 citations), who published 15 articles and is one of the top 10 cited authors in this field (Table 1). Although belonging to a smaller cluster, this author is considered one of the most influential in his field, with a significant number of citations and a significant impact on research. The interpretation of these data suggests that the identified authors with a high number of citations are important figures in their research field and have contributed significantly to the advancement and development of their field.

4.3 *Institutional Co-author Analysis*

Using VOSviewer software and citation analysis, we identified the top 10 institutions that have significantly contributed to sustainability research in the banking sector. These institutions were selected based on the large number of papers published in this area and their research impact, as shown in Table 2. Accompanying the

Table 1 Important authors by publications and citations

Ranking	Authors	Publications
1	Kirikkaleli, Dervis	23
2	Ooi, Mark K.J	15
3	Ji, Ruabo	14
4	Hare, Jonathan A.	13
5	Auld, Tony D.	12
6	Friedland, Kevin D.	10
7	Hagedorn, Mary	10
8	Powell, Eric N.	10
9	Kotlik,Petr	9
10	Searle, Jeremy B.	9
Ranking	**Authors**	**Citations**
1	Kirikkaleli, Dervis	1346
2	Suding,Katharine N.	951
3	Dixon, Kingsley W.	848
4	Poschlod,Peter	833
5	Thompson, Ken	827
6	Ooi, Mark K.J	772
7	Walk, Jeffrey I.	710
8	Hare, Jonathan A.	702
9	Dafermos, Yannis	546
10	Battiston, Stefano	547

Source: Own processing, using data provided by Web of Science Core Collection Database 2024, available at: https://0c10qjxkk-y-https-www-webofscience-com.z.e-nformation.ro/wos/woscc/basic-search (accessed in May 2024)

Table 2 Top 10 institutions in the dataset by number of affiliated publications

Ranking	Institutions	Publications	Citations
1	Chinese Academy of Sciences	136	2821
2	National Oceanic and Atmospheric Administration	77	3026
3	US Geological Survey Institution	65	2651
4	Woods Hole Oceanographic Institution	54	2019
5	University of the Chinese Academy of Sciences	46	786
6	University of Oxford	43	1232
7	University of Cambridge	42	375
8	University of Western Australia	42	1072
9	University of Cape Town	40	1605
10	University of Hamburg	40	2389

Source: Own processing, using data provided by Web of Science Core Collection Database 2024, available at: https://0c10qjxkk-y-https-www-webofscience-com.z.e-nformation.ro/wos/woscc/basic-search (accessed in May 2024)

main investigations on this topic, these institutions provided relevant and valuable results, standing out as leaders in the field. As shown in Fig. 4, applying a minimum threshold of 25 published papers, we obtained a ranking consisting of 19 clusters reflecting the contribution of institutions to climate change research in banking. In this ranking, only 833 institutions met this threshold, out of 6024, underlining their commitment and impact in this growing field. These reviews highlight the importance and contribution of institutions in sustainability research in the banking sector, highlighting research directions and collaboration between different entities to address challenges and opportunities in these critical areas.

The bibliometric analysis reveals that the Chinese Academy of Sciences is a significant player in researching and publishing papers on sustainability practices in the banking sector. With 136 papers published in this field, the Chinese Academy of Sciences stands out as one of the leading contributors to the academic literature on sustainability in the banking industry. By being the most prominent and most representative node in cluster 11, also referred to as "green lacustrine", the Chinese Academy of Sciences demonstrates a high level of expertise and commitment in this field. Cluster 11, being represented by the colour green in the network analysis, and gathering 48 items, this cluster also includes the institution University of the Chinese Academy of Sciences (affiliated to the Chinese Academy of Sciences), which has a total of 46 items. These findings suggest that the Chinese Academy of Sciences significantly impacts the development and promotion of sustainability practices in the banking sector. Through their prolific publications and leadership position in

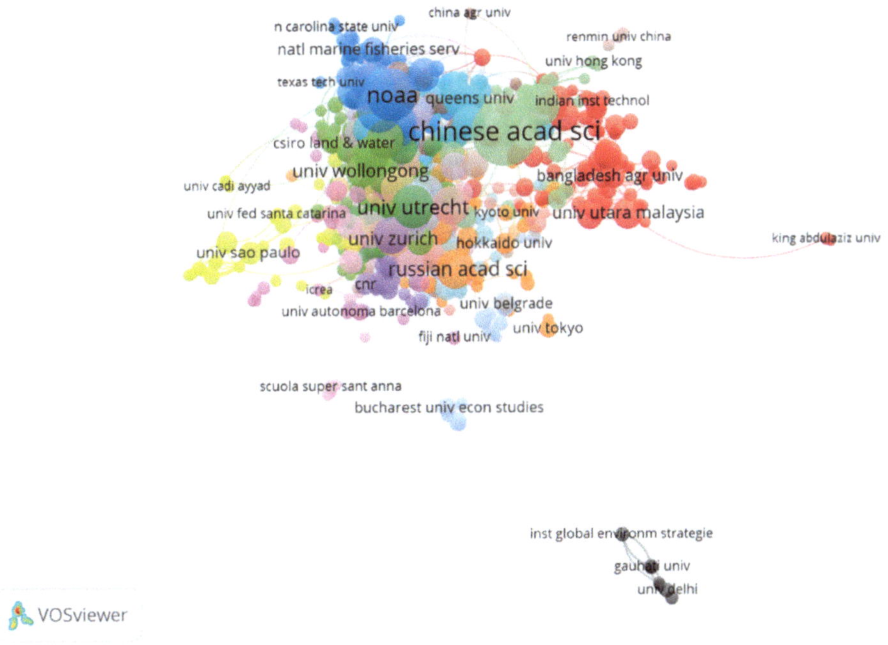

Fig. 4 Institutional co-author network. (Source: Own processing in VOSviewer)

cluster 11, they influence and guide the direction of research in this vital area for the sustainable future of the banking industry.

Another institution, the National Oceanic and Atmospheric Administration (NOAA), significantly contributes to the field of sustainability practices in the banking sector. With a total of 77 published articles and the most citations, numbering 3026, NOAA stands out as a key player in this field. As the largest node in cluster 3, represented by the blue colour in the network analysis, NOAA demonstrates that it has a central and influential position within this cluster. In this cluster, the Woods Hole Oceanographic Institution (WHOI) is also part of a total of 54 articles and 2019 citations. Given the central position of NOAA in the cluster and the importance of WHOI in the same context, it is likely that there will be significant interactions between these two institutions. These interactions may include research collaboration, data and information sharing, and participation in joint projects and initiatives related to environmental sustainability in the banking sector. Equally, with an imposing number of papers in the field of sustainability is the US Geological Survey Institution, which presents 65 papers and 2651 citations.

Cluster 19 (black) is represented by the following nodes: Gauhati University of India, Institute for Global Environmental Strategies, University of Delhi and Wildlife Institute of India. The fact that Cluster 19 shows no links to the other clusters suggests that the institutions in this cluster are relatively isolated regarding the network of interactions represented in the network analysis. This may indicate a distinct specialisation or focus of these institutions in the area of sustainability in the banking industry that differentiates them from the other clusters.

4.4 Analysis of Co-author Countries

We can examine the structure of research collaboration networks in a specific field through country co-author analysis. This analysis allows us to identify the origins of research teams and the network relationships between the authors involved. In Fig. 5, the nodes represent countries, and the distance and thickness between them reflect the degree of collaboration. The thresholds set include the minimum number of 4 papers from a country and the minimum number of citations to be considered. Based on these thresholds, only 112 of the 168 countries analysed met the conditions. In the process, countries were ranked according to the number of publications and the ranking was also based on citations over the last 24 years. Similar to the work carried out among the scientific community, we sought to identify and highlight opportunities for discovering new information and knowledge through collaboration (Beaver, 2001).

This narrow selection provides a more detailed and relevant insight into research collaboration networks. Through the country co-authorship analysis, trends in collaboration between influential countries were identified and assessed, providing opportunities for discovering new information and knowledge through collaboration. The restricted selection of participating countries resulted in 10 clusters, thus

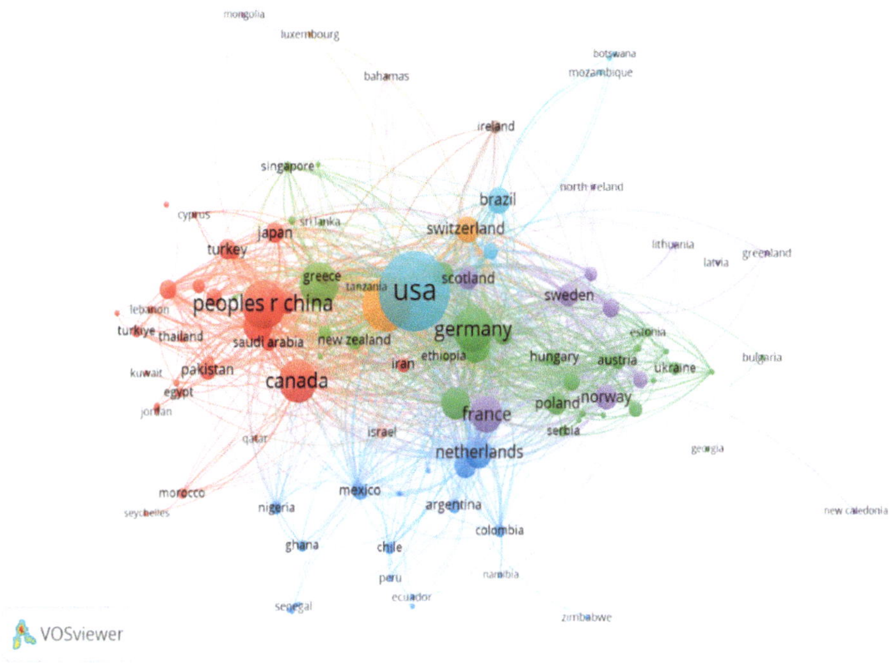

Fig. 5 Network of co-author countries. (Source: Own processing in VOSviewer)

providing a more detailed and relevant insight into collaborative research networks in the field under study.

And the top 10 countries in Table 3 represent highly ranked publications. According to the results, the USA were recognised as a leader in the research (1391 papers) that investigated and promoted environmental safeguards in the banking sector. The presence of strong links between the USA and countries in other clusters, such as China, Turkey, Japan, Canada, Egypt, Singapore, the Netherlands, France, Germany, Switzerland, Nigeria, the Bahamas, etc., may indicate a high level of research collaboration and interaction between the United States and these countries. This could suggest significant exchanges of knowledge, resources, and experience between the United States and the other mentioned countries, which would contribute to the development and advancement of the research field globally.

Next in the ranking, the United Kingdom (659) highlighted its efforts in assessing and managing sustainable measures in the banking sector, contributing to developing a comprehensive framework for implementing mitigation strategies. The fact that Cluster 7 (orange) includes only 3 countries—the United Kingdom, Switzerland and Luxembourg—suggests that these countries are engaged in a specific type of research or have some specialisation in that field. However, it is notable that although it is a smaller cluster, the UK shows 1012 strong links. This indicates significant collaboration and interaction between the UK and the other countries in the cluster and beyond.

Table 3 Prolific countries by number of publications and ranked citations

Ranking	Countries	Publications	Citations
1	USA	1391	42,053
2	England	659	20,602
3	Germany	462	12,749
4	France	297	8146
5	People's Republic of China	536	9517
6	Australia	438	13,554
7	Italy	244	5455
8	Netherlands	221	5486
9	Canada	412	11,202
10	Spain	232	6350

Source: Own processing, using data provided by Web of Science Core Collection Database 2024, available at: https://0c10qjxkk-y-https-www-webofscience-com.z.e-nformation.ro/wos/woscc/basic-search (accessed in May 2024)

Germany (the central node in the green cluster) was also noted for contributing to developing sustainability practices in the banking sector, being involved in 462 publications. Twenty three countries represent cluster 2 (green), and significant size nodes are the following countries: Italy, Russia, Serbia, Iceland, Estonia, Ukraine, Austria, Belarus, Croatia, Hungary, Georgia, Spain, Bulgaria, Poland and others. Also, other major European countries in this cluster indicate an extensive network of research collaboration in the European region.

The largest node in the red cluster-China, the emerging economy country, was included in the top 5 countries involved in collaborative research networks, suggesting a significant development and a valuable contribution to the academic field. This observation indicates China's ability to collaborate and produce scientific knowledge and the recognition and appreciation of its contribution to global research. The red cluster, consisting of 31 countries, includes diverse countries from different world regions, such as Canada, Turkey, Japan, Morocco, Egypt, Pakistan, Cyprus, Tunisia, Bangladesh and others.

This geographical diversity may indicate a global interest and commitment to research collaboration among these countries. These countries are likely to be involved in exchanges of expertise, resources and information in various research areas.

5 Practical Solutions for Better Environmental Sustainability Management in the Banking Sector

Several strategic and operational solutions could be implemented to better manage environmental sustainability in banking. First, banks could integrate environmental, social and governance (ESG) criteria into all decision-making processes, including risk assessment and capital allocation. This would involve a detailed analysis of the environmental impact of projects financed and prioritization of those that contribute to sustainability objectives, such as reducing carbon emissions or promoting renewable energy.

Developing and promoting green financial products, such as green loans, green bonds or sustainable investment funds, would also encourage companies and individuals to adopt more environmentally friendly practices. These products could include, for example, favourable lending conditions for projects that positively impact the environment, such as energy-efficient buildings or green infrastructure.

Another vital aspect is adopting an internal sustainability policy to govern the bank's operational activities. This could include measures to reduce the bank's carbon footprint, such as energy efficiency, using renewable resources, reducing paper consumption and promoting teleworking to mitigate environmental impact.

Collaboration with other financial institutions, governments and non-governmental organisations can also boost sustainability efforts. Through partnerships, banks can share best practices and technological innovations and participate in joint initiatives that promote global sustainability. This also includes supporting public policies that favour the transition to a green economy and active engagement in international debates and forums on climate change and sustainability. Through training programs and awareness-raising campaigns, banks could cultivate a culture of sustainability that influences both internal behaviour and relationships with customers and business partners.

In addition to *traditional solutions*, several innovative approaches can significantly contribute to more effective sustainability management in the banking sector. One such solution is adopting blockchain technologies to increase transparency and accountability in financing sustainability projects. Blockchain enables the tracking and verifying of capital flows to green projects, thus ensuring that funds are used as intended and directly contribute to environmental goals.

Implementing artificial intelligence (AI) and big data analytics can also be crucial in assessing environmental risks and identifying sustainable investment opportunities. AI can be used to model the potential impact of climate change on investment portfolios, enabling banks to anticipate risks and adjust their strategies accordingly. At the same time, big data analytics can facilitate the identification of emerging trends and consumer behaviours that favour sustainability, providing banks with valuable information to develop financial products and services tailored to these needs.

Another innovative aspect is the integration of sustainability-related performance metrics into compensation and incentive schemes for banking employees. By linking financial rewards to achieving sustainability targets, banks can motivate employees to adopt and promote green practices in their day-to-day work. This can include, for example, incentives for employees who develop innovative financial solutions that support the transition to a low-carbon economy.

In addition to the abovementioned solutions, other innovative approaches can support more effective sustainability management in the banking sector. One such solution is the development of sustainability assessment and reporting frameworks based on measurable and standardised impact indicators. These frameworks enable banks to accurately assess and communicate the social and environmental impacts of the projects they finance, thus facilitating comparability and transparency between different financial institutions. Standardised reporting also helps to increase the confidence of investors and other stakeholders in banks' commitment to sustainability.

Banks can also explore the creation of thematic investment funds that focus exclusively on projects and companies that promote sustainability and green innovation. These funds can attract investors who are concerned about social and environmental impact while offering competitive financial returns. Banks can play a key role in accelerating the transition to a sustainable economy by directing capital into green sectors.

Another innovative aspect is the use of crowdfunding to support sustainability initiatives. Banks can develop crowdfunding platforms dedicated to green projects, thus allowing the public to get directly involved in financing green initiatives. This type of financing expands access to capital for sustainable projects and raises public awareness of the importance of green investments.

Finally, banks can also collaborate with FinTech to develop innovative digital solutions to support sustainability management. For example, mobile apps that monitor customer spending's environmental impact or facilitate investments in sustainable projects can play an essential role in promoting responsible financial practices.

In conclusion, sustainability management in the banking sector can be improved by adopting innovative solutions that harness technology, promote transparency and accountability, and actively involve employees and the public in efforts to transition to a sustainable economy. These approaches support environmental objectives and strengthen banks as leaders in promoting sustainable development.

6 Conclusion

The comprehensive bibliometric analysis of sustainability publications within the banking sector reveals significant trends, connections, and developments in terms such as "climate change" and "banking sector." The study underscores the importance and relevance of sustainability strategies in banking, reflecting a paradigm shift toward more socially and ecologically responsible practices. Notably, the rising number of publications since 2005 indicates increasing interest and acknowledgement of the banking sector's role in addressing climate change and promoting sustainability.

The analysis of keyword co-occurrence networks shows that climate change is a central theme, closely linked with other important concepts like environmental impact, sustainable finance, and risk management. The frequent co-occurrence of terms related to climate change, CO_2 emissions, and sustainability highlights the banking sector's recognition of its environmental impact and potential contributions to mitigation efforts. Identified clusters within the network indicate a range of focus areas, from the relationship between environmental performance and financial stability to integrating advanced technologies such as artificial intelligence in fostering sustainable banking practices. This diversity underscores the complexity of sustainability issues in the banking sector, necessitating a comprehensive and interdisciplinary approach.

The study makes several *significant contributions* to our understanding of the role of the banking sector in tackling climate change. On the one hand, it provides a detailed and well-grounded overview of emerging research trends and directions in this area, highlighting the growing academic interest in sustainability in the financial sector. Therefore, the study identifies not only the main research themes, such as climate risk assessment and the development of sustainable financial instruments, but also highlights the diversity of contributions from different disciplines, including economics, finance, and environmental management.

On the other hand, this study's novelty lies in applying a rigorous bibliometric methodology, which synthesises and analyses a vast amount of data and provides a deep understanding of how the terms "climate change" and "banking sector" have evolved in literature. Moreover, the study introduces a detailed analysis of the co-occurrence of the keywords, authorship network, institutions and countries involved in the most sustainable banking practices, providing a new perspective on the thematic interconnectedness and relationships between relevant concepts in the field, and also between authors from different institutions and the most diverse countries.

Thus, the study contributes not only to consolidating existing knowledge but also to guiding future academic and practical efforts towards a more nuanced and applicable understanding of the banking sector's role in combating climate change.

Practical implications. The findings have a profound impact on both the banking industry and policymakers. The prominence of climate change and sustainability in banking suggests a need for continued efforts to integrate environmental, social, and governance (ESG) criteria into financial decision-making. Banks must acknowledge their potential to drive positive environmental outcomes through responsible lending, investment practices, and green finance initiatives. For policymakers, the study highlights the important role of regulatory frameworks in supporting sustainable practices in the banking sector. Effective policies can incentivise banks to adopt sustainable finance practices, thereby contributing to broader environmental and social goals. The growing focus on sustainability in banking also points to the importance of fostering collaboration among financial institutions, regulatory bodies, and other stakeholders to develop a cohesive and impactful sustainability strategy.

Recommendations for policymakers. To promote sustainability in the banking sector, policymakers should strengthen regulatory frameworks to support and mandate the integration of ESG criteria in banking operations. Also, collaboration between banks, regulatory bodies, and stakeholders is essential to develop innovative solutions and best practices for sustainable banking. Moreover, supporting the development and adoption of advanced technologies can enhance sustainable banking practices by improving efficiency, risk management, and creating new opportunities for sustainable investments. Raising public awareness about the importance of sustainable finance and the banking sector's role in addressing climate change is nevertheless essential. Establishing mechanisms to monitor and evaluate the progress of sustainability initiatives within the banking sector is necessary to identify gaps, measure impact, and ensure that sustainability goals are being met effectively. By implementing these strategies, policymakers and banking institutions can foster a more sustainable and resilient financial system, contributing to environmental protection, social responsibility, and long-term economic stability.

The study provides a solid basis for *future exploration* of the intersection between banking and climate change. One of the future research directions could include a deeper analysis of the effectiveness of different sustainable financial instruments implemented by banking institutions. Future research could investigate the long-term impact of these instruments on reducing carbon emissions and promoting sustainable business practices. Another promising avenue would be to explore the impact of climate change on financial risks and how the banking sector can develop predictive models and risk management strategies that incorporate environmental factors. In addition to these directions, future research could benefit from expanding bibliometric analysis to include non-academic sources, such as banks' sustainability reports, which could provide additional insights into how financial institutions perceive and address climate challenges.

Limitations of the study. Despite the valuable contributions of this study, there are some limitations worth mentioning. One of the main limitations is the dependence on the available bibliometric data sources, which might exclude some relevant papers or contributions from the grey literature or sources not indexed in the main academic databases. This may lead to a partial perspective on the current state of research in the field. The analysis also focused mainly on works published in the last two decades, which may limit our understanding of the historical evolution of the banking sector's involvement in climate change.

Given the findings and limitations identified, policymakers must encourage closer collaboration between academia, financial institutions, and regulators to improve the collection and access to relevant data on sustainability in the banking sector. Policies that support interdisciplinary research and facilitate the implementation of innovative financial solutions that respond to climate challenges are also recommended. Integrating practitioners' perspectives and empirical research into policy formulation can contribute to developing more effective and responsive strategies for the global economy.

Acknowledgments This work was supported by a grant from the Romanian Ministry of Research, Innovation and Digitalization, the project with the title "*Economics and Policy Options for Climate Change Risk and Global Environmental Governance*" (CF 193/28.11.2022, Funding Contract no. 760078/23.05.2023), within Romania's National Recovery and Resilience Plan (PNRR)—Pillar III, Component C9, Investment I8 (PNRR/2022/C9/MCID/I8)—Development of a program to attract highly specialised human resources from abroad in research, development, and innovation activities.

Competing Interests The authors have no conflicts of interest to declare relevant to this chapter's content.

Ethics Approval This study is a bibliometric analysis and did not involve any studies with human participants or animals performed by any of the authors. Therefore, ethical approval was not required.

References

Abuzahra, A., & Nobanee, H. (2022). *Sustainability in the banking sector*. Retrieved March 24, 2024, from https://www.researchgate.net/publication/358395690_Sustainability_in_the_Banking_Secto

Albuquerque, R., Koskinen, Y., Yang, S., & Zhang, C. (2020). Resiliency of environmental and social stocks: An analysis of the exogenous COVID-19 market crash. *The Review of Corporate Finance Studies, 9*(3), 593–621.

Allet, M., & Hudon, M. (2015). Green microfinance: Characteristics of microfinance institutions involved in environmental management. *Journal of Business Ethics, 126*(3), 395–414.

Alogoskoufis, S., Dunz, N., Emambakhsh, T., Lagarde, M., Martin, R., Papadopoulos, V., & Kadri, F. (2021). *ECB economy-wide climate stress test: Methodology and results* (European Central Bank Occasional paper, no. 281). European Central Bank.

Ameli, N., Kothari, S., & Grubb, M. (2021). Misplaced expectations from climate disclosure initiatives. *Nature Climate Change, 11*(11), 917–924. https://doi.org/10.1038/s41558-021-01174-8

Auer, B. R., & Schuhmacher, F. (2016). Do socially (ir)responsible investments pay? New evidence from international ESG data. *Quarterly Review of Economics and Finance, 59*, 51–62. https://doi.org/10.1016/j.qref.2015.07.002

Bachelet, M. J., Becchetti, L., & Manfredonia, S. (2019). The green bonds premium puzzle: The role of issuer characteristics and third-party verification. *Sustainability, 11*(4), 1098.

Baker, M., Bergstresser, D., Serafeim, G., & Wurgler, J. (2018). *Financing the response to climate change: The pricing and ownership of U.S. green bonds* (National Bureau of Economic Research Working paper, no. 25194). National Bureau of Economic Research.

Ball, R., & Dirk, T. (2005). *Bibliometric analysis data, facts and basic methodological knowledge: Bibliometrics for scientists, science managers, research institutions and universities* (Vol. 12). Research Center Jülich.

Bank of England. (2022). *Results of the 2021 Climate Biennial Exploratory Scenario*. Bank of England Report, May 2022.

Bank of England Prudential Regulation Authority (PRA). (2019). *Supervisory Statement SS3/19: Enhancing banks' and insurers' approaches to managing climate-related financial risks*. Bank of England.

Bank of Japan. (2021). *Implementation of climate-related financing measures*. Press Release, June 2021.

Basel Committee on Banking Supervision (BCBS). (2021). *Climate-related risk drivers and their transmission channels*. Bank for International Settlements.

Basel Committee on Banking Supervision (BCBS). (2022). *Principles for the effective management and supervision of climate-related financial risks*. Bank for International Settlements.

Bătae, O. M., Dragomir, V. D., & Feleagă, L. (2021). The relationship between environmental, social, and financial performance in the banking sector: A European study. *Journal of Cleaner Production, 290*, 125791. https://doi.org/10.1016/j.jclepro.2021.125791

Battiston, S., Mandel, A., Monasterolo, I., Schütze, F., & Visentin, G. (2017). A climate stress-test of the financial system. *Nature Climate Change, 7*(4), 283–288.

Beaver, D. D. (2001). Reflections on scientific collaboration (and its study): Past, present, and future. *Scientometrics, 52*(3), 365–377. https://doi.org/10.1023/A:1014254214337

Boffo, R., & Patalano, R. (2020). *ESG investing: Practices, progress and challenges*. OECD Publishing.

Bolton, P., Després, M., Pereira da Silva, L. A., Samama, F., & Svartzman, R. (2020). *The green swan: Central banking and financial stability in the age of climate change*. Bank for International Settlements and Banque de France.

Boyack, K., Klavans, R., & Börner, K. (2005). Mapping the backbone of science. *Scientometrics, 64*(3), 351–374. Retrieved March 24, 2024 from http://www.springerlink.com/index/Q282602257566164.pdf

Busch, T., Bauer, R., & Orlitzky, M. (2016). Sustainable development and financial markets: Old paths and new avenues. *Business & Society, 55*(3), 303–329.

Caldecott, B. (2018). *Stranded assets and the environment: Risk, resilience and opportunity*. Routledge.

Campiglio, E., Dafermos, Y., Monnin, P., Ryan-Collins, J., Schotten, G., & Tanaka, M. (2018). Climate change challenges for central banks and financial regulators. *Nature Climate Change, 8*(6), 462–468.

Carney, M. (2015). Breaking the tragedy of the horizon – Climate change and financial stability. *Speech by Mark Carney*, Governor of the Bank of England, at Lloyd's of London, 29 September 2015.

Chahine, S., & Tannir, L. (2010). On the social and financial effects of the transformation of microfinance NGOs. *VOLUNTAS: International Journal of Voluntary and Nonprofit Organizations, 21*(3), 440–461.

Chava, S. (2014). Environmental externalities and cost of capital. *Management Science, 60*(9), 2223–2247.

Climate Bonds Initiative (CBI). (2022). *Green bond market summary 2021*. Climate Bonds Initiative Report, March 2022.

Cobo, M., López-Herrera, A. G., Herrera-Viedma, E., & Herrera, F. (2012). SciMAT: A new science mapping analysis software tool. *Journal of the American Society for Information Science and Technology, 63*(8), 1609–1630. https://doi.org/10.1002/asi

Coleton, A., Brucart, M. F., Gutierrez, P., Tennier, F. L., & Moor, C. (2020). Sustainable finance market practices. Retrieved from https://www.eba.europa.eu/sites/default/documents/files/document_library/Sustainable%20finance%20Market%20practices.pdf

Collier, B., Lippert, C., & Schaper, J. (2022). Natural disasters and credit default risk: Evidence from Hurricane Katrina. *Journal of Financial Intermediation, 50*, 100945.

Contreras, L., et al. (2019). *Financial institutions as catalysts for sustainable development: A framework for green finance*. [Journal/Report].

Cosma, S., et al. (2022). *ESG integration and bank stability in the European banking sector*. [Journal/Report].

Crane, D. (1972). *Invisible colleges: Diffusion of knowledge in scientific communication*. The University of Chicago Press.

D'Orazio, P., & Popoyan, L. (2019). Fostering green investments and tackling climate-related financial risks: Which role for macroprudential policies? *Ecological Economics, 160*, 25–37.

De Solla Price, D. J. (1965). Networks of scientific papers: The pattern of bibliographic references indicates the nature of the scientific research front. *Science, 149*(3683), 510–515.

Delis, M. D., Kok, C., & Ongena, S. (2019). Climate risk and financial stability in the banking sector: Evidence from the pricing of syndicated loans. *Ecological Economics, 164*, 106315.

Dikau, S., & Volz, U. (2021). Central bank mandates, sustainability objectives and the promotion of green finance. *Ecological Economics, 184*, 107022.

Doran, N. M., Manta, A. G., Badareu, G., Berceanu, D., Doran, M. D., & Manta, F. L. (2025). Could e-commerce activities drive to climate change mitigation? Novel evidence from panel quantile regression model. *Journal of Business Economics and Management, 26*(2), 255–276. https://doi.org/10.3846/jbem.2025.23604

Ehlers, T., & Packer, F. (2017). Green bond finance and certification. *BIS Quarterly Review, 2017*, 89–104.
Ehlers, T., Packer, F., & de Greiff, K. (2020). Understanding the green bond market. In *The future of banking* (pp. 127–140). Banque de France.
Engle, R. F., Giglio, S., Kelly, B., Lee, H., & Stroebel, J. (2020). Hedging climate change news. *The Review of Financial Studies, 33*(3), 1184–1216.
European Banking Authority (EBA). (2021). *Report on management and supervision of ESG risks for credit institutions and investment firms*. EBA/REP/2021/18.
European Central Bank. (2020). *Guide on climate-related and environmental risks*. Retrieved April 2, 2024, from https://www.bankingsupervision.europa.eu/ecb/pub/pdf/ssm.202011finalguideonclimate-relatedandenvironmentalrisks~58213f6564.ro.pdf
European Central Bank (ECB). (2020). *Guide on climate-related and environmental risks: Supervisory expectations relating to risk management and disclosure*. ECB, November 2020.
European Central Bank (ECB). (2021a). *Financial Stability Review* (June 2021), Special feature on climate-related financial risks. ECB.
European Central Bank (ECB). (2021b). *ECB presents action plan to include climate change considerations in its monetary policy strategy*. ECB Press Release, 8 July 2021.
European Central Bank (ECB). (2022). *2022 Climate risk stress test – Report on the results*. ECB, July 2022.
European Commission. (2018). *Action Plan: Financing sustainable growth*. European Commission (COM/2018/097).
European Commission. (2019). *Regulation (EU) 2019/2088 on sustainability-related disclosures in the financial services sector (SFDR)*. Official Journal of the EU, L 317/1.
Ferrari, A., & Landi, V. (2024). Whatever it takes to save the planet? Central banks and unconventional green policy. *Macroeconomic Dynamics, 28*(2), 299–324. https://doi.org/10.1017/S1365100523000032
Financial Stability Board (FSB). (2020). *The implications of climate change for financial stability*. Financial Stability Board Report, November 2020.
Flammer, C. (2021). Corporate green bonds. *Journal of Financial Economics, 142*(2), 499–516.
Forcadell, F. J., & Aracil, E. (2017). European banks' reputation for corporate social responsibility. *Corporate Social Responsibility and Environmental Management, 24*(1), 1–14. https://doi.org/10.1002/csr.1378
Forcella, D., & Hudon, M. (2016). Green microfinance in Europe. *Journal of Business Ethics, 135*, 445–459. https://doi.org/10.1007/s10551-014-2476-0
Friede, G., Busch, T., & Bassen, A. (2015). ESG and financial performance: Aggregated evidence from more than 2000 empirical studies. *Journal of Sustainable Finance & Investment, 5*(4), 210–233.
Garfield, E. (1979). Is citation analysis a legitimate evaluation tool? *Scientometrics, 1*(4), 359–375. https://doi.org/10.1007/BF02019306
Gherţescu, C., Manta, A. G., Bădîrcea, R. M., Duţă, A. G., & Zaharia, A. L. (2024). Bibliometric analysis of ESG in banking: What are the key directions and collaborations behind the research? *Finance – Challenges of the Future, 24*(26), 66–79.
Gherţescu, C., Manta, A. G., & Bădîrcea, R. M. (2025). Smart agriculture and technological innovation: A bibliometric perspective on digital transformation and sustainability. *Agriculture, 15*(13), 1388. https://doi.org/10.3390/agriculture15131388
Gianfrate, G., & Peri, M. (2019). The green advantage: Exploring the convenience of issuing green bonds. *Journal of Cleaner Production, 219*, 127–135.
Global Compact. (2004). *Who cares wins: Connecting financial markets to a changing world*. Retrieved March 23, 2024, from https://www.unepfi.org/fileadmin/events/2004/stocks/who_cares_wins_global_compact_2004.pdf
Görgen, M., Jacob, A., Nerlinger, M., Riordan, R., Rohleder, M., & Wilkens, M. (2020). Carbon risk. *Available at SSRN 2930897* (later published in *Journal of Asset Management, 22*(2), 96–109, 2021).

Goss, A., & Roberts, G. S. (2011). The impact of corporate social responsibility on the cost of bank loans. *Journal of Banking & Finance, 35*(7), 1794–1810.

Gregory, A., Tharyan, R., & Whittaker, J. (2014). Corporate social responsibility and firm value: Disaggregating the effects on cash flow, risk and growth. *Journal of Business Ethics, 124*(4), 633–657. https://doi.org/10.1007/s10551-013-1898-5

Gutiérrez-Nieto, B., Serrano-Cinca, C., & Molinero, C. (2009). Social efficiency in microfinance institutions. *The Journal of the Operational Research Society, 60*(1), 104–119. https://doi.org/10.1057/palgrave.jors.2602539

Honohan, P. (2019). Should monetary policy take account of risks to financial stability? *Economic & Social Review, 50*(2), 261–276.

IFRS Foundation. (2022). IFRS S2 Climate-related Disclosures. *International Financial Reporting Standards*. https://www.ifrs.org/issued-standards/ifrs-sustainability-standards-navigator/ifrs-s2-climate-related-disclosures/

Imam, S., Kong, D., & Shah, S. Z. A. (2019). Do socially responsible firms provide superior financial performance? A trivariate analysis. *Business & Society, 58*(2), 255–286.

International Capital Market Association (ICMA). (2020). *Climate transition finance handbook*. ICMA, December 2020.

Islam, M. J., Roy, S. K., Miah, M., & Das, S. K. (2020). A review on corporate environmental reporting (CER): An emerging issue in the corporate world. *Canadian Journal of Business and Information Studies, 2*(3), 45–53.

Jung, J., Herbohn, K., & Clarkson, P. (2018). Carbon risk, carbon risk awareness and the cost of debt financing. *Journal of Business Ethics, 150*(4), 1151–1171.

Kling, G., Volz, U., Murinde, V., & Ayas, S. (2021). The impact of climate vulnerability on firms' cost of capital and access to finance. *World Development, 137*, 105131.

Krogstrup, S., & Oman, W. (2019). *Macroeconomic and financial policies for climate change mitigation: A review of the literature*. IMF working paper, no. 19/185.

Lagarde, C. (2021). Climate change and central banking. *Speech by Christine Lagarde*, President of the ECB, at the International Climate Change Conference, July 2021.

Lins, K. V., Servaes, H., & Tamayo, A. (2017). Social capital, trust, and firm performance: The value of corporate social responsibility during the financial crisis. *Journal of Finance, 72*(4), 1785–1824.

Manta, A. G., Doran, N. M., Bădîrcea, R. M., Badareu, G., & Țăran, A. M. (2023). Does the implementation of a Pigouvian tax be considered an effective approach to address climate change mitigation? *Economic Analysis and Policy, 80*, 1719–1731. https://doi.org/10.1016/j.eap.2023.11.002

Manta, A. G., Doran, N. M., Hurduzeu, G., et al. (2024a). Is there a direct benefit of using electronic commerce and electronic banking in mitigating climate change? *Climatic Change, 177*, 152. https://doi.org/10.1007/s10584-024-03809-z

Manta, A. G., Bădîrcea, R. M., Doran, N. M., Badareu, G., Gherțescu, C., & Popescu, J. (2024b). Industry 4.0 transformation: Analysing the impact of artificial intelligence on the banking sector through bibliometric trends. *Electronics, 13*(9), 1693. https://doi.org/10.3390/electronics13091693

Manta, A. G., Gherțescu, C., Bădîrcea, R. M., Manta, L. F., Popescu, J., & Lăpădat, C. V. M. (2025a). How does the interplay between banking performance, digitalization, and renewable energy consumption shape sustainable development in European Union countries? *Energies, 18*(3), 571. https://doi.org/10.3390/en18030571

Manta, L. F., Manta, A. G., & Gherțescu, C. (2025b). Decoding digital synergies: How mechatronic systems and artificial intelligence shape banking performance through quantile-driven method of moments. *Applied Sciences, 15*(10), 5282. https://doi.org/10.3390/app15105282

Meghişan-Toma, G. M., Puiu, S., Florea, N., Meghişan, F., Bădîrcea, R., & Manta, A. (2022). Sustainable transformation of Romanian companies through Industry 4.0, green production and environment commitment. *Amfiteatru Economic, 24*(59), 46–60.

Miralles-Quirós, M. M., Miralles-Quirós, J. L., & Redondo-Hernández, J. (2019). The impact of environmental, social, and governance performance on stock prices: Evidence from the banking industry. *Corporate Social Responsibility and Environmental Management, 26*(3), e1759. https://doi.org/10.1002/csr.1759

MSCI. (2021). *Foundations of climate investing: How equity markets have priced climate transition risks*. MSCI Research Report, February 2021.

Nagy, Z., Cogan, D., & Sinnreich, D. (2013). *Optimizing environmental, social, and governance factors in portfolio construction*. MSCI ESG Research.

Nagy, Z., Kassam, A., & Lee, L. E. (2015). *Can ESG add alpha?* MSCI ESG Research. Retrieved March 23, 2024, from http://www.iccr.org/sites/default/files/resources_attachments/research_insight_can_esg_add_alpha.pdf

Network for Greening the Financial System (NGFS). (2019). *First comprehensive report: A call for action – Climate change as a source of financial risk*. NGFS, April 2019.

Network for Greening the Financial System (NGFS). (2020). *NGFS Climate Scenarios for central banks and supervisors*. NGFS, June 2020.

Network for Greening the Financial System (NGFS). (2021). *Adapting central bank operations to a hotter world: Reviewing some options*. NGFS occasional paper, March 2021.

Nizam, E., Ng, A., Dewandaru, G., Nagayev, R., & Nkoba, M. A. (2019). The impact of social and environmental sustainability on financial performance: A global analysis of the banking sector. *Journal of Multinational Financial Management, 49*, 35–53.

Nollet, J., Filis, G., & Mitrokostas, E. (2016). Corporate social responsibility and financial performance: A non-linear and disaggregated approach. *Economic Modelling, 52*(Part B), 400–407.

Novy-Marx, R. (2013). The other side of value: The gross profitability premium. *Journal of Financial Economics, 108*(1), 1–28. https://doi.org/10.1016/j.jfineco.2013.01.003

OECD. (2017). *Investing in Climate, Investing in Growth*. OECD Publishing.

Ongena, S., Schopfer, A., & Wagner, A. F. (2019). Carbon emissions and the bank lending channel. Available at SSRN 3342547 (later published in *Journal of Banking & Finance, 121*, 105969, 2020).

Pástor, Ľ., Stambaugh, R. F., & Taylor, L. A. (2021). Sustainable investing in equilibrium. *Journal of Financial Economics, 142*(2), 550–571.

Poenaru, M. M., Manta, L. F., Gherţescu, C., & Manta, A. G. (2025). Shaping the future of horticulture: Innovative technologies, artificial intelligence, and robotic automation through a bibliometric lens. *Horticulturae, 11*(5), 449. https://doi.org/10.3390/horticulturae11050449

Pomering, A., & Dolnicar, S. (2006). Customers' sensitivity to different measures of corporate social responsibility in the Australian banking sector. Proceedings of the Australia and New Zealand Marketing Academy Conference (ANZMAC), Brisbane, Queensland, 4–6 December 2006. https://ro.uow.edu.au/commpapers/242

Price, D. J. d. S. (1965). Networks of scientific papers. *Science, 149*(3683), 510–515.

Ramelli, S., Wagner, A., Zeckhauser, R., & Ziegler, A. (2018). Investor rewards to climate responsibility: Evidence from the 2016 US Presidential Election. *Environmental and Resource Economics, 71*(3), 613–654.

Reboredo, J. C. (2018). Green bond and financial markets: Co-movement, diversification and price spillover effects. *Energy Economics, 74*, 38–50.

Reboredo, J. C., & Ugolini, A. (2020). Price connectedness between green bond and financial markets. *Journal of International Financial Markets, Institutions and Money, 68*, 101233.

Reboredo, J. C., & Ugolini, A. (2022). Climate-related risks and bank financial performance: An empirical analysis. *International Review of Financial Analysis, 80*, 101976.

Schoenmaker, D., & Schramade, W. (2019). *Principles of Sustainable Finance*. Oxford University Press.

Smith, A., & Nobanee, H. (2020). *Artificial intelligence: In banking – A mini-review*. https://doi.org/10.2139/ssrn.3539171.

Sommer, F. (2020). Sustainable finance: An overview. In M. La Torre & H. Chiappini (Eds.), *Contemporary issues in sustainable finance: Financial products and financial institutions* (pp. 1–22). Palgrave Macmillan. https://doi.org/10.1007/978-3-030-38829-2_1

Stauropoulou, A., & Sardianou, E. (2019). Understanding and measuring sustainability performance in the banking sector. *IOP Conference Series: Earth and Environmental Science, 362*(1), 012128. https://doi.org/10.1088/1755-1315/362/1/012128

Task Force on Climate-related Financial Disclosures (TCFD). (2017). *Final report: Recommendations of the Task Force on Climate-related Financial Disclosures.* Financial Stability Board.

Tran, Y. (2014). CSR in the banking sector: A literature review and new research directions. *International Journal of Economics, Commerce and Management, 2*(11), 1–22.

United Nations Environment Programme – Finance Initiative (UNEP FI). (2018). *Rethinking impact to finance the SDGs: A position paper and call to action.* UNEP Finance Initiative.

United Nations Framework Convention on Climate Change (UNFCCC). (2015). *Paris Agreement.* United Nations.

Vermeulen, R., Schets, E., Lohuis, M., Kölbl, B., Jansen, D. J., & Heeringa, W. (2021). The heat is on: A framework for measuring financial stress under disruptive energy transition scenarios. *Ecological Economics, 190*, 107205.

Vilanova, M., Lozano, J. M., & Arenas, D. (2009). Exploring the nature of the relationship between CSR and competitiveness. *Journal of Business Ethics, 87*(Suppl. 1), 57–69. https://doi.org/10.1007/s10551-008-9812-2

Wellalage, N., & Kumar, V. (2021). Environmental performance and bank lending: Evidence from unlisted firms. *Business Strategy and the Environment, 30*(7), 3309–3329. https://doi.org/10.1002/bse.2805

World Bank. (2020). *World Bank Group Action Plan on climate change adaptation and resilience 2019–2025.* World Bank.

Zerbib, O. D. (2019). The effect of pro-environmental preferences on bond prices: Evidence from green bonds. *Journal of Banking & Finance, 98*, 39–60. https://doi.org/10.1016/j.jbankfin.2018.10.012

Zupic, I., & Čater, T. (2015). Bibliometric methods in management and organization. *Organizational Research Methods, 18*(3), 429–472. https://doi.org/10.1177/1094428114562629

Carbon Pricing and Emissions Trading: Financial Mechanisms for Driving Climate Action and Sustainable Finance

Chi-Wei Su, Pirtea Marilen Gabriel, Ran Tao, and Rongrong Dong

1 Introduction

The twenty-first-century climate catastrophe has gained international attention. "Climate change" denotes the aggregate of all climatic modifications caused by elevated atmospheric levels of greenhouse gases (GHG). Carbon dioxide (CO_2) is the primary greenhouse gas, constituting roughly 76% of world GHG emissions (Metcalf, 2019). Shifting to low-carbon economic growth through emission reduction measures has become a global consensus and is also a necessary condition for the continuous progress of mankind (Jia & Lin, 2020), thus governments of various countries implement specific climate policies following the Paris Agreement and commit to taking prompt and practical actions (Su et al., 2025). Among these, carbon pricing is generally considered the most economically effective means for reducing greenhouse gas emissions (Tvinnereim & Mehling, 2018). Carbon pricing is an economic mechanism

C.-W. Su
Capital Market Research Institute, School of Economics, Department of Finance, Qingdao University, Qingdao, China
e-mail: cwsu7137@qdu.edu.cn; su.chih@e-uvt.ro

R. Dong (✉)
School of Economics, Department of Finance, Qingdao University, Qingdao, China
e-mail: dongrongrong@qdu.edu.cn

P. M. Gabriel
Faculty of Economics and Business Administration, Finance, Business Information Systems and Modelling Department, West University of Timisoara, Timişoara, Romania
e-mail: marilen.pirtea@e-uvt.ro

R. Tao
Doctoral School of Economics and Business Administration, West University of Timisoara, Timişoara, Romania
e-mail: ran.tao10@e-uvt.ro

C.-W. Su et al. (eds.), *Green Finance and the Challenges of Climate Change*, Sustainable Finance, https://doi.org/10.1007/978-3-032-07293-1_12

that assigns explicit value to carbon emissions, internalising environmental costs into economic decision-making to incentivise emission reductions (Solarin, 2025). Carbon pricing has become a cornerstone of climate policy, functioning through internalising greenhouse gas emission externalities.

1.1 Economic Foundations of Carbon Pricing

Carbon pricing, grounded in externality theory and market efficiency, offers a scalable solution to climate mitigation. Its theoretical foundation lies in environmental economics and market-based approaches, which aim to rectify market failures and promote cost-effective emission reductions. The economic rationale for carbon pricing rests on three fundamental principles. The first one is externality and Pigouvian taxation. The core justification for carbon pricing lies in addressing negative externalities, as polluters currently do not account for the complete social costs of emissions, such as climate change and public health effects. Pigou introduced the concept of levying a tax equivalent to emissions' marginal social cost (now termed a Pigouvian tax) as a mechanism to reconcile private and social costs, thereby achieving socially optimal pollution abatement. Carbon taxes implement the Pigouvian principle by assigning a direct price per metric ton of carbon dioxide emissions, thereby making climate externalities visible in market transactions, modifying the behaviour of both producers and consumers (Nordhaus, 2013). The second one is the Coase Theorem and tradable permits. Coase posited that delineated property rights and minimal transaction costs could achieve efficient pollution outcomes through voluntary bargaining between parties as an alternative to taxation. Emissions trading systems like the European Union Emissions Trading System (EU ETS) operationalise this approach by establishing a fixed emissions cap while permitting market-driven allowance trading among participants. This market-based mechanism enables cost-effective emission reductions by allowing firms with lower marginal abatement costs to sell surplus allowances to those facing higher reduction costs. The third one is cost-effectiveness and dynamic efficiency. Carbon pricing achieves cost-minimised abatement by letting market actors choose between paying the carbon price or investing in cleaner technologies. The meta-analysis by Gillingham and Stock (2018) demonstrates that carbon pricing stimulates low-carbon technology innovation, thereby accelerating long-term decarbonisation processes. The double dividend hypothesis posits that revenue-neutral carbon taxation, mainly when the revenues are utilised to reduce distortionary taxes such as labour taxes, can produce both environmental enhancements and economic efficiency improvements.

1.2 Carbon Pricing Measures

Carbon pricing can be divided into two methods: carbon tax and emissions trading system (ETS). The carbon tax directly levies financial fees on energy production and consumption activities, while the ETS achieves total control by establishing an emissions trading platform. Although these two carbon pricing measures are implemented differently, they are crucial in addressing climate risks.

Carbon Tax

A carbon tax is a fixed price imposed on each ton of carbon dioxide emissions. This policy plan enables the emissions of enterprises to be determined by the market (Sumner et al., 2011), which will bring significant economic and environmental benefits to society (Marron & Toder, 2014). On the one hand, it can effectively reduce the emissions of greenhouse gases such as carbon dioxide through emission costs and slow down the process of global warming (Pan et al., 2024). On the other hand, implementing the carbon tax policy can effectively curb the use of fossil fuels, promote the development of the renewable energy market (such as wind and solar energy) and drive green technological innovation. Although carbon taxes have many advantages, they also show some limitations at the same time, such as inconsistent coverage, which can easily lead to inequality. Moreover, elevated tax rates may impair industrial competitiveness, potentially resulting in carbon leakage through production relocation. Carbon taxation has emerged as a policy instrument for nations pursuing stringent climate objectives.

Emissions Trading System

The theoretical basis of the Emissions Trading System (ETS) originates from Coase's theorem. This theorem indicates that as long as a trading market for externalities (such as greenhouse gas emissions) can be established and the transaction costs are low enough, no matter how the initial property rights are allocated, the market participants can eventually reach the Pareto optimal solution through autonomous negotiation. This theory provides a theoretical basis for the government to solve the market failure caused by environmental negative externalities by clarifying the definition of emission rights (Coase, 1960).

The Emissions Trading System (ETS), known as the "Cap-and-Trade" mechanism, is a market-based environmental policy tool. This system provides enterprises with flexible emission reduction approaches by setting an upper limit on the total amount of greenhouse gas emissions and introducing emission quotas as tradable commodities into the market, while ensuring the realisation of the overall emission reduction targets (Solarin, 2025). From a legal perspective, the right to discharge pollutants can be regarded as a limited right of use related to environmental resources. The government is the ultimate proprietor of this special commodity based on the idea of treating the environment as a commodity. This system allocates quantitative emission rights to enterprises, enabling them to obtain legal emission quotas. It incorporates these rights into the market trading system, converting carbon emission rights into standardised commodities. The carbon emissions trading system plays a significant role in society and the economy (Hu et al., 2022). On the one hand, ETS ensures environmental benefits by controlling total emissions. The carbon pricing mechanism in emissions trading creates powerful economic incentives, driving corporate investment in clean technology while stimulating low-carbon innovation and fostering growth in sustainable finance. Despite their advantages, carbon markets exhibit several structural weaknesses (Borghesi et al., 2015). For instance, the regulatory process of the emissions trading system is complex and requires a well-established MRV, triggering market manipulation. Furthermore, economic recession or excessive allocation of quotas may lead to a sharp drop in carbon prices, raising the risk of carbon price fluctuations, which requires supporting measures to maintain a reasonable carbon price.

The carbon emission trading system is a trading market that uses emission rights as commodities for circulation, which can significantly reduce carbon dioxide emissions. At present, this policy plan has been implemented in many countries and regions around the world. The EU Emissions Trading System (EU ETS) is the first attempt at an emissions trading system globally. Subsequently, several countries, including the United States and China, successfully expanded and implemented this policy plan through localisation and innovations (Betsill & Hoffmann, 2011).

The European Union Emissions Trading System (EU ETS) is generally regarded as the first emissions trading system in history. Since its launch in 2005, the EU ETS has undergone four development stages, each of which has been adjusted and expanded in response to previous issues. The first stage (2005–2007) is the experimental stage, when the operation mechanism of the carbon market is tested to accumulate experience for the subsequent stages (Convery & Redmond, 2007). The second stage (2008–2012) is a transitional period, connecting with the Kyoto Protocol and strengthening the intensity of emission reduction. During this stage, the upper limit is uniformly set at the EU level, reducing by 6.5% compared with 2005. Furthermore, this carbon emissions trading system includes the aviation industry within its coverage. The EU unilaterally set free emission allowances at 85% of airlines' original emissions, with this ratio further reduced to 82% by 2013. Airlines must pay a "green toll" for any excess emissions. The regulation also stipulated penalties for non-compliance: airlines refusing to adhere would face fines of €100 (approximately $130) per ton of excess emissions. They could be banned from operating within the EU. The third stage (2013–2020) is structural reform, with the system shifting from decentralised national planning to unified management by the European Union (Koch et al., 2014). The current fourth stage (2021–2030) aims to implement stricter emission reduction targets and expand the auction mechanism.

The EU Emissions Trading System has contributed to a 21% reduction in covered emissions since 2005. The EU ETS demonstrates the effectiveness of market-based approaches to emission reduction, providing valuable experience for expanding cap-and-trade mechanisms to address climate change. Additionally, the carbon emissions trading market interacts with the financial industry, forming a virtuous cycle and promoting the development of the carbon finance industry in the European Union. More importantly, the successful implementation of this system has enhanced the EU's leadership position in the new round of international climate negotiations. However, some EU member states criticised the carbon trading system for failing to curb emissions effectively. At the same time, many non-EU countries remained hesitant to adopt similar schemes due to concerns over carbon leakage and industrial competitiveness concerns (Prentice, 2013). Despite initial challenges such as over-allocation and price volatility, the EU ETS became a model for subsequent systems (Ellerman & Buchner, 2007).

North America has adopted multiple regional emissions trading systems as market-based climate solutions, with the Western Climate Initiative (WCI) emerging as the most significant. This cap-and-trade program, jointly administered by California (since 2013) and Quebec (since 2014), regulates emissions across key sectors such as electricity generation, industrial operations, and transportation. The emission reduction effect of WCI is remarkable. Regional Greenhouse Gas Initiative (RGGI), the first market-based obligatory carbon trading system in the US power industry, is also essential. Jointly established by the Northeastern and Atlantic states, it aims to reduce carbon dioxide emissions in the power sector through a "cap-and-trade" mechanism. RGGI has decreased carbon dioxide emissions dramatically, establishing an example for other US states and boosting cross-regional climate policy collaboration. China launched the world's most extensive carbon emissions trading system in July 2021, marking a crucial step in achieving its dual carbon goals: peaking emissions by 2030 and achieving carbon neutrality by 2060. For China, establishing a unified national carbon emissions trading market is an essential institutional innovation that uses market mechanisms to reduce greenhouse gas emissions, promote the green low-carbon transformation and improve air quality (Chang et al., 2020). It is also an important policy tool for strengthening ecological civilisation construction and fulfilling international emission reduction commitments. By the end of 2024, the cumulative trading volume of quotas in the national carbon emission trading market will reach 630 million tons, with a cumulative trading value of 43.033 billion yuan. These data show that the carbon emissions market operates stably and orderly. Furthermore, this system can enhance the carbon reduction awareness of high-emission enterprises and promote the transformation to a low-carbon economy.

The carbon emissions trading system requires key factors that are essential to its operation. Implementing the carbon emissions trading system requires scientific policy and institutional design, including coverage, total quota allocation, relevant laws and regulations, etc. These are the prerequisite conditions to ensure its smooth implementation. Secondly, a sound market operation mechanism is also a factor that helps stimulate the enthusiasm of enterprises to participate in carbon dioxide reduction. Furthermore, strict data management and monitoring are also crucial to the carbon emission system. High-quality greenhouse gas emission data is the foundation for the smooth operation of the carbon emissions trading system. These factors are essential for environmental protection, economic development and technological progress (Zhang et al., 2020).

Although carbon taxes and emissions trading systems are different policy tools, their complementarity in promoting decarbonisation is gaining increasing recognition. First, their coverage areas are complementary (Lee et al., 2008). Carbon tax is more suitable for decentralised emission sources with lower regulatory difficulty (such as transportation, construction fields and small and medium-sized industrial enterprises), because it can be implemented efficiently relying on existing tax channels. In contrast, the Emissions Trading System (ETS) is more suitable for controlling significant centralised emission sources (such as the power, steel, cement and other industries), and such emission sources need to establish a complete monitoring, reporting and verification (MRV) mechanism to ensure the accuracy of data. This complementarity can optimise the efficiency of emission reduction, reduce policy costs, and enhance the effectiveness of the overall carbon pricing mechanism. Although both carbon trading and carbon tax have strong emission reduction capacity, the relative emission reduction efficiency of the carbon tax is higher than that of carbon trading (Ahmad et al., 2024). Second, the two are complementary in terms of price stability. As a price floor mechanism, the carbon tax stabilises the carbon market by preventing price collapses, providing policy flexibility while reducing the cost of industrial transformation. By establishing a price ceiling, the carbon emissions trading mechanism alleviates the potential economic impacts of carbon levies, retains emission reduction incentives and furnishes a stable investment signal for the low-carbon shift. The combined effect of the two can stabilise carbon prices, reduce risks and enhance confidence simultaneously, thus promoting long-term low-carbon investment (Zhang et al., 2022). Through the complementary elaboration of the two aspects, it can be recognised that the carbon pricing policy is of great significance for the low-carbon economy and the transformation of the energy structure.

2 Literature Review

Some scholars believe that a series of climate policies, such as carbon pricing, play an important role in climate action and sustainable finance. Solarin (2025) indicates that carbon pricing policies create a market price for carbon, spurring innovation, improving energy efficiency, and promoting renewable energy development. Gugler et al. (2023) highlight that even moderate carbon taxes can drive substantial emissions reductions, reinforcing the viability of market-based climate policies. Dorband et al. (2019) assert that climate change mitigation, domestic revenue generation, and economic inequality reduction can be achieved simultaneously, even in low- and middle-income country contexts. Moessner (2025) proposes that the prices of the carbon emissions trading system and carbon taxes have little impact on inflation, suggesting concerns about substantial inflationary impacts should not constrain the implementation of carbon pricing as a mechanism to accelerate the transition to net-zero emissions. Steinebach et al. (2021) point out that governments are more likely to adopt carbon pricing policies when both their trading partners and competitor nations in shared export markets have implemented similar measures. Linsenmeier et al. (2023) prove that emission reductions resulting from carbon pricing policy diffusion may exceed domestic mitigation effects for many nations.

However, there are also opposing opinions, arguing that the effect of carbon pricing policies is limited. Green (2021) states that although carbon pricing has been a central topic in climate policy debates, empirical evidence on its effectiveness remains limited. Only a very small part of studies have rigorously assessed its impact on emissions reductions, with most focusing on European cases, suggesting that carbon pricing, in its current implementations, has had a modest effect on lowering emissions. Abrell et al. (2022) evidence that tax-induced impacts exhibit temporal heterogeneity due to variations in relative fuel prices, showing that higher carbon taxes yield greater emissions reductions and higher costs, depending on these price dynamics. Känzig (2023) shows that carbon policy shocks raise energy prices, cut emissions, and boost green innovation, while simultaneously depressing economic activity and disproportionately affecting lower-income households.

To sum up, different scholars have different viewpoints. Therefore, this paper explores the importance of climate policies through empirical research.

3 Data and Methodology

Sustainable finance integrates environmental, social and governance (ESG) into the financial decision-making process (Edmans & Kacperczyk, 2022). Its main goal is not only to achieve economic growth, but also to protect the ecological environment and promote social progress. Before elaborating on the promoting effect of carbon pricing policies on sustainable finance, this chapter conducts a specific empirical analysis of the relationship between climate policy stability and sustainable development.

For this study, we chose the monthly data from January 2013 to December 2024 to explore the Granger causal relationship between climate policy uncertainty (CPU) and sustainability uncertainty (ESGU) in the US. Since 2013, the nineteenth session of the Conference of the Parties (COP 19) has served as an important platform for addressing worldwide climate change issues. Furthermore, President Obama announced a series of climate action plans in 2013. This part employs the Climate Policy Uncertainty Index (CPU) (Gavriilidis, 2021) and the Sustainability Uncertainty Index (ESGU) (Ongan et al., 2025) to analyse the relationship between the climate policy and sustainable development in the United States. Both of these indices are derived from the Economic Policy Uncertainty database. Furthermore, since economic policies are closely related to CPU and ESGU. Therefore, it can be determined that the Economic Policy Uncertainty (EPU) (Davis, 2016) may impact the mutual influence between CPU and ESGU. In that case, we select it as a control variable to guarantee the robustness of the outcomes in Eq. (2). Notably, the correlation between CPU and ESGU exhibits complex dynamics that EPU further influences. As demonstrated above, the interaction between CPU and ESGU exhibits time-varying variation. The traditional full-sample method fails to capture the complex Granger causality between these variables adequately. Thus, we employ an advanced sub-sample analysis to explore their dynamic interaction better.

3.1 Bootstrap Full-Sample Granger Causality Test

While the conventional vector autoregressive (VAR) method can capture the mutual relation between time series, it requires standard normality assumptions for both variables and the VAR process (Sims, 1980). In cases of non-compliance, we should avoid using the VAR model to recognise the link among time series variables due to concerns about its reliability. Shukur and Mantalos (2004) introduce critical values based on the residual bootstrap (RB) method to enhance the Granger causality test reliability and prevent erroneous outcomes. Likewise, they demonstrate that the size and power properties can amend the likelihood ratio (*LR*) tests in a finite sample (Shukur & Mantalos, 2000). Using *RB*-based modified-*LR* tests, this study investigates Granger causality between CPU and ESGU within a bivariate VAR system, as specified in Eq. (1):

$$Z_t = \gamma_0 + \gamma_1 Z_{t-1} + \ldots\ldots + \gamma_s Z_{t-s} + \mu_t \tag{1}$$

where the lag length s is selected by the sequential modified-*LR* test statistic and the Akaike Information Criterion (AIC), then this standard can seize an optimal selection. The bivariate VAR (s) is specified with $Z_t = (CPU_t, ESGU_t)'$, where Z is partitioned into CPU and ESGU components. Furthermore, given that CPU and ESGU may be linked to global economic policy uncertainty (EPU), which could influence their interaction, we include EPU as a control variable in the empirical analysis. The extended VAR (s) process is presented in Eq. (2):

$$\begin{pmatrix} CPU_t \\ ESGU_t \end{pmatrix} = \begin{pmatrix} \gamma_{10} \\ \gamma_{20} \end{pmatrix} + \begin{pmatrix} \gamma_{11}(L) & \gamma_{12}(L) & \gamma_{13}(L) \\ \gamma_{21}(L) & \gamma_{22}(L) & \gamma_{23}(L) \end{pmatrix} \begin{pmatrix} CPU_t \\ ESGU_t \\ EPU_t \end{pmatrix} + \begin{pmatrix} \mu_{1t} \\ \mu_{2t} \end{pmatrix} \tag{2}$$

where $\mu_t = (\mu_{1t}, \mu_{2t})'$ represents a white-noise process. $\gamma_{ij}(\mathrm{L}) = \sum_{k=1}^{s} \gamma_{ij,k} L^k$, where $i = 1, 2$, $j = 1, 2, 3$ and L means a lag operator, we have $L^k Z_t = Z_{t-k}$.

In terms of Eq. (2), we test the original assumption that CPU does not Granger-cause ESGU ($\gamma_{21,k} = 0$) can be tested. This hypothesis should be rejected if CPU exerts significant influences on ESGU and conversely. Similarly, the opposite original hypothesis ESGU does not Granger-cause CPU ($\gamma_{12,k} = 0$) can also be rejected if ESGU exerts significant influences on CPU.

3.2 Parameter Tability Test

The preceding analysis relies on the assumption of structural stability in the VAR framework, which frequently fails to hold in empirical applications. Accordingly, the test may yield unreliable results when parameters exhibit structural instability across the sample period. To assess structural breaks, we implement the parameter stability tests (*Sup-F*, *Ave-F*, *Exp-F*), proposed by Andrews (1993) and Andrews and Ploberger (1994). The first approach can identify the sudden structural breaks in time series variables and the VAR system, while the subsequent tests assess gradual coefficient evolution over time. Additionally, the L_c statistics test (Nyblom, 1989; Hansen, 1992) is utilized to test for random walk behavior in the parameters. Evidence of structural changes in the parameters would imply a non-constant Granger causal relation between CPU and ESGU. Consequently, these findings render conventional full-sample method may inadequate, requiring instead a sub-sample technique to properly identify the dynamic Granger causal link.

3.3 Bootstrap Sub-sample Rolling-Window Granger Causality Test

Following Balcilar et al. (2010), we utilize a rolling-window sub-sample technique to analyze the time-varying mutual relation between time series. By dividing the complete series into fixed-length segments that scroll through the sample period. Window width choice is a complex matter, smaller windows may yield biased estimates, whereas larger windows decrease the scrolls times. To resolve this methodological issue, we employ a rolling window technique that progresses sequentially through the full sample based on Pesaran and Timmermann (2005) demonstration that time-dependent VAR parameters require window widths of at least 20 periods. The specific procedure can be classified into three steps. Firstly, assuming that the extent of the time series is N and the width is m. Then, the end of each small section is m, m + 1, ..., N, and we can obtain N-m + 1 sub-samples. Secondly, every small part could acquire an interrelation via using the *RB*-based revised-*LR* method. Thirdly, we implement a bootstrap sub-sample rolling-window procedure to chronologically aggregate the *p*-values and likelihood ratio (*LR*) statistics across temporal samples, obtaining robust results. Furthermore, the average values of estimated results ($N_b^{-1} \sum_{s}^{k=1} \gamma_{21,k}^{*}$ and $N_b^{-1} \sum_{s}^{k=1} \gamma_{12,k}^{*}$) indicate the influences from CPU to ESGU and the influences from ESGU to CPU. Here, N_b shows the number of bootstrap repetitions, while $\gamma_{21,k}^{*}$ and $\gamma_{12,k}^{*}$ are the VAR (s) parameters. Consistent with Balcilar et al. (2010), this study constructs 90% confidence intervals with 5th and 95th quantiles as lower and upper bounds, respectively.

4 Results and Discussion

This section employs the ADF, PP and KPSS tests, to examine the stationary of the above series. The outcomes are presented in Table 1, and it could be observed that the first difference of CPU and ESGU have no unit roots, both of them are

Table 1 The results of unit root tests

		ADF	PP
Levels	CPU	−3.247**	−5.167***
	ESGU	−2.752*	−2.763*
First difference	CPU	−12.577***	−40.278***
	ESGU	−11.226***	−11.571***

Source: Authors' calculations
Notes: ***, ** and * are the significance at 1%, 5% and 10% levels

stationary. Then, the VAR models can be conducted effectively to perform the required Granger causality tests based on the results.

On the basis of Eq. (2), this investigation establishes a traditional VAR process to conduct the full-sample test, and then explores the Granger causal relation between CPU and ESGU. In terms of AIC, we selected 2 as the optimal lag order. Table 2 reports the results of the full-sample test, the *p*-values point out that ESGU is also not a Granger cause of CPU.

Nevertheless, the above estimation supposes that the VAR model does not exist structural breaks. Thereby, the test may be not adequate when it comes to the non-constant parameters. We, therefore, employ *Sup-F*, *Ave-F* and *Exp-F* tests to examine the parameter stability, and also use the L_c statistics test to examine whether the parameters follow a random walk process. Thus, parameter stability tests in Table 3 reveal a dynamic linkage between CPU and ESGU, suggesting full-sample approach is inappropriate in this context. In view of structural changes, we therefore employ a bootstrap sub-sample rolling-window approach to examine the dynamic causal link between these two series variables.

Furthermore, appropriate window width determination is also essential to empirical results. If the window width is considerable, the estimation results will be more uniform, and abrupt structural changes may go undetected. A narrow window width will provide an increased number of outliers and erratic outcomes, leading to incorrect estimates within the sample. We adopt a 24-month rolling window width to enhance estimation correctness, consistent with Pesaran and Timmermann (2005). Subsequently, we can examine the original assumption that CPU is not Granger cause of ESGU (ESGU is not Granger cause of CPU). Additionally, we can determine the direction of causal effects from CPU to ESGU (effects from ESGU on CPU).

Table 2 The full sample Granger causality test

Tests	H_0:CPU does not Granger cause ESGU		H_0:ESGU does not Granger cause CPU	
Bootstrap *LR* test	Statistics	*p*-values	Statistics	*p*-values
	9.759***	0.005	0.127	0.925

Source: Authors' calculations
Notes: The research counts *p*-values by using 1000 bootstrap repetitions

Table 3 The parameter stability test

	CPU		ESGU		VAR (*s*) process	
Tests	Statistics	*p*-values	Statistics	*p*-values	Statistics	*p*-values
Sup-F	30.375***	0.000	15.085	0.139	17.215	0.532
Ave-F	10.460**	0.019	6.261	0.232	7.817	0.753
Exp-F	11.439***	0.000	4.503	0.173	6.177	0.475
L_c					2.026	0.202

Source: Authors' calculations
Notes: *** and ** are the significance at 1% and 5% levels

From Figs. 1 and 2, we are able to assess the *p*-values and direction of the impacts from CPU to ESGU. The results suggest that CPU exerts positive effects on ESGU in times of 2015: M11-2016: M8, 2016: M11-2017: M4, 2019: M9-2020: M3 at the 10% significance level.

Moving on to the sub-sample period of 2015: M11-2016: M8, it has been observed that there is a positive causality from climate policy uncertainty to sustainability uncertainty. In December 2015, the Paris Agreement was adopted at the Paris Climate Change Conference, marking a new stage in global climate governance. Governments have begun to adjust domestic policies to fulfil their emission reduction commitments, but the uncertainties regarding policy directions, implementation intensity and timetables have significantly increased. The fluctuations during this policy adjustment period have pushed up the uncertainty of climate policies and further affected market expectations, leading to an increase in sustainability uncertainty. Moreover, this policy shift has shrunk traditional energy markets (e.g., coal

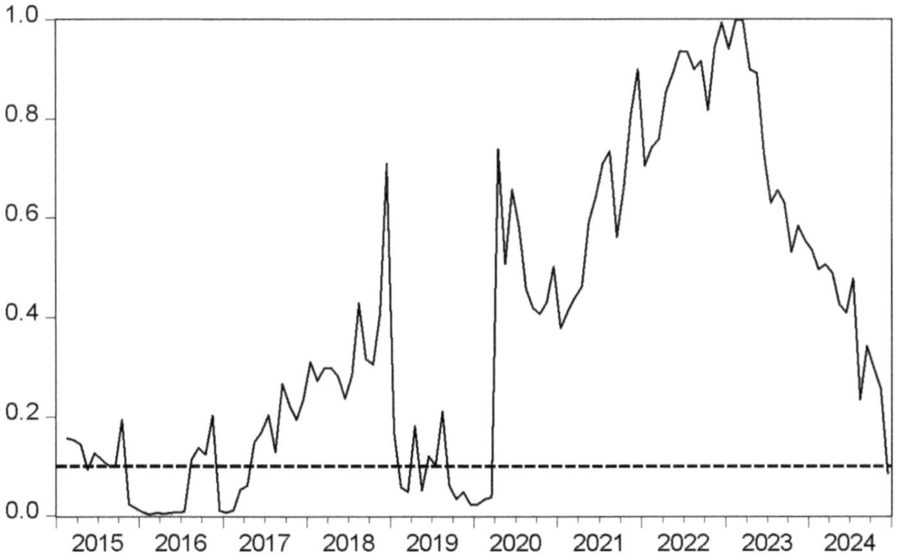

Fig. 1 Bootstrap *p*-values of rolling test statistic testing the null hypothesis that CPU does not Granger cause ESGU. (Source: Authors' calculations)

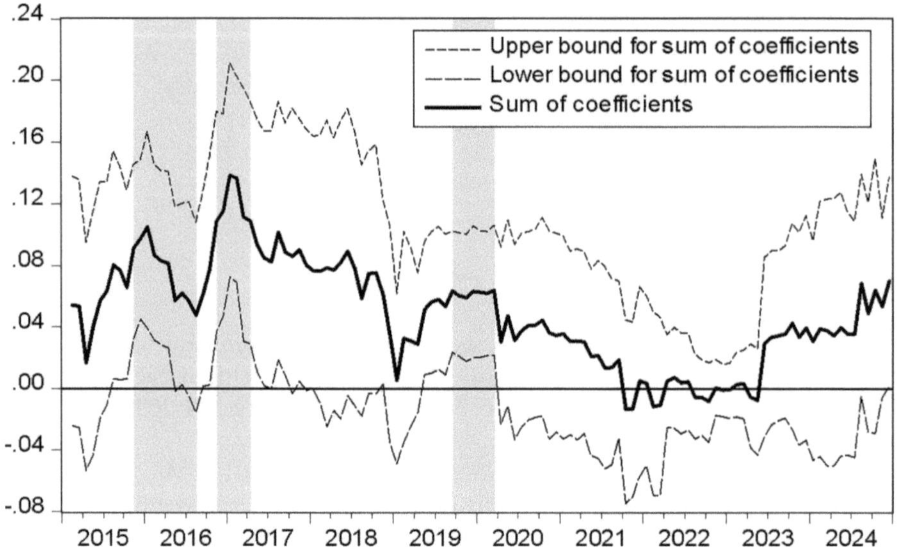

Fig. 2 Bootstrap estimates of the sum of the rolling-window coefficients for CPU impact on ESGU. (Source: Authors' calculations)

and oil), while the underdeveloped renewable energy sector has induced short-term economic volatility, thereby exacerbating sustainability challenges.

During the sub-period from 2016: M11 to 2017: M4, empirical evidence demonstrates a significant positive causal relationship from climate policy uncertainty to sustainability uncertainty. On the one hand, the inconsistency in the execution of climate policy among nations have resulted in heightened regulatory uncertainty, thereby delaying investment choices by firms. On the other hand, the volatility in energy prices and the labour market resulting from policy changes can exacerbate financial strain on conventional energy companies, ultimately causing instability in the financial market and undermining the sustainability of growth.

Similarly, a positive causal relationship from climate policy uncertainty to sustainability uncertainty can be observed during 2019: M9-2020: M3, when significant progress has also been made in global climate governance, including the United Nations Climate Action Summit in September 2019 and the full implementation stage of the Paris Agreement in early 2020. During this period, governments of various countries frequently adjusted the direction of their climate policies. For instance, the European Union launched the "Green New Deal", and China proposed preliminary discussions on the "dual carbon" goals. Such frequent adjustments in policy direction have led to more unstable market expectations for the future sustainable development path, thereby pushing up the sustainability uncertainty. Furthermore, international climate finance flows, such as green bonds and climate adaptation funds, are significantly influenced by policy signals. Policy uncertainty tends to divert funding toward projects with proven outcomes rather than those most in need of support. This unbalanced allocation of funds may have magnified the uncertainty of global sustainable development.

4.1 Functions of Carbon Pricing in Sustainable Finance

It is well known that carbon pricing policies are of great significance to sustainable finance. It can provide funds for the country and enterprises to reduce greenhouse gas emissions, promote the green transformation of a low-carbon economy, and effectively improve the ecological environment. Firstly, carbon pricing obtains funds by increasing the costs of traditional fossil energy sources (such as oil and coal) to encourage enterprises to accelerate the innovation of clean energy technologies, promote the rapid development of a low-carbon economy (Weng et al., 2022), and achieve long-term emission reduction targets. Secondly, through the emissions trading market and green bonds (Heine et al., 2019), carbon pricing policies can significantly reduce the uncertainty of consumer investment, increase expected returns, and encourage enterprises to be more actively involved in the low-carbon transformation process. Furthermore, carbon pricing can also reshape the capital flow process and optimise the efficiency of resource allocation through price signals (Michaelowa et al., 2020). Finally, it can also help provide financial assistance to certain regions and countries with low economic levels to deal with and handle climate risks, such as extreme weather and global sea level rise. In brief, this policy tool simultaneously enhances climate resilience and the momentum of low-carbon transformation. Carbon pricing offers financial incentives for sustainable financing, compelling firms to transition to a low-carbon economy and fostering sustainable growth. Ultimately, the carbon price serves as an efficacious economic instrument.

4.2 Applications of Carbon Credits and Carbon Offsets in Sustainable Finance

Carbon credit, also known as carbon right, refers to the carbon emission measurement unit obtained by a country or enterprise to reduce carbon emissions by increasing energy use efficiency, reducing pollution or reducing development under the conditions certified by the United Nations or an organisation recognised by the United Nations, and thus can be entered into the carbon trading market (Trouwloon et al., 2023). Carbon offset, or carbon neutrality compensation, generally refers to the total amount of carbon dioxide or greenhouse gas emissions directly or indirectly produced by a country, enterprise, product, activity or individual within a specific period. Through afforestation, energy conservation, emission reduction, and other forms, the carbon dioxide or greenhouse gas emissions produced by oneself are offset to achieve positive and negative offsets and reach relatively "zero emissions" (Qin et al., 2022).

The carbon credit mechanism allows enterprises or individuals to offset emissions by purchasing carbon credits. It provides an effective financing method for sustainable development, especially for developing countries with funding gaps (Otundo Richard, 2024). The carbon credit mechanism finances climate initiatives via voluntary and compliance markets, addresses the funding deficit, and simultaneously encourages the implementation of low-carbon technologies (such as solar and wind energy) through market pricing to foster green employment and economic sustainability (Martielli et al., 2024). Within the emissions trading system, regulated entities must adhere to the quota management system and mitigate their excess emissions by acquiring carbon credits (Liu et al., 2025). The carbon offset mechanism enables enterprises to invest in sustainable projects such as forest carbon sinks and biogas recovery, generating significant environmental benefits while achieving emission neutrality. Carbon credits and carbon offsets can stimulate organisations that generate emissions to implement major emission reduction strategies by assessing emission reduction activities and establishing market-driven incentive mechanisms, effectively accelerating the transition to a low-carbon economy. Furthermore, Although the carbon pricing system can significantly reduce carbon emissions, some limitations exist (Xu et al., 2023). The principal concern pertains to the debate around environmental integrity, specifically that the actual impact of emission reductions may be consistently overvalued (Trencher et al., 2024). Secondly, duplicate calculation is prevalent. Local carbon markets, national carbon taxes and international aviation carbon offset mechanisms may repeatedly price the same emission source. Third, the issue of fairness and benefit distribution is another challenge for carbon pricing policies. Some carbon projects have not been fully consulted with residents, affecting the protection of community rights and interests. More importantly, developed countries have not achieved emission reduction by purchasing carbon credits at low prices to offset their emissions, ignoring fairness among countries (Aldy & Stavins, 2012). To enhance the effectiveness of carbon pricing in promoting sustainable finance, policymakers must take corresponding measures. First, policymakers should expand the coverage of policies and establish a stable and gradually strengthened carbon pricing mechanism. Furthermore, countries set carbon prices consistent with the goals of the Paris Agreement by a carbon tax or a carbon emissions trading system to enhance the price signal. Second, governments can link carbon pricing with the classification standards of sustainable finance to guide funds to flow towards low-carbon projects. Carbon revenue subsidises the research and development of renewable energy and energy transition, enhancing social acceptance. Third, develop carbon financial derivatives, including carbon futures and options, to encourage innovation in financial products. The aforementioned policy initiatives can significantly improve the efficacy of the carbon price system.

5 Conclusions

The above empirical results indicate that the uncertainty of climate policies can positively affect the uncertainty of sustainability, which is of great significance for climate action and sustainable development. This indicates that we should choose appropriate climate policies and implement them, which play an important role in climate action and the transformation to a low-carbon economy.

5.1 Limitations of Carbon Pricing

Although carbon pricing has significant advantages, its actual effect is constrained by multiple factors (Feindt et al., 2021).

Political Resistance

The progression of carbon pricing systems faces considerable political opposition, mainly from conflicts of interest and electoral pressures (Ewald et al., 2022). They often hinder the implementation of carbon pricing policies through legislative obstruction and public opinion wars. First, traditional energy enterprises usually have strong political contribution capabilities and can use their contributions to threaten policymakers and hinder the implementation of carbon pricing policies (Narassimhan et al., 2022). Secondly, during political elections, carbon pricing may increase electricity and oil prices, causing dissatisfaction among voters and directly affecting policy implementation. Furthermore, in fossil energy-producing areas, politicians may openly oppose carbon pricing in order to retain votes (Nazari et al., 2024). The most typical example among them is that Australia abolished the fixed carbon tax policy in 2014, making it the first country in the world to abolish the pricing of carbon emissions. In fact, the carbon tax has always been the focus of political struggles among various political parties. On the carbon tax issue, the control over whether to prioritise the economy or environmental protection has always been in the hands of politicians (Crowley, 2017). This case demonstrates that carbon pricing policy is subject to resistance from political actors.

Economic Competitiveness Concerns

Although carbon pricing can help promote the shift to a low-carbon economy, it may also have a negative impact on economic competitiveness, especially on energy-intensive industries and export-oriented economies. On the one hand, carbon pricing increases the cost of using fossil energy, raising the production costs of industries that rely on high-carbon energy sources such as coal and oil (such as steel, cement, and chemical industries) and countries, weakening their market competitiveness. Furthermore, the carbon cost may put domestic goods at a significant price disadvantage in the international market, impacting the export industry and reducing international trade competitiveness. On the other hand, implementing carbon pricing policies is prone to causing carbon leakage problems. If a country or region implements strict carbon pricing alone, enterprises may shift production to countries with lenient carbon policies, resulting in the hollowing out of local industries. Still, global carbon emissions have not decreased (Branger & Quirion, 2014). Moreover, implementing carbon pricing policies will likely trigger an economic recession in the short term, suppress investment and consumption, and delay economic growth (Grubb et al., 2022).

Need for Global Coordination

Carbon pricing policies vary significantly among different countries and regions, such as different coverage areas and inconsistent carbon prices, which leads to carbon leakage, that is, high-emission enterprises shift their production to regions with low carbon costs, weakening the overall global emission reduction effect and fair competition (Böhringer et al., 2022). The effectiveness of carbon pricing policies is highly dependent on global coordination. If countries continue to act unilaterally, it may lead to low efficiency in reducing emissions, intensifying trade conflicts, and exacerbating global inequality. On the one hand, the entire international community should promote the interconnection of carbon markets and reduce trade frictions (Nachtigall et al., 2021). On the other hand, international organisations such as the International Monetary Fund and the World Bank can provide corresponding platforms to strengthen multilateral cooperation mechanisms and establish a carbon pricing mechanism with unified rules, technology sharing and fair mechanisms (Thube et al., 2021). The measures above may significantly improve the efficacy of carbon pricing mechanisms, collaboratively advance the progression of a low-carbon economy, and facilitate the green transformation of the economy.

5.2 *Policy Suggestions*

In view of the above limitations, the effectiveness of carbon pricing policies can be enhanced through relevant supplementary policies to facilitate the transition to a low-carbon economy. Subsidies for renewable energy (such as electricity price subsidies, investment subsidies and tax incentives) enhance the market competitiveness of clean energy by reducing its cost (Nicolini & Tavoni, 2017). When carbon pricing is underincentivize due to price fluctuations or limited coverage, subsidies can provide stable support to promote the development of clean energy (Murray & Maniloff, 2015). Furthermore, taxing fossil fuels is also an essential supplementary policy. It can increase the cost of using fossil fuels and promote the achievement of emission reduction targets (Mahdavi et al., 2022). Implementing these supplementary policies can compensate for the limitations of carbon pricing policies, effectively mitigating global climate risks.

Policy Combination

On the one hand, implementing policies can enhance the economy's overall efficiency and mitigate climate risks. Carbon pricing provides a stable expectation of emission reduction costs and ensures a minimum incentive for emission reduction. Carbon emissions trading adjusts the carbon price through market mechanisms and can achieve flexible emission reduction under total control. On the other hand, the coordinated implementation of policies can make up for the shortcomings of a single policy. Carbon taxes can cover industries not covered by the carbon emissions trading system, avoiding blind spots in emission reduction. The carbon emissions trading system optimises resource allocation through market transactions. The combination of the two can form a more comprehensive carbon cost coverage.

International Cooperation

With the increasing severity of climate issues, strengthening international cooperation is the inevitable way to take climate action. First, building a multilateral cooperation platform. To foster international policy alignment, a multilateral carbon pricing forum should be established to harmonise national carbon taxation and emissions trading schemes, thereby minimising cross-border trade disputes. Establishing a global carbon market governance committee is necessary to supervise the market operation, prevent carbon leakage problems and achieve climate goals. Second, establishing interconnected carbon market systems. Through international negotiations, unified carbon pricing standards and trading rules should be developed, with pilot regional market linkages, such as cooperation between the EU and China's carbon markets, to leverage market mechanisms in reducing global abatement costs. This approach would enhance efficiency while mitigating the risk of carbon leakage and associated policy inefficiencies. Third, fostering technological collaboration and green supply chain synergies. An international clean technology fund should be established to support joint research and development of critical technologies such as renewable energy and carbon capture. Simultaneously, carbon footprint certification should be implemented across transnational supply chains, with tariff incentives for low-carbon products to accelerate corporate green transition. Fourth, Scaling Up Climate Finance for Just Transition. Climate action requires enhanced cooperation. Developed countries should support capacity building in developing countries, provide climate financing, and ensure a fair transition. Only through in-depth international cooperation can the effectiveness of global climate governance be achieved.

Strengthening Public Participation and Transparency

Carbon pricing is a key policy to address the low-carbon transformation, and its effectiveness is highly dependent on public participation and social equity. To achieve this, policymakers must implement three key mechanisms: On the one hand, introduce compulsory emissions reporting requirements for high-impact industries, coupled with open-access data portals that enable real-time public oversight. On the other hand, the government should establish multi-level public participation mechanisms, such as community co-governance platforms. In short, enhancing public participation and transparency can improve policy effectiveness, reduce implementation resistance, and promote long-term emission reduction and economic growth.

This chapter explores the role and limitations of carbon pricing and emissions trading systems in climate action and sustainable finance, providing substantial policy implications for policymakers, enterprises and investors. By elucidating these market-based instruments' fundamental principles, operational mechanisms, and measurable impacts, the chapter will enable readers to contribute more effectively and advocate for the global shift to a low-carbon economy.

References

Abrell, J., Kosch, M., & Rausch, S. (2022). How effective is carbon pricing?—A machine learning approach to policy evaluation. *Journal of Environmental Economics and Management, 112*, 102589.

Ahmad, M., Li, X. F., & Wu, Q. (2024). Carbon taxes and emission trading systems: Which one is more effective in reducing carbon emissions?—A meta-analysis. *Journal of Cleaner Production, 476*, 143761.

Aldy, J. E., & Stavins, R. N. (2012). The promise and problems of pricing carbon: Theory and experience. *Journal of Environment and Development, 21*, 152–180.

Andrews, D. W. K. (1993). Tests for parameter instability and structural change with unknown change point. *Econometrica, 61*, 821–856.

Andrews, D. W. K., & Ploberger, W. (1994). Optimal tests when a nuisance parameter is present only under the alternative. *Econometrica, 62*, 1383–1414.

Balcilar, M., Ozdemir, Z. A., & Arslanturk, Y. (2010). Economic growth and energy consumption causal nexus viewed through a bootstrap rolling window. *Energy Economics, 32*, 1398–1410.

Betsill, M., & Hoffmann, M. J. (2011). The contours of "cap and trade": The evolution of emissions trading systems for greenhouse gases. *Review of Policy Research, 28*, 83–106.

Böhringer, C., Fischer, C., Rosendahl, K. E., & Rutherford, T. F. (2022). Potential impacts and challenges of border carbon adjustments. *Nature Climate Change, 12*, 22–29.

Borghesi, S., Cainelli, G., & Mazzanti, M. (2015). Linking emission trading to environmental innovation: Evidence from the Italian manufacturing industry. *Research Policy, 44*, 669–683.

Branger, F., & Quirion, P. (2014). Climate policy and the 'carbon haven' effect. *WIREs Clim Change, 5*, 53–71.

Chang, S., Yang, X., Zheng, H., Wang, S., & Zhang, X. (2020). Air quality and health co-benefits of China's national emission trading system. *Applied Energy, 261*, 114226.

Coase, R. H. (1960). The problem of social cost. *The Journal of Law & Economics, 3*, 1–44.

Convery, F. J., & Redmond, L. (2007). *Market and price developments in the european union emissions trading scheme* (Vol. 1).
Crowley, K. (2017). Up and down with climate politics 2013–2016: The repeal of carbon pricing in Australia. *Wiley Interdisciplinary Reviews: Climate Change, 8*, e458.
Davis, S. (2016). *An index of global economic policy uncertainty*. National Bureau of Economic Research.
Dorband, I. I., Jakob, M., Kalkuhl, M., & Steckel, J. C. (2019). Poverty and distributional effects of carbon pricing in low-and middle-income countries–A global comparative analysis. *World Development, 115*, 246–257.
Edmans, A., & Kacperczyk, M. (2022). Sustainable finance. *Review of Finance, 26*, 1309–1313.
Ellerman, A. D., & Buchner, B. K. (2007). The european union emissions trading scheme: Origins, allocation, and early results. *Review of Environmental Economics and Policy, 1*(1), 66–87.
Ewald, J., Sterner, T., & Sterner, E. (2022). Understanding the resistance to carbon taxes: Drivers and barriers among the general public and fuel-tax protesters. *Resource and Energy Economics, 70*, 101331.
Feindt, S., Kornek, U., Labeaga, J. M., Sterner, T., & Ward, H. (2021). Understanding regressivity: Challenges and opportunities of european carbon pricing. *Energy Economics, 103*, 105550.
Gavriilidis, K. (2021). *Measuring climate policy uncertainty*. Available at SSRN 3847388.
Gillingham, K., & Stock, J. H. (2018). The cost of reducing greenhouse gas emissions. *Journal of Economic Perspectives, 32*, 53–72.
Green, J. F. (2021). Does carbon pricing reduce emissions? A review of ex-post analyses. *Environmental Research Letters, 16*(4), 043004.
Grubb, M., Jordan, N. D., Hertwich, E., Neuhoff, K., Das, K., Bandyopadhyay, K. R., van Asselt, H., Sato, M., Wang, R., Pizer, W. A., & Oh, H. (2022). Carbon leakage, consumption, and trade. *Annual Review of Environment and Resources, 47*, 753–795.
Gugler, K., Haxhimusa, A., & Liebensteiner, M. (2023). Carbon pricing and emissions: Causal effects of Britain's carbon tax. *Energy Economics, 121*, 106655.
Hansen, B. E. (1992). Tests for parameter instability in regressions with I(1) processes. *Journal of Business & Economic Statistics, 10*, 321–335.
Heine, D., Semmler, W., Mazzucato, M., Braga, J. P., Flaherty, M., Gevorkyan, A., Hayde, E., & Radpour, S. (2019). Financing low-carbon transitions through carbon pricing and green bonds. *Vierteljahrshefte Zur Wirtschaftsforschung, 88*, 29–49.
Hu, Y.-J., Yang, L., Duan, F., Wang, H., & Li, C. (2022). A scientometric analysis and review of the emissions trading system. *Energies, 15*, 4423.
Jia, Z., & Lin, B. (2020). Rethinking the choice of carbon tax and carbon trading in China. *Technological Forecasting and Social Change, 159*, 120187.
Känzig, D. R., 2023. *The unequal economic consequences of carbon pricing*. National Bureau of Economic Research w31221.
Koch, N., Fuss, S., Grosjean, G., & Edenhofer, O. (2014). Causes of the EU ETS price drop: Recession, CDM, renewable policies or a bit of everything?—New evidence. *Energy Policy, 73*, 676–685.
Lee, C. F., Lin, S. J., & Lewis, C. (2008). Analysis of the impacts of combining carbon taxation and emission trading on different industry sectors. *Energy Policy, 36*, 722–729.
Linsenmeier, M., Mohommad, A., & Schwerhoff, G. (2023). Global benefits of the international diffusion of carbon pricing policies. *Nature Climate Change, 13*(7), 679–684.
Liu, X., Zhang, W., Liu, M., & Han, J. (2025). Green finance & carbon neutrality: Strategies and policies for a sustainable future. *Frontiers in Environmental Science, 13*, 1560927.
Mahdavi, P., Martinez-Alvarez, C. B., & Ross, M. L. (2022). Why do governments tax or subsidize fossil fuels? *The Journal of Politics, 84*, 2123–2139.
Marron, D. B., & Toder, E. J. (2014). Tax policy issues in designing a carbon tax. *American Economic Review, 104*, 563–568.

Martielli, F., Cruz, T. F. G., & Salvi, A. (2024). Financial tools for sustainable entrepreneurship: Exploring the impact of carbon credits on financial outcomes. In *Research Society on Entrepreneurial Motivation* (pp. 47–50).

Metcalf, G. E. (2019). On the economics of a carbon tax for the United States. *Brookings Papers on Economic Activity, 2019*, 405–484.

Michaelowa, A., Michaelowa, K., Shishlov, L., & Brescia, D. (2020). Catalysing private and public action for climate change mitigation: The world bank's role in international carbon markets. *Climate Policy, 21*, 120–132.

Moessner, R. (2025). Effects of carbon pricing on inflation. *Climate Policy*, 1–14.

Murray, B. C., & Maniloff, P. T. (2015). Why have greenhouse emissions in RGGI states declined? An econometric attribution to economic, energy market, and policy factors. *Energy Economics, 51*, 581–589.

Nachtigall, D., Ellis, J., Peterson, S., & Thube, S. (2021). *The economic and environmental benefits from international co-ordination on carbon pricing: Insights from economic modelling studies* (OECD Environment Working Papers No. 173), OECD Environment Working Papers.

Narassimhan, E., Koester, S., & Gallagher, K. S. (2022). Carbon pricing in the US: Examining state-level policy support and federal resistance. *Politics and Governance, 10*.

Nazari, M., Asgary, A., Nezami, I. M., & Ghayourisales, S. (2024). From resistance to resilience: A comprehensive bibliometric analysis of carbon pricing public acceptance. *Energy Research & Social Science, 107*, 103340.

Nicolini, M., & Tavoni, M. (2017). Are renewable energy subsidies effective? Evidence from Europe. *Renewable and Sustainable Energy Reviews, 74*, 412–423.

Nordhaus, W. (2013). *The climate casino: Risk, uncertainty, and economics for a warming world*. Yale University Press.

Nyblom, J. (1989). Testing for the constancy of parameters over time. *Journal of the American Statistical Association, 84*, 223–230.

Ongan, S., Gocer, I., & Işık, C. (2025). Introducing the new ESG-based sustainability uncertainty index (ESGUI). *Sustainable Development, 33*, 4457–4467.

Otundo Richard, M., 2024. Carbon credit concept and Africa's sustainable development; an empirical review.

Pan, J., Cross, J. L., Zou, X., & Zhang, B. (2024). To tax or to trade? A global review of carbon emissions reduction strategies. *Energy Strategy Reviews, 55*, 101508.

Pesaran, M. H., & Timmermann, A. (2005). Small sample properties of forecasts from autoregressive models under structural breaks. *Journal of Econometrics, Modelling structural breaks, 129*, 183–217.

Prentice, J. (2013). Carbon leakage and competitiveness under the EU ETS. *European Energy and Environmental Law Review, 22*, 132–140.

Qin, M., Su, C.-W., Zhong, Y., Song, Y., & Lobonţ, O.-R. (2022). Sustainable finance and renewable energy: Promoters of carbon neutrality in the United States. *Journal of Environmental Management, 324*, 116390.

Shukur, G., & Mantalos, P. (2000). A simple investigation of the granger-causality test in integrated-cointegrated VAR systems. *Journal of Applied Statistics, 27*, 1021–1031.

Shukur, G., & Mantalos, P. (2004). Size and power of the RESET test as applied to systems of equations: A bootstrap approach. *Journal of Modern Applied Statistical Methods, 3*, 370–385.

Sims, C. A. (1980). Macroeconomics and reality. *Econometrica, 48*, 1–48.

Solarin, A. S. (2025). Carbon pricing mechanisms for reducing greenhouse gas emissions and encouraging sustainable industrial practices. *World Journal of Advanced Research and Reviews, 25*, 001–024.

Steinebach, Y., Fernández-i-Marín, X., & Aschenbrenner, C. (2021). Who puts a price on carbon, why and how? A global empirical analysis of carbon pricing policies. *Climate Policy, 21*(3), 277–289.

Su, C. W., Song, X. Y., Dou, J., & Qin, M. (2025). Fossil fuels or renewable energy? The dilemma of climate policy choices. *Renewable Energy, 238*, 121950.

Sumner, J., Bird, L., & Dobos, H. (2011). Carbon taxes: A review of experience and policy design considerations. *Climate Policy., 11*, 922–943.

Thube, S., Peterson, S., Nachtigall, D., & Ellis, J. (2021). The economic and environment benefits from international co-ordination on carbon pricing: A review of economic modelling studies. *Environmental Research Letters, 16*, 113002.

Trencher, G., Nick, S., Carlson, J., & Johnson, M. (2024). Demand for low-quality offsets by major companies undermines climate integrity of the voluntary carbon market. *Nature Communications, 15*, 6863.

Trouwloon, D., Streck, C., Chagas, T., & Martinus, G. (2023). Understanding the use of carbon credits by companies: A review of the defining elements of corporate climate claims. *Global Challenges, 7*, 2200158.

Tvinnereim, E., & Mehling, M. (2018). Carbon pricing and deep decarbonisation. *Energy Policy, 121*, 185–189.

Weng, Z., Ma, Z., Xie, Y., & Cheng, C. (2022). Effect of China's carbon market on the promotion of green technological innovation. *Journal of Cleaner Production, 373*, 133820.

Xu, L., Solangi, Y. A., & Wang, R. (2023). Evaluating and prioritizing the carbon credit financing risks and strategies for sustainable carbon markets in China. *Journal of Cleaner Production, 414*, 137677.

Zhang, Y., Li, S., Luo, T., & Gao, J. (2020). The effect of emission trading policy on carbon emission reduction: Evidence from an integrated study of pilot regions in China. *Journal of Cleaner Production, 265*, 121843.

Zhang, Y., Qi, L., Lin, X., Pan, H., & Sharp, B. (2022). Synergistic effect of carbon ETS and carbon tax under China's peak emission target: A dynamic CGE analysis. *Science of the Total Environment, 825*, 154076.

Public-Private Partnerships Investment in Energy and Carbon Emissions: Multi-Country Evidence

Kaihua Wang and **Wenqian You**

1 Introduction

In the context of global warming, problems including the rise in sea levels and frequent extreme weather have become increasingly prominent. Carbon emissions have exacerbated the greenhouse effect and are the cynosure of international attention (Li et al., 2025; Kashif et al., 2025). According to relevant data, prior to the Industrial Revolution, the atmospheric concentration of CO_2 was likely 280 parts per million (ppm), and by 2024, this figure had surpassed 410 ppm, with an ongoing upward trend (Yoro & Daramola, 2020). These trends pose significant threats to human survival and development. Against the backdrop of economic and social growth, advancing carbon reduction has become a crucial challenge that needs to be addressed (Zhang et al., 2025). As we all know, energy is essential for social progress and economic development, and both daily life and economic production are inseparable from energy. However, the extensive utilisation of traditional energy has released enormous amounts of carbon dioxide emissions, accounting for 73.2% globally (Jiang et al., 2024). Energy investment is crucial in addressing carbon release (Boulanouar et al., 2024). The direction and structure of energy investments directly determine carbon emission levels during the generation and utilisation processes (Erdas et al., 2025). Traditional energy investments primarily target traditional energy like coal, petroleum, and natural gas (Liu et al., 2020). Long-standing fossil fuel dependence has led to the conventional energy fraction within global energy use being more than 80% (Liu et al., 2023; Li et al., 2023). As global energy demand continues to grow, if energy investments remain overly reliant on fossil fuels, the issue of carbon emissions will become even more severe (Zhang, 2024). IRENA's 2021 report, "World Energy Transitions Outlook: 1.5 °C Pathway" to align with the emission reduction vision set by the COP26 conference, renewable

K. Wang (✉) · W. You
Qingdao University, Qingdao, Shandong Province, China

C.-W. Su et al. (eds.), *Green Finance and the Challenges of Climate Change*, Sustainable Finance, https://doi.org/10.1007/978-3-032-07293-1_13

energy's proportion in the entiretyof primary energy provision need surge from 14% in 2018 to 74% by 2050 to effectively curb the trend of greenhouse gas emissions (Chen & Zhang, 2024). However, public-private partnership investment in energy (PPIE) has significantly contributed to achieving the goals of COP26. Between 2013 and 2014, the private sector contributed approximately 87% of the total financing for renewable energy, reaching $223 billion and $270 billion during 2013–2014 and 2015–2016, respectively (Rabbani et al., 2025).

In energy investment, the private sector is gradually becoming an indispensable force. As the global economy grows and market mechanisms improve, the private capital engagement in energy investment keeps rising. Germany and Canada, among other developed countries, have implemented robust policy structures and financial mechanisms to entice private capital into green energy. (Zhuang et al., 2025). With a relatively perfect institutional framework, sufficient financial resources and advanced technology reserves, developed countries can often promote the development of massive renewable energy projects through public-private cooperation in the energy field, improve the energy mix, and effectively decrease carbon emission intensity. In contrast, implementing public-private partnerships (PPP) within the energy domain in countries with lower- and middle-income levels has been uneven because of multiple constraints such as capital shortage, technical barriers and imperfect institutions (Chen, 2025). In 2022, global investment in energy transition technologies totalled $1.3 trillion, an unprecedented amount (IRENA, 2023). However, if the 1.5 °C target is to be achieved by 2050, annual investment will have to be more than quadrupled (IRENA, 2023). There is still a long way to go to increase the private sector participation in energy investment and scale up its efforts. Besides, some projects may increase carbon emission pressure due to excessive reliance on traditional energy sources. However, the influence of private sector involvement in energy investment on carbon emissions is complex. It can lead to positive and increased emissions reductions due to short-term gains. On one hand, private sector investments in renewable and clean energy technologies can help optimise the energy mix and reduce carbon emissions. Conversely, if private sector investments in traditional high-carbon energy sectors continue to rise, they could further exacerbate the carbon emission problem (Cheng et al., 2021a). Therefore, studying the connection between private sector participation in energy investment and carbon emissions is of great practical significance.

This research aims to deeply study the correlation between PPIE and carbon emission, reveal the specific impact pathway of PPIE on carbon emission through empirical analysis, and provide key theoretical support and practical guidance for global climate change response and sustainable development goals. Theoretically, there are still many gaps and controversies regarding the influence of private sector participation in energy investment on carbon emissions. Although some literature has addressed the association between energy investment and carbon emissions, there is a deficiency in systematic and detailed investigations of the private sector's specific role, influence, and pathways. This study will integrate theories from economics, environmental science, and statistics to construct a comprehensive analytical framework. It aims to delve into the driving factors behind private sector energy investment decisions and how these investments directly or indirectly affect carbon emissions through energy structure adjustment, technological innovation, and market competition. This helps enrich and improve the theoretical frameworks of climate and energy economics and fills gaps in relevant research areas. Additionally, it provides new perspectives and methods for future scholars to examine further the complex connection between energy markets and the environment. For example, by comparing the patterns of private energy investment with carbon emission data in various nations and regions, future studies can disclose common patterns and unique characteristics of private sector energy investment and carbon emission under varying levels of economic development, policy environments, and resource endowments, thus providing richer and more accurate empirical evidence for theoretical research. This article may offer potential marginal contributions in several ways. First, the study employs empirical analysis to comprehensively and deeply examine the connection between public-private energy investment and carbon emissions. By collecting relevant data from multiple countries over a long period, the study ensures that its findings are broadly representative and reliable. The analysis uses a multiple linear regression model, with PPIE as the primary explanatory variable and carbon emissions as the dependent variable. It also incorporates mediation variables, such as renewable energy consumption (REC) and energy use (EU), to accurately measure the impact paths of PPIE on carbon emissions. Second, from the research perspective, this article transcends the limitations of previous studies that focused on single countries or regions by including multiple countries. This multi-country approach allows for a more comprehensive consideration of how differences in the extent of economic growth, population scale, and resource endowments across different countries influence the relationship between PPIE and carbon emissions, thereby revealing more universal and general patterns.

The rest of this research is structured as follows: Section 2 encompasses a comprehensive review of the relevant literature. Section 3 introduces the theoretical basis. Section 4 presents the data and model frameworks. Section 5 delves into the empirical analysis. Section 6 summarises the conclusions and policy implications.

2 Literature Review

2.1 *Studies on the Factors Affecting Carbon Emissions*

Carbon emissions are affected by a combination of complex factors, and many researchers have extensively analysed them from different viewpoints. First, the level of technology plays an essential role in the issue of carbon emissions. He et al. (2025) revealed the contribution of green technology diversification to industrial carbon emission efficiency, but the positive effect will weaken as the quantile increases. Cheng et al. (2025) found that green technological advancements and the transformation of industrial frameworks significantly alleviate the carbon emission problem and show a robust spatial spillover effect. Danish et al. (2025) stated that green technology innovation drives CO_2 emission reduction by reducing pollution levels. Cao et al. (2025) noted the important effect of AI technology on reducing carbon emissions and that the application of AI to energy management will have a more considerable emission reduction effect. Second, industrial structure also serves as a vital factor in carbon emissions. Substantial disparities exist in energy intensity and carbon emission rates across industries. Hu et al. (2023) found that the influence of rationality of industrial structure on carbon emissions is asymmetrical, and that a more rational industrial structure will enhance the quality of the environment in the extended period, and vice versa in the short period. Zheng et al. (2023) pointed out that upgrading industrial structure can markedly boost the linkage network of carbon emission industries and alleviate climate warming. Duan et al. (2025) identified the elements that affect the peak of CO_2 emissions from the socioeconomic development perspective, and the economic scale and industrial outcomes are central to the carbon peaking process. Third, the energy mix is also a key factor affecting carbon emissions. Fossil fuels are major contributors to carbon emissions because they release large amounts of carbon dioxide during combustion. In contrast, clean energy sources produce little or no carbon emissions during operation. Wang and Lei (2024) pointed out that renewable energy progress can effectively mitigate carbon intensity and influence the adjacent areas via spatial spillovers, both over the near and long run. Xia et al. (2025) found that the innovation of hydrogen energy technology positively affects reducing pollution and carbon emissions synergistically. Zhang (2025) highlighted that carbon emissions drive the progress of renewable energy, reducing carbon emissions, and that government programs can accelerate the transition to effective climate action.

2.2 The Correlation Between Public-Private Partnerships and Carbon Emissions

Existing literature has yielded some results in exploring the link between private sector participation in energy investment and carbon emissions. Jiemin and Chen (2022) studied the impact of PPIE, energy consumption, and innovation on China's carbon emissions. They found that PPIE increases carbon emissions in the short term but has a positive effect over the long term. Similarly, Zhang (2024) explores the impact of public-private energy cooperation in China on CO_2 emissions, highlighting that global public-private partnerships in the energy sector can substantially lower carbon emissions and attain sustainable development goals. Kirikkaleli and Adebayo (2021) found that renewable energy consumption contributes to lowering consumption-based CO_2 emissions and that PPP energy investment also positively impacts the long term. Alnour et al. (2024) found that an increase in PPP in the transport sector would reduce transport-based CO_2 emissions in the short run, yet the influence would be less significant in the long term. Raghutla et al. (2024)

investigated the effect of PPIE on environmental performance in six global investment countries and found that PPP in the energy department inhibited carbon emissions. Raghutla and Kolati (2023) found that PPP investments in the energy sector will boost the flourishing of renewable energy, and governments must build relationships with the private enterprise at the level of renewable energy funding needs. Balsalobre-Lorente et al. (2024) explored the intermediary function of carbon policy, the circular economy and private partnerships in industrial development and found that public-private partnerships show negligible sustainability impacts on industrial development. Lu et al. (2022) pointed out that China's green finance mechanism has considerably contributed to reducing carbon emissions, and that public-private collaborations in the energy industry need to be more impactful in helping to lower the level of carbon emissions. Yang et al. (2022) examined the interplay from PPP to CO_2 emission levels under the framework of the environmental Kuznets curve (EKC) assumptions, suggesting that governments should establish more public-private partnerships in environmentally friendly fields and industries.

3 Theoretical Basis

The engagement of the private group in energy investment, especially green PPP projects, has a multifaceted mechanism for promoting carbon emission reduction. These mechanisms cover multiple levels, such as technological innovation, energy structure optimisation and energy efficiency improvement. First, the private sector has a strong incentive and resources to invest in low-carbon technologies. For example, large private equity firms (PEs) will push their portfolio energy companies to increase their R&D efforts in renewable energy technologies, energy preservation technologies, etc. Developing these technologies will improve energy performance and significantly reduce carbon emissions (Cheng et al., 2021b). In addition, the private sector can quickly apply new technologies to practical projects, such as funding renewable energy projects like solar power plants and wind farms, directly minimising the need for traditional fossil fuels. With the support and resources of governments, green PPP projects can hasten the promotion and application of new technologies on a broader scale, thus further expanding the influence of low-carbon technologies (Yin & Qamruzzaman, 2024).

Second, private sector investment can help improve the proportion of renewable energy in the energy supply. By participating in green PPP projects, private companies can work with the government to jointly create and operate renewable energy projects, including solar, wind and hydro, thereby optimising the energy mix (Qamruzzaman & Karim, 2023). In the meantime, the involvement of the private group can diversify energy supply and reduce reliance on a single high-carbon energy source. For example, by investing in the construction of biomass power generation projects and geothermal energy utilisation projects, we will increase the diversity of energy supply and reduce carbon emissions. This diversified energy supply structure not only improves the stability of the energy system but also lends significant backing to the achievement of the carbon emission reduction goals.

Third, in the energy investment process, private enterprises will pay attention to cost savings and energy efficiency to pursue economic benefits and market competitive advantages. For example, the use of high-efficiency motors, boilers, and other equipment in industrial production can significantly enhance energy efficiency and decrease energy waste and carbon emissions. Furthermore, the private sector can invest in advanced energy management systems that optimise energy efficiency through instantaneous management and control of energy consumption (Carbonara & Pellegrino, 2018). These systems can help companies better manage energy demand and reduce carbon emissions per unit of product or service, thereby driving carbon reduction at the micro level.

4 Data and Econometrics Method

Based on data from 13 countries from 1999 to 2021, this paper examines the contribution of public-private investment in the energy domain (PPIE) on carbon emissions. The selection of sample countries, which includes European countries Russia and Turkey and Asian countries China, India, Indonesia, the Philippines, Malaysia and Thailand, as well as Latin American countries Mexico, Brazil, Peru, Argentina and Colombia, is influenced by the availability of data from the World Bank's Private Sector Participation in Infrastructure (PPI) project database. The sample cycle covers the performance of PPIE from its inception to its rapid growth. From 1999 to 2000, the scale of private group involvement in energy investment was relatively limited, and the state began to introduce some policies to encourage private capital to enter the energy sector, which laid a certain policy foundation for private sector participation in energy investment. Since 2011, the global focus on clean energy has been increasing, and renewable energy investment has shown a rapid growth trend. The private sector actively develops and constructs new energy projects such as solar, wind and hydropower. For example, China's investment in wind power increased significantly in 2020. This paper's independent and dependent variables are total CO_2 emissions, excluding land use, land-use change and forestry (LULUCF) and private sector participation in energy investment (current US$). A lot of literature already studies these two variables at the national level (Kirikkaleli & Adebayo, 2021; Yang et al., 2022). Many countries are used in this study to expand the generalizability of the early research results. In terms of control variables, this study selects four factors that may affect carbon emissions. Economic growth (GDP) is usually accompanied by a rise in production capacity and the expansion of industrialisation, generating a surge in energy consumption, thus the CO_2 emission problem becomes more severe (Otim & Mutumba, 2025). The greater the total population (TP), the higher the carbon emissions will be increased by the daily activities needed for life, such as transportation, cooking, heating, etc., as well as the improvement of infrastructure (Kumar & Sen, 2025). Increasing the share of renewable electricity output (RE) reduces traditional energy consumption in electricity production and can improve environmental quality (Balsalobre-Lorente et al., 2018). Industrial value added (IND) indicates the scale of industrialisation, and the expansion of the manufacturing and industrial sectors is greatly reliant upon traditional energy consumption, causing a vast portion of air pollution and carbon emissions (Huang et al., 2023). Besides, RE is the fraction of renewable electricity in the overall electricity supply, and IND is the percentage of industrial value added to GDP.

This study chooses renewable energy consumption (REC) and energy use (EU) as mediation variables for the mechanism checking. The increase in REC will diminish carbon emissions by decreasing the proportion of traditional energy and promoting innovation and investment in renewable energy to expand its use further and reduce carbon emissions. All variable data in this article are from the World Bank database.

As illustrated by the above discussion, the empirical model is established as follows.

$$CO_{2it} = f\left(PPIE_{it}, GDP_{it}, TP_{it}, IND_{it}, RE_{it}\right) \tag{1}$$

Where i, t, CO_2, PPIE, GDP, TP, IND and RE indicate the country, time, carbon emissions, private-public-partnership in energy investment, economic growth, total population, industrial value added and energy use, respectively. After converting absolute indicators such as CO_2, PPIE, GDP and TP into natural logarithms, we use the log-linear framework to simulate the empirical equation of carbon emission increase.

$$\begin{aligned} \ln CO_{2it} = \beta_{1i} \ln PPIE_{it} + \beta_{2i} \ln GDP_{it} + \beta_{3i} \ln TP_{it} + \beta_{4i} IND_{it} \\ + \beta_{5i} RE_{it} + \phi_{it} \end{aligned} \tag{2}$$

Where β_{ai}, $a = 1, 2, 3, 4, 5$ represent the resilience of CO_2 emissions associated with public-private investments in energy, economic growth, total population, industrial value added and renewable electricity output, respectively. The ϕ_{it} stands for error term.

5 Empirical Results and Analysis

Table 1 provides the findings from descriptive statistics. The average of CO_2 is 2.55, and the standard deviation is 0.61, indicating that carbon emissions vary widely between countries. Some countries dominated by heavy industry have high carbon emissions resulting from the large consumption of energy in their production processes, especially the dependence on high-carbon fossil fuels, which can reach a maximum of 4.10. With a minimum value of 1.43, these countries with lower

Table 1 Summary statistics

VarName	Mean	SD	Min	Max
lnco$_2$	2.547	0.609	1.43	4.10
lnppie	8.913	0.713	6.48	10.54
lngdp	11.728	0.508	10.70	13.25
lntp	8.074	0.525	7.35	9.15
ind	32.739	7.179	18.19	48.53
re	32.225	24.643	4.89	90.04
lnrec	1.232	0.368	0.30	1.70
lneu	3.084	0.292	2.59	3.75

carbon emissions may have achieved better results in energy portfolio optimisation, adoption of energy-saving and emission reduction technologies, or industrial structure optimisation. The mean value of PPIE is 8.913, reflecting a certain level of activity in the private sector in the field of energy investment. The standard deviation of 0.71 shows notable differences in the countries' private sector energy investment scale. The private sector may have more financial and technical resources in some countries with relatively developed economies and a high degree of openness in energy markets. It can play a greater role in energy investment, with a larger investment scale, up to a maximum of 10.54. In some countries with underdeveloped economies or unstable energy policies, the private sector may face problems such as a shortage of funds and high investment risks, resulting in a small scale of energy investment, with a minimum value of only 6.48. The average and standard deviation of GDP are 11.728 and 0.510, respectively, indicate that the degree of economic development varies from country to country. The mean value of TP is 8.074, the standard deviation is 0.525, and the highest and lowest are 9.15 and 7.35, respectively, demonstrating that there were variations in population size among the sample countries. The average of IND is 32.739. The utmost and least values of IND can reach 48.53 and 18.19, respectively, meaning that the percentage of industrial added value in GDP fluctuates among countries. The mean value of RE is 32.225, and the standard deviation is 24.643, reflecting significant differences in the percentage of renewable electricity output in total electricity output across countries. The mean of REC is 1.233 with a standard deviation of 0.368, indicating that REC fluctuated in different countries and periods. The mean value of EU is 3.084, and the standard deviation is 0.292, suggesting that the important difference in energy use between countries is relatively small. Table 2 presents outcomes of the correlation analysis. They indicate that PPIE, GDP, TP and IND are positively related to CO_2 emissions, while RE shows a negative relationship with CO_2 emissions. In addition, an increase in GDP and TP appears to promote PPIE, with RE positively correlating with PPIE and IND negatively correlating with PPIE. The findings from the correlation analysis reveal no obvious multicollinearity among the variables.

Refer to Table 3 for the outcomes of the baseline regression. From the regression results, the coefficient of PPIE is −0.019 and is significant at the 1% *p*-value, revealing that PPIE has a markedly negative correlation with carbon emissions.

Table 2 Correlation matrix

	lnco2	lnppie	lngdp	lntp	ind	re
lnco2	1					
lnppie	0.258***	1				
lngdp	0.868***	0.384***	1			
lntp	0.848***	0.339***	0.756***	1		
ind	0.208***	−0.173***	−0.0370	0.134**	1	
re	−0.447***	0.0290	−0.198***	−0.251***	−0.500***	1

Table 3 The baseline regression between PPIE and CO_2 emissions

	lnco2
lnppie	−0.0187***
	(−4.1331)
lngdp	0.2503***
	(15.9749)
lntp	1.0941***
	(9.2175)
ind	0.0027**
	(2.2894)
re	−0.0002
	(−0.4628)
_cons	−9.1375***
	(−10.6345)
N	299
adj. R^2	0.818

Specifically, a unit growth in private energy investment would reduce total carbon output by 0.019 units. This result supports the view that increased private sector investment in energy could help reduce carbon output and suggests that PPIE can indeed significantly reduce CO_2 emissions when excluding the interference of other control variables. The result agrees with the conclusions of a wide array of existing research. (Shahbaz et al., 2020; Jiemin & Chen, 2022). The coefficient of GDP is 0.250, which shows significance at the 1% p-value, reflecting a strongly positive connection between economic growth and carbon emissions. During economic growth, with the expansion of production scale, the improvement of infrastructure construction, and the upgrading of residents' living standards, the energy demand is increasing, especially the demand for fossil energy, causing a surge in carbon emissions. In some emerging economies, the expansion of industrial activities accompanies rapid economic growth, and a vast quantity of fossil energy, such as coal and oil, is consumed, leading to an upturn in total carbon emissions. The positive correlation between TP and CO_2 is also significant at the 1% p-value, and the estimated coefficient equals 1.094. As the population gradually expands, the energy and electricity required for production and daily life and the consumption of various carbon-intensive products show an upward trend, boosting a jump in carbon emission volume. The coefficient value of IND is 0.002, which is significantly positive at the 5% p-value, implying that a rise in IND contributes to a higher carbon emission intensity. When IND rises, the contribution to economic growth from industry and manufacturing becomes greater, which often consumes too much energy and emits large amounts of pollutants. RE does not significantly affect carbon emissions in this study, probably because other variables cover its impact, and future research can focus on the interrelationship between these two variables.

To confirm the reliability of the regression results, we use the substitution of explanatory variables method for a robustness test. Table 4 exhibits the results. For the explanatory variable, the ratio of PPIE to GDP is used to replace PPIE. This substitution can more comprehensively reflect the status and effect of private energy investment in the economy from a perspective of relative proportion, avoiding the limitations caused by absolute value comparisons of investment amounts due to differences in national economic scale. Although the level of PPIE in some small economies may be relatively low, its share of GDP may be relatively high, and its influence on carbon emissions should not be overlooked. The findings reveal that the coefficient of PPIE/GDP remains inversely correlated and is markedly at the 1% p-value, indicating that private energy investment continues to have a robust beneficial influence on carbon emissions. The direction and statistical significance of the coefficients of the control variables also commonly correspond to the original regression results, further validating the stability of the model.

To further explore the influence path of PPIE to alleviate carbon emissions, we selected two mediating variables, REC and EU, for the mechanism test. The regression outcome is documented in Table 5. REC notably positively impacts PPIE at the 1% p-value, implying that the increase in PPIE promoted the development of REC. First, relying solely on government spending and investment to achieve universal access to renewable energy is unrealistic, and rural and poor areas are often

Table 4 The robustness test

	lnco2
lnppiegdp	−0.2132***
	(−3.9907)
lngdp	0.2370***
	(15.4302)
lntp	1.0881***
	(9.1402)
ind	0.0026**
	(2.2380)
re	−0.0002
	(−0.4208)
_cons	−8.9364***
	(−10.3227)
N	299
adj. R2	0.817

Table 5 Mechanism test

	(1)	(2)
	lnrec	lneu
lnppie	0.0186***	−0.0159***
	(2.6786)	(−3.9976)
lngdp	−0.2120***	0.2474***
	(−8.8261)	(18.0123)
lntp	0.5701***	0.0271
	(3.1332)	(0.2608)
ind	−0.0026	0.0026**
	(−1.4136)	(2.4715)
re	0.0023***	0.0011***
	(3.1832)	(2.5963)
_cons	−1.0391	−0.0121
	(−0.7890)	(−0.0161)
N	299	299
adj. R2	0.260	0.709

overlooked. By promoting the active participation of the private sector in energy investment, it is possible to expand the sources of financing for renewable energy projects and reduce the investment risk of long-term projects (Ha, 2025). According to recent trends, the private sector may be more disposed to channel funds into renewable energy and clean energy technology in the process of energy investment, which promotes the optimisation and upgrading of the energy mix, so that the share of clean energy in the energy consumption structure gradually increases, thereby effectively decreasing reliance on high-carbon fossil energy and reducing carbon emissions (Hu et al., 2018). Some private companies have invested heavily in solar and wind power generation projects, and with the increase and scale of these clean energy power generation projects, the energy structure has been optimised, and carbon emissions have been reduced accordingly. Second, the private sector can obtain financing for R&D and innovative renewable energy technologies, and they may be more willing to take risks than the robust government investment. Adequate financial support will significantly promote solar, wind and bioenergy development and utilisation. As a result, the REC is further enhanced, the proportion of traditional energy use decreases, thereby reducing the carbon discharge generated by unit energy consumption. The regression coefficient connecting EU with PPIE is −0.016 and is markedly at the 1% *p*-value, indicating that PPIE reduced energy use to some extent. There are two potential reasons for this. First, private sector investment in energy has supported energy-efficient equipment and technological retrofits. With private capital allocation to energy projects, industrial and manufacturing companies can use more energy-efficient equipment in their production and accelerate technological innovation in energy use to enhance energy efficiency and minimise energy loss. Second, PPIE helps optimise the structure of energy consumption. Private sector energy investment can support the transformation and modernisation of energy-intensive industries and reduce the energy demand of these industries. Overall, PPIE helps achieve more efficient energy consumption, reducing total energy use and mitigating carbon emissions.

6 Conclusion

Countries are facing serious threats, such as climate change and deteriorating environmental quality. Public-private partnerships in the energy domain are becoming increasingly active and are seen as essential support for the strategy of achieving carbon neutrality. Through an empirical analysis of the sample countries, this study investigates the connection between PPIE and carbon emissions and draws the following key conclusions.

As for the empirical analysis, the results of multiple linear regression models show that private capital in energy investment significantly negatively impacts carbon emissions. Specifically, for each 1 unit growth in private sector energy investment, total carbon emissions will be reduced by 0.019 units. This shows that from the perspective of the overall sample countries, the private sector has effectively promoted the optimisation of the energy mix and reduced the reliance on high-carbon fossil energy in the energy investment process by actively participating in renewable energy and clean energy projects, thereby reducing carbon emissions. GDP, TP and IND also exhibit a significant positive interrelation with carbon emissions, which reflects that in the process of human social development, economic growth, population expansion, and industrialisation have promoted the growth of energy demand, which is often accompanied by increased carbon emissions. When the energy mix has not yet achieved a fundamental transformation, the extensive consumption of traditional fossil energy has increased carbon emissions. From the view of the transmission channel of the impact, the role of PPIE in reducing carbon emissions via the intermediary variables REC and EU is notable. PPIE has spurred a boom in investment in renewable energy and technological innovation. Subsequently, the growth of REC significantly suppresses carbon emissions, fully reflecting the key role of renewable energy in carbon neutrality strategies. In addition, PPIE substantially decreases carbon emissions through enhancing energy efficiency and reducing energy use. Through the R&D and the application of advanced energy-saving technologies, energy efficiency can be improved, and efficient energy use can effectively reduce carbon emissions. We use the ratio of PPIE to GDP to substitute the independent variables for a robustness test, and the findings are consistent with the benchmark regression; thus, the empirical analysis in this paper is robust.

Drawing from the study's conclusions, this paper further puts forward the following policy recommendations to foster the private-public partnership investment in green energy and mitigate carbon emissions. From the government's perspective, it is necessary to construct and improve relevant policies to encourage the private sector to promote investment in renewable energy technologies. On the one hand, the government can provide direct investment subsidies for private enterprises investing in renewable energy projects, such as the hydrogen energy technique, thereby reducing the initial investment costs of enterprises. In addition, the government should reinforce the guidance and monitoring of private sector energy investment. It is essential to develop clear energy development strategies and plans, specifying the target share of renewable and clean energy in the energy mix, to provide a clear investment orientation for the private sector. Moreover, the authorities should enhance the approval and supervision of energy projects, strictly control investment in high-carbon energy projects, and encourage and support the development of low-carbon and zero-carbon energy projects. In the enterprise realm, private enterprises should proactively assume social responsibility and expand investment in the R&D of green energy technologies. Enterprises should also increase investment in key technologies such as photovoltaic systems and hydrogen energy utilisation. Technological innovation can help enterprises reduce the investment cost of green energy projects, improve energy efficiency, and boost the competitive potential of firms in the green energy market. Finally, countries should also strengthen investment cooperation in green energy projects. Private enterprises should be encouraged to invest in international green energy projects. By investing in, constructing, and operating overseas green energy projects, optimal allocation of resources can be achieved, and the global energy structure can be propelled towards a green and low-carbon transformation. Private enterprises from some developed countries can bring advanced technologies and management experience to developing countries and collaborate with local enterprises to build green energy projects, realising mutual benefits and win-win outcomes.

There are still some shortcomings in this study, which also point out the direction for future research. In terms of data, although panel data from multiple countries were collected in this study, there are still some limitations to the completeness of the data. Future research can further strengthen cooperation with national statistical agencies and international organisations to obtain more comprehensive and accurate data to improve the reliability of research. From the perspective of model construction, although the effect of GDP, population, and renewable energy electricity is considered in this paper, the relationship from PPIE to carbon emissions may be more complex in reality. Thus, the model may not fully capture all the impact mechanisms. Some potentially important variables, such as energy prices and energy subsidy policy effects, may be missing from the model. Future research can try to introduce more variables and build more complex and comprehensive models to delve deeper into the interconnectedness between the two variables.

Acknowledgments This work was supported by a grant from the Romanian Ministry of Research, Innovation and Digitalization, the project with the title "*Economics and Policy Options for Climate Change Risk and Global Environmental Governance*" (CF 193/28.11.2022, Funding Contract no. 760078/23.05.2023), within Romania's National Recovery and Resilience Plan (PNRR)—Pillar III, Component C9, Investment I8 (PNRR/2022/C9/MCID/I8)—Development of a program to attract highly specialised human resources from abroad in research, development, and innovation activities.

References

Alnour, M., Awan, A., & Hossain, M. E. (2024). Towards a green transportation system in Mexico: The role of renewable energy and transport public-private partnership to curb emissions. *Journal of Cleaner Production, 442*, 140984.

Balsalobre-Lorente, D., Shahbaz, M., Roubaud, D., & Farhani, S. (2018). How economic growth, renewable electricity and natural resources contribute to CO_2 emissions? *Energy Policy, 113*, 356–367.

Balsalobre-Lorente, D., Shah, S. A. R., & Huseynova, R. (2024). Do circular economy, public-private partnership and carbon policy manage the environmental stress? Developed countries' situation under the Prism of COP27. *Heliyon, 10*(13), e33532.

Boulanouar, Z., Essid, L., & Omri, A. (2024). Achieving carbon neutrality in emerging markets: The dual impact of energy transition investments on economic growth and carbon emissions. *International Review of Economics & Finance, 96*, 103709.

Cao, Q., Chi, C., & Shan, J. (2025). Can artificial intelligence technology reduce carbon emissions? A global perspective. *Energy Economics, 143*, 108285.

Carbonara, N., & Pellegrino, R. (2018). Public-private partnerships for energy efficiency projects: A win-win model to choose the energy performance contracting structure. *Journal of Cleaner Production, 170*, 1064–1075.

Chen, Q. (2025). Energy policy uncertainty, financial development and public-private investment in BRIC nations. *Finance Research Letters, 71*, 106417.

Chen, Z., & Zhang, S. (2024). The spatial correlation between green investment and energy-structure optimization: Evidence from China. *Energy Strategy Reviews, 53*, 101391.

Cheng, G., Zhao, C., Iqbal, N., Gülmez, Ö., Işik, H., & Kirikkaleli, D. (2021a). Does energy productivity and public-private investment in energy achieve carbon neutrality target of China? *Journal of Environmental Management, 298*, 113464.

Cheng, C., Ren, X., Dong, K., Dong, X., & Wang, Z. (2021b). How does technological innovation mitigate CO_2 emissions in OECD countries? Heterogeneous analysis using panel quantile regression. *Journal of Environmental Management, 280*, 111818.

Cheng, G., Shi, Y., & Zhou, H. (2025). Spatial statistical analysis of the effects of green technology innovation and industrial structure upgrading on carbon emission intensity. *International Review of Economics & Finance, 102*, 104163.

Danish, Hassan, S. T., & Khan, I. (2025). Achieving net-zero carbon emission targets in OECD countries: The role of the energy transition, institutional quality, and green technological innovation. *Gondwana Research, 144*, 20–32.

Duan, Z., Duan, H., Chen, S., Gao, Z., & Zhang, L. (2025). Carbon dioxide emissions peaks depend on changes in economic scale and industrial structure. *Structural Change and Economic Dynamics., 75*, 94–107.

Erdas, M. L., Caglar, A. E., Uche, E., & Ahmed, Z. (2025). A new frontier in understanding the dynamics of environmental sustainability in the context of finance and low carbon energy investment: Evidence from artificial intelligence and fourier approach. *Energy, 315*, 134419.

Ha, L. T. (2025). Rethinking energy security in the condition of a rise in public–private partnership in energy: Fresh insights from new estimation. *Energy Strategy Reviews, 60*, 101781.

He, S., Xu, S., & Zou, A. X. (2025). Does green technological diversification impact industrial carbon emissions efficiency? The role of technological specialisation. *Journal of Innovation and Knowledge, 10*(3), 100730.

Hu, H., Xie, N., Fang, D., & Zhang, X. (2018). The role of renewable energy consumption and commercial services trade in carbon dioxide reduction: Evidence from 25 developing countries. *Applied Energy, 211*, 1229–1244.

Hu, L., Yuan, W., Jiang, J., Ma, T., & Zhu, S. (2023). Asymmetric effects of industrial structure rationalization on carbon emissions: Evidence from thirty Chinese provinces. *Journal of Cleaner Production, 428*, 139347.

Huang, Y., Kuldasheva, Z., Bobojanov, S., Djalilov, B., Salahodjaev, R., & Abbas, S. (2023). Exploring the links between fossil fuel energy consumption, industrial value-added, and carbon emissions in G20 countries. *Environmental Science and Pollution Research, 30*(4), 10854–10866.

International Renewable Energy Agency (IRENA). (2023). *World Energy Transitions Outlook 2023*. IRENA. Retrieved from https://www.irena.org/Digital-Report/World-Energy-Transitions-Outlook-2023. Accessed May 2025.

Jiang, W., Zhang, Y., & Wang, K.-H. (2024). Analyzing the connectedness among geopolitical risk, traditional energy and carbon markets. *Energy, 298*, 131411.

Jiemin, H., & Chen, W. (2022). The impact of private sector energy investment, innovation and energy consumption on China's carbon emissions. *Renewable Energy, 195*(C), 1291–1299.

Kashif, U., Abbas, S., Kousar, S., & Lu, H. (2025). Linking of bio-energy and carbon neutrality: Navigating economic policy uncertainty and climate change policy in the USA. *Energy, 324*, 136012.

Kirikkaleli, D., & Adebayo, T. S. (2021). Do public-private partnerships in energy and renewable energy consumption matter for consumption-based carbon dioxide emissions in India? *Environmental Science and Pollution Research, 28*(23), 30139–30152.

Kumar, S., & Sen, R. (2025). Are larger or denser cities more emission efficient? Exploring the nexus between urban household carbon emission, population size and density. *Applied Energy, 377*, 124500.

Li, Z. Z., Su, C.-W., Moldovan, N.-C., & Umar, M. (2023). Energy consumption within policy uncertainty: Considering the climate and economic factors. *Renewable Energy, 208*, 567–576.

Li, L., Yu, L., Li, R., Zhou, X., Zhang, N., & Meng, Q. (2025). Carbon emission accounting and carbon neutrality strategies at universities: A case study from Guangzhou, China. *Building and Environment, 281*, 113210.

Liu, R., He, L., Liang, X., Yang, X., & Xia, Y. (2020). Is there any difference in the impact of economic policy uncertainty on the investment of traditional and renewable energy enterprises? – A comparative study based on regulatory effects. *Journal of Cleaner Production, 255*, 120102.

Liu, X., Adebayo, T. S., Ramzan, M., Ullah, S., Abbas, S., & Olanrewaju, V. O. (2023). Do coal efficiency, climate policy uncertainty and green energy consumption promote environmental sustainability in the United States? An application of novel wavelet tools. *Journal of Cleaner Production, 417*, 137851.

Lu, Q., Farooq, M. U., Ma, X., & Iram, R. (2022). Assessing the combining role of public-private investment as a green finance and renewable energy in carbon neutrality target. *Renewable Energy, 196*, 1357–1365.

Otim, J., & Mutumba, G. S. (2025). Carbon emissions, economic governance and economic growth. *Next Research, 2*(2), 100200.

Qamruzzaman, M., & Karim, S. (2023). Does public-private investment augment renewable energy consumption in BIMSTEC nations? Evidence from symmetric and asymmetric assessment. *Energy Strategy Reviews, 49*, 101169.

Rabbani, M. R., Kiran, M., & Shaikh, Z. H. (2025). Financing the future: Insights into sustainable energy investments through scientific mapping and meta-analysis. *Discover Sustainability, 6*(1), 1–21.

Raghutla, C., & Kolati, Y. (2023). Public-private partnerships investment in energy as new determinant of renewable energy: The role of political cooperation in China and India. *Energy Reports, 10*, 3092–3101.

Raghutla, C., Malik, M. N., Hameed, A., & Chittedi, K. R. (2024). Impact of public-private partnerships investment and FDI on CO_2 emissions: A study of six global investment countries. *Journal of Environmental Management, 360*, 121213.

Shahbaz, M., Raghutla, C., Song, M., Zameer, H., & Jiao, Z. (2020). Public-private partnerships investment in energy as new determinant of CO_2 emissions: The role of technological innovations in China. *Energy Economics, 86*, 104664.

Wang, Y., & Lei, T. (2024). Influencing mechanisms of renewable energy development on carbon emission intensity in China. *Journal of Environmental Management, 372*, 123402.

Xia, X., Luo, T., & Sun, H. (2025). Investigating the impact of hydrogen energy technology innovation on synergistic abatement of pollution and carbon emissions. *International Journal of Hydrogen Energy, 120*, 602–612.

Yang, Q., Huo, J., Saqib, N., & Mahmood, H. (2022). Modelling the effect of renewable energy and public-private partnership in testing EKC hypothesis: Evidence from methods moment of quantile regression. *Renewable Energy, 192*, 485–494.

Yin, C., & Qamruzzaman, M. (2024). Empowering renewable energy consumption through public-private investment, urbanization, and globalization: Evidence from CS-ARDL and NARDL. *Heliyon, 10*(4), e26455.

Yoro, K. O., & Daramola, M. O. (2020). CO_2 emission sources, greenhouse gases, and the global warming effect, Chapter 1. In M. R. Rahimpour, M. Farsi, & M. A. Makarem (Eds.), *Advances in carbon capture*. Woodhead Publishing.

Zhang, J. (2024). Energy access challenge and the role of fossil fuels in meeting electricity demand: Promoting renewable energy capacity for sustainable development. *Geoscience Frontiers, 15*(5), 101873.

Zhang, Z. (2025). Exploring the nexus: Climate change concerns, renewable energy, and carbon emissions. *Journal of Environmental Management, 376*, 124413.

Zhang, Y., Kou, M., & Qiao, M. (2025). Revisiting the impact of trade on carbon emissions within the context of global value chains: Empirical evidence from belt and road countries. *Journal of Cleaner Production, 511*, 145619.

Zheng, Y., Tang, J., & Huang, F. (2023). The impact of industrial structure adjustment on the spatial industrial linkage of carbon emission: From the perspective of climate change mitigation. *Journal of Environmental Management, 345*, 118620.

Zhuang, F., Han, H., Ahsan, M., & Iqbal, A. (2025). Transitioning ecological footprint: Private investment trends in renewable energy projects among G10 nations. *Renewable Energy, 251*, 123335.

Financial Development, Environmental Policies, and Green Innovation: Heterogeneous Drivers of Renewable Energy Transition in OECD Countries

Muhammad Umar, Adnan Safi, Manal Yunis, and Rahil Irfan

1 Introduction

The global transition towards renewable energy represents a challenge confronting policymakers worldwide, as countries strive to balance economic development with environmental sustainability and climate change mitigation. According to statistics, renewable energy capacity has reached record levels with approximately 3372 GW deployed globally in 2022, indicating an annual growth rate of approximately 9%, showing accelerating adoption across developed and developing economies (IRENA, 2023). However, achieving the targets established under the Paris Agreement necessitates a substantial scaling of investment flows, specifically requiring a tripling of annual clean energy investments from current levels of $1.8 trillion to $4.5 trillion by 2030 (IEA, 2023). This investment gap of $2.7 trillion annually is not only a financial constraint but also challenges whether existing financial systems, regulatory frameworks, and technological innovation

M. Umar
Department of Finance and Accounting, Adnan Kassar School of Business, Lebanese American University, Beirut, Lebanon
e-mail: muhammad.umar@lau.edu.lb

A. Safi (✉)
School of Economics, Qingdao University, Qingdao, Shandong, China
e-mail: adnansafi@qdu.edu.cn

M. Yunis
Department of Information Technology and Operations Management, Adnan Kassar School of Business, Lebanese American University, Beirut, Lebanon
e-mail: myunis@lau.edu.lb

R. Irfan
Department of Business Administration, Air University Multan Campus, Multan, Pakistan

C.-W. Su et al. (eds.), *Green Finance and the Challenges of Climate Change*, Sustainable Finance, https://doi.org/10.1007/978-3-032-07293-1_14

mechanisms possess the capacity to mobilise capital at the requisite pace and scale for limiting global temperature increases to 1.5 °C (IEA, 2023; Umar et al., 2024).

The relationship between financial development and renewable energy has emerged as a significant area of research, particularly following the Paris Agreement of 2015. Financial development plays an important role in renewable energy adoption, with several studies showing that it enhances renewable energy, though the magnitude of effects may vary across countries and development stages (Shahbaz et al., 2021). This heterogeneity in financial development is also observed in emerging countries, indicating that financial markets-related development exerts a stronger influence on renewable energy consumption than financial institutions-related development and overall financial development (Habiba & Xinbang, 2023). However, the mechanisms through which financial development influences renewable energy vary across countries and economic development levels. Kim and Park (2016) examined 30 countries during 2000–2013 and showed that developed financial markets facilitate renewable energy sectors primarily through enhanced access to external financing. On the other hand, empirical evidence from China demonstrates that green financial development negatively affected renewable energy investment efficiency, particularly through bank lending mechanisms, generating a coefficient of −0.0017 (He et al., 2019). The contrasting finding suggests that the effectiveness of financial development depends on the alignment between financial instruments and the renewable energy sector, while environmental policies play a complementary role in shaping these financial mechanisms and determining their ultimate impact on energy transition.

Environmental policies have evolved from direct regulatory measures to market-based mechanisms that work in conjunction with financial systems to influence energy investment patterns (Stavins, 2003). Liu et al. (2023) argued that the combination of environmental regulations and financial development produces interactive effects that exceed their individual contributions to the renewable energy transition. The interaction term between these variables demonstrates both statistical significance and economic relevance, indicating that environmental policies effectively channel financial resources towards renewable energy. Furthermore, evidence from the BRICST region shows that environmental regulations showed the most substantial marginal influence on renewable energy consumption, surpassing the direct effects of financial development and shadow economy (Zhang & Razzaq, 2022). Additionally, environmental governance quality significantly moderates the relationship between digital financial inclusion and renewable energy technology innovation (Wei et al., 2024). The effectiveness of renewable energy transitions is strengthened by integrated policy frameworks that combine green finance, environmental regulations, and innovation support mechanisms. Environmental regulations play a critical role in facilitating innovation, particularly when these regulations are integrated with comprehensive policy frameworks, including privacy standards and emission taxation mechanisms et consistent with the prevailing privacy regime (Pan et al., 2023). Moreover, climate change policies have significantly increased green patent applications, with non-market-based instruments (i.e., R&D subsidies and technology-support instruments) exhibiting larger effects compared to market-based mechanisms, particularly in competitive economies and sectors characterised by minimal financial constraints (Bettarelli et al., 2023). These findings indicate that energy transitions require comprehensive policies rather than isolated interventions.

The existing literature has overlooked the influence of financial development, environmental policies, and green innovation on renewable energy development. Moreover, previously conducted studies have employed mean-based approaches that assume uniform effects, neglecting the heterogeneous impacts occurring at different levels of renewable energy development. Furthermore, existing literature has focused primarily on renewable energy consumption or investment flows while overlooking production capabilities as a distinct dimension that represents technological readiness and infrastructure development. Therefore, the present study contributes to the existing literature by examining the impact of financial development, environmental policy stringency, and green innovation on renewable energy production capabilities and consumption patterns in OECD countries from 1990 to 2022. The distinction between production capacity and consumption provides comprehensive insights into energy transition dynamics that are often obscured in aggregate analyses. We have also incorporated trade openness and natural resource rents as control variables that ensures the robustness of estimated effects for the primary determinants of renewable energy. This integrated approach gives us evidence-based insights for designing differentiated policy strategies that leverage financial markets, regulatory frameworks, and innovation systems to accelerate renewable energy transitions, while accounting for heterogeneous effects across different stages of development.

2 Literature Review

2.1 Financial Development and Renewable Energy

The relationship between financial development and renewable energy deployment has attracted considerable attention in recent empirical literature. Fotio et al. (2022) analysed the impact of financial integration on renewable energy capacity in 16 Sub-Saharan African countries. Their results showed that enhanced financial integration increases renewable energy capacity in both the long and short run, though the effect varies across countries and levels of capacity. While financial integration supports growth in renewable energy, it does not guarantee diversification of the energy mix, and financial development within domestic systems is negatively associated with renewable energy production. These findings suggest that the benefits of financial integration are positive but heterogeneous and may depend on the specific structure of financial linkages and policies. Financial development comprises multiple dimensions that affect renewable energy outcomes through different channels. Wang and Zhao (2022) analysed the interplay of renewable energy investment efficiency, shareholder control, and green financial development in China. They showed that green finance development helps alleviate investment shortages but can also hinder renewable energy investment efficiency through its influence on bank loan issuance, with intermediary effects from short- and long-term loans accounting for 9.38% of the overall impact. The study also highlighted that ownership structures and shareholder control are significantly connected to investment efficiency and the effect of green finance, with some risks that green finance development may increase under-investment in renewable energy firms.

Financial development and renewable energy consumption demonstrate varying temporal dynamics across regions and methodologies. Pata et al. (2022) employed a Fourier and wavelet based quantile causality analysis to examine the U.S. data and showed that financial development, particularly the depth and access to financial markets, significantly promotes renewable energy consumption at higher quantiles in the medium and long run. Mukhtarov and Mikayilov (2023) examined the relationship between financial development and renewable energy consumption in Poland, showing that both financial development and economic growth favour the transition to renewable energy. Their results suggested that continued improvements in financial sector depth can help increase renewable energy investment, broaden access to modern energy services, and reduce reliance on fossil fuels, thereby supporting Poland's energy transition and potentially alleviating energy poverty.

2.2 *Environmental Policies and Renewable Energy*

Environmental policy stringency influences renewable energy through different mechanisms. Tan et al. (2022) revealed an inverted U-shaped relationship between environmental regulations and renewable energy investment in China, finding that both command-and-control and market-incentive regulations initially promote investment but become counterproductive beyond optimal intensity levels. This non-linear relationship is mediated by enterprise costs, with heterogeneous effects observed between state-owned and non-state-owned enterprises. Similarly, Wang et al. (2024) demonstrated that China's new energy demonstration city construction policy significantly enhances renewable energy technology innovation, with stronger effects in eastern regions and areas with developed digital finance. Moreover, similar evidence emerges from OECD countries, where Shahzad et al. (2024) find that environmental regulations significantly enhance clean energy usage when supported by effective governance measures, while Bashir et al. (2022) conversely report that environmental regulations hinder renewable energy consumption in OECD economies.

Katircioglu and Katircioglu (2023) confirm long-term effects of environmental taxation on renewable energy stock prices in Turkey, demonstrating how regulatory variations significantly influence energy stock performance and investment flows.

Recent developments in financial technology offer new pathways for regulatory effectiveness, as Sreenu (2024) shows how fintech and green bonds promote clean energy generation in India when supported by proactive regulatory frameworks. Moreover, Lu et al. (2020) highlighted feed-in tariffs as particularly successful policy instruments across multiple countries. These findings collectively suggest that environmental regulations' effectiveness depends on policy, implementation intensity, and governance quality.

2.3 *Green Innovation and Renewable Energy*

Green innovation emerges as an important factor that enhances renewable energy development. Behera et al. (2024) showed that green innovation enhances renewable energy use in European Union nations. However, Alfalih (2025) revealed a threshold-dependent relationship through panel discrete threshold regression analysis of the top greenest countries, finding that green innovation positively affects renewable energy generation only in countries with lower levels of green innovation, while exhibiting negative effects in countries with higher green innovation levels. This threshold effect suggests potential diminishing returns or saturation points in the green innovation-renewable energy nexus. Similarly, Dogan et al. (2025) establish a positive relationship between green technology and renewable energy innovation in BRICS countries using FMOLS and DOLS methods, emphasising the need for policymakers to promote green technology advancement to enhance renewable energy innovation.

The mechanisms through which green innovation influences renewable energy are multifaceted, involving financial channels, research and development activities, and emission reduction pathways. Solarin et al. (2022) provide compelling evidence that renewable energy innovation significantly enhances renewable energy production across all quantiles in BRICS countries from 1993 to 2018, with stronger effects observed in countries with smaller renewable energy production per capita. Long et al. (2023) demonstrate that green credit policy significantly promotes the total amount, quality, and quantity of green innovation in renewable energy enterprises through reduced green agency costs and increased R&D spending. Furthermore, Zhu and Zhang (2023) identify renewable energy development as an essential mediating mechanism through which green technology innovation achieves emission reduction, based on panel data from 30 Chinese provinces. Wang et al. (2023) findings showed that innovation for renewable energy, supported by green finance, effectively reduces greenhouse gas emissions in G7 countries. These studies collectively underscore the critical role of green technological innovation as both a direct driver and indirect facilitator of renewable energy development through various institutional and financial pathways.

3 Methodology and Data

3.1 Data and Variables

This study employs a panel dataset covering selected OECD countries from 1990 to 2022. The OECD country sample is ideal for examining renewable energy transitions as these nations collectively represent a substantial share of the global energy supply (IEA, 2023) and have implemented diverse policy approaches to promote renewable energy development. The temporal coverage captures critical periods of renewable energy development, including the post-Kyoto Protocol era, the 2008 financial crisis and subsequent recovery, and recent acceleration in clean energy investments following the Paris Agreement.

The dependent variables capture two distinct dimensions of renewable energy development. First, Renewable energy consumption (REE) is measured as the percentage of total final energy consumption from renewable sources, encompassing hydro, solar, wind, biomass, and geothermal energy. This indicator reflects the actual utilisation of renewable energy to meet final energy demand across all sectors. Second, Renewable energy capabilities (REcap) are measured by the share of electricity generated from renewable sources, representing the sector's supply-side capacity and infrastructure. The distinction between consumption and capabilities is crucial as it allows examination of both demand-side adoption and supply-side development of renewable energy systems (World Bank, 2024).

Financial development (FD) is measured using domestic credit to the private sector as a percentage of GDP (World Bank, 2024). This indicator primarily reflects the depth of bank-based financial systems, which are a key channel for funding renewable energy projects. Environmental policy stringency (EnvPol) is measured using the OECD Environmental Policy Stringency Index, a comprehensive indicator that aggregates market-based and non-market environmental policy instruments across multiple domains, including climate change, air pollution, and biodiversity protection. Green innovation (GIN) is environment-related technology taken from the OECD database. Natural resource rents (NR) are obtained from the World Bank's World Development Indicators. Trade openness (TO) shows the total trade in goods and services (exports + imports) as a percentage of GDP, taken from the World Bank.

3.2 Empirical Models

Based on the theoretical framework and existing literature, we specify two empirical models to examine the determinants of renewable energy outcomes:

Model 1 (Renewable Energy Consumption): $

$$REE_{\{it\}} = \alpha_0 + \beta_1 FD_{\{it\}} + \beta_2 EnvPol_{\{it\}} + \beta_3 TO_{\{it\}} + \beta_4 NR_{\{it\}} + \beta_5 GIN_{\{it\}} + \mu_i + \varepsilon_{\{it\}}$$

Model 2 (Renewable Energy Capabilities):

$$REcap_{it} = \alpha_0 + \gamma_1 FD_{it} + \gamma_2 EnvPol_{it} + \gamma_3 TO_{it} + \gamma_4 NR_{it} + \gamma_5 GIN_{it} + \mu_i + \varepsilon_{it}$$

Where subscripts i and t denote country and time, respectively, μ_i represents country-specific effects, and ε_{it} is the error term. The coefficients β and γ capture the marginal effects of each explanatory variable on renewable energy consumption and capabilities, respectively.

3.3 Methodological Approach

First, we provide descriptive analysis for the variables, including normality analysis. Given the interconnected nature of global energy markets and policies across OECD countries, we test for cross-sectional dependence using Pesaran's (2004) CD test. Subsequently, we examine slope heterogeneity using the Pesaran and Yamagata (2008) test, which evaluates whether the relationships between variables differ across countries. To examine the variables' stationarity properties, we employ the Im et al. (2003) method, which accounts for cross-sectional dependence by including cross-sectional averages.

For cointegration testing, we utilise the Westerlund (2005) error-correction-based panel cointegration test, which examines whether error correction exists for individual panel members and the panel as a whole. The variance ratio statistic tests the null hypothesis of no cointegration against the alternative of cointegration for at least some cross-sectional units.

The primary empirical analysis employs the Method of Moments Quantile Regression (MMQR) approach developed by Machado and Silva (2019), which provides several advantages for this study's context. Unlike traditional mean regression methods, MMQR allows examination of how the relationships between variables differ across the conditional distribution of renewable energy outcomes, revealing potential heterogeneities and non-linearities.

The MMQR estimator offers several key advantages: (i) it accounts for individual heterogeneity through fixed effects while allowing these effects to vary across quantiles; (ii) it provides a complete characterisation of the conditional distribution rather than focusing solely on average effects; (iii) it is robust to outliers and non-normal distributions; and (iv) it accommodates both location and scale shifts, revealing whether explanatory variables affect not only the level but also the dispersion of outcomes. We employ bootstrap quantile regression following Koenker (2005) as a robustness check.

4 Results and Discussion

Table 1 presents the descriptive statistics for all variables employed in this study. The renewable energy consumption (REE) exhibits a mean value of 0.912 with a standard deviation of 0.476, indicating moderate variability across the sample countries. The negative skewness (−0.404) suggests that the distribution is left-tailed, with more countries clustered toward higher renewable energy consumption levels. Similarly, renewable energy capabilities (REcap) show a higher mean value of 1.14 but with greater variability (SD = 0.559), reflecting heterogeneous production capacities across nations. The negative skewness (−0.395) and high kurtosis (2.483) for REcap indicate significant outliers, likely representing countries with either exceptionally high or low renewable energy production capabilities.

Financial development (FD) demonstrates relatively low variability (SD = 0.231) around a mean of 1.907, suggesting that sample countries have achieved comparable levels of financial system development. Environmental policy stringency (EnvPol) exhibits the highest relative variability among the main explanatory variables, with a coefficient of variation exceeding the mean values, indicating substantial heterogeneity in regulatory approaches across countries. The normality test results, as indicated by the Chi-square statistics, reject the null hypothesis of normal distribution for all variables ($p < 0.05$), justifying the use of quantile regression methods that are robust to non-normal distributions.

The cross-sectional dependency (CD) test results presented in Table 2(a) reveal evidence of cross-sectional dependence across the variables except green innovation (GIN). The CD statistics for REE (35.340), REcap (28.06), and EnvPol (41.740) are highly significant, with average absolute correlations exceeding 0.70. The only exception is green innovation (GIN), which shows no significant cross-sectional dependence (CD = 0.780), suggesting that innovation activities remain largely country-specific despite globalisation trends.

The slope heterogeneity tests (Table 2b) strongly reject the null hypothesis of homogeneous slopes for both models, with delta statistics of 12.969 and 11.884, respectively. This finding corroborates the theoretical expectation that the relationships between financial development, environmental policies, and renewable energy outcomes vary significantly across countries due to differences in institutional quality, economic structures, and energy market characteristics.

Table 1 Descriptive statistics

Variables	Mean	Std. Dev.	Min	Max	Skew.	Kurt.	Chi2	Chi2-Pvalue
REE	.912	.476	−.398	1.788	−.404	2.646	16.443	0.000
REcap	1.14	.559	−.238	1.999	−.395	2.483	21.613	0.000
FD	1.907	.231	1.206	2.327	−.746	3.14	27.154	0.000
EnvPol	.274	.283	−1.079	.689	−1.52	5.909	126.125	0.000
TO	.263	.05	.142	.363	−.273	2.58	12.431	0.002
NR	−.661	.789	−2.095	1.126	.349	2.174	49.419	0.000
GIN	−.006	.116	−.699	.428	−.648	6.53	68.652	0.000

Table 2 Cross-sectional dependency and slope homogeneity

a) CD Analysis			
Variable	**CD-STAT**	**P-Value**	**abs(corr)**
REE	35.340	0.000	0.752
FD	23.570	0.000	0.497
EnvPol	41.740	0.000	0.814
TO	28.340	0.000	0.612
NR	15.020	0.000	0.424
Recap	28.06	0.000	0.687
GIN	0.780	0.433	0.322
b) Slope Heterogeneity Analysis			
Models		**STAT**	**P-value**
REE FD EnvPol TO NR GIN	Delta	12.969	0.000
	Adj.	15.661	0.000
REcap FD EnvPol TO NR GIN	Delta	11.884	0.000
	Adj.	14.350	0.000

Table 3 Unit root analysis

	At level	First difference
	STAT	*STAT*
REE	−0.104	−5.536
REcap	−0.529	−6.289
FD	−2.318***	
EnvPol	−1.570	−6.194
TO	−1.274	−4.962
NR	−2.585***	
GIN	−2.503***	

Note: *** $p < 0.01$, ** $p < 0.05$, * $p < 0.1$

The unit root test results (Table 3) reveal mixed integration orders among the variables. While REE, REcap, EnvPol, and TO are integrated of order one I(1), requiring first differencing to achieve stationarity, FD, NR, and GIN appear stationary at levels I(0).

Westerlund cointegration tests (Table 4) provide strong evidence of long-run relationships in both models. The variance ratio statistics of −2.579 and −2.699 for Models 1 and 2, respectively, indicate that the variables move together in the long run, supporting the theoretical framework that renewable energy development is fundamentally linked to financial development, policy frameworks, and technological progress. This finding is particularly important as it validates the examination of long-run equilibrium relationships despite short-run dynamics and adjustment processes.

The MMQR results for Model 1 (Table 5) reveal that financial development's impact on renewable energy consumption exhibits significant heterogeneity across the conditional distribution. At the lower quantile (Q25), financial development

Table 4 Westerlund cointegration with trend

Models		STAT	P-value
REE FD EnvPol TO NR GIN	Variance Ratio	−2.579	0.005
REcap FD EnvPol TO NR GIN	Variance Ratio	−2.699	0.003

Table 5 MMQR Model-1

DV: RE	loc	sca	Q25	Q50	Q75	Q90
FD	0.272**	0.387***	−0.0646	0.364***	0.627***	0.794***
	(0.128)	(0.0804)	(0.178)	(0.122)	(0.108)	(0.117)
EnvPol	0.455***	−0.203***	0.632***	0.407***	0.269***	0.182**
	(0.0878)	(0.0553)	(0.122)	(0.0830)	(0.0746)	(0.0801)
TO	2.583***	2.286***	0.592	3.125***	4.682***	5.669***
	(0.527)	(0.332)	(0.748)	(0.517)	(0.444)	(0.483)
NR	0.150***	0.00683	0.144***	0.151***	0.156***	0.159***
	(0.0243)	(0.0153)	(0.0333)	(0.0226)	(0.0207)	(0.0221)
GIN	1.435***	−0.0373	1.468***	1.427***	1.401***	1.385***
	(0.234)	(0.147)	(0.321)	(0.218)	(0.200)	(0.213)
Constant	−0.327	−0.941***	0.493	−0.549*	−1.191***	−1.597***
	(0.342)	(0.215)	(0.476)	(0.326)	(0.290)	(0.312)

*Note: Standard errors in parentheses. *** $p < 0.01$, ** $p < 0.05$, * $p < 0.1$*

shows an insignificant effect (−0.0646), suggesting that countries with low renewable energy consumption face structural barriers that financial development alone cannot overcome. However, the effect becomes increasingly positive and significant at higher quantiles, reaching 0.794 at Q90. This pattern indicates that financial development's effectiveness in promoting renewable energy consumption depends critically on existing renewable energy infrastructure and market maturity, consistent with the findings of Pata et al. (2022) who documented that financial development significantly promotes renewable energy consumption at higher quantiles in the medium and long run, emphasising that the finance-renewable energy relationship varies by level and time horizon. This finding suggests that while financial development generally promotes renewable energy adoption, it may also exacerbate inequalities between leading and lagging countries, confirming Zhao and Zhao's (2024) concerns about innovation inequality in renewable energy development.

For renewable energy capabilities (Model 2, Table 6), financial development demonstrates uniformly positive and significant effects across all quantiles, with coefficients ranging from 0.620 at Q25 to 1.230 at Q90. The stronger and more consistent effects on capabilities compared to consumption suggest that financial development primarily facilitates renewable energy infrastructure investment and capacity building, with actual utilisation depending on additional factors such as grid integration and demand-side management. This distinction between capacity and consumption effects provides new insights into the mechanisms through which financial systems influence energy transitions. Figure 1 visually gives the results for Model 1.

Table 6 MMQR Model-2

Dv: REcap	loc	sca	Q25	Q50	Q75	Q90
FD	0.826***	0.267***	0.620***	0.845***	1.089***	1.230***
	(0.146)	(0.0859)	(0.173)	(0.146)	(0.152)	(0.173)
EnvPol	0.295***	−0.143**	0.405***	0.285***	0.155	0.0794
	(0.0991)	(0.0585)	(0.118)	(0.0991)	(0.104)	(0.118)
TO	3.031***	2.714***	0.934	3.223***	5.702***	7.132***
	(0.562)	(0.332)	(0.693)	(0.601)	(0.588)	(0.643)
NR	0.0871***	0.0380**	0.0577*	0.0897***	0.124***	0.144***
	(0.0284)	(0.0167)	(0.0336)	(0.0283)	(0.0297)	(0.0337)
GIN	1.041***	0.361**	0.762**	1.067***	1.397***	1.587***
	(0.266)	(0.157)	(0.315)	(0.265)	(0.278)	(0.316)
Constant	−1.231***	−0.777***	−0.631	−1.286***	−1.995***	−2.405***
	(0.374)	(0.221)	(0.447)	(0.377)	(0.391)	(0.442)

Note: Standard errors in parentheses. *** p < 0.01, ** p < 0.05, * p < 0.1

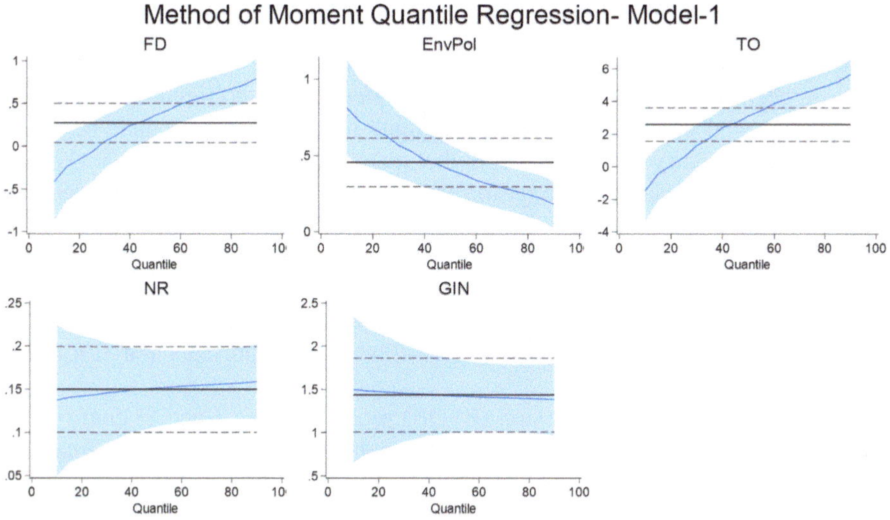

Fig. 1 MMQR model 1

Environmental policy stringency exhibits contrasting patterns between renewable energy consumption and capabilities. For consumption (Model 1), the effect is strongest at lower quantiles (0.632 at Q25) and diminishes at higher quantiles (0.182 at Q90), though remaining significant throughout. This inverse relationship suggests that stringent environmental policies are most effective in countries with currently low renewable energy consumption, where regulatory push can overcome market failures and incumbent resistance. The negative scale parameter indicates that environmental policies reduce outcome dispersion, promoting convergence in renewable energy consumption across countries.

For capabilities (Model 2 Table 6), environmental policy effects follow a similar declining pattern across quantiles but lose significance at higher quantiles (Q75 and Q90). This finding suggests that market forces and technological momentum may substitute for regulatory pressure in countries with already-developed renewable energy capabilities. These results align with Tan et al.'s (2022) findings of an inverted U-shaped relationship between environmental regulations and renewable energy investment, supporting the notion of optimal policy stringency levels beyond which additional regulations yield diminishing returns.

Green innovation emerges as a consistently strong driver of both renewable energy consumption and capabilities across all quantiles. For consumption (Model 1), the coefficients range from 1.385 to 1.468, with minimal variation across the distribution. This uniformity suggests that technological innovation provides universal benefits regardless of current renewable energy development levels, contrasting with the heterogeneous effects of financial development and environmental policies. The capabilities model shows similar patterns but with increasing effects at higher quantiles (0.762 at Q25 to 1.587 at Q90), indicating that innovation's impact on production capacity is amplified in countries with more developed renewable energy sectors. Alfalih's (2025) findings revealed that green innovation enhances renewable energy generation.

Trade openness demonstrates strong positive effects on both renewable energy consumption and capabilities, with coefficients increasing across quantiles. The effects are particularly pronounced at higher quantiles (5.669 and 7.132 at Q90 for Models 1 and 2, respectively), suggesting that trade's benefits for renewable energy are greatest in countries already integrated into global clean energy value chains. Natural resource rents show consistently positive effects across all specifications, contradicting the resource curse hypothesis for renewable energy development. The positive coefficients (ranging from 0.0577 to 0.159) suggest that resources can facilitate renewable energy transitions. Figure 2 visually represents the results of Model 2.

The bootstrap quantile regression (BSQR) results (Tables 7 and 8) confirm the MMQR findings while revealing some additional insights. The effects of financial development remain positive and significant at higher quantiles but show instability at Q25, with a negative coefficient in Model 1 (−0.225) that lacks statistical significance. This instability at lower quantiles reinforces the interpretation that financial development requires complementary conditions to promote renewable energy in lagging countries effectively. Figure 3 visually represents Model 1 results obtained using bootstrap quantile regression.

Environmental policy effects remain robust across both methods, though with some variation in magnitude. The bootstrap results show stronger effects at lower quantiles (0.851 at Q25 in Model 1), indicating that regulations may have a more pronounced impact in countries with lower levels of renewable energy development, potentially contributing to convergence across countries. Green innovation effects remain strong and significant across most specifications, though showing some instability at Q90 in the bootstrap results, possibly due to small sample sizes at extreme quantiles. Figure 4 visually represents model-2 results obtained using bootstrap quantile regression.

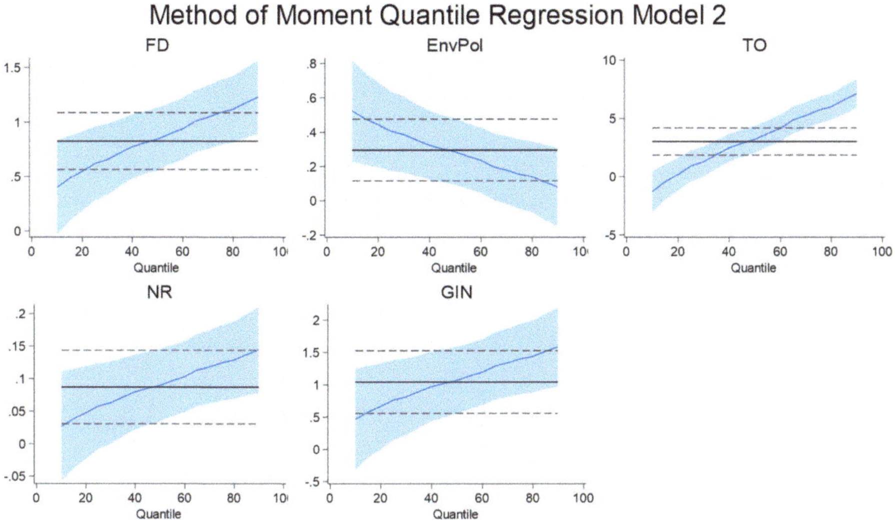

Fig. 2 MMQR Model-2 (REcap)

Table 7 Bootstrap Quantile Regression Analysis Model 1

DV: RE	Q25	Q50	Q75	Q90
FD	−0.225	0.620***	0.642***	0.191***
	(0.215)	(0.208)	(0.121)	(0.0566)
EnvPol	0.851***	0.226**	0.277***	0.331***
	(0.174)	(0.0958)	(0.0435)	(0.0267)
TO	−0.499	3.654***	5.129***	5.000***
	(0.675)	(0.789)	(0.238)	(0.245)
NR	0.162***	0.121***	0.164***	0.205***
	(0.0216)	(0.0437)	(0.0306)	(0.0185)
GIN	1.180***	1.697***	1.434***	0.353*
	(0.422)	(0.185)	(0.354)	(0.182)
Constant	1.008*	−1.133**	−1.322***	−0.279**
	(0.550)	(0.572)	(0.283)	(0.128)

Note: Standard errors in parentheses. *** $p < 0.01$, ** $p < 0.05$, * $p < 0.1$

5 Conclusion and Implications

This study examined the heterogeneous effects of financial development, environmental policy stringency, and green innovation on renewable energy in the OECD countries from 1990 to 2022. We employed the Method of Moments Quantile Regression (MMQR) approach across quantiles to capture heterogeneous effects on renewable energy consumption and production capabilities. The distinction between

Table 8 Bootstrap Quantile Regression Analysis Model 2

REcap	Q25	Q50	Q75	Q90
FD	0.514**	1.168***	0.902***	0.248
	(0.252)	(0.216)	(0.187)	(0.161)
EnvPol	0.624***	0.0827	0.183	0.153
	(0.190)	(0.0672)	(0.124)	(0.0979)
TO	−0.318	4.265***	5.647***	5.475***
	(1.083)	(0.716)	(0.395)	(0.416)
NR	0.0294	−0.00227	0.127***	0.132***
	(0.0399)	(0.0310)	(0.0338)	(0.0139)
GIN	0.512	1.254***	1.281***	0.241
	(0.691)	(0.353)	(0.413)	(0.391)
Constant	−0.233	−2.155***	−1.615***	−0.136
	(0.697)	(0.554)	(0.375)	(0.376)

Standard errors in parentheses. *** $p < 0.01$, ** $p < 0.05$, * $p < 0.1$

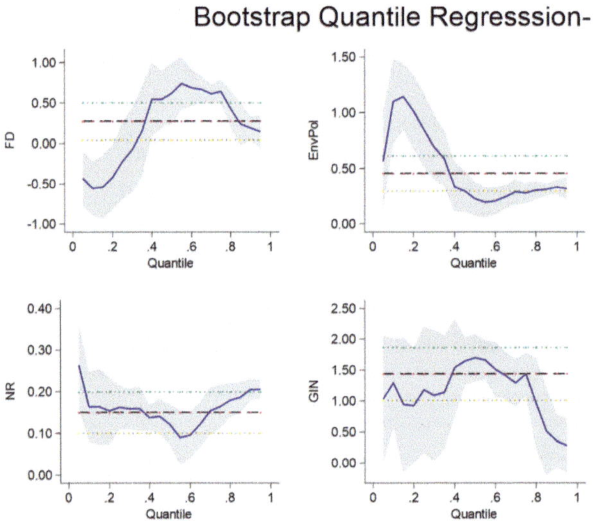

Fig. 3 BSQR Model 1

renewable energy capabilities and consumption as outcome variables provided a comprehensive overview of the multifaceted nature of energy transitions.

The empirical findings reveal three key insights. First, financial development exhibits pronounced heterogeneity in its effects on renewable energy outcomes. While countries with low renewable energy adoption (lower quantiles) show negligible or negative responses to financial development, the effects become increasingly positive and significant at higher quantiles, reaching 0.794 for consumption and 1.230 for capabilities at the 90th percentile. This pattern suggests that financial

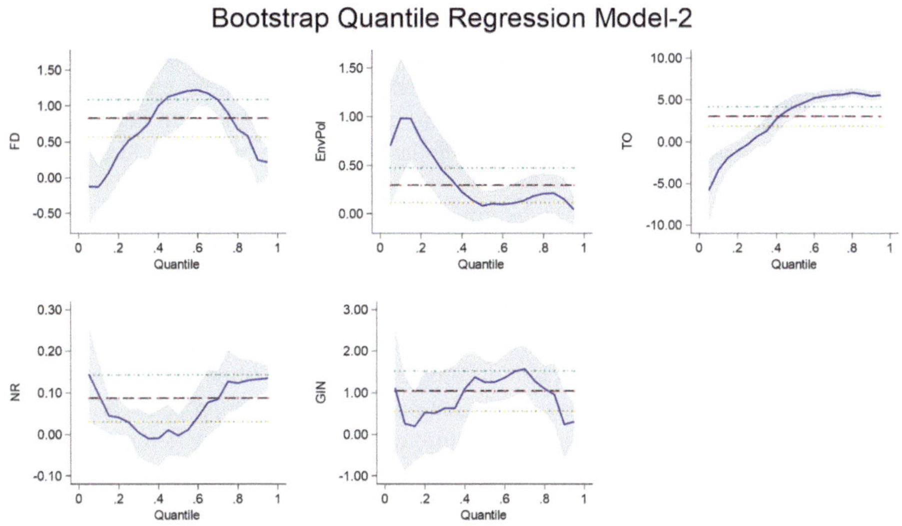

Fig. 4 BSQR Model 2

development alone is insufficient to initiate renewable energy transitions but becomes increasingly effective once basic market infrastructure and institutional frameworks are established. Compared to consumption, the stronger effects on capabilities indicate that financial systems primarily facilitate supply-side investments in renewable energy infrastructure rather than directly influencing demand-side adoption.

Second, environmental policy stringency emerges as a key convergence mechanism, exhibiting the most potent effects at lower quantiles (0.632 at Q25) and diminishing impacts at higher quantiles (0.182 at Q90) for renewable energy consumption. The negative scale parameter confirms that environmental policies reduce cross-country disparities in renewable energy adoption. These findings have important implications for international climate cooperation, suggesting that coordinated environmental regulations can help narrow the renewable energy gap between leading and lagging nations while maintaining progress in advanced economies.

Third, green innovation demonstrates consistent positive effects across the entire distribution, with coefficients ranging from 1.385 to 1.468 for consumption and showing even stronger impacts on capabilities. Unlike the heterogeneous effects of financial development and environmental policies, green innovation provides universal benefits that transcend country-specific barriers. These findings underscore the public good nature of green technologies and highlights the importance of international technology transfer mechanisms and collaborative innovation networks in accelerating global renewable energy transitions.

The policy implications of these findings are substantial. For countries at early stages of renewable energy development, the results suggest that environmental regulations should be the primary policy instrument, complemented by targeted support for green innovation. Financial sector reforms alone are unlikely to catalyse renewable energy adoption without accompanying regulatory frameworks and technological capabilities. For countries with established renewable energy sectors, deepening financial markets and maintaining innovation momentum become relatively more important, while additional regulatory stringency may yield diminishing returns.

The interactive nature of these determinants also points toward the need for policy coherence and sequencing. This study's findings support a staged approach where initial emphasis on environmental regulations creates market demand, followed by financial sector development to channel investments efficiently, all while maintaining consistent support for green innovation throughout the transition process. International organisations and development agencies should tailor their support programs accordingly, recognising that one-size-fits-all approaches will likely be ineffective given the heterogeneous nature of renewable energy determinants.

Acknowledgement Declaring Use of AI for Language Improvement: We acknowledge the use of an LLM for language enhancement purposes. All text has been reviewed by the authors for accuracy and content validity.

Competing Interests The authors have no conflicts of interest to declare relevant to this chapter's content.

References

Alfalih, A. A. (2025). Assessing the impact of green technological innovation and human capabilities on renewable energy supply: Evidence from the top-greenest countries. *Sustainable Energy Technologies and Assessments, 73*, 104152. https://doi.org/10.1016/j.seta.2024.104152

Bashir, M. F., Benjiang, M. A., Bashir, M. A., Radulescu, M., & Shahzad, U. (2022). Investigating the role of environmental taxes and regulations for renewable energy consumption: Evidence from developed economies. *Economic Research-Ekonomska Istraživanja*. https://www.tandfonline.com/doi/abs/10.1080/1331677X.2021.1962383

Behera, P., Behera, B., & Sethi, N. (2024). Assessing the impact of fiscal decentralisation, green finance and green technology innovation on renewable energy use in European Union countries: What is the moderating role of political risk? *Renewable Energy, 229*, 120715. https://doi.org/10.1016/j.renene.2024.120715

Bettarelli, L., Furceri, D., Pizzuto, P., & Shakoor, N. (2023). *Environmental policies and innovation in renewable energy*. International Monetary Fund.

Dogan, B., Nketiah, E., Ghosh, S., & Nassani, A. A. (2025). The impact of green technology on renewable energy innovation: Fresh pieces of evidence under the role of research & development and digital economy. *Renewable and Sustainable Energy Reviews, 210*, 115193. https://doi.org/10.1016/j.rser.2024.115193

Fotio, H. K., Nchofoung, T. N., & Asongu, S. A. (2022). Financing renewable energy generation in SSA: Does financial integration matter? *Renewable Energy, 201*, 47–59. https://doi.org/10.1016/j.renene.2022.11.047

Habiba, U., & Xinbang, C. (2023). The contribution of different aspects of financial development to renewable energy consumption in E7 countries: The transition to a sustainable future. *Renewable Energy, 203*, 703–714. https://doi.org/10.1016/j.renene.2022.12.088

He, L., Liu, R., Zhong, Z., Wang, D., & Xia, Y. (2019). Can green financial development promote renewable energy investment efficiency? A consideration of bank credit. *Renewable Energy, 143*, 974–984. https://doi.org/10.1016/j.renene.2019.05.059

IEA. (2023). *Executive summary – Net zero roadmap: A global pathway to keep the 1.5 °C goal in reach – Analysis*. https://www.iea.org/reports/net-zero-roadmap-a-global-pathway-to-keep-the-15-0c-goal-in-reach/executive-summary

Im, K. S., Pesaran, M. H., & Shin, Y. (2003). Testing for unit roots in heterogeneous panels. *Journal of Econometrics, 115*(1), 53–74. https://doi.org/10.1016/S0304-4076(03)00092-7

IRENA. (2023). *Renewable capacity statistics 2023*. International Renewable Energy Agency. https://www.irena.org/Publications/2023/Mar/Renewable-capacity-statistics-2023

Katircioglu, S., & Katircioglu, S. (2023). The effects of environmental taxation on stock returns of renewable energy producers: Evidence from Turkey. *Renewable Energy, 208*, 311–323. https://doi.org/10.1016/j.renene.2023.03.089

Kim, J., & Park, K. (2016). Financial development and deployment of renewable energy technologies. *Energy Economics, 59*, 238–250. https://doi.org/10.1016/j.eneco.2016.08.012

Koenker, R. (2005). *Quantile Regression*. Cambridge University Press.

Liu, W., Shen, Y., & Razzaq, A. (2023). How renewable energy investment, environmental regulations, and financial development derive renewable energy transition: Evidence from G7 countries. *Renewable Energy, 206*, 1188–1197. https://doi.org/10.1016/j.renene.2023.02.017

Long, Y., Yang, B., & Liu, L. (2023). Can green credit policy promote green innovation in renewable energy enterprises: Evidence from China. *Environmental Science and Pollution Research, 30*(41), 94290–94311. https://doi.org/10.1007/s11356-023-29041-3

Lu, Y., Khan, Z. A., Alvarez-Alvarado, M. S., Zhang, Y., Huang, Z., & Imran, M. (2020). A critical review of sustainable energy policies for the promotion of renewable energy sources. *Sustainability, 12*(12), Article 12. https://doi.org/10.3390/su12125078

Machado, J. A. F., & Santos Silva, J. M. C. (2019). Quantiles via moments. *Journal of Econometrics, 213*(1), 145–173. https://doi.org/10.1016/j.jeconom.2019.04.009

Mukhtarov, S., & Mikayilov, J. I. (2023). Could financial development eliminate energy poverty through renewable energy in Poland? *Energy Policy, 182*, 113747. https://doi.org/10.1016/j.enpol.2023.113747

Pan, W., Cao, H., & Liu, Y. (2023). "Green" innovation, privacy regulation and environmental policy. *Renewable Energy, 203*, 245–254. https://doi.org/10.1016/j.renene.2022.12.025

Pata, U. K., Yilanci, V., Zhang, Q., & Shah, S. A. R. (2022). Does financial development promote renewable energy consumption in the USA? Evidence from the Fourier-wavelet quantile causality test. *Renewable Energy, 196*, 432–443. https://doi.org/10.1016/j.renene.2022.07.008

Pesaran, M. H. (2004). *General diagnostic tests for cross section dependence in panels* (SSRN scholarly paper 572504). https://doi.org/10.2139/ssrn.572504

Pesaran, M. H. H., & Yamagata, T. (2008). Testing slope homogeneity in large panels. *Journal of Econometrics, 142*(1), 50–93. https://doi.org/10.1016/j.jeconom.2007.05.010

Shahbaz, M., Topcu, B. A., Sarıgül, S. S., & Vo, X. V. (2021). The effect of financial development on renewable energy demand: The case of developing countries. *Renewable Energy, 178*, 1370–1380. https://doi.org/10.1016/j.renene.2021.06.121

Shahzad, K., Raza, H. A., & Shahbaz, M. (2024). Environmental regulations' impact on clean energy consumption: Under the assistance of governance measures. *Renewable Energy, 237*, 121766. https://doi.org/10.1016/j.renene.2024.121766

Solarin, S. A., Bello, M. O., & Tiwari, A. K. (2022). The impact of technological innovation on renewable energy production: Accounting for the roles of economic and environmental factors using a method of moments quantile regression. *Heliyon, 8*(7). https://doi.org/10.1016/j.heliyon.2022.e09913

Sreenu, N. (2024). The impact of Fintech and green bonds on the Indian renewable energy production. *Renewable Energy, 237*, 121807. https://doi.org/10.1016/j.renene.2024.121807

Stavins, R. N. (2003). Chapter 9 – Experience with market-based environmental policy instruments. In K.-G. Mäler & J. R. Vincent (Eds.), *Handbook of environmental economics* (Vol. 1, pp. 355–435). Elsevier. https://doi.org/10.1016/S1574-0099(03)01014-3

Tan, X., Zhang, Q., Zeng, A., & Xing, X. (2022). Impact of environmental regulations on the investment level of renewable energy enterprises. *China Population Resources and Environment, 32*(07), 127–136.

Umar, M., Ji, X., Safi, A., & Afshan, S. (2024). Decentralisation, institutional quality, and carbon neutrality: Unraveling the nexus in China's pursuit of sustainable development. *Economic Analysis and Policy, 82*, 1238–1249. https://doi.org/10.1016/j.eap.2024.05.008

Wang, B., & Zhao, W. (2022). Interplay of renewable energy investment efficiency, shareholder control and green financial development in China. *Renewable Energy, 199*, 192–203. https://doi.org/10.1016/j.renene.2022.08.122

Wang, Z., Teng, Y.-P., & Xie, L. (2023). Innovation for renewable energy and energy related greenhouse gases: Evaluating the role of green finance. *Sustainable Energy Technologies and Assessments, 57*, 103279. https://doi.org/10.1016/j.seta.2023.103279

Wang, W., Wang, J., & Wu, H. (2024). Assessing the potential of energy transition policy in driving renewable energy technology innovation: Evidence from new energy demonstration city pilots in China. *Economic Change and Restructuring, 57*(5), 160. https://doi.org/10.1007/s10644-024-09748-0

Wei, D., Ahmad, F., Chandio, A. A., & Khan, I. (2024). Digital financial inclusion role to promote renewable energy technology innovation in Chinese prefectural cities: Moderating role of environmental governance. *Renewable Energy, 237*, 121704. https://doi.org/10.1016/j.renene.2024.121704

Westerlund, J. (2005). New simple tests for panel cointegration. *Econometric Reviews, 24*(3), 297–316. https://doi.org/10.1080/07474930500243019

World Bank. (2024). World development indicators. https://databank.worldbank.org/reports.aspx?source=World-Development-Indicators

Zhang, R. J., & Razzaq, A. (2022). Influence of economic policy uncertainty and financial development on renewable energy consumption in the BRICST region. *Renewable Energy, 201*, 526–533. https://doi.org/10.1016/j.renene.2022.10.107

Zhao, X., & Zhao, J. (2024). Digital finance and inequality in renewable energy technology innovation. *Energy & Environment, 35*(7), 3673–3702. https://doi.org/10.1177/0958305X231171352

Zhu, P.-H., & Zhang, K. (2023). How does income and green technology innovation influence the emissions reduction effect of renewable energy: Evidence from Chinese provincial data. *Environmental Science and Pollution Research, 30*(29), 74056–74069. https://doi.org/10.1007/s11356-023-27677-9

Challenges and Barriers to Scaling Green Finance: Overcoming Policy, Market, and Climate Change Impediment

Badareu Gabriela, Nicoleta Mihaela Doran, and Firu Mihai Alexandru

1 Introduction

As global concern about climate change grows and the pressure to meet sustainability goals intensifies, green finance has become a key focus in academic discussions, policy planning, and financial sector strategies. At its core, green finance refers to a wide range of financial instruments, regulatory measures, and market-driven approaches aimed at steering both public and private capital toward environmentally sustainable initiatives (OECD, 2017). It plays a key role in supporting the transition to a low-carbon economy by financing renewable energy, sustainable infrastructure and climate change adaptation projects, while contributing to broader decarbonisation efforts. (Campiglio, 2016; MacNeil & Kokoric, 2021).

There is growing academic interest in green finance. While initially, research focused on instruments such as green bonds or ESG assets, the field is currently multidisciplinary and located at the intersection of climate policy, economics, and financial regulation. (Taghizadeh-Hesary & Yoshino, 2019; D'Orazio, 2022). The state of knowledge in the field highlights that progress in green finance is hindered by inadequate government policy. (Volz et al., 2015; Monasterolo, 2020). In addition, recent research emphasises that Institutions and the regulatory framework play a crucial role in transitioning to a sustainable economy. At the same time, there is a need for innovative financial instruments to overcome structural obstacles. Thus, green bonds, ESG funds, and public-private guarantee schemes are proving to be

B. Gabriela (✉) · N. M. Doran
Faculty of Economics and Business Administration, University of Craiova, Craiova, Romania
e-mail: gabriela.badareu@edu.ucv.ro; nicoleta.doran@edu.ucv.ro

F. M. Alexandru
Faculty of Economics and Business Administration, Doctoral School in Economic Sciences "Eugeniu Carada", University of Craiova, Craiova, Romania
e-mail: firu.mihai.j3t@student.ucv.ro

C.-W. Su et al. (eds.), *Green Finance and the Challenges of Climate Change*, Sustainable Finance, https://doi.org/10.1007/978-3-032-07293-1_15

concrete financial mechanisms that allow the alignment of markets' economic interests with climate and social objectives. Furthermore, integrating sustainability into monetary policy and financial supervision marks a major conceptual transformation in which climate risks become central financial stability factors and are addressed in an integrated manner with economic policies.

The current challenges in green finance research are because we have a sector that, although booming, still suffers from a lack of coherence and coordination. Although academic interest has increased significantly, studies are often thematically dispersed or focused on specific cases, without building an integrated picture at a systemic level. In this context, closer collaboration between the institutions involved, central banks, regulators, investors, and a harmonisation of standards, becomes crucial in order to mobilise the capital necessary for the green transition efficiently. While individual studies offer valuable insights into specific instruments (such as Flammer, 2021 on green bonds) or regional experiences (like Cui et al., 2020 in the Chinese context), comprehensive and data-driven syntheses that trace the field's evolution, spotlight key contributors, and link theory to real-world application remain scarce.

Moreover, few studies address climate risk, public support mechanisms, and institutional innovation within green finance literature from a systemic perspective (Battiston et al., 2017).

Therefore, this chapter responds to the need for a structured bibliometric assessment that captures the intellectual evolution of green finance as an academic domain. Using data from the Web of Science Core Collection (2009–2025), this study employs tools such as VOSviewer and bibliometric mapping to analyse the temporal distribution of publications, co-authorship networks, keyword co-occurrence, and the citation impact of the most influential works in the field.

The study is guided by several key objectives:

- To identify the most cited authors, journals, and institutions contributing to green finance research;
- To analyse the thematic structure of the literature through keyword clustering;
- To track how topics such as "policy barriers", "market barriers", and "climate change" are integrated into the discourse over time;
- To highlight collaboration patterns and knowledge production networks across countries and institutions.

By consolidating existing knowledge and detecting emerging trends, this study contributes to the academic understanding of green finance and the practical development of policy instruments and financial mechanisms supporting the ecological transition. It also responds to recent calls for a more integrated and strategic approach to sustainable finance (NGFS, 2019; EU High-Level Expert Group on Sustainable Finance, 2018).

The results will provide valuable insights for the academic community, policymakers, investors, and institutions engaged in sustainable finance. By presenting a structured and data-driven overview of green finance research, this study offers a foundation for future interdisciplinary work and policy innovation that can accelerate the ecological transition and address the complex barriers that continue to impede its implementation.

The chapter is structured as follows: section 1 analyzes the theoretical framework of green finance; section 2 describes the bibliometric methodology; section 3 presents the results of scientific mapping; and the final section offers conclusions and future research directions.

2 Methodology and Data

In recent years, bibliometric analysis has gained increasing relevance as a method for understanding the evolution of academic research, both in terms of content and structure. This approach enables researchers to identify trends in publication output, collaborative networks among authors or institutions, and emerging thematic directions within a given field (Ellegaard & Wallin, 2015; Badareu et al., 2024). According to Donthu et al. (2021), bibliometric methods offer a systematic framework for managing large volumes of scientific data, facilitating the identification of influential contributors and tracing the conceptual development of academic disciplines. Although this methodology was once underutilized in environmental and financial studies, its adoption has grown rapidly—driven by the technological advancement of visualization tools such as VOSviewer and Gephi, as well as enhanced access to comprehensive academic databases like Web of Science and Scopus (Aria & Cuccurullo, 2017; Donthu et al., 2019; Linnenluecke et al., 2017; Verma & Gustafsson, 2020; Badareu et al., 2025).

In the context of green finance, bibliometric analysis provides a structured lens through which the development of scholarly discourse on sustainable financial practices can be assessed, while also highlighting persistent barriers that hinder broader implementation. This study aims to identify the key obstacles confronting green finance, with a particular focus on regulatory challenges, market inefficiencies, and climate-related risks. To ensure the relevance and focus of the dataset, the search strategy combined the term "green finance" with at least one of the following keywords: "policy," "market," or "climate change." This allowed for the selection of a representative body of literature situated at the intersection of financial systems, sustainability, and institutional dynamics.

The final dataset consists of 1726 publications, spanning from 2009—when the first article on green finance was indexed in Web of Science—through to 2025, which was still ongoing at the time of analysis. VOSviewer (version 1.6.18) was employed to create visual maps of co-authorship networks, institutional collaboration, and keyword co-occurrence. In these visualizations, node size indicates frequency (such as the number of publications or keyword appearances), while the connecting lines reflect thematic or collaborative relationships. Color-coded clusters denote distinct thematic groups and research directions.

This bibliometric approach allowed for a detailed exploration of how academic discourse around green finance has evolved over time, shedding light on dominant topics, influential scholars, and emerging areas of interest. Co-citation and keyword co-occurrence analyses revealed strong conceptual linkages among themes such as financial innovation, regulatory constraints, and the integration of climate-related risks into financial decision-making (Ravikumar et al., 2015; Wang et al., 2010). The findings present a coherent and evidence-based overview of the current academic landscape, while also identifying the institutional and systemic challenges that continue to impede the scaling of green financial instruments and practices.

Ultimately, this study contributes to the expanding body of literature seeking to map the knowledge architecture of green finance and to inform the development of more effective policies and financial mechanisms capable of addressing the obstacles posed by regulatory inertia, market dysfunctions, and the increasing volatility associated with climate change.

3 Result

This article aims to provide a comprehensive analysis of the academic literature on green finance by examining four key dimensions: the yearly evolution of publications in the field, the identification of the most influential authors and their contributions, the detection of emerging research themes through keyword co-occurrence analysis, and the evaluation of the most highly cited works that have shaped the scientific discourse, particularly those addressing the key barriers to scaling green finance—whether policy-related, market-driven, or linked to climate change.

To address the research questions stated in the introduction, the analysis is based on data extracted from the Web of Science database and is supported by a comprehensive bibliometric methodology. As illustrated in Fig. 1, the analysis follows a structured approach, which includes:

1. mapping the annual evolution of scientific output on green finance,
2. examining the co-occurrence of key terms related to "green finance" and at least one of "policy", "market", or "climate change",
3. analysing the co-authorship networks to identify major contributors and collaborative clusters, and
4. highlighting the most cited publications that provide critical insights into the systemic challenges faced in scaling green finance initiatives.

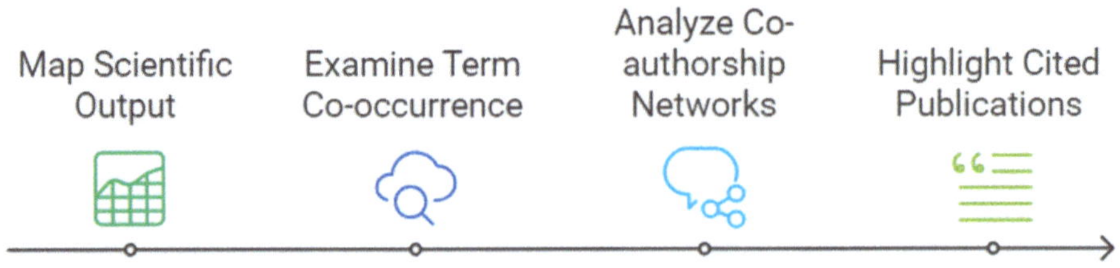

Fig. 1 Structured Methodological Framework for the Bibliometric Analysis of Green Finance Research. (Source: Own Processing, 2025)

3.1 Annual Trends in Green Finance Research

The publication evolution in the field of green finance reflects an accelerated growth trajectory, signalling a recent scientific maturation and a steadily increasing concentration of research interest in this interdisciplinary area. Drawing from Web of Science data, the first documented publication dates back to 2009, with a single article (see Fig. 2). Between 2009 and 2016, the publication rate remained modest, with fewer than five papers published annually. This slow start can be explained by the emerging nature of the green finance concept at the time, which lacked both a clearly defined theoretical framework and institutional consolidation.

This initial stagnation correlates with the absence of a robust regulatory framework and the conceptual difficulty of integrating environmental considerations into traditional financial instruments. Beginning in 2017, a modest resurgence in academic interest was observed, followed by a consistent upward trend in 2019. This moment aligns with growing international debates on climate change, increasing mobilisation of institutional actors toward green investments, and introducing new EU regulations on the taxonomy of sustainable activities.

A turning point is reached in 2022, with 283 papers published, accounting for 16.39% of the total output analysed. This increase is quickly surpassed in the following years: 460 publications in 2023 (26.65%) and an absolute peak in 2024 with 519 articles (30.07%). This publication boom is likely driven by global urgencies such as the European Green Deal (European Commission, 2018), the 2030 Agenda for Sustainable Development, a wave of ESG (Environmental, Social, and Governance) regulations, and growing pressure to accelerate the transition to a low-carbon economy. Simultaneously, it reflects the institutionalisation of green finance as a field where academic research contributes actively to public policy and strategic investment decision-making.

As for 2025, although the year is still ongoing, it already accounts for 244 publications (14.13%), indicating a sustained upward trend and confirming the continued relevance of this topic. This momentum suggests that green finance has become

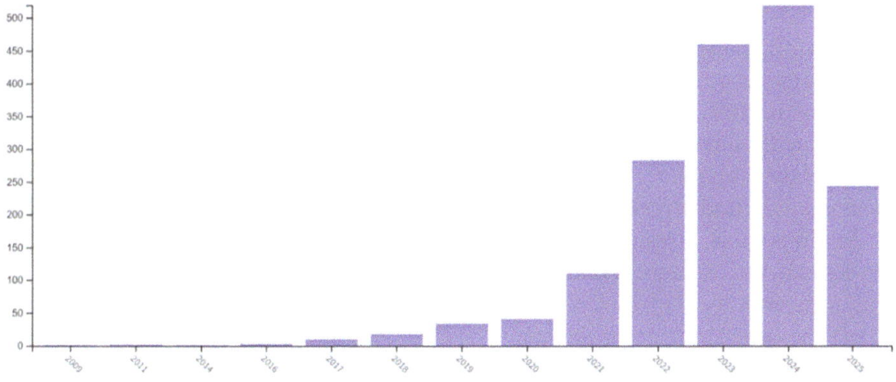

Fig. 2 Evolution of Scientific Publications on Green Finance. (Source: Web of Science, 2025)

increasingly aligned with global priorities such as sustainable development, economic resilience, and climate justice. The recent dynamic reflects a response to ongoing environmental and economic crises and result from consolidating a new conceptual and methodological framework within sustainable finance. The exponential growth in scholarly output highlights academia's increasing engagement in supporting sustainable development goals through green financial instruments and responsible investment practices.

Further, the chronological analysis of publication volume reveals the transformation of green finance from a niche topic into a mature, high-impact field of study with real-world implications for both academic research and policymaking. This trend is expected to continue as governments, financial institutions, and the scientific community collaborate more closely to meet the demands of the ecological transition. The field's evolution underscores a growing global awareness that sustainability is no longer optional but a fundamental condition for long-term economic viability.

3.2 Keyword Analysis in Green Finance Research

The co-occurrence analysis of keywords (with a minimum threshold of 5 occurrences) reveals a dense and diversified thematic landscape, where 452 terms out of a total of 4751 appear frequently enough to outline consistent research directions (see Fig. 3). This semantic concentration highlights the intellectual core of the scientific literature concerning the intersection between fiscal policy, sustainability, and the green transition. In particular, a set of highly frequent terms stands out, such as climate change (101), CO2 emissions (164), green finance (1008), carbon emissions (57), policy (160), sustainability (96), and renewable energy (231), which underline the centrality of concerns related to climate change, regulatory frameworks, and green financial instruments.

In addition to these dominant themes, the analysis reveals strong semantic connections between concepts such as innovation, investment, efficiency, and economic growth, pointing to a clear research orientation toward identifying mechanisms by which public policies can stimulate ecological innovation and sustainable development. At the same time, the frequent occurrence of terms like carbon neutrality (33), green bonds (109), and environmental regulation (39) suggests increased interest in corrective fiscal policies designed to support climate objectives. Thus, the literature consistently emphasises the need to integrate financial mechanisms into environmental policy frameworks in a coherent and results-oriented manner.

An important aspect is the identification of recurring obstacles, expressed through keywords such as constraints (48), barriers (18), uncertainty (20), climate policy uncertainty (13), governance (47), and regulations (15). These terms directly reflect the central theme of the study — Overcoming Policy, Market, and Climate Change Impediments — and confirm that a significant part of the literature is focused on understanding and overcoming systemic barriers to the green transition.

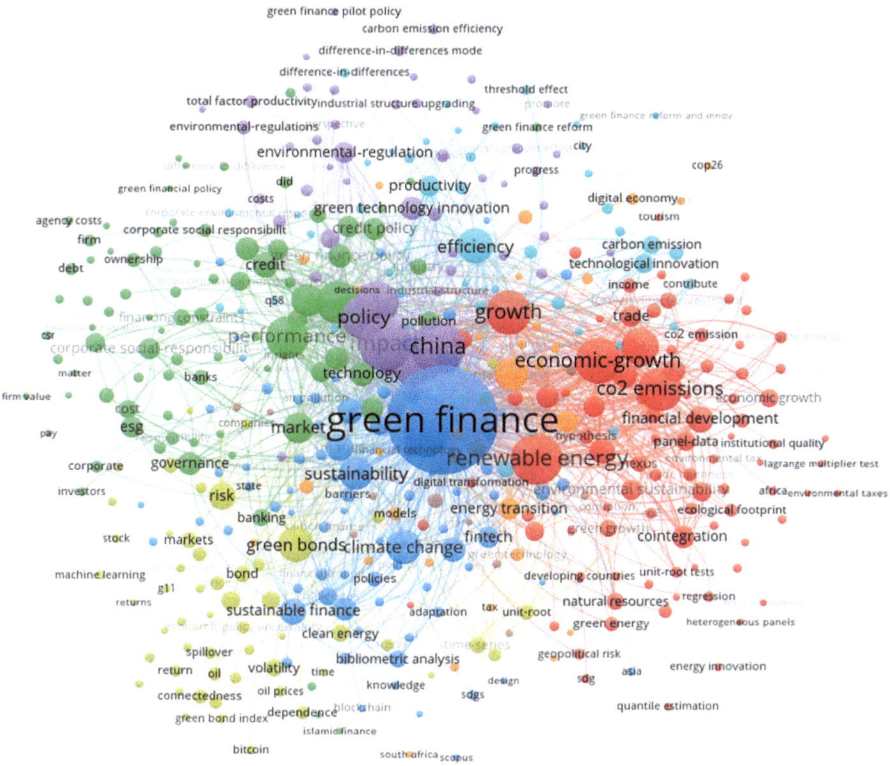

Fig. 3 Keyword Co-Occurrence Network in Green Finance Research. (Source: Own Data Processing Using VOSviewer, 2025)

The links between these terms and those associated with energy transition, such as renewable energy development, energy transition, and green innovation, reflect an integrative approach that sees barriers not only as administrative limits but also as expressions of an institutional and economic infrastructure that is not yet fully adapted.

Moreover, a thematic network emerges that brings together normative dimensions (policy, regulation), technological aspects (blockchain, AI, green technology), and social components (climate justice, corporate social responsibility), suggesting a growing interdisciplinary character of recent research (Network for Greening the Financial System, 2019). The frequent appearance of concepts such as environmental quality, decarbonization, financial inclusion, and green credit also indicates an expansion of interest toward concrete implementation and financing tools relevant to achieving climate goals and supporting a more equitable and resilient economy.

To sum up, the co-occurrence analysis of keywords confirms the existence of a mature and coherent conceptual core, centred on the relationship between public policy, ecological innovation, and green financial instruments. At the same time, the constant presence of terms indicating institutional or market barriers shows that the specialised literature is deeply engaged in identifying ways to overcome these impediments. Thus, in order to support the green transition and respond effectively to climate challenges, a systemic approach is essential—one that combines strategic regulation, sustainable financing, and innovation support within an adaptive and forward-looking governance framework.

3.3 Influential Contributors in Green Finance Research: Citation Trends and Academic Impact

Against the backdrop of intensified research efforts surrounding the transition to a sustainable economy, bibliometric indicators such as co-authorship patterns, publication volume, and citation counts have become essential tools for understanding scientific productivity and academic influence. These metrics allow not only the identification of the most active scholars but also those whose work has significantly shaped global academic discourse.

The number of publications attributed to an author often reflects sustained involvement in a given field, whereas citation counts indicate the extent to which the academic community acknowledges the relevance and impact of their contributions. In the field of green finance—marked by its interdisciplinary nature and shaped by regulatory developments, market dynamics, and climate-related concerns—bibliometric analysis offers valuable insights. It helps to pinpoint thought leaders, highly cited researchers, and key collaborative networks that influence research priorities and policy design.

Thus, bibliometric analysis transcends mere quantitative assessment and becomes a strategic approach for uncovering high-impact research, fostering academic collaboration, and supporting the advancement of knowledge in climate finance and sustainability-oriented economics.

As illustrated in Fig. 4, Farhad Taghizadeh-Hesary emerges as a leading figure in green finance, with 27 publications and 2188 citations. His work examines how financial mechanisms and regulatory frameworks can facilitate the low-carbon transition, especially in developing economies. A major research focus includes mobilizing private capital and expanding the green bond market.

Chien-Chiang Lee follows closely, with 16 publications and 1766 citations. His research explores the interactions between financial volatility, climate change, and the performance of green fiscal tools, applying advanced econometric techniques. With 5 publications and 946 citations, Yao Wang investigates the role of the banking sector and the development of innovative financial instruments for climate action, placing particular emphasis on integrating ESG (environmental, social, and governance) criteria into investment practices across both public and private sectors. Also with 5 publications, Ehsan Rasoulinezhad (653 citations), examines how trade policy affects the pace of energy transition and explores the macroeconomic vulnerabilities triggered by energy price volatility.

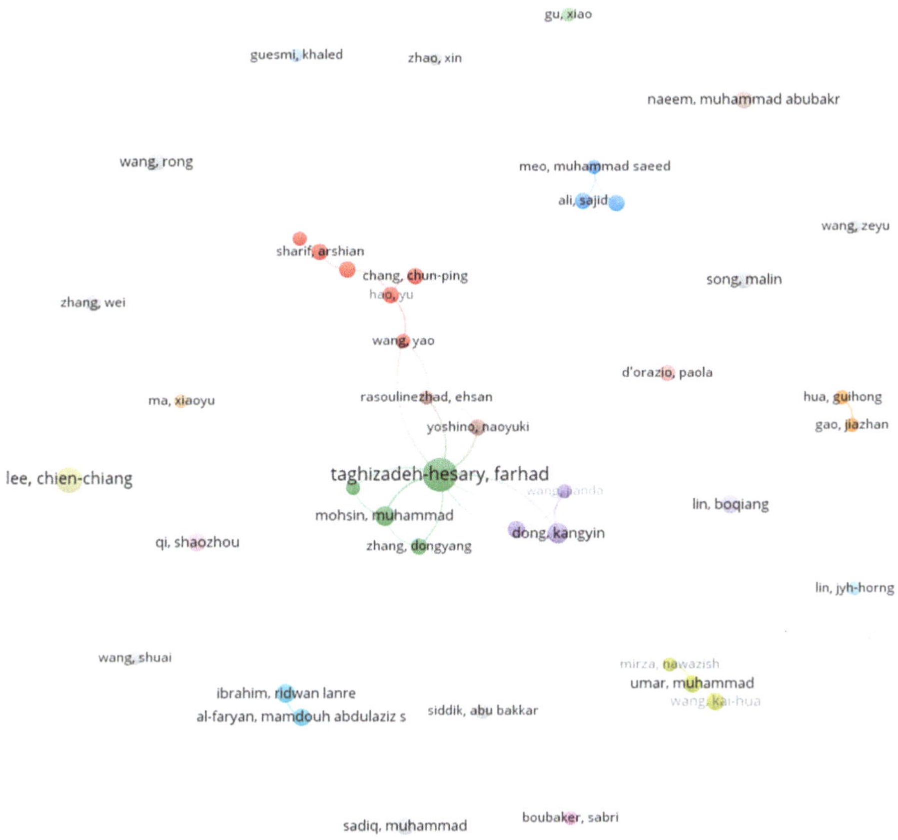

Fig. 4 Co-Authorship Network in Green Finance Research. (Source: Own Data Processing Using VOSviewer, 2025)

Muhammad Mohsin, with 9 publications and 687 citations, focuses on the intersection between digital transformation, energy transition, and financial sustainability, offering policy-driven insights aimed at reducing reliance on fossil fuels. Malin Song (7 publications, 658 citations) analyzes investor behavior in relation to energy efficiency, developing models to encourage private sector engagement in green investments and to understand how market volatility influences confidence in sustainable assets.

Research conducted by Muhammad Sadiq (7 publications, 556 citations) and Arshian Sharif (6 publications, 608 citations) sheds light on psychological and behavioral aspects of sustainable investing, emphasizing how individual perceptions and social influences contribute to ecologically responsible financial decisions. Muhammad Irfan (6 publications, 618 citations) investigates public acceptance of green technologies, underlining the importance of environmental awareness and education campaigns in shaping consumer behavior and fostering support for sustainability-driven innovation.

In the realm of financial governance, Naoyuki Yoshino (6 publications, 476 citations) and Dongyang Zhang (6 publications, 474 citations) explore how monetary policy and regional financial systems can be leveraged to support sustainable investment, highlighting the systemic nature of green finance and the necessity for integrated regulatory approaches.

Yu Hao (6 publications, 477 citations) offers empirical evidence on the link between corporate financial performance and greenhouse gas emissions, advocating for fiscal reforms that support decarbonization and embed sustainability into corporate reporting practices.

Chun-Ping Chang and Muhammad Umar, each with 6 and 7 publications respectively (403 citations each), focus on the macroeconomic consequences of energy policy, calling for the alignment of environmental objectives with broader economic strategies to ensure long-term, sustainable growth.

Muhammad Saeed Meo (5 publications, 399 citations) concentrates on the adoption of green fiscal instruments in developing countries, arguing for tailored policy solutions that reflect regional development challenges and socio-economic diversity. Muhammad Abubakr Naeem (7 publications, 362 citations) addresses the growing concern over climate-related systemic risks in global financial markets, contributing to the development of predictive models for anticipating ecological disruptions and improving financial system resilience.

In the area of green investment allocation, Abu Bakkar Siddik (6 publications, 349 citations) and Sabri Boubaker (5 publications, 342 citations) examine the role of corporate governance in enhancing organizational performance during the ecological transition. Paola D'Orazio (6 publications, 329 citations) explores the intersection of banking regulation and financial stability in the decarbonization context, stressing the importance of integrating climate-related risks into financial supervision and risk assessment frameworks. Kangyin Dong (10 publications, 255 citations) contributes by developing integrated models that account for both economic and environmental risks, offering a data-driven foundation for sustainable public policy and resilient financial systems.

Finally, Wasim Iqbal (5 publications, 261 citations) and Nawazish Mirza (5 publications, 240 citations) explore the compatibility between green fiscal reforms and economic competitiveness, advocating for policy harmonization at both European and global levels to ensure coherence, policy alignment, and long-term sustainability.

The findings of the analysis point to the existence of a core group of scholars who have exerted significant influence on the development of research in green finance, as evidenced by both their high publication output and citation impact. Farhad Taghizadeh-Hesary stands out as the most prolific author within the dataset, noted for his comprehensive approach to green financing mechanisms, with a particular focus on their implementation in emerging economies. At the same time, Chien-Chiang Lee offers valuable contributions through his examination of the interplay between financial instability and environmental fiscal policies, relying on rigorous econometric tools to substantiate his empirical findings.

Moreover, Yao Wang underlines the strategic importance of the banking sector and green bond instruments in facilitating the ecological transformation. Muhammad Mohsin contributes by exploring how digitalisation interacts with clean energy initiatives and broader economic sustainability goals. In turn, Paola D'Orazio draws attention to the role of financial regulation in safeguarding stability during the decarbonisation process. Collectively, these scholarly efforts expand the theoretical understanding of green finance and provide actionable insights for policymakers, reinforcing the value of academic research as a cornerstone for advancing a sustainable and climate-resilient economic model.

3.4 *Top-Cited Studies on Green Finance: Academic Influence and Thematic Relevance*

This analysis identifies and examines the 10 most cited scientific papers in the Web of Science database to trace the evolution and consolidation of the field of green finance: Overcoming Policy, Market, and Climate Change Impediment. These publications represent the most impactful contributions to the academic discourse, with citation frequency indicating their influence in shaping research trajectories and informing public policy development. By focusing on these benchmark studies, we aim to outline how green finance is addressed at the intersection of environmental, economic, and institutional concerns.

The bibliometric evidence (see Fig. 5—Keyword Co-occurrence, threshold $\geq$50) reveals a high concentration of articles that surpass the relevance threshold of 50 citations, confirming the scientific maturity and growing interest in this field (see Fig. 5). However, to provide a structured and focused synthesis, this review concentrates on the top 10 most cited works. These studies converge around critical themes such as the role of green finance in reducing CO_2 emissions, promoting innovation, improving energy efficiency, and addressing systemic barriers through policy and market-based mechanisms.

The study "How does green finance affect green total factor productivity? Evidence from China" (Lee & Lee, 2022) represents a key contribution to understanding the role of green finance in overcoming structural barriers related to policy, market failures, and climate change. With a notable 742 citations, this study is currently recognised as the most impactful publication in the domain of green finance, based on data from the Web of Science. Relying on a panel data analysis covering 30 provinces in China over the period 2006–2018, the authors provide empirical evidence that the expansion of green finance contributes positively to improvements in green total factor productivity. The impact of green finance appears to be especially significant in regions characterized by high pollution levels, limited civic participation, and more advanced socio-economic development. These findings suggest that green finance has the potential to mitigate deeper institutional and market inefficiencies. Moreover, the strategic implementation of targeted public policies

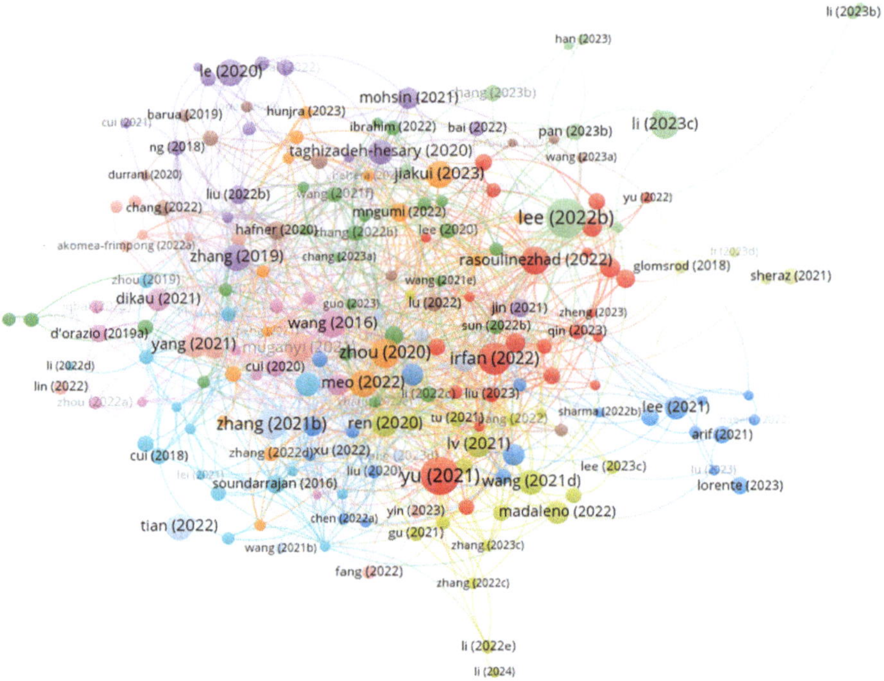

Fig. 5 Most Cited Publications in Green Finance: Key Contributions to the Field. (Source: Own Data Processing Using VOSviewer, 2025)

further enhances its effectiveness, reinforcing the idea that green finance is not only a financial tool but also a catalyst for ecological transformation and sustainable, long-term growth.

A significant contribution to the green finance literature is the study by Yu et al. (2021), titled "Demand for green finance: Resolving financing constraints on green innovation in China", which ranks as the second most cited publication in the field, with 714 citations. Based on robust empirical data, the research highlights that limited access to financial resources represents a major barrier to environmentally oriented innovation among Chinese firms, particularly during the period from 2001 to 2017. The study reveals that despite demonstrating strong innovation capacity, private enterprises encounter more significant challenges than state-owned firms in securing green funding. While green finance policies appear to alleviate some of these barriers, the analysis highlights persistent inequities in credit access. To mitigate these issues, the authors advocate for greater state support directed at privately owned companies, improved transparency through stricter disclosure regulations, and the development of a unified assessment framework under the supervision of the China Banking Regulatory Commission. Overall, the study reinforces the role of green finance as a pivotal mechanism for dismantling structural constraints and promoting inclusive, innovation-driven responses to climate change.

The study titled "Influence mechanism between green finance and green innovation: Exploring regional policy intervention effects in China" (Irfan et al., 2022), cited 491 times, is recognised as the third most impactful contribution within the thematic area Green Finance: Overcoming Policy, Market, and Climate Change Impediments. Drawing on comprehensive regional data from China (2010–2019), the authors provide compelling evidence that green finance exerts a strong and consistent influence on green innovation outcomes. Notably, regions designated as pilot zones for green financial reform exhibit enhanced effects, demonstrating the importance of localised policy measures in driving innovation.

The research identifies three key pathways through which green finance stimulates innovation: shaping the industrial structure, supporting economic growth, and increasing investment in research and development. These insights underscore that the success of sustainable innovation initiatives depends on access to green finance and the broader institutional and economic environment in which such efforts occur. The authors recommend reinforcing regional policy frameworks, fostering environments that support innovation, and improving the allocation of green financial resources to accelerate ecological transformation.

The study "A bibliometric analysis on green finance: Current status, development, and future directions" by Zhang et al. (2021), which ranks fourth among the most cited articles within the theme Green Finance: Overcoming Policy, Market, and Climate Change Impediments (with 430 citations), offers a valuable methodological contribution to the field. The paper provides a comprehensive mapping of the specialized literature, analyzing current trends and identifying relevant opportunities for future research in a domain characterized by rapid and ongoing development. By analysing influential authors, institutions, and thematic patterns, the paper reveals the interdisciplinary expansion of green finance studies and the persistent lack of a unified conceptual framework. Their analysis delineates the most influential lines of inquiry and proposes a solid conceptual foundation to support green finance's theoretical and practical development. The study's findings offer valuable guidance for shaping a coherent scientific framework to sustain the ecological transition in the face of climate and institutional challenges.

The study "Role of green finance in improving energy efficiency and renewable energy development" (Rasoulinezhad & Taghizadeh-Hesary, 2022), ranked 5th in the field of Green Finance: Overcoming Policy, Market, and Climate Change Impediment, with 372 citations, highlights the essential role of green finance in supporting the energy transition. By analysing data from 10 major economies, the authors demonstrate that green bonds significantly contribute to reducing CO_2 emissions and promoting investment in renewable energy, even if short-term causal relationships are not identified. The study emphasises the need for long-term public policy support to attract private capital to green projects, especially in the post-pandemic economic context. Thus, the paper reinforces the idea that green finance can overcome market and policy barriers, positioning itself as a strategic tool in addressing climate change.

The ranking of landmark studies in the field Green Finance: Overcoming Policy, Market, and Climate Change Impediment is followed by the paper "Green finance, fintech and environmental protection: Evidence from China" (Muganyi et al., 2021), with 364 citations. This study examines the impact of China's green finance policies on industrial emissions using panel data from 290 cities (2011–2018) and the Semi-parametric Difference-in-Differences (SDID) method. The results show that green finance policies have significantly reduced emissions, while fintech development has supported investments in environmental protection. The paper highlights the complementary role of financial technology in enhancing market-based tools for the green transition and recommends expanding green financial products and encouraging fintechs to engage in environmental objectives actively.

The study titled "The role of green finance in reducing CO_2 emissions: An empirical analysis" (Meo & Karim, 2022), ranked seventh among the most cited works within the thematic area Green Finance: Overcoming Policy, Market, and Climate Change Impediments (with 365 citations), presents robust empirical evidence on the crucial role of green finance in lowering carbon emissions. Employing the Quantile-on-Quantile Regression (QQR) approach, the authors examine the relationship between green financial flows and CO_2 emissions across 10 leading economies engaged in ecological transition. This advanced method allows for a nuanced interpretation of how the interaction between these variables evolves under different market conditions and across multiple quantile levels. The findings consistently reveal a negative correlation, indicating that as green finance expands, carbon emissions tend to decline—reinforcing its strategic importance in mitigating climate change.

However, the strength of this relationship varies depending on the phase of the economic cycle (expansion or contraction) and the specific institutional and macroeconomic context of each country. These variations underscore the need for adaptive green finance strategies tailored to national and regional realities. In this regard, green finance emerges as a versatile and effective instrument in addressing policy shortcomings and correcting structural market inefficiencies.

Ranked eighth, the article "Fostering green development with green finance: An empirical study on the environmental effect of green credit policy in China" (Zhang et al., 2019) offers a rigorous empirical evaluation of how targeted financial policies influence corporate environmental performance. The authors use panel data covering 945 publicly listed firms from 30 Chinese provinces during the period 2004–2017, assessing the impact of the Green Credit Policy (GCP) introduced in 2012. By treating the GCP as a quasi-natural experiment and employing the difference-in-differences (DID) method, the study finds that although the policy initially provided financial incentives to energy- and emission-intensive firms, it progressively discouraged investment in these sectors. The results demonstrate notable environmental improvements—particularly in eastern and western provinces—through reductions in sulfur dioxide emissions and wastewater discharge, illustrating the capacity of well-designed public policies to strengthen green finance's effectiveness in addressing market failures and promoting sustainable development.

A subsequent study by the same authors, published in 2021 under the same title, is ranked ninth, with 360 citations. This follow-up analysis reinforces the earlier findings, showing that green credit instruments can successfully channel financial resources toward environmentally responsible sectors. Using the same dataset and DID methodology, the authors demonstrate that the GCP has had a lasting deterrent effect on capital allocation to high-emission, high-energy sectors ("two high" industries) and has significantly contributed to reducing industrial pollution. The policy's impact was more pronounced among large state-owned enterprises compared to small and medium-sized firms. Regionally, it proved more effective in eastern and western provinces, while having a weaker influence in central regions. These findings highlight the importance of tailoring green finance interventions to regional and institutional contexts, confirming the essential role of such instruments in supporting climate-compatible economic transitions.

Ranked tenth, the article "Nexus between green finance, non-fossil energy use, and carbon intensity: Empirical evidence from China" (Ren et al., 2020) provides valuable insights into the interrelated dynamics between green finance development, non-fossil energy consumption, and carbon intensity. By constructing a composite index to capture green finance performance and applying a Vector Error Correction Model (VECM), the authors show that progress in green financial instruments—combined with increased reliance on clean energy—contributes significantly to lowering carbon intensity.

Nonetheless, the study also reveals a negative feedback mechanism: high carbon intensity can discourage sustainable investment and slow the uptake of clean technologies. This highlights a vulnerability in the absence of consistent and supportive policy frameworks. The authors call attention to shortcomings in current regulation and advocate for strengthening green finance mechanisms, implementing more effective carbon trading systems, and providing stronger incentives for non-fossil energy use. These policy measures are essential for overcoming institutional inertia and market distortions, thereby enabling green finance to more effectively support the ecological and economic transformation.

A broader review of the most cited works in the thematic area Green Finance: Overcoming Policy, Market, and Climate Change Impediments reveals the evolution of a mature interdisciplinary research field. Central themes include the role of green finance in reducing CO_2 emissions, driving green innovation, and facilitating the energy transition. The literature emphasizes the importance of targeted public interventions and specialized financial instruments—such as green bonds and green credit—in achieving these objectives.

Collectively, the core body of research affirms that green finance is far more than a mechanism for allocating capital efficiently. It functions as a key lever for structural reform, helping economies adapt to climate imperatives and enabling the financial sector to become an active agent in the ecological transition. These influential contributions offer actionable insights for designing coherent, forward-looking policies capable of transforming financial systems in line with sustainability goals.

4 Conclusions

The bibliometric analysis highlights a rapidly expanding field of research that has transitioned from a marginal subdomain to a central focus within contemporary economic and financial literature. The chronological evolution of publication volume shows a significant increase starting in 2015, corresponding to a key global moment—the adoption of the Paris Agreement and the consolidation of the Sustainable Development Goals (SDGs). This context has stimulated the scientific community's interest in identifying financial solutions adapted to the ecological transition, reinforcing green finance as a field with real-world impact on public policy and investment decisions.

From a theoretical standpoint, the keyword co-occurrence analysis reveals the existence of a mature conceptual core, built around the interaction between public policy, green financial instruments (such as green bonds and green loans), and ecological innovation. This core is complemented by a range of secondary themes that reflect structural concerns such as institutional barriers, regulatory uncertainty, and market dysfunctions. Thus, the specialised literature proposes an interdisciplinary framework in which insights from environmental sciences, governance theory, and climate policy complement traditional economic concepts. The result consolidates a complex research field capable of integrating ecological challenges into existing financial and economic models.

From a practical perspective, the contributions of the most highly cited authors—among whom Farhad Taghizadeh-Hesary, Chien-Chiang Lee, Yao Wang, Muhammad Mohsin, and Paola D'Orazio stand out—offer significant value through the development of public policy and financial regulation strategies adapted to the current ecological context. Their studies emphasise the need for institutional reform, support for green innovation, and efficient capital allocation through specialised financial instruments. Moreover, recent literature increasingly reflects the need for a systemic approach that combines strategic regulation, innovation support, and active private sector involvement in financing the climate transition.

The results of the analysis provide a solid framework for developing strategic recommendations. First, fostering active interdisciplinary collaboration among economists, climate scientists, financial experts, and policymakers is essential to ensure the development of coherent, interconnected, and sustainable public policies. Second, green finance policies should be tailored to the specific characteristics of each region, taking into account their level of economic development and institutional capacity. In addition, supporting applied research and assessing the tangible impact of green financial instruments should become strategic priorities, helping to build investor confidence and mobilize broader participation from economic actors.

Regarding future research directions, bibliometric findings highlight several promising avenues. On one hand, comparative analyses between countries or regions are needed to identify best practices in the implementation of green finance. On the other hand, integrating emerging technologies—such as blockchain, artificial intelligence, and FinTech solutions—into the design of green financial products opens important opportunities for improving efficiency and enhancing transparency in the sector. Moreover, deeper exploration is needed into the psychological and behavioral aspects of financial decision-making related to sustainability, including how climate risk is perceived and how investor preferences evolve.

In conclusion, the bibliometric analysis shows that green finance has moved beyond its early framing as a public policy instrument or niche innovation, becoming a strategic pillar of the economic transition toward sustainability. Scientific research plays a vital role in this process by offering strong theoretical foundations and practical solutions to support an ecological transition that is balanced, equitable, and effective.

Acknowledgments This work was supported by a grant from the Romanian Ministry of Research, Innovation and Digitalization, the project with the title *"Economics and Policy Options for Climate Change Risk and Global Environmental Governance"* (CF 193/28.11.2022, Funding Contract no. 760078/23.05.2023), within Romania's National Recovery and Resilience Plan (PNRR)—Pillar III, Component C9, Investment I8 (PNRR/2022/C9/MCID/I8)—Development of a program to attract highly specialised human resources from abroad in research, development, and innovation activities.

Competing Interests The authors have no conflicts of interest to declare relevant to this chapter's content.

References

Aria, M., & Cuccurullo, C. (2017). Bibliometrix: An R-tool for comprehensive science mapping analysis. *Journal of Informetrics, 11*(4), 959–975. https://doi.org/10.1016/j.joi.2017.08.007

Badareu, G., Doran, M. D., Firu, M. A., Olaru, S. M., & Doran, N. M. (2024). Financial development and climate change: A detailed bibliometric investigation. *Economies, 12*(288).

Badareu, G., Cârstina, S., Militaru, F., Siminică, M. I., & Cîrciumaru, D. (2025). What is the current status of research on the impact of digitalization in medicine? A bibliometric analysis. *Healthcare, 13*(93). https://doi.org/10.3390/healthcare13020093

Battiston, S., Mandel, A., Monasterolo, I., Schütze, F., & Visentin, G. (2017). A climate stress-test of the financial system. *Nature Climate Change, 7*(4), 283–288. https://doi.org/10.1038/nclimate3255

Campiglio, E. (2016). Beyond carbon pricing: The role of banking and monetary policy in financing the transition to a low-carbon economy. *Ecological Economics, 121*, 220–230. https://doi.org/10.1016/j.ecolecon.2015.03.020

D'Orazio, P. (2022). Green macro-financial policies in emerging markets: Signals and patterns of recent developments. *Environmental Economics and Policy Studies, 24*(1), 1–23. https://doi.org/10.1007/s10018-021-00301-z

Donthu, N., Kumar, S., & Pattnaik, D. (2019). Forty-five years of journal of business research: A bibliometric analysis. *Journal of Business Research, 109*, 1–14. https://doi.org/10.1016/j.jbusres.2019.10.039

Donthu, N., Kumar, S., Mukherjee, D., Pandey, N., & Lim, W. N. (2021). How to conduct a bibliometric analysis: An overview and guidelines. *Journal of Business Research, 133*, 285–296. https://doi.org/10.1016/j.jbusres.2021.04.070

Ellegaard, O., & Wallin, J. A. (2015). The bibliometric analysis of scholarly production: How great is the impact? *Scientometrics, 105*(3), 1809–1831. https://doi.org/10.1007/s11192-015-1645-z

European Commission. (2018). *Final report of the high-level expert group on sustainable finance*. Publications Office of the European Union. https://ec.europa.eu/info/publications/180131-sustainable-finance-report_en

Flammer, C. (2021). Corporate green bonds. *Journal of Financial Economics, 142*(2), 499–516. https://doi.org/10.1016/j.jfineco.2021.05.014

Irfan, M., Zhao, Z.-Y., Li, H., Rehman, A., & Ali, S. (2022). Influence mechanism between green finance and green innovation: Exploring regional policy intervention effects in China. *Technological Forecasting and Social Change, 182*, 121882. https://doi.org/10.1016/j.techfore.2022.121882

Lee, C.-C., & Lee, C.-C. (2022). How does green finance affect green total factor productivity? Evidence from China. *Energy Economics, 107*, 105863. https://doi.org/10.1016/j.eneco.2022.105863

Linnenluecke, M. K., Chen, X., Ling, X., Smith, T., & Zhu, Y. (2017). Research in finance: A review of influential publications and a research agenda. *Pacific-Basin Finance Journal, 43*, 188–199. https://doi.org/10.1016/j.pacfin.2017.04.005

MacNeil, I., & Kokoric, A. (2021). Sustainable finance and the EU green bond standard. *Law and Financial Markets Review, 15*(1), 12–24. https://doi.org/10.1080/17521440.2021.1881896

Meo, M. S., & Karim, M. Z. A. (2022). The role of green finance in reducing CO_2 emissions: An empirical analysis. *Borsa Istanbul Review, 22*(1), 169–178. https://doi.org/10.1016/j.bir.2021.08.002SCIRP+2MDPI+2SSRN+2

Monasterolo, I. (2020). Climate-related financial risk: An emerging field in climate finance. *Climate Policy, 20*(6), 752–764. https://doi.org/10.1080/14693062.2019.1704678

Muganyi, T., Yan, L., & Sun, H.-P. (2021). Green finance, fintech and environmental protection: Evidence from China. *Environmental Science and Ecotechnology, 7*, 100107. https://doi.org/10.1016/j.ese.2021.100107

Network for Greening the Financial System. (2019). *A call for action: Climate change as a source of financial risk*. https://www.ngfs.net/en/call-action

Organisation for Economic Co-operation and Development. (2017). *Investing in climate, investing in growth*. OECD Publishing. https://doi.org/10.1787/9789264273526-en

Rasoulinezhad, E., & Taghizadeh-Hesary, F. (2022). Role of green finance in improving energy efficiency and renewable energy development. *Energy Efficiency, 15*(2), 14. https://doi.org/10.1007/s12053-022-10021-4

Ravikumar, S., Agrahari, A., & Singh, S. N. (2015). Mapping the intellectual structure of Scientometrics: A co-word analysis of the journal *Scientometrics* (2005–2010). *Scientometrics, 102*(1), 929–955. https://doi.org/10.1007/s11192-014-1445-7

Ren, S., Hao, Y., & Wu, H. (2020). Nexus between green finance, non-fossil energy use, and carbon intensity: Empirical evidence from China. *Energy Policy, 144*, 111605. https://doi.org/10.1016/j.enpol.2020.111605

Taghizadeh-Hesary, F., & Yoshino, N. (2019). The way to induce private participation in green finance and investment. *ADB Institute Working Paper Series*. https://www.adb.org/publications/way-induce-private-participation-green-finance-and-investment, 31

Verma, S., & Gustafsson, A. (2020). Investigating the emerging COVID-19 research trends in the field of business and management: A bibliometric analysis approach. *Journal of Business Research, 118*, 253–261. https://doi.org/10.1016/j.jbusres.2020.06.057

Volz, U., Böhnke, J., Knierim, L., Richert, K., & Roehrl, R. A. (2015). Financing the green transformation. *UNEP Inquiry Working Paper*. https://wedocs.unep.org/handle/20.500.11822/20624

Wang, M. H., Yu, T. C., & Ho, Y. S. (2010). A bibliometric analysis of the performance of water research. *Scientometrics, 84*(3), 813–820. https://doi.org/10.1007/s11192-009-0112-z

Yu, C.-H., Wu, X., Zhang, D., Chen, S., & Zhao, J. (2021). Demand for green finance: Resolving financing constraints on green innovation in China. *Energy Policy, 153*, 112255. https://doi.org/10.1016/j.enpol.2021.112255

Zhang, D., Zhang, Z., & Managi, S. (2019). A bibliometric analysis on green finance: Current status, development, and future directions. *Finance Research Letters, 29*, 425–430. https://doi.org/10.1016/j.frl.2019.02.003

Zhang, S., Wu, Z., He, Y., & Hao, Y. (2021). Fostering green development with green finance: An empirical study on the environmental effect of green credit policy in China. *Journal of Cleaner Production, 279*, 123456. https://doi.org/10.1016/j.jclepro.2020.123456

The Role of Green Finance in Building Climate Resilience: Financing Adaptive Solutions for Vulnerable Regions

Li Zhengzheng

1 Introduction

Against the backdrop of the ongoing worldwide climate crisis, climate resilience (CR) has become more than ever a compelling necessity. CR means the collective capacity of natural ecosystems, human societies, and economic systems to sustain basic functionalities, manage systemic risk, and improve sustainable development avenues through adaptive and co-evolutionary processes under the pressures of climate change (Seidu et al., 2025). This insight emphasises an integrated system's capacities to weather, respond to, and pursue sustainable development choices in the midst of adversity, such as episodic extreme weather events (e.g., heatwaves, floods, droughts) and gradual climatic trends (e.g., sea-level rise, temperature rise). In natural systems, CR develops enhanced adaptive abilities by application of nature-based solutions such as forest protection and wetland rehabilitation. These nature restoration measures reduce the risks of natural hazards and preserve key biodiversity functions. Socioeconomically, CR operationalises risk reduction efforts against economic losses from extreme weather events into action, thereby securing people's well-being and enabling sustainable development trajectories. The Paris Agreement formally integrates adaptive capacity development in climate action strategies, emphasising climate-resilient development as central to ensuring the 2 °C temperature control objective. As a result, developing cross-sectoral and multi-scale CR systems has become a necessary means to manage risky climate uncertainties.

The impacts of global climate change on vulnerable regions have emerged as a critical topic of research in international sustainable development. Vulnerable regions that are typical of agricultural production regions, coastal regions, and

L. Zhengzheng (✉)
School of Economics, Qingdao University, Qingdao, China
e-mail: lizhengzheng@qdu.edu.cn

C.-W. Su et al. (eds.), *Green Finance and the Challenges of Climate Change*, Sustainable Finance, https://doi.org/10.1007/978-3-032-07293-1_16

desert zones are under structural threat from higher rates of frequency of weather-related phenomena. Particularly in agricultural production areas, climate change adversely impacts agricultural production systems by altering natural climatic variables such as temperature and precipitation, thereby constraining opportunities for crop diversification (Das et al., 2024). It is also significant to note the acute vulnerability of global coastal ecosystems. Global climate change and sea-level rise are accelerating coastal land erosion, compromising the sustainability of coastal resource-based livelihoods (Das, 2024). Moreover, the fact that Iraq is predominantly found within dry and semi-dry climatic zones makes it highly susceptible to climate change impacts, particularly through decreased rainfall patterns and increased warming trends (Saeed et al., 2022). Climate change exacerbates the water shortage in Iraq while increasing the frequency and magnitude of heatwaves and flash flood incidences (Rasheed et al., 2023). Thus, adaptation financing strategies should be developed because of the increased adverse impacts of climate change in vulnerable regions.

The 2016 Guidelines on the Green Financial System define green finance (GF) as financial services supporting economic activities resulting in environmental improvement, reducing climate change, and enhancing resource efficiency. This involves financing and investing, project business operations, and project risk management for environmental protection, energy conservation, clean energy, green transport, and green buildings. Specifically, GF targets optimal resource distribution (e.g., technology, capital, policy, labour, etc.) among industries to steer the flow of resources towards green and low-carbon sectors (Saeed Meo & Abd Karim, 2022). It targets lowering the finance cost for green businesses, expanding their margins, and streamlining the production and use of green technologies (Irfan et al., 2022). Besides, it facilitates the shift of industries by regulating the expansion of highly polluting sectors, streamlining industrial frameworks, promoting resource efficiency, and stimulating sustainable economic growth (Lee et al., 2023).

GF and CR have a close relationship as the two essential dimensions in addressing climate change. On the one hand, GF enhances CR capacity through channelling capital flows into highly resilient and low-carbon industries. Through the GF framework, capital support funds low-carbon projects, reduces initial capital investment costs, and enhances the green firms' sustainable development capacity (Hu et al., 2021). On the other hand, GF can indirectly facilitate CR through its intermediary function. For example, in agriculture, GF promotes agricultural total factor productivity, such as technological innovation and the efficiency of resource use (Fei, 2023), thus enhancing the ability of agriculture to adjust to extreme weather. Similarly, urban-scale research has also determined that GF reduces overall carbon emissions due to optimising industry structures and increasing energy efficiency, making cities more climate-resilient (Liu et al., 2024). On top of this, the expansion of GF can be facilitated by CR. Besides alleviating the consumption of greenhouse gases and improving environmental quality, effective CR project implementation and operation also yield consistent returns for investors. This enhances GF's market confidence and visibility, further driving its growth.

This chapter examines the pivotal role of GF in enhancing CR, focusing on its capacity to provide adaptive financing solutions for vulnerable regions through capital allocation, policy coordination, and technological innovation. First, it highlights the severity of GF and CR in vulnerable regions. We especially analyse the climate risks and challenges that representative vulnerable areas face, such as high exposure to climate hazards, low adaptive capacity, and fragile socioeconomic systems. Second, the study explains the significance of GF in building CR by elucidating its mechanisms in capital allocation, policy coordination, and technological innovation. Furthermore, by examining the effect of CPIA policy and institutions for environmental sustainability rating on the percentage of the total population with electricity access (AET), the percentage of the total population using clean fuels and cooking technologies (AFTC), and the percentage of the urban population with electricity access (AEU) in LDCs countries, this study confirms the positive role of GF on CR.

The remainder of this chapter is structured as follows. Section 2 reviews the current status and points out the challenges of CR and GF in vulnerable regions. Section 3 analyses the role of GF in CR through capital allocation, policy frameworks, and technological innovation. Section 4 provides the empirical evidence on the GF and CR in LDCs. Section 5 concludes this chapter and proposes several strategies and a future outlook for enhancing GF in fragile regions.

2 Background Review of Climate Resilience and Green Finance in Vulnerable Regions

The global climate is changing at an unprecedented rate, with far-reaching consequences for vulnerable regions worldwide. Least Developed Countries (LDCs) comprise the most climate-vulnerable population of countries in the world. The population dynamics, low human development index, threshold development capacity, and developmental challenges render nations like this more vulnerable to climate change since their human capacity or ability to adapt and recover to climate stressors is essentially limited, more so, as they depend on climate-sensitive sectors for development. The global climate change rate, rising sea levels, natural disasters, and drought can be alarming to these populations operating in sensitive and vulnerable conditions (UNDP, 2015). These regions, often characterised by their high exposure to climate hazards, low adaptive capacity, and fragile socioeconomic systems, are disproportionately affected by climate change. The impacts range from more frequent and intense extreme weather events, such as floods, droughts, and heatwaves, to long-term changes in sea-level rise, precipitation patterns, and temperature increases.

- **High Exposure to Climate Hazards:** Vulnerable regions are typically located in areas that are highly exposed to various climate hazards. For example, small island developing states (SIDS) are highly vulnerable to sea-level rise, threatening

to inundate their low-lying coastal areas, eroding beaches, and contaminating freshwater sources. Many SIDS, such as the Maldives and Tuvalu, are at the forefront of this threat, with some estimates suggesting that parts of these islands could be uninhabitable within the next few decades. In addition, regions in the tropics and sub-tropics are more prone to intense tropical cyclones. Countries in the Caribbean, like Haiti and the Dominican Republic, regularly experience the devastating impacts of these storms. These cyclones bring strong winds, heavy rainfall, and storm surges, which can cause widespread damage to infrastructure, homes, and agricultural land, leading to significant economic losses and displacement of populations.
- **Low Adaptive Capacity:** Various factors often limit the adaptive capacity of vulnerable regions. Economically, many of these regions are developing countries with limited financial resources. They struggle to invest in the necessary infrastructure, technology, and human capital required to adapt to climate change. For instance, in sub-Saharan African countries, a large proportion of the population lives in poverty, and governments have limited budgets to allocate towards climate adaptation measures. This limited budgetary investment restricts their ability to build flood-resilient structures, create water management infrastructure, or implement early warning mechanisms for extreme weather (Jayles et al., 2022). Also, these regions might lack the required technological and institutional capacity. A shortage of access to advanced climate predictive technologies in some rural South Asian areas is also a problem, which is indispensable for highly informed agricultural planning and disaster readiness. Weakly governed institutions, suffering from incapacities in policy implementation, also hamper the effective delivery of climate adaptation programs (Johnson et al., 2025).
- **Socioeconomic Fragility:** Vulnerable regions' weak socioeconomic systems rely heavily on climate-sensitive industries, such as agriculture and fisheries. Agriculture is the backbone of the economy and provides employment for a large percentage of the population in most of Africa. However, changed climate patterns like increased droughts and unpredictable rainfall are considerably impacting crop yields. Smallholder farmers, with no means of investing in climate-smart agriculture practices, are particularly vulnerable. In addition, these regions may have high levels of population density, especially in coastal areas or urban slums. For example, Bangladesh has a high population density in flood-exposed coastal regions. The high population densities in these regions contribute to climate-related disasters' impacts because of the challenges in acquiring appropriate shelter, food provision, and medical attention after a disaster. Social imbalances, especially gender discrimination, also contribute to vulnerability in certain groups in these regions. Women, in general, possess fewer resources and decision-making powers and are consequently more vulnerable to the impacts of climate change.

Climate resilience projects unavoidably have gigantic initial construction costs and greater risks (Gabriel et al., 2016), a typical scenario in vulnerable regions. Compared to those with high adaptive potential, such regions are doubly vulnerable along natural ecosystems and socio-developmental aspects, significantly elevating the cost of putting up such ventures. The natural systems approach detects exposed regions, such as coastal lowlands and drylands, by the heightened frequency and magnitude of extreme climatic events, compounded by ecologically formed structures with highly constrained climatic hazard resilience. Consequently, climatic resilience programs in such cases must be normative to more stringent technical specifications and operating practices, thereby escalating material procurement, construction complexity, and long-term maintenance expenditures. From a social perspective, vulnerable regions typically suffer from lagging public services and infrastructure. For instance, infrastructure such as energy, water conservancy, and transportation is typically outdated and highly susceptible to the destruction caused by extreme weather (Meregillano & Delina, 2023). Further, public services like healthcare and emergency management have poor coverage. With the impacts and stresses brought about by climate change, they can only resort to stopgap measures, and the cost and amount of effort in developing long-term resilience are higher (Nazari et al., 2025). In addition, climate change's non-linear and unpredictable character makes it challenging to predict extreme events, thereby rendering traditional risk assessment models somewhat useless when used for resilience projects. At the same time, the environment's vulnerability could also lead to a chain reaction wherein one disaster can cause the contagion and failure of systemic risks, thus further intensifying the risks.

Furthermore, vulnerable regions often confront multi-dimensional financial exclusion constraints, stemming from monolithic economic structures, inadequate infrastructure, and unmitigated climate risk exposures, which collectively restrict their access to formal financial markets. Specifically, the economies of vulnerable regions are often highly dependent on climate-sensitive industries such as agriculture and fishing. These industries are particularly vulnerable to extreme weather conditions (Diallo & Tapsoba, 2022), resulting in significant fluctuations in the income of local residents. Consequently, financial institutions, driven by risk management imperatives, tend to classify these regions as high-risk cohorts, imposing stricter eligibility criteria for insurance and credit products or even withdrawing services entirely. Secondly, the financial infrastructure in disaster-prone areas remains underdeveloped, characterised by sparse banking networks and high operational maintenance costs. This has led to a lack of motivation for financial institutions to establish their presence in this region, resulting in lower coverage and convenience of financial services in the area. Finally, due to the scarcity of climate monitoring data, financial institutions find it difficult to accurately assess risks, prompting them to escalate interest rates or insurance premiums. These factors have led to a low penetration rate of financial services in vulnerable regions, which will further exacerbate these regions' financial risks and economic vulnerability, pushing them further away from the financial market.

Technical barriers also impede the penetration of green finance into vulnerable regions (Adegboyo et al., 2025). On the one hand, the climatic conditions in vulnerable regions impose stringent technical specifications on green technology deployment, creating significant implementation barriers. For instance, the high temperature and humidity in tropical regions necessitate better anti-ageing properties for photovoltaic panels, while wind turbines in areas with frequent sandstorms require special anti-abrasion designs; otherwise, they will shorten the equipment's lifespan and increase maintenance costs. On the other hand, due to the lack of a technical training system in vulnerable regions, they cannot meet the professional maintenance needs required for introduced green technologies (such as intelligent microgrids, green irrigation), thus making many projects difficult to sustain and eventually leading to paralysis. The increase in failed project cases will, in turn, reduce investor confidence and the speed of technology introduction.

Regarding institutional frameworks, financial systems in vulnerable regions are typically constructed around traditional industries, such as agriculture and fossil fuels. This structure results in a scarcity of technological reserves and institutional experience necessary for a green transition (Chendo, 1994). Moreover, constrained by limited fiscal resources, local governments often face conflicts between short-term economic interests and long-term climate objectives. These contradictions undermine the capacity and willingness of governments to advance the construction of green governance systems and regulatory frameworks, thereby impeding the smooth entry and effective application of green finance in these regions. This predicament is manifested explicitly in the following aspects: An incomplete legal system leads to chaotic market entry standards. The absence of statutory definitions for "green finance" and "climate resilient projects" creates room for subjective discretion in project certification, fostering the risk of "greenwashing" and eroding investors' trust in the regional market(Li & Deng, 2025). Secondly, fragmented government functions exacerbate the inefficiency of policy implementation. Unclear divisions of responsibilities among environmental, financial, and fiscal regulatory authorities often result in conflicting approval processes or regulatory vacuums, significantly reducing the efficiency of project advancement. Lastly, lagging regulatory capabilities constrain the precision of risk assessment. Technical shortcomings and data deficiencies force institutions to rely on international templates, resulting in policies that lack local specificity.

Specific social equity issues emerge during the construction of climate-resilient projects, which essentially stem from the misalignment of "cost bearing" and "benefit distribution" across temporal and spatial dimensions. These issues are specifically manifested in the following aspects: Firstly, intergenerational equity is imbalanced. Climate-resilient projects, such as flood control projects and drought-resistant agricultural transformation, often require massive upfront investments. However, the short-term costs may be shouldered by vulnerable groups of the present generation, while the benefits will only materialise after decades. This creates a temporal mismatch where the current generation's sacrifice is exchanged for the future generation's welfare. Moreover, the absence of an intergenerational compensation mechanism further intensifies this equity dilemma. Secondly, there's the issue of regional equity. Vulnerable regions typically suffer from low fiscal revenue and poor credit ratings, which leads to high financing costs and makes market-based financing difficult. Due to factors such as assessment and allocation criteria (Wilkinson et al., 2023), the allocation of international climate finance often exhibits a "visibility preference". For instance, multilateral development banks' climate funds are predominantly allocated to high-emission areas and a few relatively wealthy countries, with less going to highly vulnerable and low-emission nations (Xie et al., 2023). Consequently, the Matthew effect in resource allocation emerges. Higher-income areas are attractive to funds through green finance, generating a positive climate resilience cycle. The most vulnerable areas, however, are those that are most in need of protection but are in a negative cycle of increased vulnerability and progressively more difficult financing. Lastly, there are also equity conflicts among different groups. Applying climate-resilient projects can restrict traditional livelihoods and degrade traditional cultural landscapes. This leads to intense conflicts of interest claims among different groups in the process of project advancement, thereby further speeding up social equity issues. Additionally, in climate governance, there needs to be a balance between climate resilience and environmental sustainability as well.

Climate resilience is interested in adaptation and rebounding (Yi et al., 2025), i.e., surviving climate catastrophes and transcending food security issues. Environmental sustainability is, however, interested in restraining and mitigating, i.e. low-carbon emissions and promoting ecological preservation. There are paradoxes in promoting the building of both. On one hand, the technology that climate resilience actions are targeted at is incompatible with low-carbon technology. Thus, enhancing regional climate resilience can increase carbon emissions directly and generate other issues, lowering environmental sustainability. For example, concrete sea walls are very effective against storm surges, but cement manufacture is a high-carbon dioxide process that somewhat negates the disaster prevention benefit of the project. Conversely, contradictions exist within the assessment system due to space and time mismatches. For instance, constructing dams along the Mekong River has increased flood control ability upstream and is hence more flood climate resistant. However, it has led to soil erosion in the downstream delta and reduced mangrove carbon sequestration ability. This establishes a conflict of balance: sacrificing sustainability for temporary safety, and rejecting present threats for the benefit of planning in the future.

3 Understanding the Role of Green Finance in Climate Resilience

3.1 Capital Allocation

As a pivotal policy instrument for reconciling environmental governance with economic development, GF plays a critical role in enhancing CR through its capital allocation mechanism. Its core role is demonstrated in guiding capital toward sectors with CR benefits through market-based mechanisms while constructing a systemic risk hedging framework. From the perspective of capital allocation pathways, green financial systems make the long-term costs of climate risks explicit through tools such as environmental information disclosure, climate stress testing, and differentiated pricing, prompting financial institutions to reassess the resilience value of their portfolios. For instance, the issuance rates of green bonds are typically lower than those of conventional bonds, with such price signals incentivising investors to prioritise support for CR projects like flood control infrastructure upgrades and ecological restoration (Xu & Yang, 2025). Regarding resource allocation efficiency, GF enhances the bankability of climate-resilient projects through risk-sharing mechanisms. Financial instruments like catastrophe bonds and parametric climate insurance products quantify climate risks into tradable financial contracts. The United States National Flood Insurance Program exemplifies this approach, where risk dispersion mechanisms enable coastal communities to access post-disaster reconstruction financing. Similarly, Shenzhen's catastrophe insurance pool and Shanghai Songjiang's rice high-temperature index insurance pilot project exemplify how financial mechanisms can turn climate risks into manageable financial costs, enhancing the disaster resistance of agricultural and urban systems.

3.2 Policy Coordination

The policy coordination function of GF for CR is mainly reflected in the synergy of policy objectives and policy instruments. On the one hand, remarkable synergistic enhancement mechanisms exist between GF and CR construction at the policy objective level, as the two form systematic policy synergies through goal coupling and effect stacking. From the dual dimensions of mitigation and adaptation, GF policies can not only guide low-carbon technological innovation through capital allocation to directly serve greenhouse gas emission reduction goals (Manta et al., 2024), but also promote investments in adaptive infrastructure such as disaster early warning systems and ecological restoration projects via climate risk pricing mechanisms, thereby enhancing the resilience of economic and social systems to climate disasters. In addition, countries have systematically embedded CR indicators into the assessment systems of financial institutions by establishing green financial regulatory frameworks, effectively guiding micro-subjects' behaviours to align with macro climate goals. Such policy coordination constructs a self-reinforcing dynamic cycle of "policy guidance, market response, and target iteration" through market mechanisms, substantially enhancing climate policy implementation effectiveness. This institutional architecture concurrently provides dual structural support for achieving the Paris Agreement's temperature control targets and developing climate-adaptive societies. On the other hand, the policy coordination role of GF in CR is further manifested in the synergy of policy instruments. Specifically, green bonds and green credit complement each other through maturity mismatch optimisation and sectoral targeting. The former supports large-scale resilience projects such as coastal protection zones and climate monitoring systems by leveraging their long-term and low-cost characteristics (Xu & Yang, 2025), while the latter directs offers preferential low interest rates and higher credit quotas to the green ecological industry, addressing short-term CR financing gaps (Fan & Gao, 2024). This combination of GF instruments fulfils cross-cycle financing needs for CR development, contributing to climate change mitigation.

3.3 Technological Innovation

Finance development provides a solid foundation for green technological innovations to achieve through financial market extension and reduced capital costs, critical objectives required for the operational deployment of green technology (Jarijari et al., 2025). GF is the linchpin of financial development, lowering the cost of finance for environmentally friendly projects to facilitate sustainable investment while enhancing the appeal of green technology innovations (Taghizadeh-Hesary & Yoshino, 2020). Moreover, GF fosters green innovation abilities by spurring corporate research and development with financial support mechanisms. More specifically, its environmental regulation role channels capital investment to environmentally friendly activities, such as efforts to strengthen incentive mechanisms for corporate investment (Lei et al., 2025). Therefore, GF can encourage green technological innovation by alleviating financing constraints and enhancing the scale of research and development investments. Indeed, green technological innovation reduces environmental and ecological degradation (Musah et al., 2022). Green technological innovations are renewable energy systems, energy-efficient appliances, sustainable agricultural methods, and pollution control technology (Acheampong et al., 2019; Qudrat-Ullah & Nevo, 2021). On the one hand, green technology innovations have considerably minimised fossil fuel dependence by creating renewable energy systems, such as photovoltaic, wind power generation, and hydroelectric power generation (Jarijari et al., 2025). Empirical data prove that using solar power in sub-Saharan African nations such as Kenya and Ghana has successfully mitigated emissions by approximately 400 million kilograms of carbon dioxide (CO_2). Green activities have also enhanced atmospheric conditions and delivered socioeconomic advantages to host communities (Zafar et al., 2021). On the other hand, agricultural green technology innovations, such as precision agriculture and organic cultivation methods, realise the maximum efficacy in resource application through mitigating synthetic agrochemical usage while at the same time increasing mechanisms of soil conservation and hydrological regulation (Mensah et al., 2021). This way, GF acts as a catalytic driver towards encouraging eco-innovative technological improvement, eventually increasing the resilience of climatic systems through a more resilient adaptive governance framework.

4 Empirical Analysis: Examining the Effect of Green Finance on Climate Resilience in Vulnerable Regions

4.1 Data Description

To confirm the effect of green finance on climate resilience in vulnerable regions, we conducted an empirical examination in vulnerable regions. Our sample comprises 19 LDCs countries that, based on the World Bank's classification, were categorized in the low-income group from 2005 to 2023 (for a complete list of countries, see Table A1 in the appendix).

We select three indicators—the percentage of the total population with electricity access (AET), the percentage of the total population using clean fuels and cooking technologies (AFTC), and the percentage of the urban population with electricity access (AEU)—to proxy the CR (Barkat et al., 2023). AET directly reflects a region's electricity popularization degree and is key for assessing energy accessibility. A higher AET implies better-developed power infrastructure, which enhances society's climate change-coping capacity. AFTC focuses on a clean energy structure. Widespread clean energy use reduces greenhouse gas emissions and indoor air pollution, which boosts climate resilience. AEU highlights urban electricity coverage. As economic and population hubs, cities' stable power supply is critical for extreme-weather resilience, ensuring normal traffic, medical, and communication system operation. Overall, these three indicators provide a macro-level view of energy supply popularisation and cleanliness while emphasising urban energy-use features. This comprehensive approach effectively measures regional climate resilience. An increase in the level of these variables implies a rise in CR. For the explanatory variables, we focus on GF, which is measured as CPIA policy and institutions for environmental sustainability rating (ESR). Other controlled variables are the Gini coefficient (GINI), the urban population (URB), and financial development level (FD), which is proxied by domestic credit to the private sector as a share of GDP. The variable list, variable definitions, and data sources are shown in Table A2 in the Appendix.

Table 1 shows the descriptive statistics of the variables. AFTC, AET, and AEU, which represent the population's access to clean energy and electricity, have means

Table 1 Descriptive statistics

Variable	Mean	Std. dev.	Min	Max
AFTC	7.551	15.261	0.1	65.6
AET	27.056	18.223	2.7	83.6
AEU	61.329	21.224	11.9	100
ESR	3.172	0.590	2	4.5
GINI	39.967	5.582	29.6	56.2
URB	30.753	11.888	9.375	64.472
FD	13.001	7.901	1.201	48.534

of 7.551, 27.056, and 61.329, respectively. ESR, the key independent variable measuring the development of GF, averages 3.172. The control variables, GINI, URB, and FD, have distinct means and standard deviations. To ensure data integrity, we addressed missing values using an interpolation technique. In addition, each variable's minimum and maximum values differ greatly, indicating a wide range of variation across the variables. These statistics provide essential insights into the variables' central tendencies, dispersion, and ranges, offering a foundation for further analysis.

4.2 Model and Methodology

We employ panel data models with fixed and random effects as our primary estimation methods to examine how GF impacts CR. The specific model formulation is presented as follows:

$$Y_{jt} = \beta_0 + \beta_1 ESR_{jt} + \beta_2 C_{jt} + \mu_j + \varepsilon_{jt} \tag{1}$$

where Y_{jt} represents the three measures of CR defined earlier, j denotes the country (j = 1, …, 109) and t denotes the time (t = 2005, …, 2023). ESR_{jt} are the CPIA policy and institutions for environmental sustainability rating by country j in year t. C_{jt} represents the set of control variables. μ_j is a vector of country dummies which captures unobserved country-specific fixed effects and ε_{jt} refers to the idiosyncratic error term.

4.3 Results and Discussion

Table 2 employs panel fixed and random effects models as baseline estimates, showcasing the impact of the CPIA policy and institutions for environmental sustainability rating on access to clean fuels and cooking technologies (AFTC). The Hausman test results in Table 2 indicate that the fixed effects estimator is suitable for our model (over random effects). Hence, the subsequent discussion focuses solely on the results of the fixed effects.

Table 2 shows that the coefficient of the CPIA policy and institutions for environmental sustainability rating is positive and statistically significant in all estimations. This indicates that a 1 unit increase in CPIA policy and institutions for environmental sustainability rating is associated with a 1.398 percentage point increase in access to clean fuels and cooking technologies. Thus, GF plays a role in reducing CR in developing and low-income countries. A plausible explanation for this outcome is that a higher CPIA environmental sustainability rating often implies that the region or country has a more robust policy framework and more proactive initiatives in place for environmental protection and sustainable development. Governments

Table 2 Access to clean fuels and technologies for cooking (% of population)

Variable	Fixed effects	Fixed effects	Fixed effects	Fixed effects	Random effects
	(1)	(2)	(3)	(4)	(5)
ESR	1.985*** (3.46)	1.476*** (2.52)	1.153** (1.96)	1.398*** (2.36)	1.299*** (2.19)
GINI		−0.245*** (−3.38)	−0.208*** (−2.86)	−0.202*** (−2.79)	−0.217*** (−3.01)
URB			0.250*** (2.84)	0.387*** (3.75)	0.354*** (3.67)
FD				−0.135*** (−2.49)	−0.123*** (−2.32)
Constant	1.427 (0.78)	12.794*** (3.35)	4.684 (0.99)	1.183 (0.24)	2.972 (0.52)
Observations	361	361	361	361	361
R-squared	0.934	0.932	0.933	0.936	0.921
Country FE	Yes	Yes	Yes	Yes	No
Hausman Test					10.74
Prob.					0.030

Note: Standard errors in parentheses, *** $p < 0.01$, ** $p < 0.05$

will likely introduce clear energy transition policies, clean fuel subsidy programs, and incentives encouraging sustainable practices. This policy direction can redirect resources toward clean fuels and cooking technologies, enhancing people's access to them. Therefore, a positive correlation exists between environmental sustainability rating and access to clean fuels and cooking technologies. Regarding the control variables, the Gini index and financial development coefficients are negative and statistically significant in the fixed effects model. This means these factors negatively impact access to clean fuels and cooking technologies. Conversely, the urban population proportion coefficient is positive and significant at the 1% level, indicating that a larger urban population enhances access to clean fuels and cooking technologies, thereby strengthening urban climate resilience.

Table 3 reports the results using the second proxy for CR, namely, access to electricity. From the results in Table 3, environmental sustainability has a positive and significant impact on electricity access in all estimations. Specifically, a one-unit increase in environmental sustainability leads to a 2.674 percentage point increase in electricity access. This phenomenon can be explained in multiple ways. From the policy and regulation perspective, higher ratings imply clear and proactive environmental policies and strictly enforced environmental regulations. This drives the energy structure towards a green and low-carbon transition, creating space for green energy electricity development. Regarding resource allocation, the rating upgrade guides funds into the green energy sector. This accelerates grid upgrades and improves power infrastructure. It sends positive market signals regarding market development and business investment, attracting more enterprises and investors to the green energy electricity sector. These factors interact and jointly boost the electricity supply. Therefore, it is evident that FG has a positive impact on

Table 3 Access to electricity (% of population)

Variable	Fixed effects	Fixed effects	Fixed effects	Fixed effects	Random effects
	(1)	(2)	(3)	(4)	(5)
ESR	7.609*** (6.13)	5.877*** (4.76)	2.540*** (3.04)	2.674*** (3.16)	2.979*** (3.38)
GINI		−0.831*** (−5.44)	−0.456*** (−4.43)	−0.453*** (−4.39)	−0.528*** (−4.96)
URB			2.585*** (20.76)	2.660*** (18.06)	2.204*** (16.38)
FD				−0.074 (−0.96)	0.063 (0.82)
Constant	2.921 (0.74)	41.550*** (5.16)	−42.312*** (−6.31)	−44.230*** (−6.32)	−29.957*** (−3.90)
Observations	361	361	361	361	361
R-squared	0.775	0.769	0.952	0.955	0.850
Country FE	Yes	Yes	Yes	Yes	No
Hausman Test					57.25
Prob.					0.000

Note: Standard errors in parentheses, *** $p < 0.01$

CR. Regarding the control variables, the urban population coefficient has a significantly positive effect on electricity supply in the fixed effects model. This suggests a positive link between urban population growth and increased electricity supply. The increasing electricity demand can explain this phenomenon as the urban population grows. Conversely, the Gini coefficient has a significantly negative impact on electricity supply. This can be explained by the fact that a higher Gini coefficient implies significant income inequality. Low-income groups lack sufficient funds to purchase electrical equipment and cover the high upfront costs of electricity access. Considering the uneven distribution of demand and limited market expansion in certain areas, power companies have reduced their willingness to invest in construction and upgrading electricity infrastructure.

Subsequently, we investigate the relationship between environmental sustainability and electricity supply in urban areas. The results in Table 4 show that environmental sustainability ratings positively affect urban electricity supply. Specifically, a one-unit increase in GF can lead to a 5.642 percentage point increase in CR. It is interesting to compare the results in Tables 3 and 4. The coefficient of environmental sustainability rating's impact on urban electricity supply is larger, nearly three times that of the total electricity supply. This clearly shows that environmental sustainability has a greater influence on urban areas than rural regions. Therefore, urban areas seem to benefit more from environmental sustainability, as their electricity supply usually surpasses that of rural regions.

The analysis across Tables 2, 3, and 4 consistently demonstrates environmental sustainability ratings' positive and significant impact on energy access and climate resilience in LDCs countries. Specifically, a higher CPIA environmental sustainability rating is associated with increased access to clean fuels, cooking

Table 4 Access to electricity, urban areas (% of urban population)

Variable	Fixed effects	Fixed effects	Fixed effects	Fixed effects	Random effects
	(1)	(2)	(3)	(4)	(5)
ESR	11.781*** (8.78)	9.401*** (7.24)	6.315*** (6.33)	5.642*** (5.71)	5.992*** (5.90)
GINI		−1.142*** (−7.11)	−0.795*** (−6.46)	−0.813*** (−6.75)	−0.907*** (−7.42)
URB			2.391*** (16.07)	2.014*** (11.71)	1.534*** (10.19)
FD				0.371*** (4.11)	0.518*** (5.87)
Constant	23.943*** (5.58)	77.010*** (9.09)	−0.565 (−0.07)	9.064 (1.11)	24.531*** (2.85)
Observations	361	361	361	361	361
R-squared	0.775	0.753	0.949	0.934	0.818
Country FE	Yes	Yes	Yes	Yes	No
Hausman Test					31.66
Prob.					0.000

Note: Standard errors in parentheses, *** $p < 0.01$

technologies, and improved electricity access. This relationship is interpreted as evidence that robust environmental policies and sustainable development initiatives drive resource allocation toward clean energy solutions, enhancing energy infrastructure and market participation. Control variables such as the Gini index and financial development show negative impacts on energy access, highlighting the barriers posed by income inequality and underdeveloped financial systems. Conversely, urban population growth positively influences electricity supply, reflecting higher demand and infrastructure development in urban areas. Notably, environmental sustainability substantially impacts urban electricity supply more than rural areas, suggesting that urban regions benefit more significantly from sustainability efforts. The findings underscore the critical role of environmental sustainability in advancing energy access, particularly in urban contexts. Overall, the results presented in Tables 2, 3, and 4 strongly support the important role of GF in enhancing CR.

5 Conclusion, Implications and Outlook

5.1 Conclusion of the Chapter

A review of the preceding analysis reveals that numerous obstacles and challenges persist in developing green finance in vulnerable regions. For instance, climate-resilient projects in these regions often exhibit characteristics of high costs and high risks, which can hinder their access to the green finance market. Meanwhile, vulnerable regions commonly suffer from weak technological capabilities, inadequate reservoirs of professional knowledge, and imperfect governance and regulatory frameworks. These problems lead to technological bottlenecks and management dilemmas for climate-resilient projects during the planning, construction, and post-maintenance phases. Beyond the constraints imposed by the regions' conditions, social equity and interest balance issues cannot be overlooked. These include the imbalance between cost-sharing and benefit distribution and the complex trade-offs between prioritising the enhancement of resilience to extreme weather events and safeguarding the sustainability of the development environment.

Green finance can be pivotal in building climate resilience across multiple dimensions, including capital allocation, policy coordination, and technological innovation. At the level of capital allocation, green finance leverages market mechanisms, channels funds into climate-resilient sectors, and enhances the financial ability of climate-resilient projects through risk-sharing mechanisms. Regarding policy coordination, green finance and climate resilience construction exhibit synergistic effects in policy objectives and policy instruments. Policy objectives guide relevant technological innovations and infrastructure investments, while policy instruments such as green bonds meet the financing needs for developing climate-resilient projects. At the technological innovation level, green finance incentivises enterprises to engage in green technology research and development by reducing financing costs, promoting green transformation and enhancing climate resilience. This multidimensional synergy provides crucial support for developing climate-resilient projects in vulnerable regions.

We also confirm the effect of green finance on climate resilience by investigating the relationship between these two indicators in vulnerable regions. The empirical results demonstrate green finance's positive and significant impact on climate resilience in LDCs countries. Specifically, a higher CPIA environmental sustainability rating is associated with increased access to clean fuels, cooking technologies, and improved electricity access. This relationship is interpreted as evidence that robust environmental policies and sustainable development initiatives drive resource allocation toward clean energy solutions, enhancing energy infrastructure and market participation.

5.2 Implications and Strategies for Enhancing Green Finance in Vulnerable Regions

Government policy plays an important role in driving green finance in vulnerable regions. They guide green financial resources to vulnerable regions through strategic planning, standard setting, and innovative tools. Firstly, from the strategic planning aspect, governments need to launch special relief policies in the vulnerable regions, such as preferential tax relief, tax-exempt interest on green bonds, and subsidies on finance. These measures can reduce the financing cost of climate-resilient projects and facilitate their development (Xu et al., 2022). In addition, governments have to encourage improved regulatory regimes and standard setting. Governments have to develop region-based standards for green project classification so that one would know what economic activities can go hand-in-hand with climate resilience goals. Aside from this, mandatory environmental information disclosure has to be introduced. Financial institutions and companies must disclose their exposure to climate risk in finance, investment, and business operations. This promotes market transparency and the prevention of greenwashing and fraud behaviours. Governments can also use new financial products to improve the financing of climate-resilient projects. Governments can use fiscal tools to mobilise capital, such as investment and guarantee funds, to reduce investors' risks. Furthermore, they can issue market-based instruments like climate resilient bonds and parametric insurance (e.g., crop insurance with automatic payment triggered by rainfall) to mitigate the impact of extreme weather (Adegboyo et al., 2025).

International collaboration enables vulnerable regions to overcome resource constraints. It establishes resource reallocation stages and capacity-building forums through multilateral frameworks, thus addressing the loopholes of vulnerable regions in finance and technology. From a fund mobilisation point of view, international collaboration channels capital into climate-resilient initiatives through diverse funding instruments and risk-sharing facilities. For example, the Green Climate Fund (GCF) invests financial resources in adaptation to climate risk and climate risk mitigation, intending to support the development of green technology in vulnerable areas (Markandya et al., 2015). Debt-for-nature swaps are also innovative, repurposing debt repayment pressure as a stimulus for climate action. For example, Ecuador has secured debt relief to protect the Galapagos Islands' biodiversity. Multilateral development banks, as major climate finance providers (Xie et al., 2023), combine long-term low-interest loans, grants, and technical cooperation. These favourable financing terms lower the initial investment threshold for climate projects in vulnerable regions. Regarding capacity building, international cooperation accelerates vulnerable regions' technological and managerial progress through technology transfer agreements and knowledge sharing networks. For example, triangular cooperation mechanisms, which involve technical assistance from developed countries, drive technology diffusion via tripartite collaboration. International research alliances, such as the West African Science Service Centre on Climate Change and Adapted Land Use (WASCAL), employ the regional climate modelling centres to create risk assessment tools for vulnerable regions. These initiatives overcome local research resource limitations through data sharing, systematically enhancing climate prediction and disaster response capabilities.

Capacity building enhances stakeholders' understanding and practical abilities in green finance and climate-resilient projects. It includes raising awareness and developing professional skills. Firstly, education is vital for filling the knowledge gap in climate action. Schools should develop local curriculum materials linked to regional climate risks in primary and secondary education. Through participatory teaching like sandbox simulations, the goal is for adolescents to connect climate adaptation with incremental improvements from an early cognitive stage. Higher education institutions can offer green finance courses and conduct interdisciplinary research based on local characteristics. This approach cultivates professional talents and generates local solutions. Secondly, professional skills development should be guided by the needs of vulnerable regions. For communities susceptible to extreme weather, the focus should be on training practical skills for constructing and maintaining climate-resilient projects. For local governments, simplified climate budgeting tools and community-led disaster early warning response procedures should be developed. Entrepreneurs need to build capacity for carbon sink projects based on local natural resources. Moreover, green financial products should be utilised to reduce barriers for entrepreneurs in technology applications and market entry.

Technological support focuses on three key aspects: introducing applicable technologies, promoting local innovation, and establishing a technology-finance collaboration mechanism. First, as previously mentioned, green technology transfer from developed to vulnerable regions can be facilitated via international collaboration and the establishment of low-cost technology licensing pools. This enables vulnerable regions to access climate-friendly technology patents under favourable terms. In addition, vulnerable regions should focus on technological innovation and the application of digital tools. For example, climate technology labs can be set up in disaster-prone areas to develop targeted solutions based on local needs. Digital tools like blockchain and big data can also be utilised to track the flow of green funds and optimise resource allocation. This leads to the design of more innovative and more credible green financial products. Lastly, to achieve technology finance integration, technology-driven funds can be set up to support the commercialisation of clean technologies. Guarantee mechanisms can also be introduced to reduce financial institutions' concerns regarding adopting new technology. This facilitates the smoother penetration of green finance into vulnerable regions.

The private sector is key in implementing green finance strategies in vulnerable regions through financial instruments like bonds and insurance (Motlagh et al., 2024). Firstly, it offers sustainable funding for projects. Private financial entities, including banks, insurance companies, and investment funds, utilise green bonds and climate finance to support green projects, effectively filling the public funding gap in these initiatives (Yan & Haroon, 2023). Secondly, the private sector manages and shares risks through innovative products. Index insurance and climate derivatives are tools that can mitigate the business risks caused by extreme climate events. This helps alleviate investors' concerns and attracts more capital to green projects. Lastly, the private sector contributes to the development of market mechanisms. Taking voluntary carbon markets as an example, carbon trading activities convert carbon emission rights into tradable economic assets. This generates revenue streams for green projects such as forest protection and renewable energy. At the same time, market price signals guide the optimal allocation of resources toward low-carbon sectors. As a result, economic systems' greening and operational efficiency in vulnerable regions are enhanced (Deknatel & van der Loos, 2025).

In strengthening green finance strategies for vulnerable regions, successful public-private partnerships integrate policy support from the public sector with funding and technological resources from the private sector, effectively advancing green finance and climate-resilient projects. The successful models of this cooperation mainly include the following types: Firstly, the mixed financing model, where public funds or international aid funds bear the initial losses and subordinated debts, while private capital holds the priority debt, thereby reducing the investment risks of the private sector and stimulating private capital to participate in high-risk green projects. Secondly, climate insurance and risk-sharing mechanisms enhance the efficiency of claims settlement through parametric insurance and expand the underwriting capacity of private insurance companies through public reinsurance, thereby increasing insurance coverage and reducing the impact of extreme climate events on economic activities in vulnerable regions. Thirdly, the technology sharing and capacity building model combines technology and institutions. Through technology transfer and training, private enterprises help vulnerable regions enhance their ability to apply green technologies. At the same time, the public sector provides policy support and infrastructure, forming a sustainable knowledge spillover effect. Fourthly, the green supply chain and market linkage model internalises externalities through procurement agreements, premium acquisitions, etc. For example, Nestlé, in collaboration with the Vietnamese government, promoted drought-resistant coffee varieties and provided technical training, achieving a 30% increase in production under the support of public certification standards, demonstrating the practical value of complementarity between market mechanisms and public regulations.

5.3 Outlook

Looking ahead, green finance demonstrates significant potential for further innovation. In terms of products and services, the innovative development of financial instruments such as climate adaptation bonds and climate insurance can restructure the risk-return profile, thereby channelling capital towards climate-resilient sectors. On the technological front, digital tools like big data and artificial intelligence can leverage machine learning to optimise investment logic, reshape the analytical framework of green finance, and enhance operational efficiency. Blockchain technology, conversely, can ensure the alignment of capital flows with climate objectives, thereby reducing regulatory compliance costs.

In addition, given the global externalities of climate change, developing green finance necessitates deepening international cooperation (Ferrari Minesso & Pagliari, 2023). Concerning standard harmonisation, the continuous growth of cross-border capital flows has placed higher demands for greater international mutual recognition on information disclosure and taxonomy standards in green finance. Future international cooperation should establish a unified standard framework to mitigate information asymmetry and guide the effective allocation of cross-border capital towards climate-resilient sectors. Regarding market integration, international cooperation is a crucial link connecting regional green finance markets and constructing a global climate investment and financing network. To this end, it is essential to promote policy coordination in cross-border payments and tax incentives, break down regional market barriers, and unlock the liquidity of green financial products.

Acknowledgments This work was supported by a grant from the Romanian Ministry of Research, Innovation and Digitalization, the project with the title "*Economics and Policy Options for Climate Change Risk and Global Environmental Governance*" (CF 193/28.11.2022, Funding Contract no. 760078/23.05.2023), within Romania's National Recovery and Rcsilience Plan (PNRR)—Pillar III, Component C9, Investment I8 (PNRR/2022/C9/MCID/I8)—Development of a program to attract highly specialised human resources from abroad in research, development, and innovation activities.

Competing Interests The authors have no conflicts of interest to declare relevant to this chapter's content.

Appendix

Table A1 List of countries

Low-income countries (19)				
Burkina Faso	Burundi	Central African	Republic	Chad
Congo, Dem. Rep.	Ethiopia	Gambia, The	Guinea	Madagascar
Malawi	Mali	Mozambique	Niger	Rwanda
Sierra Leone	Sudan	Yemen, Rep.	Uganda	

Table A2 Variables definition

Variables	Code	Source
Access to clean fuels and technologies for cooking (% of population)	AFTC	World Development Indicators (WDI)
Access to electricity (% of population)	AET	World Development Indicators (WDI)
Access to electricity, urban (% of urban population)	AEU	World Development Indicators (WDI)
CPIA policy and institutions for environmental sustainability rating (1 = low to 6 = high)	ESR	World Development Indicators (WDI)
Gini index	GINI	World Development Indicators (WDI)
Energy intensity level of primary energy (MJ/$2017 PPP GDP)	EI	World Development Indicators (WDI)
Urban population (% of total population)	URB	World Development Indicators (WDI)
Domestic credit to private sector (% of GDP)	FD	World Development Indicators (WDI)

References

Acheampong, A. O., Adams, S., & Boateng, E. (2019). Do globalization and renewable energy contribute to carbon emissions mitigation in Sub-Saharan Africa? *Science of the Total Environment, 677*, 436–446.

Adegboyo, O. S., Yang, L., Ndayambaje, J. d. D., & Gakuru, E. (2025). The impact of climate change on conflict: Does financial development matter? *Economic Analysis and Policy, 86*, 1261–1284.

Barkat, K., Alsamara, M., & Mimouni, K. (2023). Can remittances alleviate energy poverty in developing countries? New evidence from panel data. *Energy Economics, 119*, 106527.

Chendo, M. A. C. (1994). Towards sustainable renewable energy technology in Africa. *Energy Conversion and Management, 35*(12), 1173–1190.

Das, S. (2024). Determinants of livelihood vulnerability to climate change among the agrarian community: Empirical evidence from the coastal region of Indian Sundarban biosphere reserve. *Regional Studies in Marine Science, 78*, 103777.

Das, S., Sharma, K. K., Majumder, S., Das, D., & Chowdhury, I. R. (2024). Spatio-temporal variation and relationship between agricultural efficiency and irrigation intensity in a semi-arid region of India. *Regional Sustainability, 5*(2), 100144.

Deknatel, N., & van der Loos, A. (2025). The intangible technological innovation system: The role and influence of voluntary and compliance carbon markets on carbon dioxide removal in the European Union. *Energy Research & Social Science, 119*, 103851.

Diallo, Y., & Tapsoba, R. (2022). *Climate shocks and domestic conflicts in Africa*. International Monetary Fund.

Fan, Q., & Gao, W. (2024). Climate risk and financial stability: The mediating effect of green credit. *Finance Research Letters, 65*, 105558.

Fei, W. L. (2023). Can green finance enhance the resilience of agricultural development: Mediation effect based on total factor productivity. *Operations Research and Fuzziology, 13*(06), 6893–6907.

Ferrari Minesso, M., & Pagliari, M. S. (2023). No country is an Island. International cooperation and climate change. *Journal of International Economics, 145*, 103816.

Gabriel, C.-A., Kirkwood, J., Walton, S., & Rose, E. L. (2016). How do developing country constraints affect renewable energy entrepreneurs? *Energy for Sustainable Development, 35*, 52–66.

Hu, G., Wang, X., & Wang, Y. (2021). Can the green credit policy stimulate green innovation in heavily polluting enterprises? Evidence from a quasi-natural experiment in China. *Energy Economics, 98*, 105134.

Irfan, M., Razzaq, A., Sharif, A., & Yang, X. (2022). Influence mechanism between green finance and green innovation: Exploring regional policy intervention effects in China. *Technological Forecasting and Social Change, 182*, 121882.

Jarijari, T., Deka, A., & Cavusoglu, B. (2025). The synergistic role of natural resources, green finance and green technological innovations in promoting environmental sustainability in sub-Saharan Africa. *Resources Policy, 105*, 105598.

Jayles, B., Cheong, S. A., & Herrmann, H. J. (2022). Interactions between communities improve the resilience of multicultural societies. *Physica A: Statistical Mechanics and its Applications, 607*, 128164.

Johnson, M., Ballantyne, A., Graham, J., Holden, Z., Hoylman, Z., Jensco, K., Ketchum, D., Kimball, J., & Mitchell, J. (2025). An ecosystem resilience index that integrates measures of vegetation function, structure, and composition. *Ecological Indicators, 171*, 113076.

Lee, C. C., Wang, F., Lou, R., & Wang, K. (2023). How does green finance drive the decarbonization of the economy? Empirical evidence from China. *Renewable Energy, 204*, 671–684.

Lei, H., Gao, R., Ning, C., & Sun, G. (2025). Green finance and corporate green innovation. *Finance Research Letters, 72*, 106577.

Li, Y., & Deng, Z. (2025). Can environmental regulation inhibit enterprises' greenwashing? Evidence from implementing the New Environment Protection Law. *Finance Research Letters, 81*, 107476.

Liu, X., Cui, J., Wu, Y., Yue, X., Shen, J., & Guo, P. (2024). The nexus between green finance and carbon emissions: Evidence from maturity mismatch in China. *Sustainability, 16*(10), 4319.

Manta, O., Folcut, O., Militaru, I., & Edu, T. (2024). Digitization and green finance, tools for sustainable tourism. In *Contemporary marketing and consumer behaviour in sustainable tourism* (pp. 9–27). Routledge.

Markandya, A., Antimiani, A., Costantini, V., Martini, C., Palma, A., & Tommasino, M. C. (2015). Analyzing trade-offs in international climate policy options: The case of the green climate fund. *World Development, 74*, 93–107.

Mensah, J. T., Amoah, A., & Asare, P. (2021). Environmental sustainability and economic development in Sub-Saharan Africa: The role of institutional quality. *Environmental Science and Pollution Research, 28*(6), 6699–6713.

Meo, M. S., & Abd Karim, M. Z. (2022). The role of green finance in reducing CO_2 emissions: An empirical analysis. *Borsa Istanbul Review, 22*(1), 169–178.

Meregillano, E. O., & Delina, L. L. (2023). Stronger typhoons, weaker electricity systems? A review of the impacts of extreme weather events on coastal communities and strategies for electric system resilience. *The Electricity Journal, 36*(9), 107339.

Motlagh, F., Hamideh, S., Gallagher, M., Yan, G., & van de Lindt, J. W. (2024). Bonds for disaster resilience: A review of literature and practice. *International Journal of Disaster Risk Reduction, 104*, 104318.

Musah, M., Owusu-Akomeah, M., Nyeadi, J. D., Alfred, M., & Mensah, I. A. (2022). RETRACTED ARTICLE: Financial development and environmental sustainability in West Africa: Evidence from heterogeneous and cross-sectionally correlated models. *Environmental Science and Pollution Research, 29*(8), 12313–12335.

Nazari, R., Karimi, M., Nikoo, M. R., Khoshkonesh, A., & Museru, M. L. (2025). Can small towns survive climate change? Assessing economic resilience and vulnerability amid major storms. *Journal of Cleaner Production, 498*, 145158.

Qudrat-Ullah, H., & Nevo, C. M. (2021). The impact of renewable energy consumption and environmental sustainability on economic growth in Africa. *Energy Reports, 7*, 3877–3886.

Rasheed, N. J., Al-Khafaji, M. S., & Alwan, I. A. (2023). Investigation of rivers planform change in a semi-arid region of high vulnerability to climate change: A case study of Tigris River and its tributaries in Iraq. *Regional Studies in Marine Science, 68*, 103233.

Saeed, F. H., Al-Khafaji, M. S., & Al-Faraj, F. A. (2022). Spatiotemporal hydroclimatic characteristics of arid and semi-arid river basin under climate change: A case study of Iraq. *Arabian Journal of Geosciences, 15*(14), 1260.

Seidu, S., Chan, D. W., & Debrah, C. (2025). A novel climate resilience implementation model for the construction industry: An international perspective. *Building and Environment, 274*, 112793.

Taghizadeh-Hesary, F., & Yoshino, N. (2020). Green finance and green technological innovation in emerging markets: Evidence from Sub-Saharan Africa. *Journal of Financial Stability, 46*, 100720.

United Nations Development Programme (UNDP). (2015). *Responding to climate change in least developed countries*. UNDP.

Wilkinson, E., Panwar, V., Pettinotti, L., Cao, Y., Corbett, J., & Bouhia, R. (2023). *A fair share of resilience finance for Small Island developing states: Closing the gap between vulnerability and allocation*. ODI Report.

Xie, L., Scholtens, B., & Homroy, S. (2023). Rebalancing climate finance: Analysing multilateral development banks' allocation practices. *Energy Research & Social Science, 101*, 103127.

Xu, X., & Yang, L. (2025). Green bond financing and regional ecological resilience: The role of financial agglomeration mechanisms. *Finance Research Letters, 80*, 107355.

Xu, N., Kasimov, I., & Wang, Y. (2022). Unlocking private investment as a new determinant of green finance for renewable development in China. *Renewable Energy, 198*, 1121–1130.

Yan, J., & Haroon, M. (2023). Financing efficiency in natural resource markets mobilizing private and public capital for a green recovery. *Resources Policy, 85*, 103841.

Yi, R., Chen, Y., & Chen, A. (2025). Evaluating urban climate resilience in the Yangtze River Delta urban agglomeration: A novel method integrating the DPSIR model and Sustainable Development Goals. *Journal of Environmental Management, 376*, 124517.

Zafar, A., Wang, L., & Hussain, S. (2021). Innovation and research in green technologies: The role of government funding and policy in Sub-Saharan Africa. *Journal of Cleaner Production, 292*, 125928.

Sustainable Agriculture Financing: Green Finance for Climate-Smart Agriculture

Bădîrcea Roxana Maria, Manta Alina Georgiana, Gherţescu Claudia, and Olaru Mihail

1 Introduction

In the twenty-first century, sustainable agriculture has emerged as a cornerstone of global development strategies, addressing the complex interplay between food security, environmental sustainability, and climate resilience. The rising pressures of climate change, coupled with land degradation, water scarcity, and biodiversity loss, have underscored the need to transform agricultural systems toward more sustainable and climate-smart models (FAO, 2021; Liu et al., 2023). In this context, financing mechanisms play a critical role in facilitating this transition, ensuring that agricultural practices not only enhance productivity but also contribute to the broader goals of environmental conservation and climate change mitigation (Huang & Wang, 2014; Howden et al., 2007).

Sustainable agriculture financing refers to the mobilisation and allocation of financial resources in support of farming practices that maintain the economic viability of farms while ensuring environmental sustainability and social equity (Fatemi & Fooladi, 2013). With agriculture accounting for approximately 24% of global greenhouse gas (GHG) emissions when including land use change (IPCC, 2019), the sector has significant potential and responsibility to contribute to global climate goals (Poenaru et al., 2025). Climate-smart agriculture (CSA), a concept developed by the Food and Agriculture Organisation (FAO), is designed to address these

B. R. Maria (✉) · M. A. Georgiana
Faculty of Economics and Business Administration, University of Craiova, Craiova, Romania
e-mail: roxana.badircea@edu.ucv.ro; alina.manta@edu.ucv.ro

G. Claudia · O. Mihail
Doctoral School in Economic Sciences "Eugeniu Carada", Faculty of Economics and Business Administration, University of Craiova, Craiova, Romania
e-mail: ghertescu.claudia.x6e@student.ucv.ro; mihai.olaru@noarimgroup.ro

C.-W. Su et al. (eds.), *Green Finance and the Challenges of Climate Change*, Sustainable Finance, https://doi.org/10.1007/978-3-032-07293-1_17

challenges by promoting agricultural practices that increase productivity, enhance resilience, and reduce emissions (Lipper et al., 2014; Desalegn et al., 2024).

However, the widespread adoption of CSA practices often requires substantial upfront investments in technology, knowledge transfer, infrastructure, and capacity building costs that are frequently prohibitive for smallholder farmers and marginalised agricultural communities (Zhang et al., 2025; Manta et al., 2025a). Thus, green finance, defined as financing investments that provide environmental benefits in the broader context of environmentally sustainable development, becomes essential in bridging this financial gap (UNEP, 2016; Manta et al., 2025b). Green finance mechanisms such as green bonds, concessional loans, sustainability-linked loans, and blended finance have been increasingly deployed to fund CSA initiatives globally (OECD, 2022). These instruments support environmental objectives and offer opportunities for financial institutions to diversify their portfolios and align with emerging Environmental, Social, and Governance (ESG) criteria (Gherțescu et al., 2024).

The concept of green finance in agriculture is intrinsically tied to broader frameworks such as the United Nations Sustainable Development Goals (SDGs), particularly Goal 2 (Zero Hunger), Goal 13 (Climate Action), and Goal 15 (Life on Land). To achieve these goals, financing systems must evolve to incentivise sustainable land use, promote renewable energy in farming, reduce chemical inputs, and invest in ecosystem services (World Bank, 2020; Doran et al., 2025). However, existing financial systems are often ill-equipped to address the specific needs of sustainable agriculture (Rahman & Hossain, 2024; Robinson, 2024). Conventional banking models may not adequately consider the long-term environmental returns or the systemic risks of climate change, leading to underinvestment in green agricultural projects (GCF, 2019; Gong & Zhang, 2025).

A significant barrier to scaling sustainable agriculture financing is the perceived high risk and low return of agricultural investments, especially in developing countries (Aznar-Sánchez et al., 2019; Sarker et al., 2019). Smallholder farmers, who constitute a significant portion of the agricultural workforce in low-income regions, frequently face limited access to credit due to a lack of collateral, weak financial infrastructure, and inadequate risk mitigation tools (Baker et al., 2018; IFAD, 2021). Green finance can mitigate these challenges by introducing innovative financial instruments, de-risking mechanisms, and public-private partnerships that redistribute risk and attract private capital (Meghişan-Toma et al., 2022). For example, blended finance approaches, where concessional public funds are used to leverage private sector investment, have shown promise in financing climate-resilient agricultural projects (Convergence, 2020).

Moreover, digital technologies and financial innovation offer new avenues to scale green finance in agriculture (Wadumestrige Dona et al., 2021). Digital platforms, blockchain, satellite data, and AI-driven credit scoring are being utilised to enhance green financing schemes' transparency, efficiency, and inclusivity (Manta et al., 2024b). These technologies can reduce transaction costs, improve creditworthiness assessments, and facilitate monitoring environmental outcomes (WEF, 2021; Lee et al., 2024). For instance, digital microfinance platforms have enabled rural farmers to access climate-smart inputs and insurance products tailored to weather variability and crop cycles (Lagarde, 2021; Kling et al., 2021).

In recent years, several countries have implemented successful models of sustainable agriculture financing. For example, in Kenya, the government's collaboration with international donors led to establishing the Kenya Climate-Smart Agriculture Project (KCSAP), which provided financial and technical support to thousands of farmers in adopting resilient agricultural practices (World Bank, 2020). Similarly, green bonds in India have been used to finance solar-powered irrigation and agroforestry projects that align with agricultural productivity and climate objectives (Climate Bonds Initiative, 2021). These examples demonstrate the potential of well-designed financing frameworks to catalyse large-scale agricultural transformation (Yaron, 1994; Manta et al., 2023).

Despite these advances, the global financing gap for sustainable agriculture remains substantial. Estimates suggest that an additional USD 300–350 billion annually is required to transform global food systems in a manner consistent with the Paris Agreement and the SDGs (FAO & UNDP, 2021). Bridging this gap will need more financial resources and greater alignment among stakeholders, including governments, private investors, financial institutions, farmers, and civil society organisations. Transparency, accountability, and impact measurement must also be prioritised to ensure that green finance delivers tangible environmental and social outcomes.

This study aims to analyse the scientific literature published in sustainable agriculture financing using bibliometric methods. The research aims to identify common themes, the authors with the most contributions, active institutions and collaborations between countries (Shehzad & Khan, 2024). By organising and interpreting this information, the study provides a clear picture of how the field will evolve from 2020 to 2025. The importance of this study stems from the need to understand better how funding for sustainable agriculture is addressed in scientific research. The topic is linked to current challenges such as climate change, resource management and food security. The results can be helpful for researchers, academic institutions and policymakers who want to prioritise research, foster collaboration, and support public policy or investment. The study can also be a starting point for future research, providing a documented basis for more detailed exploration of subthemes or regions (Gherțescu et al. 2025).

The research questions of this study are:

RQ1: What are the central themes and dominant conceptual relationships in the scientific literature on financing sustainable agriculture, and how do they reflect the integration of economic, environmental and managerial dimensions in the 2020–2025 period?

RQ2: How are the key concepts grouped in the literature on financing sustainable agriculture, and what thematic relationships are emerging between sustainable agriculture's economic, environmental and managerial dimensions?

RQ3: What are the characteristics of the distribution of authors, and how are collaborative networks structured in the literature on sustainable agriculture finance?

RQ4: Which academic institutions have the highest level of scientific output in sustainable agriculture finance, and how has their contribution evolved from 2020 to 2025?

RQ5: How are the international collaborative networks structured, and which countries are leading the scientific output in sustainable agriculture financing in terms of volume of publications and type of collaboration (national or international)?

The article is structured in five sections. Section 2 presents the bibliometric methodology, data sources, and article selection criteria. Section 3 presents the results and discussion of the analysis, highlighting publication trends, authors, and relevant institutions. Section 4, entitled Sustainable Financing in Contemporary Literature, analyses recent studies and assesses the impact of sustainable financing on agriculture. Section 5 formulates conclusions, main contributions, and future research directions.

2 Materials and Methods

This study conducted the bibliometric analysis using the Bibliometrix package within the RStudio environment (Version 2024.12.1 + 563), a powerful integrated development platform for the R programming language, tailored for statistical computing and data visualisation. RStudio offers desktop and server-based versions, enabling remote access through a web browser, which is particularly useful for collaborative research and large-scale data processing. Bibliometrix, a comprehensive tool for quantitative research in bibliometrics and scientometrics, provides a flexible and reproducible framework for importing, filtering, analysing, and visualising bibliographic data (Rusydiana et al., 2021).

This methodology serves as a robust approach to exploring the intellectual structure, thematic evolution, and collaborative dynamics within academic disciplines. It allows researchers to tailor bibliometric indicators and generates a variety of science maps, such as co-word networks, co-citation matrices, and thematic clusters, which are essential for interpreting patterns and identifying knowledge gaps (Aguinis et al., 2019; Manta et al., 2024a). Additionally, integrating with R ensures transparency and replicability, as the entire analytical workflow can be documented and reused for future studies.

The bibliometric analysis followed a structured, multi-step approach using the Web of Science Core Collection database and the Bibliometrix software package in the RStudio environment.

In the first step (Fig. 1), the Web of Science Core Collection database was accessed using the keyword "sustainable agriculture financing." This search aimed to identify relevant scientific works at the intersection between sustainable financing and the agricultural sector. This initial search returned 2113 results.

In the second step, filters were applied to refine the dataset. A temporal filter was used, limiting the selection to 2020–2025 to highlight recent and current research

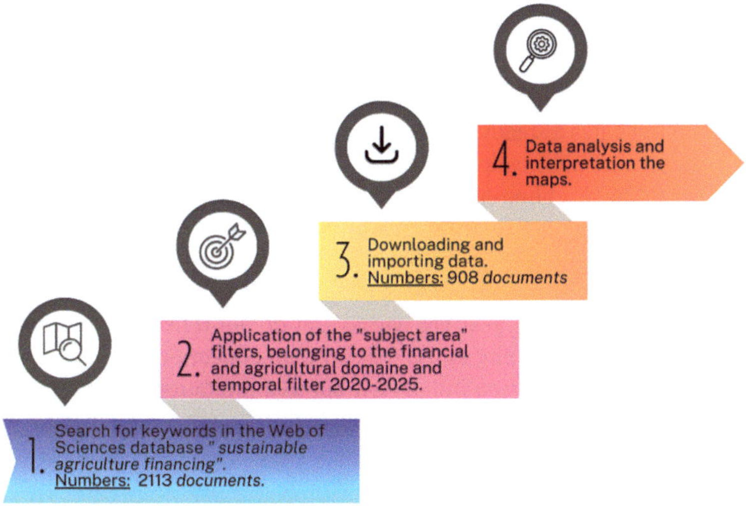

Fig. 1 Methodological steps in bibliometric analysis. (Source: own processing)

directions in the field. In addition, only papers falling into the thematic categories relevant to the present study, i.e. those belonging to the financial and agricultural domains, were selected. After applying these criteria, the final dataset was reduced to 908 scientific articles, which were considered representative of the proposed analysis.

In the next step, the 908 results were exported from the Web of Science platform as a BibTeX format file, which ensures compatibility with the Bibliometrix package and allows efficient manipulation of bibliographic metadata (title, author, year, abstract, keywords, source, etc.).

Subsequently, the actual bibliometric analysis was performed in RStudio, using the functionalities offered by Bibliometrix. These include data pre-processing (field cleaning, duplicate removal, name variant unification), and the generation of visual maps, such as keyword co-occurrence maps, author and document co-citations maps, institutional and geographical collaboration maps, and evolutionary thematic analysis.

This approach allowed the identification of central research themes, networks of collaboration between authors and institutions, and emerging trends in the literature on sustainable agriculture financing. All steps of the analysis were documented and executed in a replicable way, ensuring the transparency of the methodological process and the possibility of replication in future similar studies.

3 Results and Discussions

The trend in the publication of scientific articles in sustainable agriculture financing in 2020–2025 indicates a steady and significant increase in academic interest in this topic. The number of publications increased from 108 articles in 2020 to a peak of 227 articles in 2024, doubling the volume in just 4 years. This upward trend reflects a growing concern of the scientific community to find sustainable solutions in agriculture in the context of climate, food and economic crises.

The noticeable decrease in 2025 (with only 84 publications) can be explained by the fact that the data for this year are partial and do not reflect the full annual volume, as the year is still ongoing at the time of data collection. Therefore, it should not be interpreted as a reduction in interest, but as a limitation related to the timing of extracting information from the database.

The 2020–2025 period is relevant because it covers the years after the launch of the European Green Pact, the Paris Agreement and the new directions in the Common Agricultural Policy, a period in which global sustainability policies have intensified. At the same time, the COVID-19 pandemic has shifted research priorities and boosted reflection on the resilience of food systems, which explains the accelerated dynamics of publications. This period is therefore essential for understanding recent developments, new policy directions, and how the scientific community has responded to the challenges in agriculture through sustainable and innovative approaches (Fig. 2).

Figure 3 provides an overview of the main bibliometric indicators for the scientific literature dedicated to the financing of sustainable agriculture from 2020 to 2025. In total, 908 documents from 316 sources (journals, conferences, collective volumes, etc.) were analyzed, reflecting moderate but sustained scientific activity.

The total number of authors is 4835, indicating a high level of research engagement overall. However, only 51 papers were signed by a single author, implying that

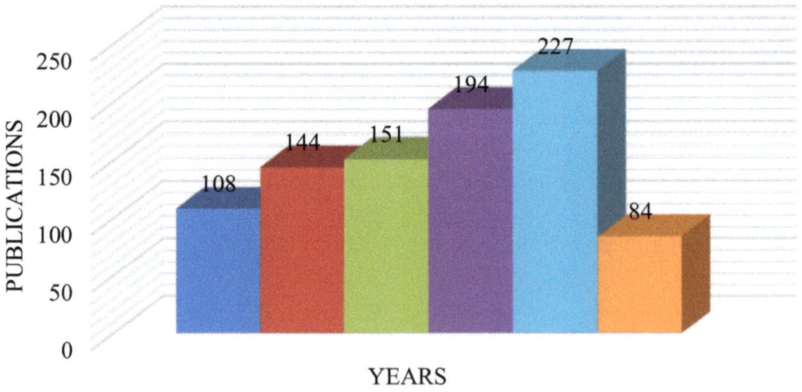

Fig. 2 The increase in the number of publications on financing sustainable agriculture between 2020 and 2025. (Source: Processed data from Web of Science Database, 2025)

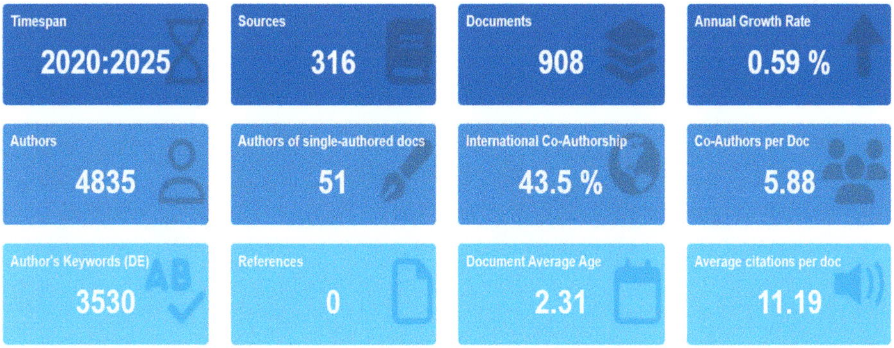

Fig. 3 General bibliometric indicators of the literature on financing sustainable agriculture (2020–2025). (Source: own processing in Bibliometrix)

the field is deeply collaborative. This is also supported by the average indicator of 5.88 co-authors per paper and the high percentage of international co-authorship, 43.5%, indicating a significant degree of transnational cooperation in knowledge generation.

The total number of keywords authors use is 3530, suggesting an important thematic diversity and a trend towards interdisciplinary exploration. However, the annual growth rate of publications is relatively low (0.59%), which may indicate either a recent plateau of interest or a stabilisation of the field after a period of expansion.

The average number of citations per paper is 11.19, indicating good scientific impact and relevance within the academic community. At the same time, the average age of the documents is 2.31 years, which means that the field is relatively new and growing, with recent publications already attracting significant citations.

3.1 Comparative Co-occurrence Network Analysis of Keywords and Word Cloud in the Literature on Sustainable Agriculture Financing

To answer research question RQ1, a bibliometric analysis was conducted to explore the thematic configuration of the sustainable agriculture financing literature. To capture both the frequency and the conceptual relationships among key terms, two distinct yet complementary visualization techniques were applied: the keyword co-occurrence network (Fig. 4) and the Word Cloud representation (Fig. 5). While the co-occurrence network reveals structural relationships and thematic clustering of concepts, the Word Cloud offers a synthetic view of term prominence across the literature. Together, these tools enabled a comprehensive examination of dominant research directions and their alignment with sustainable agriculture's economic, ecological, and managerial dimensions. This integrated approach provided a

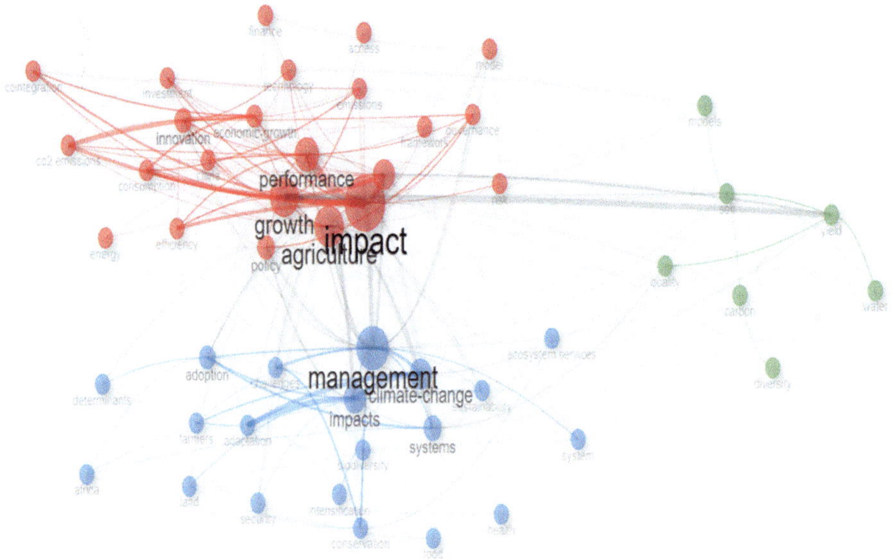

Fig. 4 Co-occurrence network of keywords in the field of "Sustainable Agriculture Financing" (2020–2025). (Source: own processing in Bibliometrix)

Fig. 5 Word Cloud is one of the most frequent keywords in the literature on sustainable agriculture financing (2020–2025). (Source: own processing in Bibliometrix)

mapping of the field's most relevant themes and more profound insight into how knowledge is organised and interconnected in current academic discourse.

In Fig. 4, the co-occurrence network was generated based on the frequency with which keywords appear together in the same publications. The result highlights a series of well-defined thematic clusters. A first cluster is centred around the terms climate change, sustainability, policy, biodiversity and adaptation, indicating a significant research concern for agriculture's environmental and regulatory dimension. A second considerable cluster includes terms such as agriculture, productivity, performance, innovation, management and systems, signalling a strong orientation towards efficiency, innovation and agricultural governance. The third cluster focuses on economic growth, investment, adoption and technology, highlighting the interest in integrating economic and technological mechanisms into sustainable agricultural processes.

This structuring of the network allows for precise observation of how the different research areas -ecology, economics, and management - interlink around the central theme of agriculture. Furthermore, the central positioning of the nodes' impact, agriculture and management indicates these concepts' pivotal role in the literature. The high connectivity of these terms suggests that most research explores the effects of various policies, technologies and climatic conditions on the performance of agricultural systems.

In contrast, Fig. 5, the Word Cloud representation, provides a synthetic view, based solely on the frequency of occurrence of key words in the entire body of literature analysed. In this picture, terms such as impact, management, growth, agriculture, performance, climate change, productivity and innovation stand out for their prominent size, confirming their central relevance in current scientific discourse. Unlike the co-occurrence network, the Word Cloud does not highlight the semantic relationships between terms. Still, it effectively shows the thematic hierarchy and the amount of interest associated with each concept.

Comparative analysis of the two visualisations highlights several key thematic convergences. Impact, agriculture, management, performance, productivity and climate change are the central themes common to both instruments, indicating a clear consensus in the scientific community on research priorities in the field of sustainable agriculture. The co-occurrence network provides a more complex perspective on the interdependencies between concepts, revealing significant connections between the ecological, economic and managerial dimensions. The Word Cloud, in contrast, highlights a consistent presence of the terms governance, adoption, systems, soil, yield and sustainability, which supports the idea that the current literature addresses production factors, agricultural policies and economic outcomes in an integrated way.

Another interesting element from Word Cloud is the presence of geographical and contextual perspectives, such as the emergence of the terms China and Africa, suggesting researchers' interest in regional case studies and analysing differences according to socio-economic context. Also, the technological dimension, illustrated by terms such as innovation, technology and investment, is recurrent in both representations, confirming the importance of technological progress in the transition towards a more sustainable agriculture.

Therefore, combining the two visual methods provides a broader and deeper understanding of the thematic landscape in the scientific literature on financing sustainable agriculture.

3.2 *Analysing Themes and Relationships Between Concepts in Research on Financing Sustainable Agriculture*

In order to answer RQ2 and better understand the structure of research on sustainable agriculture financing, two distinct but complementary visual tools were used: the thematic map, known as the Thematic Map, and the factor map, known as the Factorial Map. These graphical representations allow not only to identify dominant, emergent or marginal themes, but also to explore the semantic relationships between terms and how they are positioned within the literature. Together, the two types of analysis help to highlight the linkages between the economic, environmental and managerial components involved in the transition towards sustainable agriculture.

In Fig. 6, the thematic map organizes the key terms into four quadrants according to two key dimensions, namely the degree of centrality, which reflects the relevance of a theme to the field as a whole, and the degree of density, which indicates the level of internal development of that theme.

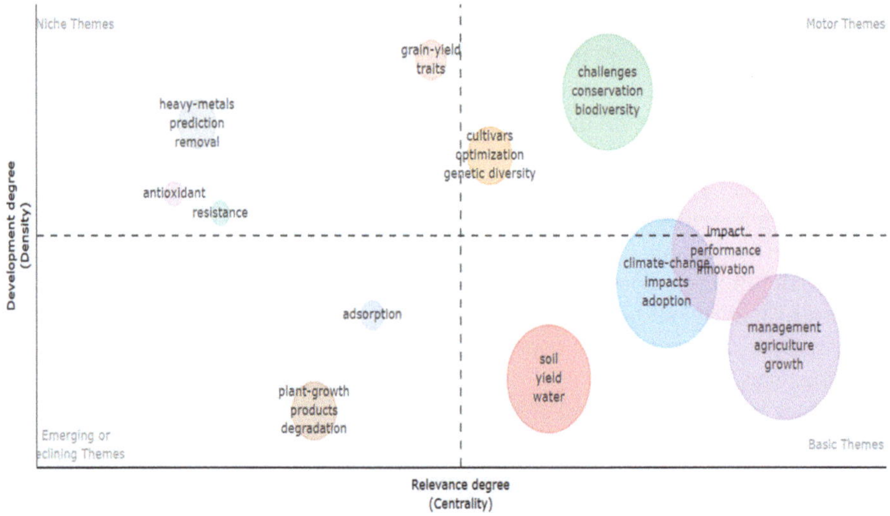

Fig. 6 Thematic map of key concepts in the literature on financing sustainable agriculture (2020–2025). (Source: own processing in Bibliometrix)

In the lower-right quadrant, considered as the core theme area, are the clusters formed around the concepts of management, agriculture and growth, climate change, impacts and adoption, as well as impact, performance and innovation. These themes appear frequently in the literature and form the conceptual foundation of the field. They are strongly interconnected, reflecting the interdependence between agricultural policies, economic performance and the need to adapt to climate change.

The concepts of challenges, conservation, and biodiversity are at the top right of the graph and are associated with the driving themes. This cluster is characterised by a central position and high density, which signals that these themes are relevant and well developed. They are frequently addressed in interdisciplinary research concerning protecting natural resources and ecosystem conservation.

On the top left, corresponding to niche topics, are terms such as heavy metals, removal, antioxidants, and resistance. These topics are well defined but relatively isolated from the core literature, giving them a specialized character with limited applicability in specific contexts, e.g., in areas affected by pollution or biological stresses.

The lower left quadrant, corresponding to emerging or declining topics, contains concepts such as adsorption, degradation, and plant-growth products. These can be interpreted either as new research directions in their infancy or as topics that have lost relevance recently. The lack of strong thematic connections with the rest of the network indicates untapped potential or a tendency towards marginalisation.

In Fig. 7, factor analysis provides another perspective by representing terms according to their semantic proximities in a two-dimensional space. The two main

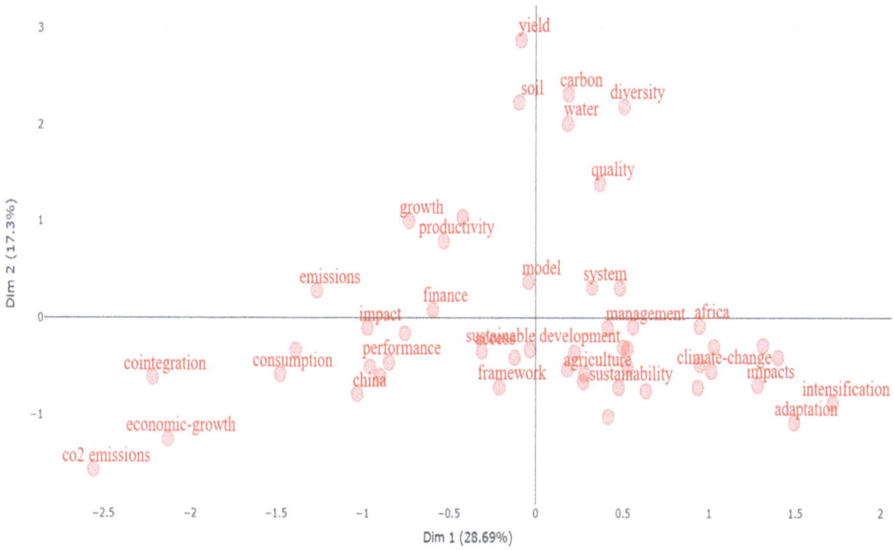

Fig. 7 Factor map of co-occurrence of terms in the literature on sustainable agriculture financing (2020–2025). (Source: own processing in Bibliometrix)

dimensions, with explanatory weights of approximately 28.7% and 17.3%, reflect the latent semantic structure of the analysed literature.

On the graph's horizontal axis, a separation is visible between concepts that refer to economics and public policy, located on the left side, and those that describe agronomic and ecological aspects on the right side. Thus, on the left side are terms such as cointegration, economic growth, consumption and finance, while on the right side are concepts such as yield, soil, carbon, water and diversity. This dissociation highlights the existence of two broad strands of research: one focusing on macroeconomic issues and the other on agricultural practices and resources.

The vertical axis seems to reflect the themes' applicability. Terms at the top, such as yield, soil, and diversity, indicate direct concerns with agricultural outcomes and performance, while terms at the bottom, such as CO_2 emissions, cointegration, and framework, suggest a more analytical approach oriented towards modeling and theoretical assessment.

Terms such as sustainability, management, agriculture, and development are located close to the center of the graph, indicating that they function as points of convergence between the different research directions. Their presence in a central area confirms that the literature on financing sustainable agriculture is anchored in fundamental, widely used, and well-integrated concepts.

Also, the terms climate change, adaptation, and intensification appear on the bottom right, suggesting a resilience-oriented and structural reform-oriented direction in agriculture, especially in the context of climate change. On the left centre, concepts such as growth, productivity, performance, and impact reinforce the managerial and economic axis of the research.

By way of comparison, the thematic map allows one to classify themes according to their global relevance and internal maturity, while factor analysis allows semantic proximity relationships between terms to be observed. Both tools converge in identifying the thematic core formed around sustainability, agriculture, performance, and adaptation. At the same time, they also highlight the existence of specialised or emerging areas that may constitute promising directions for future research.

3.3 Author Dynamics and Collaborative Structure in Research on Financing Sustainable Agriculture

Understanding the dynamics of authorship and collaborative networks is an essential step in a thorough bibliometric analysis, and it is crucial to assess how knowledge is produced and disseminated in the field of sustainable agriculture finance. Identifying the most prolific authors, assessing the distribution of scientific activity according to theoretical models such as Lotka's Law, and analysing co-authorship networks provide valuable insights into the structure of the scientific community involved in this field, thus answering RQ3.

Figure 8 presents a distribution of the most relevant authors according to the number of publications indexed in the scientific literature in 2020–2025, with the central theme of financing sustainable agriculture. This representation highlights the quantitative contribution of the leading researchers active in the field, allowing the identification of the authors dominating the scientific landscape regarding productivity.

The author with the highest visibility is Zhang Y, with a total of 12 published papers, closely followed by Zhang X with 11 publications. These figures suggest a constant and sustained involvement in the topics analysed, which gives them a central role in shaping the research directions. The next group of authors, which includes Li Y, Liu Y and Zhang L, stand out with 9 publications each, indicating a significant contribution and possible collaboration in consolidated research networks.

In the range of 7 to 8 papers, names such as Li W (8 papers), Li H, Wang L, Wang S and Zhang S appear with 7 publications each. These authors can be considered essential actors in the thematic development of the topic, although they have somewhat less influence than the field leaders.

It is worth noting that many of the authors highlighted in this graph come from Asia, especially China, which may reflect a geographical concentration of research in this region. This observation is supported by the repetitive presence of common surnames in Chinese research, such as Zhang, Li, Wang and Liu. This trend emphasises not only the strategic interest of Asian countries in sustainable agriculture but also the intensification of scientific and institutional efforts in this sector.

Figure 9 represents the distribution of authors' productivity in the field of sustainable agriculture funding analysed by applying Lotka's Law. This bibliometric

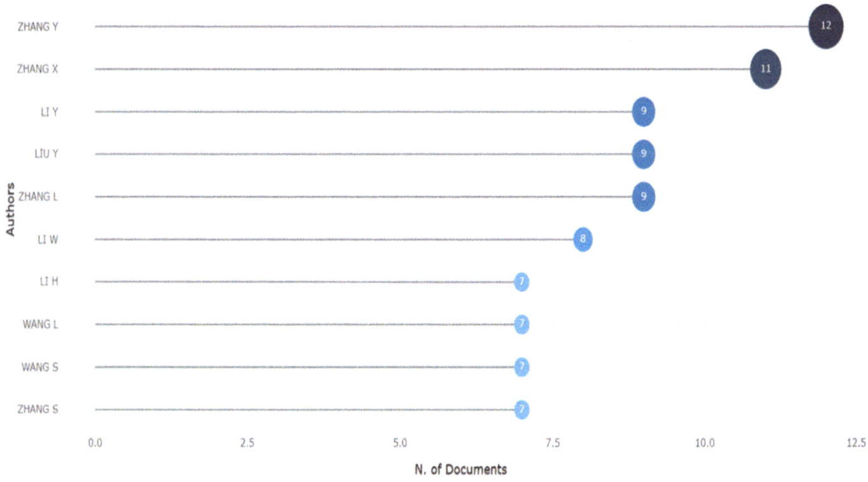

Fig. 8 Most prolific authors in sustainable agriculture financing (2020–2025). (Source: own processing in Bibliometrix)

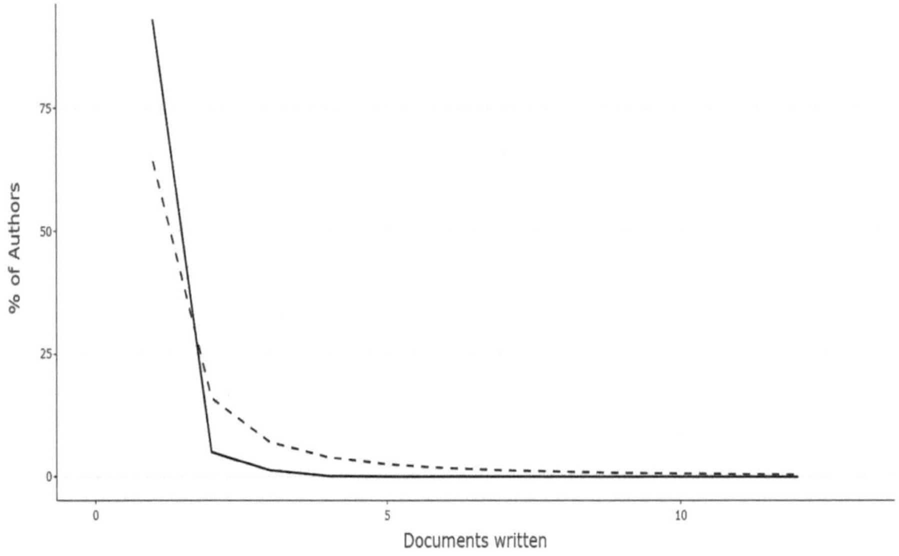

Fig. 9 Authors' productivity analysis of sustainable agriculture financing based on Lotka's Law. (Source: own processing in Bibliometrix)

law states that most authors publish a single paper, while only a few researchers contribute multiple papers, with the distribution following an inverse proportional relationship.

The graph's solid curve reflects the authors' empirical distribution according to the number of papers published. At the same time, the dashed line indicates the theoretical distribution predicted by Lotka's Law. We observe that about 80% of the authors published a single article in 2020–2025, which confirms the basic assumption of the law: publication is dominated by isolated contributions, and only a substantial minority of authors have an intense and repetitive publication activity.

The differences between the two curves, especially in the range of those who have published two or three articles, may signal a slight deviation from the ideal pattern, suggesting a slightly more pronounced tendency in this field for repeated collaboration or membership in well-defined research groups. However, the overall distribution largely follows the classic pattern of concentrating productivity around a small number of authors.

This distribution emphasises the fragmented and broadly participatory nature of sustainable agricultural finance. The involvement of a large number of single authors may reflect either the topic's cross-cutting interest across disciplines or the fact that research in this area is often conducted in applied contexts with one-off contributions.

Therefore, the analysis of author productivity through the lens of Lotka's Law confirms the predominance of occasional publications and emphasises the importance of identifying opinion leaders with sustained activity for mapping the scientific dynamics and centres of excellence in the field.

Figure 10 illustrates the collaborative network between authors contributing to the literature on financing sustainable agriculture in 2020–2025. This visual representation highlights not only the intensity of collaborations, but also the structure of research clusters, the frequency of interactions and the degree of centrality of specific authors within the network.

The densest cluster identified in the network is represented by the group of authors colored in red: Zhang Y, Zhang X, Li Y, Liu Y, and Zhang L. These authors appear connected through multiple links, indicating a well-consolidated and active collaborative network. The central positioning of these nodes also confirms that they play an essential role in the dissemination and circulation of ideas in the field. The close interactions in this cluster suggest that they belong to the same national or international research teams or collaborative projects.

A second significant cluster, marked in orange, is dominated by authors Doro G and Grabenocchi G, around whom several close co-authors gravitate. This cluster presents a very dense internal network but poorly connected with the rest of the authors in other clusters, signaling an independent, thematically, or geographically isolated research core.

On the top right, another cluster consists of authors Wang X, Wang H, Liu Z and Chen W, marked in blue, who collaborate predominantly among themselves, with no visible links to the core clusters. A similar situation is noticeable in the pink-colored cluster, which includes authors Ali M, Ali Mas, and Luo J, who suggest a distinct regional or institutional collaborative network. Likewise, the author Akzar R, colored in brown, appears to be completely isolated, with a single link, indicating a punctual participation in the literature without significant collaborations.

The presence of several loosely connected groups suggests that research in sustainable agriculture finance is fragmented, structured around distinct research centres with few visible transnational or inter-institutional links. At the same time, the

Fig. 10 Author collaboration network in sustainable agriculture financing research (2020–2025). (Source: own processing in Bibliometrix)

predominance of Asian surnames in the central cluster confirms the geographical concentration of much of the scientific output in Asia, particularly in China.

This analysis of collaboration between authors contributes to identifying nuclei of scientific influence, the degree of thematic cohesion, and opportunities for further interdisciplinary or international collaborations in future research on sustainable agriculture.

3.4 Institutional Dynamics of Scientific Output in the Field of Sustainable Agriculture Financing

Understanding how different academic institutions contribute to knowledge development is essential in the bibliometric analysis of a fast-growing field such as sustainable agriculture financing. Monitoring scientific output based on institutional affiliation makes it possible to identify centres of excellence, geographical trends and research strategies adopted by universities and institutes. In addition, the dynamics of these contributions provide a clear perspective on the evolution of scientific interest over time and the degree of involvement of institutional actors at the international level. To answer RQ4, the evolution of the number of articles published by the most active institutions over 2020–2025 will be analysed, focusing on universities that have significantly contributed to the literature.

Figure 11 shows the evolution of the scientific output of the most active academic institutions in sustainable agriculture funding from 2020 to 2025.

The most prominent institution is Sichuan Agricultural University, which shows accelerated growth, reaching 38 published articles in 2025. This upward trajectory indicates a substantial consolidation of scientific activity at the university, suggesting sustained investment in research and a possible strategic positioning in sustainable agriculture.

In second place is the Hungarian University of Agriculture and Life Sciences, with 26 publications in 2025, demonstrating a steady and increasing scientific activity throughout the analysed interval. The following institutions in the ranking are Nanjing Agricultural University with 22 papers, South China Agricultural University with 21 papers, and the University of São Paulo and Northwest A&F University, respectively, each with 19 papers in 2025. These figures suggest an increasingly active involvement of universities from Asia and a representative institution from Latin America.

It is worth noting that all six institutions analysed show a clear upward trend over the 2020–2025 period, with different paces. Sichuan Agricultural University and Nanjing Agricultural University have accelerated trajectories, while the University of São Paulo and South China Agricultural University have a more moderate but steady evolution. This diversity of trajectories suggests differences in research development strategies, available resources and institutional thematic priorities.

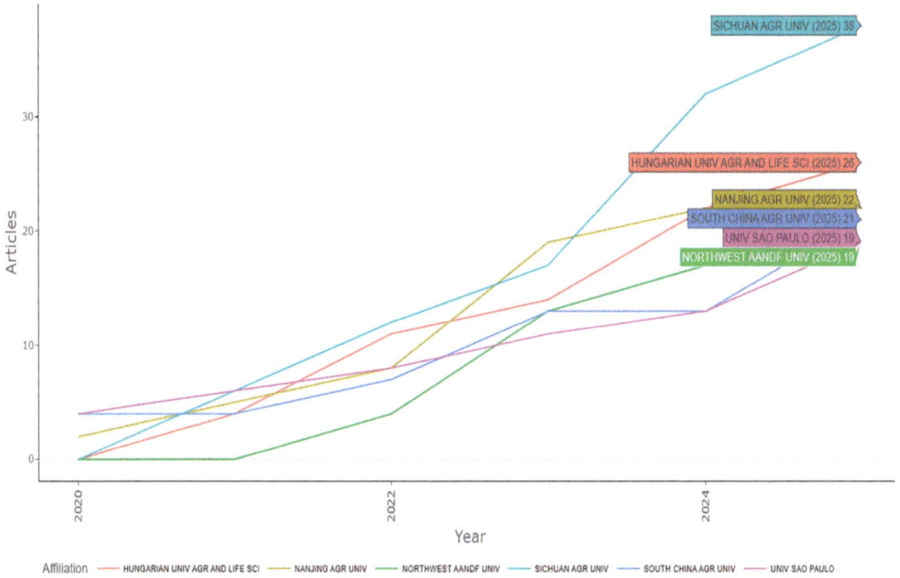

Fig. 11 Evolution of the scientific output of the leading institutions in the field of financing sustainable agriculture (2020–2025). (Source: own processing in Bibliometrix)

Geographically, the predominance of Chinese universities is evident, which confirms previous findings at the level of authors and collaborative networks. This indicates that research on sustainable agriculture is firmly anchored in Asia, particularly in China. In this region, demographic pressures and climate change are generating an urgent need for sustainable solutions in the agricultural sector. At the same time, the emergence of the Hungarian University of Agriculture and Life Sciences among the top institutions suggests an increasingly visible activation of the European space in this field.

This analysis confirms that research development in sustainable agriculture financing is marked by an intense institutional concentration, with a few reference universities functioning as scientific production and innovation centres. At the same time, trends suggest expanding contributions, both geographically and institutionally, which offer essential opportunities for strengthening international collaborations and global knowledge dissemination.

3.5 International Scientific Collaboration and Geographical Distribution of Contributions to Research on Financing Sustainable Agriculture

To answer research question RQ5, the global network of collaboration between countries and the geographical distribution of corresponding authors were analyzed, highlighting the volumes of publications and the typology of scientific collaborations (national or international) to outline the role of leading countries and their degree of integration in international academic networks in the field of sustainable agriculture financing.

Figure 12 provides a geospatial representation of the network of collaboration between countries involved in research on financing sustainable agriculture from 2020 to 2025. The map highlights both the intensity of the links between countries, illustrated by the density and thickness of the connecting lines, and the geographical distribution of scientific activity, reflected by the colouring of the participating countries.

It is clear that China occupies a central and dominant position in this global network, functioning as a significant hub of international collaboration. It is involved in numerous transnational partnerships, spanning all continents, including North America, Europe, Africa, South-East Asia and Oceania. China's central role is supported by its track record in terms of the number of publications, prolific authorship, and institutional affiliations, which confirm its position as a global leader in research in this field.

Alongside China, other countries with high visibility in this network are the United States, India, Brazil, Germany, Australia, and South Africa, which also form important nodes of collaboration. These countries exhibit multiple and globally

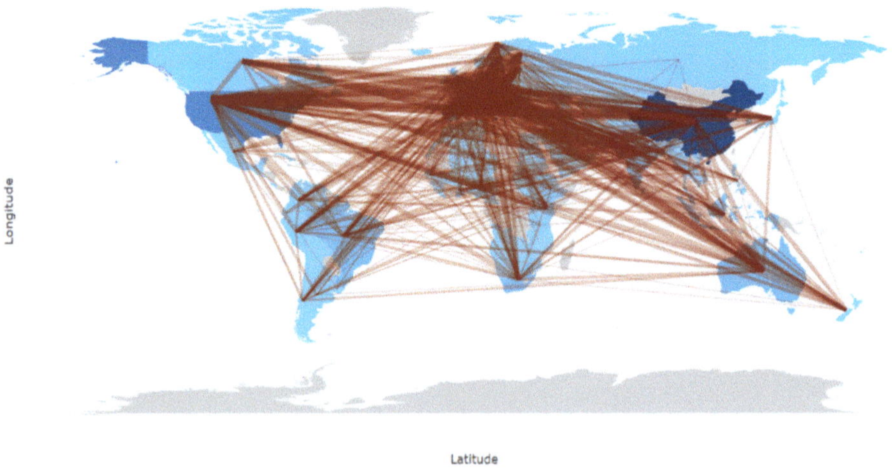

Fig. 12 Map of international research collaborations on financing sustainable agriculture (2020–2025). (Source: own processing in Bibliometrix)

distributed connections, reflecting an openness to international cooperation and active involvement in topics related to sustainable agriculture, financing, and climate change adaptation.

There are also well-defined regional networks, particularly in Asia, where collaborations between China, South Korea, Japan, Thailand and Vietnam are frequent. Similarly, in Europe, there are visible connections between countries such as Hungary, Germany, Italy and the UK, indicating the existence of European research consortia or initiatives focusing on sustainability and agriculture.

The strong presence of connections with countries in Africa, such as Nigeria, Kenya, and South Africa, suggests increased interest in scaling up applied research in climate-vulnerable contexts. At the same time, the emergence of links with countries in Latin America and Oceania indicates an increasingly globalised nature of the field.

This map confirms that research on financing sustainable agriculture is deeply internationalised, with a complex network of scientific collaborations across regional boundaries. The exchange of expertise and cooperation between developed and developing countries plays a key role in building a global scientific framework for sustainable development in agriculture.

Figure 13 shows the distribution of scientific papers according to the country of affiliation of the corresponding author, providing a key insight into scientific leadership in sustainable agriculture financing. In addition, the graph differentiates

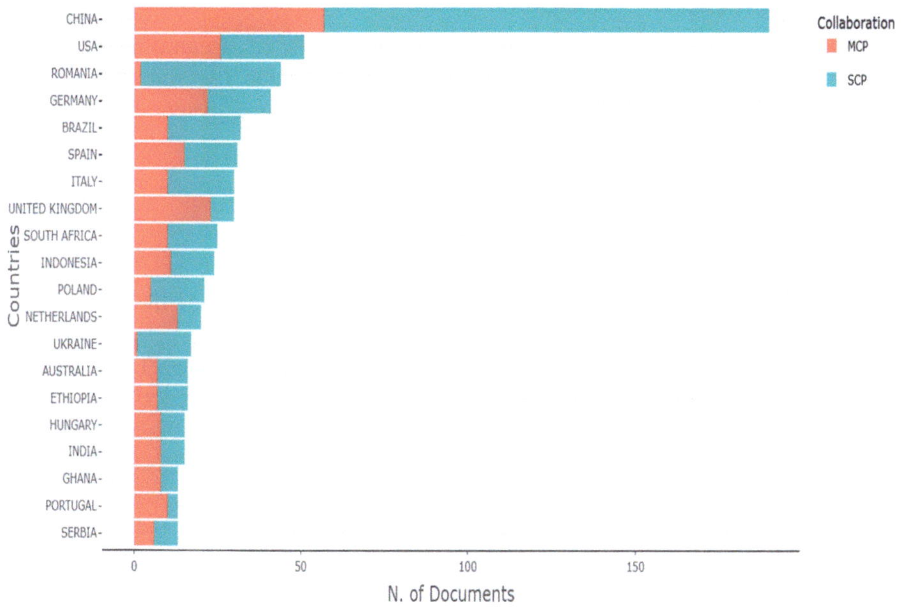

Fig. 13 Geographical distribution of corresponding authors and typology of international collaborations in research on financing sustainable agriculture (2020–2025). (Source: own processing in Bibliometrix)

between Single Country Publications (SCP) and Multiple Country Publications (MCP), allowing an assessment of the degree of openness and interconnectedness between countries.

China stands out, with a much higher number of articles than the other countries, thus consolidating its position as a global leader in terms of volume and its capacity to generate publications through international collaboration. However, the significant share of SCP-type publications indicates that much of China's scientific output is carried out within national institutions, with a predominantly domestic orientation.

The United States comes second, with a relative balance between domestic and international publications. Romania follows in third place, with a notable number of publications, indicating an increasingly active involvement of Romanian researchers in this emerging field. Romania is characterised by a predominantly collaborative profile, with a high share of publications in international partnerships.

Germany, Brazil, Spain, Italy and the United Kingdom contribute significantly to the literature, with a balanced representation between SCP and MCP, reflecting a mixed scientific production pattern based on local expertise and international networks. Countries in Africa, such as South Africa, and Asia, such as Indonesia and India, are also well represented, with a high degree of internationalisation, suggesting openness to transnational cooperation in sustainable development contexts.

Interestingly, countries such as Hungary, Ethiopia, Ghana and Serbia, although present with a smaller number of articles, have a relatively high share of international collaborations, indicating a strategy of integration into global research networks through external partnerships.

The graph, therefore, illustrates who contributes most to research on financing sustainable agriculture and how these contributions are made, highlighting patterns of scientific cooperation and the level of internationalisation of research.

4 Sustainable Financing in Contemporary Literature

Scientific literature in recent years provides growing evidence of the benefits of adequate financing for sustainable agriculture, both economically, socially and environmentally. A number of empirical studies confirm that access to credit and financing significantly stimulates the adoption of climate-smart practices and can improve farm performance. For example, Finizola e Silva et al. (2024), in a meta-analysis of CSA adoption factors in Africa, concluded that access to credit consistently increases the likelihood of adopting CSA practices (farmers with access to dedicated loans had a ~ 20% higher adoption rate). Similarly, Sisay et al. (2023), in a study of farmers in Ethiopia's Rift Valley, found that the availability of financing (credit and additional income) is one of the most important determinants of households' decisions to invest in CSA technologies, along with access to information and land (Kangogo et al., 2021).

These results support the hypothesis that financial barriers are often more critical than technological knowledge. Many farmers are aware of beneficial practices but lack the resources to implement them without financial support (Belay et al., 2022).

Furthermore, the literature shows that sustainable financing generates long-term economic benefits for rural communities (Xie & Li, 2023; Scarlet, 2024). By increasing productivity and reducing losses caused by climate disasters, agricultural incomes become more secure and can increase (Amare & Balana, 2023). Ma and Rahut (2024) summarise the results of 19 studies indicating that adopting CSA, facilitated by financing and training interventions, has increased household incomes and local economic diversification, contributing to rural poverty reduction. For example, a study of the "climate-smart villages" program shows that CSA technology adoption rates increased substantially when targeted loans and subsidies were provided, leading to increased local grain production and improved food security (Ouédraogo et al., 2019; Adesipo et al., 2020). At the same time, the resilience benefits are significant: farmers who adopted adaptive practices (irrigation, resistant crops, insurance) with financial support recovered more quickly after episodes of drought or flooding compared to control groups (Musyoki et al., 2022; Djido et al., 2022). For example, a study in Ghana (Djido et al., 2022). showed that households that received climate information services and microcredit for adaptation reduced their crop losses by half compared to those that did not receive assistance, demonstrating the role of financing in reducing vulnerability (Zheng, 2021).

One more thing that's been pointed out in the literature is the positive link between training, association, and financing (Christian et al., 2024). Farmers who are part of cooperatives or producer associations have better access to credit and inputs, which leads to higher adoption rates for sustainable practices (Abegunde et al., 2020; Kangogo et al., 2021). Community savings and credit groups have enabled the mobilization of local savings and the financing of micro-investments (wells, solar pumps, etc.), with demonstrable benefits (Zampaligré & Fuchs, 2019; Belay et al. (2022) observed that farmers with more education and experience tend to use external financing more efficiently and adopt more complex CSA technology packages, suggesting that investments in human capital (education, technical training) complement financial investments and amplify their impact. Thus, a mix of interventions such as credit, training, and organisation generates the best results in terms of sustainable transformation.

Regarding the environment, funding for regenerative practices brings benefits such as restoring soil fertility, reducing erosion and carbon storage (Khangura et al., 2023). Research on green-funded projects shows that farms that have implemented crop rotation with legumes and conservation agriculture (no-till) have seen increases in soil organic matter content and a decrease in chemical fertiliser requirements, which improves long-term profitability (Trivedi et al., 2021). Investors in carbon-positive agriculture also report generating carbon credits that can be monetised, creating an additional funding source for farmers (Henderson et al., 2022). However, the literature also points to the risk of "greenwashing" if standards are not strict, i.e., labelling conventional investments as "green" without any real impact. In this regard, transparency and rigorous impact reporting become essential (Xu et al., 2020; Falcone & Sica, 2019).

In conclusion, 2020–2024 has brought progress and persistent challenges in financing sustainable agriculture. On the one hand, there has been increased recognition of the importance of the agricultural sector in climate finance, new financial instruments have been developed, and evidence has accumulated on the effectiveness of climate-smart investments. On the other hand, the financing gap remains large, requiring additional efforts to mobilise capital and reform agricultural policies. The literature convergently emphasizes the need for a package of measures: reallocating subsidies to sustainable practices (Ricciardi et al., 2021), massively stimulating private investment (Sharma & Laurent, 2024), strengthening risk reduction mechanisms, and expanding financial inclusion in rural areas (Benni, 2024; Kaur Makkar et al., 2023). The critical mass of funding can be achieved by combining these approaches to transform global agriculture into an engine of sustainable development and climate resilience.

5 Conclusion

This study provides a comprehensive and well-structured bibliometric analysis of the scientific literature on sustainable agriculture financing in 2020–2025. Using a rigorous methodology based on visual tools and quantitative indicators, the research contributes to building a clear picture of this interdisciplinary field's structure, dynamics and development directions. Among the study's most important contributions are identifying the central themes dominating the current scientific discourse, such as investment impacts, sustainability of farming systems, climate change adaptation, resource management and innovation. The research also highlights emerging directions that are less developed but have the potential to strengthen in the future, such as biodiversity conservation, green technologies and energy efficiency in agriculture.

The integration of the thematic, co-occurrence and factor analyses allowed a layered understanding of the field by correlating between the centrality and maturity of the themes and observing semantic relationships between key concepts. In parallel, the analysis of authors' productivity and distribution according to Lotka's Law highlighted the degree of concentration of scientific production. In contrast, the network of collaboration between authors revealed the structure of the scientific community: dense in some regions and fragmented in others. At the institutional level, universities such as Sichuan Agricultural University and the Hungarian University of Agriculture and Life Sciences stood out as centres of excellence in scientific production. At the same time, the analysis of national contributions highlighted China, the United States and Romania as leaders in the publication of papers in the field of sustainable agriculture financing.

The originality of the research lies in the fact that it represents one of the few bibliometric analyses focusing exclusively on the intersection between agricultural sustainability and sector-specific financing. Unlike other studies that treat sustainable agriculture as a general topic or green finance in isolation, this paper brings an integrated perspective on an emerging but critical area for achieving the Sustainable Development Goals. The novelty also lies in the multimodal approach, which maps the thematic structure and correlates the production volume with the typology of collaborations, institutional structure and geographical distribution, thus providing a three-dimensional picture of the field.

The results are relevant for several categories of actors regarding practical applicability. For researchers and academic institutions, identifying central and emerging themes provides clear directions for further research and the development of interdisciplinary projects. Knowledge of prominent authors and institutions can facilitate establishing strategic partnerships and international consortia. For practitioners and public policy developers, the results provide an empirical basis for understanding current and emerging concerns in the field and the dynamics of the actors involved.

On this basis, clear recommendations for policymakers can be formulated. First, public policies on the transition towards sustainable agriculture should include measures to stimulate international research collaboration by supporting transnational partnerships and joint funding programs. Second, research capacity in underrepresented regions needs to be strengthened through investment in scientific infrastructure, academic mobility and the formation of regional centres of excellence. At the same time, integrating scientific research into the agricultural policy-making process can ensure a better alignment between academic innovations and the practical needs of farmers and rural communities.

In addition to these findings, the study opens essential directions for future research. One of these would be the in-depth analysis of sub-domains such as using green finance instruments in agriculture, integrating circular economy principles in agricultural production, or developing sustainable agricultural governance models. Future studies may also investigate the dynamics of scientific influence over time by applying longitudinal methods or combining bibliometric data with altimetric indicators to assess the impact of research beyond the academic sphere, in public policy or social perception. Another proper direction would be comparative analysis across regions (e.g. Europe vs. Asia or global North vs. South) to understand structural and strategic differences in approaches to agricultural sustainability.

Related to the study's limitations, it must be noted that it relied exclusively on data available in international databases such as Web of Science, which may exclude some significant contributions published in local journals or languages other than English. The selection of keywords and the analysis period (2020–2025) may also influence the results, especially in a rapidly expanding field. In addition, although visual methods provide an intuitive and synthetic understanding, they may not fully capture the theoretical complexity or the institutional and political context of the publications analysed. Also, the analysis of collaborations is based solely on the co-authorship recorded in the articles, without reflecting informal or early-stage collaborations.

In conclusion, this research provides a broad, coherent and documented picture of the scientific landscape in the field of sustainable agriculture financing, highlighting dominant directions, relevant actors and opportunities for future development. Combining thematic, institutional and geographical analysis makes the study a valuable tool for the scientific community, policymakers, and practitioners interested in sustainable development in the agricultural sector.

Acknowledgments This work was supported by a grant from the Romanian Ministry of Research, Innovation and Digitalization, the project with the title "*Economics and Policy Options for Climate Change Risk and Global Environmental Governance*" (CF 193/28.11.2022, Funding Contract no. 760078/23.05.2023), within Romania's National Recovery and Resilience Plan (PNRR)—Pillar III, Component C9, Investment I8 (PNRR/2022/C9/MCID/I8)—Development of a program to attract highly specialised human resources from abroad in research, development, and innovation activities.

Competing Interests The authors have no conflicts of interest to declare relevant to this chapter's content.

Ethics Approval This study is a bibliometric analysis and did not involve any studies with human participants or animals performed by any authors. Therefore, ethical approval was not required.

References

Abegunde, V. O., Sibanda, M., & Obi, A. (2020). Determinants of adoption of climate-smart agricultural practices among smallholder farmers: Evidence from southern Africa. *Journal of Agriculture and Environment, 16*(2), 143–158. https://doi.org/10.1007/s10806-020-09849-w

Adesipo, A., Fadeyi, O., Kuca, K., Krejcar, O., Maresova, P., Selamat, A., & Adenola, M. (2020). Smart and climate-smart agricultural trends as core aspects of smart village functions. *Sensors, 20*(21), 5977. https://doi.org/10.3390/s20215977

Aguinis, H., Ramani, R. S., & Villamor, I. (2019). The first 20 years of Organizational Research Methods: Trajectory, impact, and predictions for the future. *Organizational Research Methods, 22*(2), 463–489. https://doi.org/10.1177/1094428118786564

Amare, M., & Balana, B. (2023). Climate change, income sources, crop mix, and input use decisions: Evidence from Nigeria. *Ecological Economics*. https://doi.org/10.1016/j.ecolecon.2023.107892

Aznar-Sánchez, J. A., Piquer-Rodríguez, M., Velasco-Muñoz, J. F., & Manzano-Agugliaro, F. (2019). Worldwide research trends on sustainable land use in agriculture. *Land Use Policy, 87*, 104069. https://doi.org/10.1016/j.landusepol.2019.104069

Baker, M., Bergstresser, D., Serafeim, G., & Wurgler, J. (2018). *Financing the response to climate change: The pricing and ownership of U.S. green bonds* (National Bureau of Economic Research Working paper no. 25194). National Bureau of Economic Research.

Belay, A., Oludhe, C., Mirzabaev, A., Recha, J., Berhane, Z., Osano, P., et al. (2022). Knowledge of climate change and adaptation by smallholder farmers: Evidence from Southern Ethiopia. *Heliyon, 8*(11), e12089. https://doi.org/10.1016/j.heliyon.2022.e12089

Benni, N. (2024). *Promoting inclusive green finance in agriculture – Selected case studies*. FAO; GIZ. https://doi.org/10.4060/CB7917EN

Christian, M., Obi, A., Zantsi, S., Mdoda, L., & Jiba, P. (2024). The role of cooperatives in improving smallholder participation in agri-food value chains: A case study of one local municipality in eastern cape, South Africa. *Sustainability*. https://doi.org/10.3390/su16062241

Climate Bonds Initiative. (2021). Agriculture criteria – Climate bonds standard & certification scheme. https://www.climatebonds.net/standard/agriculture

Convergence. (2020). The State of Blended Finance 2020. https://www.convergence.finance

Desalegn, G., Tangl, A., Fekete-Farkas, M., Gudisa, G., & Boros, A. (2024). Linking policies and regulations to sustainable finance to promote urban agriculture: Evidence from micro and small businesses. *Heliyon, 10*(11), e31938. https://doi.org/10.1016/j.heliyon.2024.e31938

Djido, A., Zougmoré, R., Houessionon, P., & Ouédraogo, I. (2022). To what extent do weather and climate information services drive the adoption of climate-smart agriculture practices in Ghana? *Climate Risk Management, 32*, 100309. https://doi.org/10.1016/j.crm.2021.100309

Doran, N. M., Manta, A. G., Badareu, G., Berceanu, D., Doran, M. D., & Manta, F. L. (2025). Could e-commerce activities drive to climate change mitigation? Novel evidence from panel quantile regression model. *Journal of Business Economics and Management, 26*(2), 255–276. https://doi.org/10.3846/jbem.2025.23604

Falcone, P. M., & Sica, E. (2019). Assessing the opportunities and challenges of green finance in Italy: An analysis of the biomass production sector. *Sustainability, 11*(2), 517. https://doi.org/10.3390/su11020517

FAO. (2021). *The State of Food and Agriculture 2021. Making agrifood systems more resilient to shocks and stresses*. Food and Agriculture Organization of the United Nations. https://doi.org/10.4060/cb4476en

FAO & UNDP. (2021). *Scaling up climate finance in agriculture*. FAO. https://www.fao.org/3/cb3820en/cb3820en.pdf

Fatemi, A. M., & Fooladi, I. J. (2013). Sustainable finance: A new paradigm. *Global Finance Journal, 24*(2), 101–113. https://doi.org/10.1016/j.gfj.2013.07.006

Finizola e Silva, M., Van Schoubroeck, S., Cools, J., & Van Passel, S. (2024). A systematic review identifying the drivers and barriers to the adoption of climate-smart agriculture by smallholder farmers in Africa. *Frontiers in Environmental Economics, 3*, 1356335. https://doi.org/10.3389/frevc.2024.1356335

GCF. (2019). *Financing climate-smart agriculture: A catalyst for transformation*. Green Climate Fund. https://www.greenclimate.fund

Gherțescu, C., Manta, A. G., Bădîrcea, R. M., Duță, A. G., & Zaharia, A. L. (2024). Bibliometric analysis of ESG in banking: What are the key directions and collaborations behind the research? *Finance – Challenges of the Future, 24*(26), 66–79.

Gherțescu, C., Manta, A. G., & Bădîrcea, R. M. (2025). Smart agriculture and technological innovation: A bibliometric perspective on digital transformation and sustainability. *Agriculture, 15*(13), 1388. https://doi.org/10.3390/agriculture15131388

Gong, Q., & Zhang, R. (2025). Digital finance as a catalyst for energy transition and sustainable rural economic growth. *Finance Research Letters, 71*, 106405. https://doi.org/10.1016/j.frl.2024.106405

Henderson, B., Lankoski, J., Flynn, E., Sykes, A., Payen, F. T., & MacLeod, M. (2022). *Soil carbon sequestration by agriculture: Policy options* (OECD food, agriculture and fisheries papers no. 174). Organisation for Economic Co-operation and Development (OECD). https://doi.org/10.1787/63ef3841-en

Howden, S. M., Soussana, J. F., Tubiello, F. N., Chhetri, N., Dunlop, M., & Meinke, H. (2007). Adapting agriculture to climate change. *Proceedings of the National Academy of Sciences of the United States of America, 104*(50), 19691–19696. https://doi.org/10.1073/pnas.0701890104

Huang, J., & Wang, Y. (2014). Financing sustainable agriculture under climate change. *Journal of Integrative Agriculture, 13*(4), 698–712. https://doi.org/10.1016/S2095-3119(13)60698-X

IFAD. (2021). *Rural development report 2021: Transforming food systems for rural prosperity*. International Fund for Agricultural Development. https://www.ifad.org

IPCC. (2019). *Climate change and land: An IPCC special report on climate change, desertification, land degradation, sustainable land management, food security, and greenhouse gas fluxes in terrestrial ecosystems*. https://www.ipcc.ch

Kangogo, D., Dentoni, D., & Bijman, J. (2021). Adoption of climate-smart agriculture among smallholder farmers: Does farmer entrepreneurship matter? *Land Use Policy, 109*, 105666. https://doi.org/10.1016/j.landusepol.2021.105666

Kaur Makkar, M., Bhat, B. A., Gupta, N., & Vaid, A. (2023). Harvesting sustainable agriculture with climate finance: Review. *E3S Web of Conferences, 453*, 01042. https://doi.org/10.1051/e3sconf/202345301042

Khangura, R., Ferris, D., Wagg, C., & Bowyer, J. (2023). Regenerative agriculture – A literature review on the practices and mechanisms used to improve soil health. *Sustainability*. https://doi.org/10.3390/su15032338

Kling, G., Volz, U., Murinde, V., & Ayas, S. (2021). The impact of climate vulnerability on firms' cost of capital and access to finance. *World Development, 137*, 105131.

Lagarde, C. (2021). Climate change and central banking. *Speech by Christine Lagarde, President of the ECB, at the International Climate Change Conference*, July 2021.

Lee, C.-C., Yuan, Z., & Kang, Y. (2024). Green finance and land ecological security: A potential mechanism for sustainable development. *Economic Analysis and Policy, 84*, 1222–1241. https://doi.org/10.1016/j.eap.2024.10.003

Lipper, L., Thornton, P., Campbell, B. M., Baedeker, T., Braimoh, A., Bwalya, M., … & Hottle, R. (2014). Climate-smart agriculture for food security. *Nature Climate Change, 4*(12), 1068–1072. https://doi.org/10.1038/nclimate2437

Liu, Y., Deng, Y., & Peng, B. (2023). The impact of digital financial inclusion on green and low-carbon agricultural development. *Agriculture, 13*(9), 1748. https://doi.org/10.3390/agriculture13091748

Ma, W., & Rahut, D. B. (2024). Climate-smart agriculture: Adoption, impacts, and implications for sustainable development. *Mitigation and Adaptation Strategies for Global Change, 29*(44). https://doi.org/10.1007/s11027-024-10139-z

Manta, A. G., Doran, N. M., Bădîrcea, R. M., Badareu, G., & Ţăran, A. M. (2023). Does the implementation of a Pigouvian tax be considered an effective approach to address climate change mitigation? *Economic Analysis and Policy, 80*, 1719–1731. https://doi.org/10.1016/j.eap.2023.11.002

Manta, A. G., Bădîrcea, R. M., Doran, N. M., Badareu, G., Gherţescu, C., & Popescu, J. (2024a). Industry 4.0 transformation: Analysing the impact of artificial intelligence on the banking sector through bibliometric trends. *Electronics, 13*(9), 1693. https://doi.org/10.3390/electronics13091693

Manta, A. G., Doran, N. M., Hurduzeu, G., et al. (2024b). Is there a direct benefit of using electronic commerce and electronic banking in mitigating climate change? *Climatic Change, 177*, 152. https://doi.org/10.1007/s10584-024-03809-z

Manta, A. G., Gherţescu, C., Bădîrcea, R. M., Manta, L. F., Popescu, J., & Lăpădat, C. V. M. (2025a). How does the interplay between banking performance, digitalization, and renewable energy consumption shape sustainable development in European Union countries? *Energies, 18*(3), 571. https://doi.org/10.3390/en18030571

Manta, L. F., Manta, A. G., & Gherţescu, C. (2025b). Decoding digital synergies: How mechatronic systems and artificial intelligence shape banking performance through quantile-driven method of moments. *Applied Sciences, 15*(10), 5282. https://doi.org/10.3390/app15105282

Meghişan-Toma, G. M., Puiu, S., Florea, N., Meghişan, F., Bădîrcea, R., & Manta, A. (2022). Sustainable transformation of Romanian companies through Industry 4.0, green production and environment commitment. *Amfiteatru Economic, 24*(59), 46–60.

Musyoki, M., Busienei, J., Gathiaka, J., & Karuku, G. (2022). Linking farmers' risk attitudes, livelihood diversification and adoption of climate smart agriculture technologies in the Nyando basin, South-Western Kenya. *Heliyon, 8*, e09305. https://doi.org/10.1016/j.heliyon.2022.e09305

OECD. (2022). *Climate-smart agriculture: Policies, practices and financing for food security, adaptation and mitigation*. Organisation for Economic Co-operation and Development. https://www.oecd.org/agriculture/topics/climate-change

Ouédraogo, M., Houessionon, P., Zougmoré, R. B., & Partey, S. T. (2019). Uptake of climate-smart agricultural technologies and practices: Actual and potential adoption rates in the Climate-Smart Village site of Mali. *Sustainability, 11*(17), 4710. https://doi.org/10.3390/su11174710

Poenaru, M. M., Manta, L. F., Gherțescu, C., & Manta, A. G. (2025). Shaping the future of horticulture: Innovative technologies, artificial intelligence, and robotic automation through a bibliometric lens. *Horticulturae, 11*(5), 449. https://doi.org/10.3390/horticulturae11050449

Rahman, M. M., & Hossain, M. E. (2024). Green technology, policy and sustainable finance nexus with SDG-12: Moderating effects of stakeholder awareness. *Sustainable Futures, 8*, 100405. https://doi.org/10.1016/j.sftr.2024.100405

Ricciardi, V., Mehrabi, Z., Wittman, H., James, D., & Ramankutty, N. (2021). A multi-billion-dollar opportunity: Repurposing agricultural support to transform food systems. *Nature Food, 2*(9), 680–683. https://doi.org/10.1038/s43016-021-00395-9

Robinson, G. M. (2024). Global sustainable agriculture and land management systems. *Geography and Sustainability, 5*(4), 637–646. https://doi.org/10.1016/j.geosus.2024.09.001

Rusydiana, A., Slamet, A., As-Salafiyah, A., Djamaluddin, Y. S., & Marlina, L. (2021). Fiqh on finance: A Scientometric analysis using Bibliometrix. *Library Philosophy and Practice (e-journal), 5436*. https://digitalcommons.unl.edu/libphilprac/5436

Sarker, A. H., Bornman, J. F., & Marinova, D. (2019). A framework for integrating agriculture in urban sustainability in Australia. *Urban Science, 3*(3), 50. https://doi.org/10.3390/urbansci3030050

Scarlet, J. (2024). Sustainable investment strategies for rural development. *International Journal of Developing Country Studies*. https://doi.org/10.47941/ijdcs.2172

Sharma, D., & Laurent, S. (2024). *Transition in action: Agri-food* (Climate Bonds Initiative Report, April 2024). Climate Bonds Initiative. https://www.climatebonds.net/files/reports/agri-food-transition-in-action-160524.pdf

Shehzad, A., & Khan, K. I. (2024). Time traveling through research: Bibliometric analysis of biodiversity finance in agricultural sector for SDGs. *Journal of Agriculture and Food Research, 18*, 101485. https://doi.org/10.1016/j.jafr.2024.101485

Sisay, T., Tesfaye, K., Ketema, M., Dechassa, N., & Getnet, M. (2023). Climate-smart agriculture technologies and determinants of farmers' adoption decisions in the Great Rift Valley of Ethiopia. *Sustainability, 15*(4), 3471. https://doi.org/10.3390/su15043471

Trivedi, P., Mattupalli, C., Eversole, K., & Leach, J. E. (2021). Enabling sustainable agriculture through understanding and enhancement of microbiomes. *New Phytologist, 230*(6), 2129–2147. https://doi.org/10.1111/nph.17319

UNEP. (2016). *Green finance for developing countries: Needs, concerns and innovations*. United Nations Environment Programme. https://www.unep.org

Wadumestrige Dona, C. G., Mohan, G., & Fukushi, K. (2021). Promoting urban agriculture and its opportunities and challenges – A global review. *Sustainability, 13*(17), 9609. https://doi.org/10.3390/su13179609

WEF. (2021). *Innovation to accelerate climate-smart agriculture*. World Economic Forum. https://www.weforum.org

World Bank. (2020). *Transforming agriculture for climate resilience: A policy framework*. World Bank. https://www.worldbank.org/en/topic/agriculture/publication

Xie, X., & Li, D. (2023). Green finance helps rural revitalization under the new institutional economics. *Frontiers in Humanities and Social Sciences*. https://doi.org/10.54691/w2b3ra41

Xu, H., Mei, Q., Shahzad, F., Liu, S., Long, X., & Zhang, J. (2020). Untangling the impact of green finance on the enterprise green performance: A meta-analytic approach. *Sustainability, 12*(21), 9085. https://doi.org/10.3390/su12219085

Yaron, J. (1994). What makes rural finance institutions successful. *World Bank Research Observer, 9*, 49–70. https://doi.org/10.1093/WBRO/9.1.49

Zampaligré, N., & Fuchs, L. (2019). Determinants of adoption of multiple climate-smart adaptation practices in Sudano-Sahelian pastoral and agro-pastoral production systems. *Sustainability, 11*(18), 48431. https://doi.org/10.3390/su11184831

Zhang, W., Zhang, Y., & Mou, S. (2025). Green finance development on corporate sustainability: Evidence from China. *Finance Research Letters, 82*, 107468. https://doi.org/10.1016/j.frl.2025.107468

Zheng, W. (2021). Governance structures in sustainable investment: A case study in rural China. *Journal of Corporate Finance, 52*, 264–276. https://doi.org/10.1016/j.jcorpfin.2021.101037

The Road Ahead: Future Directions in Green Finance—Global Collaboration and Trends for Climate Action

Meng Qin, Lobonț Oana-Ramona, and Costea Florin

1 Introduction to Global Green Finance Trends

With nations prioritising sustainability and climate change agendas, green finance has been a key promoter to enhance investment in clean energy (Zhang et al., 2019). Climate-themed bond issuance is a significant driver in advancing sustainability agendas by triggering financial incentives for industries to lower their carbon footprint (Pata et al., 2025). Initially launched in 2007 by the European Investment Bank, climate-aligned bonds have since witnessed exponential global growth, becoming a crucial financial product in both developing and developed nations (Shah et al., 2025). Such green debt products represent a significant landmark in the green finance sector, enabling finance mobilisation to priority climate projects while simultaneously providing new investment avenues for emerging markets (Ngwenya & Simatele, 2020). Unlike traditional bonds that allocate funds to standard projects, green bonds are specifically targeted towards financing projects that deal with climate change mitigation, the preservation of natural resources, or ecosystem protection (Ning et al., 2023). In this context, the World Bank Group finances developing nations in its International Bank for Reconstruction and Development (IBRD) operations. It serves as the world's largest development institution and a key source of

M. Qin (✉)
School of Marxism, Qingdao University, Qingdao, China

L. Oana-Ramona
Finance, Business Information Systems and Modelling Department, Faculty of Economics and Business Administration, West University of Timişoara, Timişoara, Romania
e-mail: oana.lobont@e-uvt.ro

C. Florin
Doctoral School of Economics and Business Administration, West University of Timişoara, Timişoara, Romania
e-mail: florin.costea@e-uvt.ro

C.-W. Su et al. (eds.), *Green Finance and the Challenges of Climate Change*, Sustainable Finance, https://doi.org/10.1007/978-3-032-07293-1_18

development knowledge. Data compiled from the World Bank indicates that from 2008 to 2024, the multilateral organisation has conducted green bond issuances totalling more than 230 individual issuances in 28 currencies with a total value of more than 20 billion dollars. During the 2023 fiscal period, these environmentally-focused financial instruments facilitated 955 million dollars in project financing to support developing nations' shift toward climate-adaptive, low-emission economic models. The Climate Bond Initiative reports that the global sustainable debt market, including green, social, sustainability, and sustainability-linked instruments, has reached an aggregate value of 5.7 trillion dollars by the end of 2024. This asset class has demonstrated remarkable growth since its inception, with annual green bond issuance surging from negligible levels in 2007 to 305 billion dollars in 2020, accumulating over 1 trillion dollars in total funding. Research suggests these instruments confer reputational benefits to issuers by demonstrating environmental commitment (Guesmi et al., 2025). China has emerged as a major player in this market, with combined government and corporate green bond issuances surpassing 540 billion dollars since 2015, which is the second largest volume globally after the U.S. While China's green bond market expanded dramatically from 206.6 billion yuan to over 1 trillion yuan, it still represents a fraction of the country's conventional bond market (Yan et al., 2025). Egypt made history on September 29, 2020, by becoming the Middle East and North Africa (MENA) region's first sovereign green bond issuer. This aligns with Egypt Vision 2030's target to boost green project allocations from 14% (2020) to 30% of government investment budgets. Meanwhile, the Organisation for Economic Co-operation and Development (OECD) data reveals that developed nations achieved their collective 100 billion dollars climate finance pledge in 2022, mobilising 115.9 billion dollars for developing countries' climate initiatives.

Climate-aligned bonds and environmental, social and governance (ESG)-oriented investments demonstrate significant capacity to advance sustainable financial systems and foster environmentally conscious economic expansion (Huang, 2024). Superior ESG performance could help reduce corporate financial risks, promote corporate innovation and enhance corporate value (Wang et al., 2025). As global economies transition toward premium-quality development models, the ESG paradigm is a catalytic framework that transforms corporate social responsibility from a secondary concern into a fundamental strategic component (Liu, 2025). The size of the ESG data and related services market is expected to grow at a rate of 23% through 2025 and has likely exceeded 1.5 billion dollars in 2023. At the current pace of 630 billion dollars per year, ESG investment, as an asset class in the investment field, has started to scale up and continues to grow, according to the International Monetary Fund (IMF). The IMF collaborates with member states and global stakeholders to develop a standardised classification system for ESG-aligned financial products. Recent market research by Banque Nationale de Paris (BNP) Paribas reveals that approximately 20% of institutional investors now integrate ESG criteria into at least 50% of their equity allocation decisions, underscoring rising demand for sustainable investment strategies (Vu et al., 2025). China's ESG adoption has accelerated significantly, as demonstrated by 374 ESG-themed public funds by mid-2022, representing 241.9 billion shares and a combined net asset value of 392.5

billion yuan. Similarly, India's ESG fund assets under management (AUM) experienced remarkable growth, jumping 150% from 275 million dollars (2020) to 650 million dollars (2021) before exceeding 1 billion dollars in 2023 (Gaurav et al., 2025). Nevertheless, ESG implementation remains highly heterogeneous, with substantial variations observed cross-nationally and temporally within individual markets. At a high-level seminar hosted by the People's Bank of China and the IMF, a survey of 425 investors found that 53% believe poor ESG data quality and availability are the biggest barriers to expanding sustainable investment. Sustainability reporting has emerged as a crucial activity for companies to transparently disclose their achievement towards ESG issues to the concerned stakeholders (Ferri et al., 2023). To effectively integrate sustainability issues into asset allocation decisions, investment practitioners need sufficient and reliable ESG metrics to assess firm-level sustainability performance (Jonsdottir et al., 2022). Strong ESG credentials enhance green innovation capabilities by removing capital constraints and accelerating research investments (Lu et al., 2025). Further, external assessments of environmental, social, and governance (ESG) factors induce firms to diversify their green intellectual property portfolios, thus enhancing capital investment and innovation activities in clean technologies (Chen et al., 2023).

Blended finance instruments that blend public and private funds have emerged as a prominent solution to construct biodiversity markets and mobilise other private capital towards Sustainable Development Goals (Christiansen, 2021; Smith et al., 2022). This financing mix is crucial in development economics, using public capital to catalyse the private sector's engagement in tackling the global agenda's pressing challenges in emerging economies (Léon, 2025). Current research indicates that substantial climate-aligned funding is essential for meaningful decarbonisation. Analysis by Climate Policy Initiative reveals that Sub-Saharan Africa and South and East Asian markets, particularly eight high-potential nations, collectively represent over 360 billion dollars in renewable energy investment opportunities through 2030. China's ambitious climate roadmap targeting peak emissions by 2030 and carbon neutrality by 2060 necessitates approximately 1.4 trillion dollars in annual investments to fulfil its Paris Agreement commitments (Pindiriri & Kwaramba, 2024). Nevertheless, existing global climate finance flows remain inadequate to achieve the necessary emission reductions. While global climate financing has surged by over 100% from 2018 to 2022, current levels must expand approximately five times to meet the annual 7.4 trillion dollars requirement for achieving the 1.5 °C climate target by 2030. Consequently, substantial mobilisation efforts are required to secure climate financing at the necessary scale from both international and domestic sources. The OECD data show private capital mobilised by public climate funds, for which comparable records exist only from 2016 onwards, increasing from 14.4 billion dollars in 2021 to 21.9 billion dollars in 2022. As a new tool under the UNFCCC framework, the GCF is critical for accelerating financial investment in climate change mitigation and adaptation actions (Bowman & Minas, 2019). Doubling the GCF's ability to mobilise various financial resources, both public and private capital, while mitigating the risk involved in conventional financing streams would greatly augment the overall level of climate-related investments (Amighini et al., 2022).

2 Key Financial Instruments and Innovation

2.1 Empirical Analysis of Climate Bonds

The global climate bond market has experienced a record period of expansion (Afful-Koomson, 2015; Azhgaliyeva & Kapsalyamova, 2023; Christnacht & Mertzanis, 2025), with the Climate Bonds Initiative reporting cumulative issuance of 6.9 trillion dollars of green, social, sustainability, and sustainability-linked bonds (GSS+) through December 2024. An impressive 5.7 trillion dollars, representing 83% of this total, meets the organisation's stringent alignment criteria, demonstrating the market's increasing commitment to robust sustainability standards. Among them, green bonds still dominate the market with an issuance of 671.7 billion dollars, while social and sustainability bonds show stable growth rates of 9% and 32%, respectively. In contrast, sustainability-linked bonds face a sharp drop of 65%, reflecting continuing investor concerns over the credibility of targets. Geographically, Europe is still the dominant region, representing 45% of the total issuance (475 billion dollars), followed by Asia-Pacific (26% share) and North America (16%, especially the U.S., with 45% year-on-year expansion). Moreover, China is the largest contributor to aligned sustainability, which has expanded enormously by 310%. Regarding currency denomination, the euro (39%) and U.S. dollar (30%) continue to dominate, representing 69% of the share. In comparison, the Chinese renminbi consolidates its position as the third-most-used currency with a 7% share. Then, we will conduct a detailed analysis of the dominant green bonds:

Green bonds still flow predominantly into energy, buildings, and public transportation projects, which is shown in Fig. 1. Specifically, in the field of energy, it accounted for 56% in 2015 and is the absolute main flow direction of green bonds.

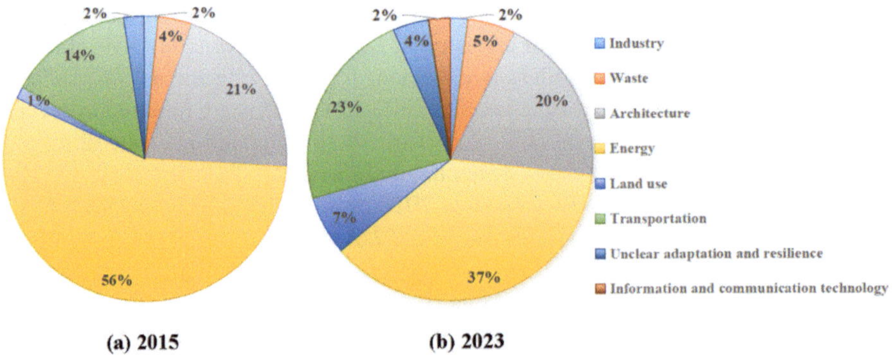

Fig. 1 The use of green bonds. (Notes: The data is obtained from the Climate Bonds Initiative)

By 2023, it drops to 37%. Although it still ranks first in proportion, its share has significantly shrunk due to the development in other fields. In the field of architecture, it accounted for 21% in 2015 and 20% in 2023, with a slight decline in proportion. However, it still holds an essential position in the overall capital allocation. In the transportation sector, the proportion has significantly increased from 14% in 2015 to 23% in 2023, reflecting a notable enhancement in the support of green bonds for green projects in the transportation sector, such as new energy vehicles and green transportation infrastructure. In the industrial sector, it accounts for 1% in 2015 and rises to 7% in 2023, indicating that green transformation projects in the industrial sector (such as green manufacturing and industrial energy conservation) are gradually receiving more financial attention. In the field of waste treatment, 4% in 2015 and 5% in 2023, with a slight increase in proportion, reflect continuous investment in projects such as waste recycling and green treatment. The land use sector accounts for 2% in 2015 and 2023, with a stable proportion, which might involve projects such as ecological protection and sustainable land development. In information and communication technology, there is no definite proportion in 2015, which accounts for 2% in 2023, indicating that green bonds have begun to get involved in green projects in this field (such as energy conservation in data centres). In unclear adaptation and resilience, it rose from 2% in 2015 to 4% in 2023, reflecting increased funding for climate adaptation and disaster resilience projects that are difficult to categorise.

Investment in green bonds has changed from a primary emphasis in the energy sector in 2015 to encompass other sectors like transportation and industry by 2023. This reveals the trend of diversifying green financial support and green developmental demands into different sectors (Qin et al., 2023). It also means that layout adjustment is needed for the green transformation in various industries in the global effort to advance carbon neutrality and sustainable development (Nhan et al., 2025; Shahbaz et al., 2025).

2.2 *Empirical Analysis of Sustainability-Linked Loans*

Sustainability-linked loans (SLLs) represent an innovative financing mechanism where loan terms (primarily interest rates) are directly tied to the borrower's achievement of predefined sustainability performance targets, including pollution reduction, carbon emissions, energy efficiency metrics, and ESG ratings (Auzepy et al., 2025). Distinct from conventional green loans and analogous to sustainability-linked bonds, SLLs offer dual advantages: unrestricted use of proceeds and financial incentives (or penalties) contingent upon sustainability performance. This structure makes SLLs applicable to any borrower committed to advancing sustainability objectives, thereby significantly expanding their potential market reach (Pohl et al., 2023; Gonzalez et al., 2025). In April 2017, the International Netherlands Group (ING) provided Philips with the world's first 1-billion-euro SLL. As a new type of green financial instrument, SLLs have witnessed accelerated development.

According to the data statistics from the Environmental Finance Data Website, from the issuance of the first SSL in 2017 to 2024, a total of 3123 SSLs have been issued globally. Global SSLs started at 4.7 billion dollars in 2017. They have developed rapidly in 2018 and 2019, with their scales increasing tenfold and threefold yearly. In 2019, they exceeded the scale of green loans for the first time. From 2021 to 2022, the annual issuance volume of SSLs has continued to expand. In 2022, they exceeded 677.6 billion dollars at a growth rate of two times. Hence, the scale and activity level in the broad green finance debt market show a year-on-year growth trend. In 2023, the global scale of SSLs slightly decreased compared to 2022 (554.4 billion dollars), but it remains an essential form of sustainable financing in the loan market. After that, global SSLs have recovered to 613.9 billion dollars, reflecting enduring appeal and strong adoption by market participants on both the borrowing and lending sides.

SLLs have gained significant traction across key industries, with utilities and financial sectors leading adoption (total proportion is close to 40%), while oil and gas, industrial manufacturing, and transportation each account for 5–8% market share. The financial sector's prominent involvement highlights its crucial role in enabling transition finance. In addition, regional deployment patterns reveal distinct sectoral focuses: North American SLLs concentrate in real estate, energy, and information technology (IT), whereas European markets prioritise energy, finance, and food industries.

Performance metrics show evolving complexity, with single-key performance indicator (KPI) structures representing 42.13% of loans, dual-KPI frameworks at 22.22%, and triple-KPI arrangements comprising 25.93%. The average number of KPIs per loan has demonstrated consistent growth, rising from 1.72 in 2019 to 2.02 in 2021. Environmental metrics dominate KPI categorisation at 51%, primarily featuring greenhouse gas (GHG) emissions and energy efficiency indicators aligned with climate goals. ESG ratings from independent evaluators form the second priority area at 22%, with social metrics at 20% for supply chain, workplace safety, and diversity. Governance indicators stand out by their absence, representing a mere 1%.

The above detailed examination shows how SLLs evolve as a multi-purpose capital instrument that stimulates quantifiable sustainability improvement in various economic sectors while remaining sensitive to regional and sectoral issues.

2.3 *Empirical Analysis of Carbon Trading Markets*

The global carbon market is a fundamental solution to climate change (Wang et al., 2022; Su et al., 2023; Kong et al., 2025), with rapid evolution and complex transformation over the recent years. By January 2025, 38 operational emissions trading systems (ETS) will be in place worldwide, covering over 10 billion tons of CO_2 equivalent, representing 19% of total greenhouse gas emissions globally. Besides, these nations possess one-third of the world population and produce 58% of the world's gross domestic product (GDP), with another 20 nations at various phases of implementing ETS. That is, the coverage of the global carbon trading market is expanding on an ongoing basis, with increasingly more countries and regions realising the need to implement this instrument for managing carbon emissions. For instance, China's national ETS is a case in point of such expansion, with plans to increase coverage in 2025 from the power sector to energy-intensive industries like steel, cement, and aluminium smelting. Such expansion increases the covered emissions from 5 to 8 billion tons of CO_2 annually, encompassing over 60% of China's emissions.

From the price dynamics and revenue perspective, 2024 has witnessed a divergence in carbon price trends across markets. Prices have declined in long-established systems like the EU, UK, California, and Washington. At the same time, China, Korea, and regional greenhouse gas emission reduction action (RGGI) are relatively stable or have increased. Auction revenues have globally amounted to approximately 70 billion dollars and remained a significant source of climate finance even as they decreased year-on-year. Further, market reforms have increasingly favoured auction-based allocation to achieve optimal efficiency and price signals, as witnessed by ongoing reforms in the EU and California systems.

The evolution of the Carbon trading market displays distinct regional characteristics shaped by local policy priorities and economic contexts, as presented in Table 1.

Additionally, the carbon trading market remains dogged by long-standing problems like carbon leakage, fragmentation and public acceptance (Pu et al., 2025; Sun et al., 2025; Xiao et al., 2025). These would have to be managed with greater international coordination to converge standards and enhance system interoperability. With advancing technology and policy further developed, carbon trading markets are set to become an even more vital part of global decarbonisation, supporting Paris Climate Agreement objectives and the net-zero transition.

Table 1 Differences in regional development

Regions	Features of development
Europe and Central Asia	EU ETS continues to develop, having recently included maritime emissions. Germany and Austria's national systems will transition to the EU ETS 2. Turkey and Ukraine move forward with ETS preparations.
North America	Subnational systems are leading in policy innovation, with California's cap-and-trade system and Canada's federal backstop at the forefront. Market linkages between Washington, California, and Quebec are under consideration.
Asia-Pacific	China's nationwide expansion of ETS is the foundation of regional development, complemented by India and Indonesia's nascent systems. Japan plans to render its voluntary ETS obligatory by 2026, with Malaysia, Philippines and Thailand considering comparable mechanisms.
Latin America	Brazil has established a national ETS, and Chile, Colombia and Mexico are accelerating implementation. The Dominican Republic is considering pilot options.

3 Role of Multilateral and National Institutions

3.1 Climate-Related Development Finance

For the multilateral institutions, the World Bank has reaffirmed its position as the leading issuer in the GSS+ market by pricing 262 sustainability bonds worth 54.3 billion dollars in 2024, thus accounting for 26.3% of overall global sustainability bond issuance. Its cumulative issuance has totalled 300 billion dollars by 2024, highlighting the priority of local currency bonds (for example, Kazakhstani tenge and Uruguayan peso) to enable the SDGs and foster inclusive investments in emerging markets (Park & Shin, 2025). Immediately following the World Bank is the International Development Association (IDA), which has released 11.1 billion dollars of sustainability bonds, focusing mainly on low-income nations under the Private Sector Window.

The Inter-American Development Bank (IDB) has been instrumental in issuing 8.1 billion dollars of sustainability bonds and facilitating innovative social finance initiatives, including BancoSol's gender bond in Bolivia of 29 million dollars that expanded microfinance opportunities to women-run businesses. Meanwhile, the Asian Development Bank (ADB) has facilitated the development of municipal green bonds in Southeast Asia, with Vadodara Municipal Corporation's inaugural green bond of 12 million dollars for water and waste management projects in India being an excellent case in point. In Europe, the European Investment Bank (EIB) is at the forefront of green finance, issuing 16.6 billion dollars of aligned green bonds, while the Agence Française de Développement (AFD) has also backed sustainability bonds, funding climate-resilient investment in Africa and Latin America. The African Development Bank (AfDB) has prioritised social bonds (4.2 billion dollars) in low-cost housing and small and medium-sized enterprises (SME) lending, even when volumes of regional green bonds fall. In addition, the Green Climate Fund (GCF) and Global Environment Facility (GEF) have delivered essential climate finance to support developing countries in realising their nationally determined contributions (NDC) ambitions.

National institutions dominate the growth of climate-related development finance, with state-sponsored and sovereign issuers dominating massive green and sustainability-linked bond issues (Altaf et al., 2025). The following table is a roundup of noteworthy activity by institution and country (Table 2):

3.2 Policy-Based Lending and Regulatory Influence

First, we analyse the policy support and regulatory framework of multilateral institutions. Multilateral development banks (MDBs) are central in shaping the sustainable debt market through policy-directed lending and regulatory frameworks. The report of the Climate Bonds Initiative illustrates how such institutions as the World

Table 2 National institutions in climate-related development finance

Institutions	Countries	2024 Issuance	Key focus areas	Notable features
Kreditanstalt für Wiederaufbau (KfW)	Germany	13.4 billion dollars (green bonds)	Domestic and international climate projects	Major contributor to Germany's 73.3-billion-dollar green bond market
China's four largest banks[a]	China	8.61 billion dollars (Q4 2024, green bonds)	Solar, wind, and public transport	Aligned with China's Green Bond Catalogue & Common Ground Taxonomy (CGT)
California Community Choice Financing Authority (CCCFA)	USA	9.1 billion dollars (green municipal bonds)	Renewable energy power purchase Agreements (PPAs)	Largest U.S. municipal green bond issuer, financing solar/wind power prepayments
French Treasury	France	8.7 billion dollars (green sovereign)	Low-carbon energy, climate adaptation	80% are allocated to ESG investors Cumulative green debt: 84.9 billion dollars
Japanese Government	Japan	10.6 billion dollars (green transformation bonds)	Low-carbon tech R&D	First sovereign transition bond certified by Climate Bonds
Thai Government	Thailand	868 million dollars (sovereign sustainability-linked bond, SLB)	GHG reduction (30% by 2030), electric vehicle (EV) adoption	First Association of Southeast Asian Nations (ASEAN) sovereign SLB, linked to emissions targets
German Government	Germany	19 billion dollars (green sovereign)	Energy transition, green infrastructure	Lead Europe's sovereign green issuance (73.3 billion dollars total)
UK Government	UK	17.3 billion dollars (green sovereign)	Clean transport, energy efficiency	Part of Europe's 101.9 billion dollars sovereign GSS+ volume in 2024
Italian Government	Italy	14.5 billion dollars (green sovereign)	Renewable energy, sustainable transport	Among the top 10 sovereign issuers globally

Notes: [a] China's four largest banks include Industrial and Commercial Bank of China Limited (ICBC), Agricultural Bank of China (ABC), Bank of China (BOC), and China Construction Bank (CCB)

Bank not only provide finance in the form of bond issuances but also technical support and framework development to aid policy development in the emerging economies. For example, the World Bank's Sustainability Bond Framework contains explicit guidelines on the use of proceeds, where projects must be aligned with the

UN SDGs and the greenhouse gas emission reduction goals of the Paris Climate Agreement. In addition, the International Finance Corporation (IFC) and the German Agency for International Cooperation (GIZ) have developed Thailand's first 'traffic light' taxonomy, which maps projects based on their environmental footprint onto red, yellow, and green labels. This practice serves as a model for Southeast Asian transition finance.

Then, we consider the role of the national regulatory agency. National governments employ tax credits, mandatory reporting, and taxonomies to mobilise capital to climate-related activities (Azhgaliyeva & Kapsalyamova, 2023; Bai et al., 2024). Specifically, the EU Taxonomy requires financial institutions to report on the environmental performance of their portfolios and explicitly excludes natural gas and nuclear energy from its green activity definition. Likewise, the People's Bank of China released the Green Bond Endorsed Projects Catalogue (2021 Edition), aligning domestic and international standards with the Common Ground Taxonomy (CGT) it has created with the EU. For instance, 74% of the green bonds issued by China's four largest banks in 2024 finance solar, wind, and public transportation projects, indicating policy backing for priority areas. Japan's Ministry of Economy, Trade, and Industry (METI) also sets an example for regulatory leadership in its green transformation strategy, which requires transition bond issuers to develop science-based decarbonisation pathways. Among them, the Climate Bonds Initiative's first sovereign transition bond to be certified finances low-carbon technology development at the nascent phase.

Some countries have addressed market bottlenecks through innovative regulatory tools. For example, Thailand's inaugural sovereign SLBs link coupon payments to GHG emissions reduction goals (a 30% reduction by 2030) and establish KPIs in its Sustainability-Linked Financing Framework. This framework has been certified by the Climate Bonds Initiative as a benchmark for emerging-market sovereign SLBs. Meanwhile, Colombia's Green Taxonomy has catalysed 9 billion dollars of green investments by training 2700 financial practitioners, demonstrating the importance of regulatory capacity-building. Nevertheless, regulatory fragmentation is a global problem.

Furthermore, the government attracts private capital through the risk-sharing mechanism. Divergences remain, for example, in the EU and China differing over whether natural gas is a transition fuel, and the U.S. municipal green bond market, as highlighted by the CCCFA 9.1 billion dollars issuance, is premised on local regulation instead of a federal regime. Governments also use risk-sharing instruments to mobilise private capital. For instance, India's Ultra Mega Solar Power Project, a public-private-partnership (PPP), mobilises private investment and adds 2.5% to solar power generation capacity. The Dutch development bank has also issued 1 billion euros of water bonds, credit guarantees by the public sector, reducing investor risks in water projects.

4 Emerging Trends and Challenges

4.1 Emerging Trends

Fintech in Green Finance

The introduction of financial technology (fintech) is transforming green finance through enhanced efficiency, transparency, and accessibility (Wan et al., 2023; Kozar & Wodnicka, 2024; Li et al., 2024a). Specifically, blockchain technology is applied to enhance traceability in green bonds to track environmental effects and reduce the possibility of greenwashing properly (Li et al., 2024b). Artificial intelligence and big data analytics also enable more precise evaluation of ESG risks and tracking of real-time sustainability metrics. Internet platforms allow broader access to green investments using crowdfunding models for renewable energy projects. Smart contracts provide automated enforcement of sustainability agreements, and distributed ledger technology facilitates carbon credit trading.

To quantitatively examine this influence, we utilise daily fintech and green finance data spanning 2018–2025 (shown in Table 3), sourced from Standard & Poor's (S&P) Global's comprehensive database (https://www.spglobal.com).

Next, we will conduct a correlation analysis of these variables; the results are shown in Table 4 and Fig. 2. It could be found that GBI is positively related to DBI, GAFI, GFPI, and DLI, with correlation coefficients above 40%. This is proof that green finance and fintech are mutually dependent, whereby the incorporation of fintech into green finance has the potential to improve operational efficiency, lower risks, and facilitate sustainable financial growth (Xu et al., 2023; Kashif et al., 2024;

Table 3 The measurement of variables

Variables	Indicators	Symbols	Descriptions
Green finance	S&P Green Bond Select Index	GBI	Tracking bonds whose earnings are used to fund environmental protection projects
Fintech	S&P Kensho global alternative finance index	GAFI	Measuring the performance of companies that focus on providing alternative financing and wealth management
	S&P Kensho Global Future Payments Index	GFPI	Measuring the performance of companies worldwide that offer products or services related to universal platforms that allow consumers to conduct transactions using digital accounts
	S&P Kensho distributed ledger index	DLI	Measuring the performance of service companies engaged in developing distributed ledger technology and blockchain, and other technologies that make these technologies operational
	S&P Kensho democratised banking index	DBI	Measuring major operators of innovative financial services, including direct lending, crowdfunding, automated wealth management, insurance services tailored to individual needs, digital currencies and related services.

Table 4 The measurement of variables

Variables	DBI	GBI	GAFI	GFPI	DLI
DBI	1.0000	0.6519	0.6524	0.9558	0.9474
GBI	0.6519	1.0000	0.4968	0.4441	0.5997
GAFI	0.6524	0.4968	1.0000	0.4660	0.5998
GFPI	0.9558	0.4441	0.4660	1.0000	0.8776
DLI	0.9474	0.5997	0.5998	0.8776	1.0000

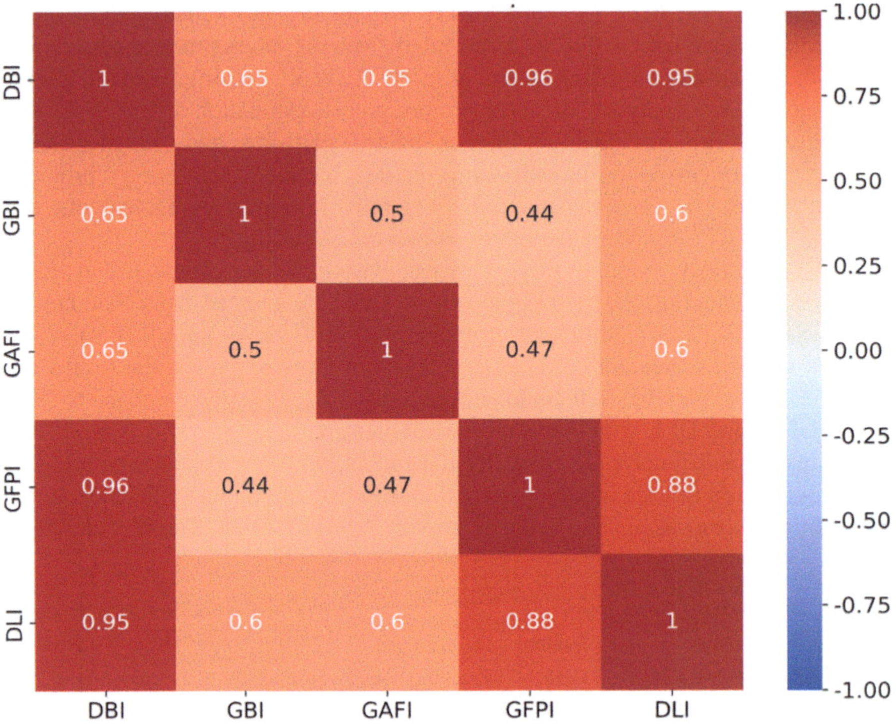

Fig. 2 The correlation heatmap

Qin et al., 2024; Dai et al., 2025). In addition, this friendly interaction implies that adopting fintech innovations into green financial systems can speed the shift to a low-carbon economy.

Regulatory regimes, however, are having a harder time keeping up with the speed at which these technological advancements are evolving, often lagging due to the complexity of balancing innovation with risk management. Though some governments have introduced sandbox frameworks to experiment with new solutions within regulated environments, regulatory fragmentation is a risk absent global coordination and adaptive policymaking. Without adaptive governance frameworks, the complete potential of fintech-enabled green finance can be limited by compliance issues and varied standards in every market.

Nature-Based Solutions

Nature-based solutions (NbS) are within sustainable finance initiatives (Gao et al., 2025; Martin et al., 2025). Financial instruments are being developed to finance ecosystem restoration, regenerative agriculture, and blue carbon projects. Biodiversity-linked bonds and debt-for-nature swaps reflect growing recognition of natural capital's financial value. Corporate and sovereign issuers are incorporating NbS into their sustainability reporting, with metrics extending beyond carbon to address water stewardship and land-use pressures. NbS has attracted approximately 200 billion dollars in financing during 2022 alone, as per the United Nations Environment Programme's report in 2023, but this is still overshadowed by more than 6 trillion dollars towards nature-degrading activities. Specifically, there are issues of standardising measurement approaches and building robust valuation frameworks for ecosystem services. The Taskforce on Nature-related Financial Disclosures (TNFD) framework and the science-based targets initiative for nature institutionalise such practices in mainstream finance.

Transition Finance

Transition finance has become a credible intermediary between conventional and sustainable economic paradigms, especially for hard-to-abate industries. Innovative financial instruments, such as sustainability-linked bonds with tiered pricing, are being utilised to align financial incentives with decarbonisation trajectories (Chen et al., 2025; Khan et al., 2025). Sectoral transition frameworks for sectors like steel, cement, and aviation are charted with near-term emissions reduction targets and long-term net-zero commitments. Banks are developing transition taxonomies to separate true transition activities from greenwashing ones. The market is also seeing an increase in just transition financing that addresses the social aspects of the energy transition. Over the past few years, several countries or regions such as Japan, the EU, Singapore, ASEAN, and China have successively formulated transition finance standards, which are shown in Table 5. Nonetheless, disagreements remain on suitable timelines, the role of natural gas, and the need to balance environmental integrity and economic viability in transition strategies.

4.2 Potential Challenges

Data Gaps

The absence of comparable and transparent data remains a key obstacle for global green finance markets. Inconsistent definitions of 'green' or 'sustainable' among countries, institutions and markets pose substantial obstacles to cross-border data comparability. For example, classification frameworks such as the EU Taxonomy

Table 5 The measurement of variables

Country/ Region	Framework type	Key features	Sector coverage
Japan	Principle-based guidelines + sectoral roadmaps	Defines key elements for disclosure but leaves specific technologies/activities to market The Ministry of Economy, Trade and Industry (METI) provides 2050 carbon-neutral roadmaps for sectors as references for transition plans	Steel, chemicals, power, gas, oil, paper, automotive, etc.
EU	Taxonomy-based (list of activities)	Includes 28 transition activities aligned with the 1.5 °C pathway Sets sector-specific CO_2 thresholds based on EU ETS benchmarks	Manufacturing, transport, energy, construction, information and communications technology (ICT)
Singapore	Taxonomy-based (green vs. transition activities)	Green activities: Fully 1.5 °C-aligned Transition activities: Do not meet 1.5 °C but significantly reduce emissions by 2030; re-evaluated post-2030	Energy, real estate, transport, industry, agriculture, ICT, circular economy, carbon capture, utilisation and storage (CCUS)
ASEAN	Taxonomy-based (3-tiered transition activities)	Classifies transition activities into: 1. 1.5 °C-aligned 2. Below 2 °C-aligned 3. Not 2 °C-aligned but with a 2-year remediation commitment	Broad coverage (reference for member economies)
China	Transition finance catalogue (project + entity-level)	Covers 4 pilot sectors (coal power, steel, construction materials, agriculture); expanding to 7 more Entity-level financing allowed with requirements on targets/ governance Technical thresholds based on domestic advanced values, 1.5 °C/2 °C-aligned	Coal power, steel, construction, agriculture, shipping (under development)

and China's Green Bond Endorsed Projects Catalogue show considerable differences in essential aspects. Reporting practices are also part of the problem, as most corporations report environmental metrics selectively, usually only reporting on carbon emissions but not on other material aspects like biodiversity footprint or water resource management. Data availability and quality are also a concern, especially in frontier markets where poorly organised ESG data collection mechanisms frustrate precise environmental footprint analysis.

The relatively recent emergence of green finance means the absence of long-term performance records, which frustrates the development of solid risk analysis and investment decisions. Technological deficiencies exacerbate the issues because the financial sector has yet to adopt blockchain and artificial intelligence-driven solutions on a large scale that can render data tracking and verification more effective

(Managi et al., 2022). In addition, there is a lack of an integrated data exchange framework between financial institutions, governments, and corporate firms, leading to complex information silos that make overall analysis problematic (Jawadi et al., 2025).

To address these issues, global collaboration is necessary to develop consistent data standards, as demonstrated by the guidelines suggested by the International Financial Reporting Standards (IFRS) Foundation's International Sustainability Standards Board (ISSB). Greater collaboration between the private and public sectors in developing open data platforms, like the European Single Electronic Format (ESEF) reporting tool, would be a great leap towards transparency. Applications of new technologies, particularly blockchain due to its traceability function, offer promising ways to guarantee data integrity and accessibility in green finance markets.

Potential Risks

The green finance sector faces a considerable challenge to its credibility and effectiveness. A widespread phenomenon within this market is greenwashing, in which financial institutions and enterprises exaggerate their environmental sustainability initiatives but fail to take substantial actions, resulting in erosion of market trust (Gacem et al., 2025; Li & Zhong, 2025; Zhao & Wang, 2025). This trend is marked by confusing sustainability claims, like the release of 'sustainable bonds' with no definite use-of-proceeds guidance, or 'net-zero' commitments while continuing to invest in fossil fuel projects. The absence of independent verification worsens it as most green bonds rely on internal decisions rather than obtaining full third-party certification, such as that offered by the Climate Bonds Standard.

Transition risks are also a perilous concern. Policy uncertainty creates uncertainty because carbon tax changes or subsidy removal can fundamentally alter project economics (Chang et al., 2025; Gaies et al., 2025). Technological obsolescence poses an imminent danger to development, particularly to emerging green technologies like hydrogen energy. Market uptake risks also contribute to the intricacy, in which poor consumer or corporate demand for green products can result in underwhelming project returns.

Financial risks introduce additional complexity to green finance (Kong et al., 2025). Liquidity risks are a concern since long-term assets inherently characterise the green bond market with minimal demand on the secondary market for trade (Kocaarslan, 2023). Credit risks manifest as policy dependence, whereby renewable energy projects, most exposed to shifting government incentives, may struggle to repay (Li et al., 2025).

Fighting these complex risks needs a comprehensive approach. First, strengthening regulatory frameworks to require solid disclosure of fund distribution and environmental effects, as the European Union's Sustainable Finance Disclosure Regulation (SFDR) has done, is a good starting point. Second, universal adoption of independent certification regimes, such as those used by the Climate Bonds Initiative

(CBI), would contribute meaningfully to market integrity. Thirdly, new risk management tools like green financial derivatives and insurance products would give investors working in this new environment greater protection.

Limitations in Current Frameworks

Existing frameworks on green finance contain numerous inherent limitations that compromise their effectiveness (Shi et al., 2025). Policy and regulatory fragmentation prevalent in key markets presents diverse challenges to implementation (Chen & Zhao, 2021). Green finance standard discrepancies in the EU, U.S., and Asia hinder cross-border investments. At the same time, even in jurisdictions with comprehensive guidelines, weak enforcement regimes cast doubt on the effectiveness of compliance measures.

Market mechanisms are incomplete, particularly in establishing credible pricing mechanisms. The so-called 'green premium' of green bonds is not universally recognised in the market, without offering any adequate return incentives to investors. The current incentive system also demonstrates imbalances, where most subsidies and tax incentives accrue to large corporations and leave small and medium enterprises with prohibitive participation costs.

There are significant institutional capacity and market awareness gaps. Most financial institutions lack the technical expertise to evaluate green projects appropriately, resulting in inefficient capital allocation. On the investor side, structural knowledge gaps result in ESG considerations being perceived as voluntary factors, rather than essential risk parameters (Zheng et al., 2023; Krueger et al., 2024).

Addressing these weaknesses requires multi-pronged solutions. International coordination through mechanisms like the Group 20 (G20) Sustainable Finance Working Group could align policy frameworks. Establishing dedicated capacity-building institutions, like the United Nations Development Programme (UNDP)'s Green Finance Labs, would drive technical capabilities in the ecosystem. Developing new financial instruments, particularly transition bonds that are tailor-made for carbon-intensive industries, could introduce much-needed flexibility in the transition to low-carbon economies.

References

Afful-Koomson, T. (2015). The Green Climate Fund in Africa: What should be different? *Climate and Development, 7*(4), 367–379.

Altaf, A., Anwar, M. A., Shahzad, U., & Bilan, Y. (2025). Exploring the nexus among green finance, renewable energy and environmental sustainability: Evidence from OECD economies. *Renewable Energy, 244*, 122589.

Amighini, A., Giudici, P., & Ruet, J. (2022). Green finance: An empirical analysis of the Green Climate Fund portfolio structure. *Journal of Cleaner Production, 350*, 131383.

Auzepy, A., Bannier, C. E., & Martin, F. (2025). Are sustainability-linked loans designed to effectively incentivize corporate sustainability? A framework for review. *Financial Management, 52*(3), 643–670.
Azhgaliyeva, D., & Kapsalyamova, Z. (2023). Policy support in promoting green bonds in Asia: Empirical evidence. *Climate Policy, 23*(4), 430–445.
Bai, X., Zhao, W., & Tian, G. (2024). ESG certification, green innovation, and firm value: A quasi-natural experiment based on SynTao Green Finance's ESG ratings. *Pacific-Basin Finance Journal, 86*, 102453.
Bowman, M., & Minas, S. (2019). Resilience through interlinkage: The green climate fund and climate finance governance. *Climate Policy, 19*(3), 342–353.
Chang, S., Qin, M., Hsueh, H.-P., & Lobont, O.-R. (2025). Could U.S.-China conflicts intensify climate transition risks? *Economic Analysis and Policy, 86*, 1592–1604.
Chen, Y., & Zhao, Z. J. (2021). The rise of green bonds for sustainable finance: Global standards and issues with the expanding Chinese market. *Current Opinion in Environmental Sustainability, 52*, 54–57.
Chen, C., Fan, M., & Fan, Y. (2023). The impact of ESG ratings under market soft regulation on corporate green innovation: An empirical study from informal environmental governance. *Frontiers in Environmental Science, 11*, 1278059.
Chen, P., Yao, X., Li, J., & Wang, Y. (2025). Does the transition finance policy facilitate lending for heavy polluters? Insights from China. *Finance Research Letters, 83*, 107625.
Christiansen, J. (2021). Fixing fictions through blended finance: The entrepreneurial ensemble and risk interpretation in the Blue Economy. *Geoforum, 120*, 93–102.
Christnacht, L., & Mertzanis, C. (2025). Foreign direct investment, technology transfer and the global issuance of green bonds. *Energy Economics, 143*, 108246.
Dai, B., Zhang, J., & Hussain, N. (2025). Policy pathways through FinTech and green finance for low-carbon energy transition in BRICS nations. *Energy Strategy Reviews, 57*, 101603.
Ferri, S., Tron, A., Colantoni, F., & Savio, R. (2023). Sustainability disclosure and IPO performance: Exploring the impact of ESG reporting. *Sustainability, 15*(6), 5144.
Gacem, S., Hervé, F., & Marsat, S. (2025). When green turns red: Is the perception of greenwashing a barrier to individual green investment? *Finance Research Letters, 82*, 107605.
Gaies, B., Chaâbane, N., Adeosun, O. A., & Sahut, J.-M. (2025). Climate transition risks, ESG sentiment and market value: Insights from the European stock market. *Energy Economics, 148*, 108605.
Gao, M., Lee, K. E., & Shamsuddin, A. S. (2025). Landscape approaches and stakeholder engagement in nature-based solutions for sustainable river floodplains: A systematic review. *Ecological Indicators, 176*, 113686.
Gaurav, A., Baishnab, K., & Singh, P. K. (2025). Intelligent ESG portfolio optimization: A multi-objective AI-driven framework for sustainable investments in the Indian stock market. *Sustainable Futures, 9*, 100832.
Gonzalez, T. A., Romero, L. C., Karmaziene, E., & Yuan, X. (2025). Green gains: The impact of REITs' environmental performance on sustainability-linked loan interest rates. *Finance Research Letters, 71*, 106415.
Guesmi, K., Makrychoriti, P., & Pyrgiotakis, E. G. (2025). Climate change exposure and green bonds issuance. *Journal of International Money and Finance, 152*, 103281.
Huang, L. (2024). Green bonds and ESG investments: Catalysts for sustainable finance and green economic growth in resource-abundant economies. *Resources Policy, 91*, 104806.
Jawadi, F., Pondie, T. M., & Cheffou, A. I. (2025). New challenges for green finance and sustainable industrialization in developing countries: A panel data analysis. *Energy Economics, 142*, 108120.
Jonsdottir, B., Sigurjonsson, T., Johannsdottir, L., & Wendt, S. (2022). Barriers to using ESG data for investment decisions. *Sustainability, 14*(9), 5157.

Kashif, M., Ullah, A., Ullah, S., & Qian, N. (2024). Towards a greener future: The impact of financial technology (FinTech) and climate finance on ecological sustainability. *Journal of Environmental Management, 370*, 122876.

Khan, S., Pinglu, C., Ullah, A., & Qian, N. (2025). Facilitating sustainable corporate growth in China: Examining the role of transition finance, green technology innovation, and ESG strategies in the path to sustainability. *Journal of Environmental Management, 389*, 126085.

Kocaarslan, B. (2023). Funding liquidity risk and the volatility of U.S. municipal green bonds during the COVID-19 pandemic. *Finance Research Letters, 58*(Part D), 104560.

Kong, X., Zheng, Y., Yang, S., & Wang, Y. (2025). Profit-driven economic expansion planning of CCUS deployment and carbon trading in carbon-neutral energy systems under liberalized market-oriented decarbonization strategies with step-based pricing. *Energy, 297*, 136762.

Kozar, Ł. J., & Wodnicka, M. (2024). FinTech and green finance: Bibliometric analysis. *Procedia Computer Science, 246*, 2080–2089.

Krueger, P., Sautner, Z., Tang, D. Y., & Zhong, R. (2024). The effects of mandatory ESG disclosure around the world. *Journal of Accounting Research, 62*(5), 1795–1847.

Léon, F. (2025). In the shadows of support: Unintended outcomes of DFIs' intermediated lending in Africa. *World Development, 191*, 106998.

Li, H., & Zhong, Y. (2025). Digital transformation of enterprises and greenwashing behavior: Evidence from China. *Finance Research Letters, 82*, 107585.

Li, S., Chen, R., Li, Z., & Chen, X. (2024a). Can blockchain help curb "greenwashing" in green finance? Based on tripartite evolutionary game theory. *Journal of Cleaner Production, 435*, 140447.

Li, Y., Chu, E., Nie, S., Peng, X., & Yi, Y. (2024b). Fintech, financing constraints and corporate green innovation. *International Review of Financial Analysis, 96*(Part A), 103650.

Li, J., Chen, C., & Wu, Z. (2025). Asymmetric credit risk pathways in China's green bond market: A conjunctural analysis using modified PFM and fsQCA. *Finance Research Letters, 83*, 107626.

Liu, F.-J. (2025). ESG performance, managerial ability and corporate investment efficiency. *Finance Research Letters, 80*, 107416.

Lu, C., Wu, C., Feng, L., Zhan, J., Shi, Y., & Chen, H. (2025). Does ESG performance enhance corporate green technological innovation? Micro evidence from Chinese-listed companies. *Sustainability, 17*(2), 636.

Managi, S., Broadstock, D., & Wurgler, J. (2022). Green and climate finance: Challenges and opportunities. *International Review of Financial Analysis, 79*, 101962.

Martin, J. G. C., Scolobig, A., Linnerooth-Bayer, J., Irshaid, J., Rodriguez, J. J. A., Fresolone-Caparrós, A., & Oen, A. (2025). The nature-based solution implementation gap: A review of nature-based solution governance barriers and enablers. *Journal of Environmental Management, 388*, 126007.

Ngwenya, N., & Simatele, M. (2020). Unbundling of the green bond market in the economic hubs of Africa: Case study of Kenya, Nigeria and South Africa. *Development Southern Africa, 37*(6), 888–903.

Nhan, D. T., Lu, W.-M., Kuo, K.-C., & Nguyen, D. T. (2025). Special economic zones and sustainable development: A study in provinces of China. *Sustainable Futures, 7*, 100864.

Ning, Y., Cherian, J., Sial, M., Alvarez-Otero, S., Comite, U., & Zia-Ud-Din, M. (2023). Green bond as a new determinant of sustainable green financing, energy efficiency investment, and economic growth: A global perspective. *Environmental Science and Pollution Research, 30*(22), 61324–61339.

Park, C.-Y., & Shin, K. (2025). The development of local currency bond markets and uncovered interest rate parity. *Journal of International Money and Finance, 154*, 103310.

Pata, U., Kartal, M., Ahmed, Z., & Sinha, A. (2025). Relationship between green bonds and carbon neutrality: Evidence from top five emitting countries' sectoral CO_2 emissions. *Financial Innovation, 11*(1), 10.

Pindiriri, C., & Kwaramba, M. (2024). Climate finance in developing countries: Green budget tagging and resource mobilization. *Climate Policy, 24*(7), 894–908.

Pohl, C., Schüler, G., & Schiereck, D. (2023). Borrower- and lender-specific determinants in the pricing of sustainability-linked loans. *Journal of Cleaner Production, 385*, 135652.
Pu, R.-Y., Liang, Q.-M., Wei, Y.-M., Yan, S.-Y., Wang, X.-Y., Li, D.-H., Yi, C., & Ji, C.-J. (2025). Impact of the China's new energy market on carbon price fluctuation risk: Evidence from seven pilot carbon markets. *Energy Strategy Reviews, 59*, 101718.
Qin, M., Zhang, X., Li, Y., & Badarcea, R. M. (2023). Blockchain market and green finance: The enablers of carbon neutrality in China. *Energy Economics, 118*, 106501.
Qin, L., Aziz, G., Hussan, M. W., Qadeer, A., & Sarwar, S. (2024). Empirical evidence of fintech and green environment: Using the green finance as a mediating variable. *International Review of Economics & Finance, 89*(Part A), 33–49.
Shah, S. S., Nakouwo, S. N., Sobirjonovna, G. M., & Khan, A. (2025). Exploring the sustainability impact of green bonds on ecological and resource capacities. *Renewable Energy, 243*, 122590.
Shahbaz, M., Barut, A., Alofaysan, H., & Çamkaya, S. (2025). The impact of inclusive finance and lifelong learning on carbon neutrality: Evidence from African countries. *Research in International Business and Finance, 78*, 102996.
Shi, P., Long, H., Li, Y., Li, X., & Wang, X. (2025). Agricultural green production efficiency within a green finance framework: Empirical evidence from China. *International Review of Financial Analysis, 97*, 103814.
Smith, J., Samuelson, M., Libanda, B., Roe, D., & Alhassan, L. (2022). Getting blended finance to where it's needed: The case of CBNRM enterprises in Southern Africa. *Land, 11*(5), 637.
Su, C. W., Pang, L. D., Qin, M., Lobonţ, O. R., & Umar, M. (2023). The spillover effects among fossil fuel, renewables and carbon markets: Evidence under the dual dilemma of climate change and energy crises. *Energy, 274*, 127304.
Sun, X., Bao, M., Ding, Y., Ouyang, X., & Song, Y. (2025). Multi-stage portfolio optimization for GenCo considering multi-correlation between electricity and carbon markets. *Applied Energy, 393*, 126052.
Vu, T. N., Lehkonen, H., Junttila, J.-P., & Lucey, B. (2025). ESG investment performance and global attention to sustainability. *The North American Journal of Economics and Finance, 75*, 102287.
Wan, S., Lee, Y. H., & Sarma, V. J. (2023). Is Fintech good for green finance? Empirical evidence from listed banks in China. *Economic Analysis and Policy, 80*, 1273–1291.
Wang, X. Q., Su, C. W., Lobonţ, O. R., Li, H., & Nicoleta-Claudia, M. (2022). Is China's carbon trading market efficient? Evidence from emissions trading scheme pilots. *Energy, 245*, 123240.
Wang, X., Tang, M., Wenya, L., & Xu, L. (2025). ESG performance and cross regional investment: Evidence from Chinese listed companies. *Finance Research Letters, 81*, 107439.
Xiao, J., Wang, Y., & Wen, D. (2025). Global climate policy uncertainty and carbon market volatility: Aggravating or mitigating across market conditions? *Economics Letters, 254*, 112441.
Xu, J., Chen, F., Zhang, W., Liu, Y., & Li, T. (2023). Analysis of the carbon emission reduction effect of Fintech and the transmission channel of green finance. *Finance Research Letters, 56*, 104127.
Yan, M., Li, X., Zhao, X., & He, Z. (2025). Liquidity and cost advantage of green bonds. *Finance Research Letters, 71*, 106433.
Zhang, D., Zhang, Z., & Managi, S. (2019). A bibliometric analysis on green finance: Current status, development, and future directions. *Finance Research Letters, 29*, 425–430.
Zhao, J., & Wang, X. (2025). Managerial myopia, technological innovation, and greenwashing: Moderating roles of financing constraints and environmental uncertainty. *Technological Forecasting and Social Change, 218*, 124210.
Zheng, J., Jiang, Y., Cui, Y., & Shen, Y. (2023). Green bond issuance and corporate ESG performance: Steps toward green and low-carbon development. *Research in International Business and Finance, 66*, 102007.